MW00837937

Sandwich Structures: Theory and Responses

Terry John Hause

Sandwich Structures: Theory and Responses

Terry John Hause
U.S. Army
Warren, MI, USA

ISBN 978-3-030-71894-7 ISBN 978-3-030-71895-4 (eBook)
https://doi.org/10.1007/978-3-030-71895-4

This Springer imprint is published by the registered company Springer Nature Switzerland AG
The registered company address is: Gewerbestrasse 11, 6330 Cham, Switzerland

This book is in honor of Dr. Liviu Librescu who sacrificed his life, on April 16th, 2007, during the Virginia Tech Shooting, to save the students in his class by holding the door shut while the other students escaped through the window. Dr. Liviu Librescu was a great mentor, colleague, and friend. May he never be forgotten for the selfless act he carried out.

Preface

Over the past several decades sandwich structures have increased in their use and applications. Just to name a few, sandwich structures have applications in supersonic/hypersonic aircraft, Cryogenic tanks, Flight and launch vehicles, trains, automobiles, naval ships, and several components found in satellites. With all of these applications it is imperative that a good understanding of these structures is in hand both from a theoretical and practical standpoint. In the future the applications will increase for the use of these structures. With the consideration of all of these applications a great deal of knowledge with regard to their behavior or response under complex loading conditions is required. Currently, there is need for a comprehensive book on sandwich structures that provides a comprehensive theoretical base upon which to build. Very few if any books on sandwich structures provide this background from a consistent standpoint. There is a plethora of texts on plates, shells, and beams. A comprehensive theory that provides a solid foundation in the subject matter for the practicing engineer, instructor, or scientist is needed. In general, the majority of technical information on sandwich structures is found in technical journals. Unfortunately, the theory varies among the various researchers. This is another reason to present a consistent understandable theory in text format. For these reasons, this text is being written.

This text will expose scientists and engineers to the most advanced state-of-the-art theory and response behavior. This text will serve mainly as a theoretical reference in conjunction with various applications. The scientist and practicing engineer will be exposed to several theoretical aspects and their solution techniques. These techniques can serve as a basis for other types of structures such as composite structures, monocoque plates and shells, thin-walled beams, etc. Also, this text will serve to better understand how certain variables can affect the design of components which are of the sandwich-type construction.

Additional benefits for the interested reader are it will provide a theoretical tool which will provide an understanding of the response and instability of advanced sandwich structures exposed to complex loading conditions. It will discuss and present both the incompressible and compressible core theories. The idea of integrating functionally graded materials into these structures are presented. A better understanding of how to alleviate unwanted detrimental effects due to various complex loading conditions by the use of the structural tailoring technique is promoted. Many aspects discussed and presented in this book is of benefit to the interested reader. In addition, it will provide a basic solid mechanics foundation which can be applied to other types of structures.

Chapter 2 presents the theoretical foundation for sandwich plates and shells with a transversely incompressible core. The theory is very comprehensive covering a broad spectrum of various specialized cases concerning both the linear and nonlinear theories with the inclusion of the geometrical imperfections. Also considered are the external loadings on the structure pertaining to the compressive/tensile edge loadings as well as the transverse pressure. The governing equations are developed via Hamilton's principle which is an enormously powerful theoretical tool which can be used for a whole host of solid mechanics problems. Chapter two sets the tone for Chaps. 3, 4, 5 and 6. Chapters 3, 4, 5 and 6 contain applications of the theory developed in Chap. 2 such as the buckling problem, post buckling, free vibration, and the dynamic response problem. Chapter 3 considers the linear governing equations developed in Chap. 2 as it pertains to the buckling problem. These equations are then solved via an immensely powerful solution technique referred to as the Extended Galerkin Method (EGM). Several validations are made to validate the theory with various other numerical results which clearly reveal the response of the structure under various material and geometrical properties. The nonlinear governing equations from Chap. 2 are applied to the post buckling problem in Chap. 4 where three cases are considered. The first case considers cross-ply laminated sandwich plates and shells. The second case considers angle-ply laminated sandwich plates, and finally angle-ply sandwich shells are considered. The governing equations from Chap. 2 are highlighted then the details of the solution technique are presented which involves the EGM and Newton's method to arrive at the post buckling response. Several results are presented considering the effect of the geometric imperfections, the effect of the curvature, the effect of the ply angle, the transverse pressure, biaxial edge loading, the end -shortening of the edges, etc.

Chapter 5 addresses the free vibration problem considering the linear governing equations from chapter two with the inclusion of the inertia term. The governing equations are solved via the EGM with validations and results which follow. The results consider the anisotropy of the face sheets, the stacking sequence of the face sheet lamina, the fiber orientation of the face sheets, the orthotropic properties of the core, in addition to other important geometrical and material properties. All of these results are highlighted with respect to the effects on the eigenfrequencies of the structure.

Chapter 6 addresses the dynamic response of sandwich plates and shells, where in contrast to the free vibration problem, time-dependent external loadings are considered for the dynamic response problem. The governing equations from Chap. 2 as it applies to the dynamic response problem are solved via closed-form techniques such as the Laplace Transform and the EGM. To generate the results several external time-dependent loads are considered such as the sonic boom, triangular pulse,

Heaviside step function, rectangular pulse, the sine pulse, a tangential traveling air blast, and the Friedlander in-Air explosive pulse, and a case for underwater blast loading. The highlighted results consider various geometrical and material parameters and their effect on the dynamic response of the structure.

Chapters 2, 3, 4, 5 and 6 are devoted to the sandwich structure with a transversely incompressible core. In Chaps. 7 and 8 attention is given to the theory of sandwich plates and shells with a transversely compressible core which captures local and global wrinkling. Two very comprehensive detailed theories are addressed where the first theory considers that the core transverse displacement is modeled with a first order power series whereas the second theory considers that the core transverse displacement is modeled with a second order power series. The latter is considered in Chap. 8 while the former is addressed in Chap. 7. Both theories are identical in their approach to the development of the governing equations where Hamilton's principle is utilized. Theory two requires more computational effort to produce the governing equations, as a result of an assumed higher order transverse core displacement. At the conclusion of both Chaps. 7 and 8 there is an application of the theory to highlight a solution technique to solving these governing equations.

In Chap. 9, a nonlinear theoretical foundation considering the first order shear deformation theory for functionally graded sandwich plates and shells is presented. Two types of functionally graded sandwich structures are considered. The first type considers the that the face sheets are functionally graded while the core is homogeneous. The second case considers the opposite scenario where the face sheets are homogeneous, and the core is functionally graded. The theory employs Hamilton's principle while considering the tangential and rotatory inertias.

It is hoped that the material covered in this book sets a foundation upon which to build from. It should be noted that although very comprehensive and detailed theories have been presented which have enormous applications, this material was not meant to cover every single case or application in which these equations could be applied. There are still several areas that need to be addressed both for the incompressible and compressible core case. The theoretical tools have been provided in this text with the idea that the many other applications for sandwich structures can now be explored.

Warren, MI, USA Terry John Hause

Acknowledgements

I would like to thank God for the inspiration, motivation, and opportunity afforded me to write this book. I would also like to thank my dear wife Marilou and my dear son Jourdan for their understanding and patience during the writing of this book.

Contents

Chapter 1
Introduction

Abstract Sandwich structures have several prominent roles in the aerospace industry such as their use in aircraft engines, the fuselage, the floor, side-panels, overhead bins, and the ceiling. These structures have many additional applications such as their use in naval ship bulkheads, deck houses, aircraft hangars, locomotive cabs, buses, satellites, automobiles, etc. With this in mind, a very intensive study of their structural performance is necessary under extreme complex loading conditions. As an introduction, a very preliminary overview of the composition of typical sandwich structures are discussed with regard to their structural elements. Secondly various applications of this type of structure are listed. A few historical applications are briefly mentioned with various types of failure modes, such as wrinkling, that sandwich structures can endure. Finally, an overview of the text is presented for the readers benefit.

1.1 Preliminary Overview

A sandwich structure is a three-layered structure composed of two facings with a thick core layer in-between. The core layer is multiple times the thickness of the facings. The facings are usually very thin and stiff which can be constructed from a variety of different kinds of materials. Some of the most common types are anisotropic laminated composites, a single-layered isotropic or orthotropic material, and functionally graded materials. Generally, the core is soft and only carries the transverse shear stresses. The other type of core is referred to as a strong core which carries both the tangential and transverse shear stresses. Typical types of core constructions are foam, honeycomb, web, and truss-type core construction. The former two are soft-core type while the latter two are of the strong-core type. Usually these type structures are exposed to in-plane and lateral loading.

Sandwich structures have several applications. Overtime their use seems to keep increasing due to advanced manufacturing techniques and the introduction of new types of materials integrated into the structure, either within the facings or within the core. Some of the aeronautical applications of sandwich structures includes aircraft engines, the fuselage, the floor, side panels, overhead bins, and the ceiling. These

T. J. Hause, *Sandwich Structures: Theory and Responses*,
https://doi.org/10.1007/978-3-030-71895-4_1

structures have many additional applications such as use in naval ship bulkheads, deck houses, aircraft hangars, locomotive cabs, buses, satellites, automobiles, etc. According to Vinson (2005), during the last decade, sandwich structures have found their way into wind energy systems. GE Energy declares there are 6900 installations worldwide, while their growth rate is up to 20% annually. According to Vinson (2005) Germany, Spain, the US, and Denmark are the leaders in wind energy installations. These structures inherently have many benefits such as large bending stiffness, providing a smoother surface finish for aerodynamic applications, provide excellent sound and thermal insulation depending on the materials used in their construction, increased strength at elevated temperatures, increased operational time, and lightweight in construction.

The application of sandwich construction has been around for several decades. The British de Havilland Mosquito bomber of World War II utilized the sandwich-type construction within the airframe. The face sheets were birch bonded to a balsa wood core. Other airplanes such as the B-58, B-70, F-111, C-5a, and many others employed this type of construction. The advantage is the high strength-to-weight ratio inherent within the structure. Various spacecraft have leveraged this type of structure such as the Apollo spacecraft, The Spacecraft LM Adapter fairings on the Centaur, and other launch vehicles. They were also used in propellant tank bulkheads.

1.2 Composition of the Sandwich Structure

Sandwich structures provide a wide range of facing and core material selections. The choice of materials depends on the application in which the structure will be utilized, the loading conditions, the availability, and the cost. As an example, in aerospace applications glass-epoxy or glass-vinyl ester are used in the facings of civil and marine structures. The core is often aluminum or Nomex honeycomb. In civil engineering applications, the core is usually a closed- or open-cell foam. Balsa core is used in ships according to Birman and Kardomateas (2018). There are four types of commonly used cores. They are foam, honeycomb, truss, and web-type core construction. Considering foam, there are several types of foam materials. The first is Polystyrene, which has a high compressive strength and resist water penetration. Then there is Phenolic, which is fire resistant, has a low density, and low mechanical properties. Next there is polyurethane. This is utilized for producing the fuselage, wing tips, and other curved parts of small aircraft. There is polystyrene which is used for airfoil shapes. Then there is Polyvinyl Chloride (PVC), which has a high compressive strength and durability and is fire resistant. Finally, there is Polymethacrylimide which is used for lightweight sandwich construction.

The second type of core is the honeycomb-type structure. The cell-type structure can be round, square, or triangular. Some of the materials used in the construction of the honeycomb core are kraft paper, thermoplastics, aluminum, steel, titanium, aramid paper, fiberglass, carbon, and ceramics. Each has their own properties and benefits depending on the application in which it will be used. The honeycomb-type core construction is used mainly in the aerospace industry. They are lightweight, flexible, fire retardant, have good impact resistance, and have the best strength-to-weight ratios. The third and fourth types are the truss and web core which are used in civil engineering applications. Besides the traditional construction, current research shows that the latest designs in sandwich construction include new core concepts in design, incorporating nanotubes and smart materials, as well as functionally graded materials.

1.3 Failure Modes of Sandwich Structures

The structural instability of a sandwich structure can appear in any number of ways depending on the loading and construction. The first type of failure mode is Intracellular Buckling or face dimpling. This is a localized form of failure which occurs when the core is not continuous. Above the core cells, the facings buckle with the cell walls acting as edge supports. As this progresses, buckling ensues. The second type of failure is face wrinkling. This usually occurs if the core is compressible. A compressible core can carry normal or extensible straining thus stresses. This mode of failure appears as short wavelengths in the facings, which involves the normal to facings straining of the core. This wrinkling can by symmetrical or asymmetrical with respect to the mid-surface of the core before being deformed. Ultimate failure usually occurs by core crushing, tensile rupture of the core, or tensile rupture of the core-to-facing bond. The third type of failure is shear crimping. This is a general instability mode where the buckle wavelength is noticeably short due to a low transverse modulus for the core. This failure mode occurs quickly and causes failure due to shearing. The fourth and fifth types are the General Instability and the Panel Instability.

With all of the applications and construction scenarios available to these type structures it is imperative that a good theoretical understanding of some of these concepts are understood. In the forthcoming chapters a comprehensive theoretical base is presented to obtain the pertinent response under various loading scenarios.

1.4 Contents of the Text

The governing equations for a sandwich plates and shells modeled with a transversely incompressible core are presented in Chap. 2. The kinematic equations such as the displacement field and the strain–displacement relationships are formulated for the case of thick doubly curved shells with the First-Order Shear Deformation theory in mind. The Green's Strain Tensor is presented with the introduction of geometric imperfections. The constitutive equations are developed based on the generalized Hooke's Law following with the developments of the equations of motion. Two forms of the equations of motion are presented. One is referred to as the mixed formulation and the other form is the displacement formulation. These equations of motion are developed via Hamilton's Principle. As a byproduct from Hamilton's Principle, the boundary conditions are simply supported and clamped boundary conditions. The pertinent governing equations are formulated for the general case of thick shells including the geometric imperfections, body forces and inertia terms, prescribed edge loadings, and the transverse loadings. In general, the body forces are neglected and considered negligible to the response of the structure. Chapters 3 and 4 provide the theoretical developments, the solution methodology, and in addition, some numerical results concerning the problem of buckling and post-buckling, respectively. In the case of buckling, which is the linearized counterpart of the post-buckling solution, validations are made against the theory with remarkable agreement. The presentation of the numerical results, in the case of post-buckling, are broken down into two categories, cross-ply and angle-ply laminated facings. Several issues are highlighted in both cases such as the effect of the geometric imperfections, the effect of the aspect ratio, the material directional properties in the face sheets, the effect of the panel face thickness, the effect of curvature, the snap-through-type behavior, the layup sequence of the facings (structural tailoring), the in-plane and tangential prescribed edge loadings, etc. In Chaps. 5 and 6, the free vibration and dynamic response of sandwich structures are investigated. The dynamic response considers various time-dependent external loadings for both sandwich plates and shells. The solution methodology leads to an eigenvalue problem which is conducive to a closed-form solution utilizing the Extended Galerkin Method (EGM) and the Laplace Transform Method (LTM) to determine both the frequency and the dynamic response. The effects of the panel curvature, anisotropy, structural tailoring, the orthotropy of the core, and the effects of damping are all considered in regard to the structural response. Various types of stimuli such as the sonic boom, the triangular pulse, Heaviside step function, the rectangular pulse, sine pulse, tangential traveling air blast, and the Friedlander in-air explosive pulse are considered. Chapter 7 presents the basic governing equations for the transversely compressible core model, whereby the core is considered extensible in the transverse direction. By modeling the core as extensible, the wrinkling phenomenon is captured which can lead to a local (face wrinkling) and global instability mode which is not captured by the incompressible core model. Chapter 8 is built upon Chap. 7 in that the core is modeled with a higher-order transverse displacement function to the second order, in contrast to Chap. 7 where the transverse displacement is considered linear. The theory is computationally more intensive but provides for an improved behavior of the extensibility of the core. Following the theoretical developments in Chap. 7, an application making use of the theory concerning the dynamic response is presented. A brief overview of how to apply and solve the equations is highlighted. In contrast, an application of the theory presented following the theoretical developments in Chap. 8 is discussed highlighting buckling and post-buckling. In both cases numerical results are omitted. Prominent authors such as Hohe and Librescu (2003, 2006) have generated and published several numerical results regarding this subject matter. The reader is referred to their publications for the results. There has been a plethora of results compiled by these top researchers where a detailed analysis of both the local (wrinkling) and global bucking and post-buckling under various geometrical and loading scenarios have been presented and discussed in sufficient detail. In Chap. 9, the state-of-the-art components of functionally graded sandwich plates and shells are discussed. An introduction to functionally graded sandwich structures is provided where the governing equations are derived for a both symmetric and asymmetric cases.

It is hoped that this text will provide some insight and promote a better understanding of sandwich structures under complex loading conditions for the practicing engineer, research scientist, or graduate student wanting to acquire knowledge with regard to these type structures.

References

Birman, V., & Kardomateas, G. A. (2018). Review of current trends in research and applications of sandwich structures. *Composites B, 142*, 221–240.

Hohe, J., & Librescu, L. (2003). A nonlinear theory for doubly curved anisotropic sandwich shells with transversely compressible core. *International Journal of Solids and Structures, 40*, 1059–1088.

Hohe, J., & Librescu, L. (2006). Dynamic buckling of flat and curved sandwich panels with transversely compressible core. *Composite Structures, 74*, 10–24.

Vinson, J. R. (2005). Sandwich structures: Past, present, and future. In O. T. Thomsen et al. (Eds.), *Sandwich structures 7: Advancing with sandwich structures and materials*. Dordrecht: Springer.

Chapter 2
Theory of Sandwich Plates and Shells with an Transversely Incompressible Core

Abstract An extremely robust geometrical nonlinear theory of initially imperfect doubly curved sandwich shells with an incompressible core is presented. The aspects of the theory consider dissimilar face sheets with imposed anisotropic laminated composite construction. The core is considered broad in its inherent properties by considering both the weak and strong core types. The influence of the geometric imperfections on the theoretical developments are also included which adds an additional level of complexity. The governing equations are simplified for various specialized cases for two formulations referred to as the mixed formulation and the displacement formulation. These formulations are then broken down individually each for plate and shell-type sandwich construction. Finally, the boundary conditions are briefly discussed.

2.1 Introduction

In this chapter, the sandwich structure is introduced with a very comprehensive and detailed presentation of the basic terminology, basic assumptions, and the governing nonlinear theory regarding the case for the transversely incompressible core. The derivation is provided in extreme detail which is then specialized for several different cases. The presentation of the governing equations commences with the highlights of the displacement field, the nonlinear Green–Lagrange strain displacement field with the introduction of geometrical imperfections along with the stress–strain relationships, based on the generalized Hooke's Law. Following in a similar vein, the principles of shallow shell theory are adopted and the concept of the stress and stress couple resultants are presented. The equations of motion are derived via an energy approach known as Hamilton's principle. The equations are built upon a very general base considering the strong core–type construction with transverse shear considered in the facings (thick facings). The equations are then reduced considering symmetry, a soft/weak core, with the adoption of the Love–Kirchhoff hypothesis where the facings are assumed thin whereby the transverse shear stresses can be neglected. Following these assumptions, the equations are further reduced

T. J. Hause, *Sandwich Structures: Theory and Responses*,
https://doi.org/10.1007/978-3-030-71895-4_2

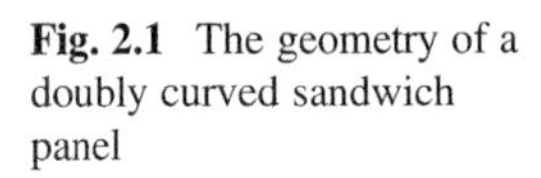

Fig. 2.1 The geometry of a doubly curved sandwich panel

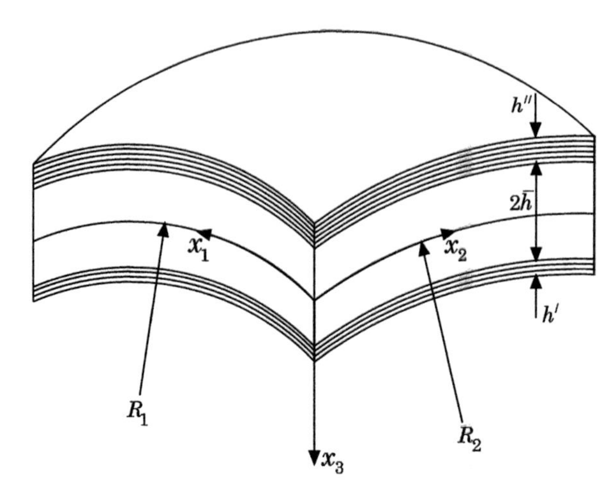

considering the linear theory for both plates and shells with applications of these equations in mind. This chapter serves as a precursor to Chaps. 3, 4, 5, and 6.

2.2 Preliminaries and Basic Assumption

Referring to Fig. 2.1, the middle surface of the core becomes the global mid-surface of the structure which is referred to an orthogonal curvilinear coordinate system x_i ($i = 1,2,3$).

The coordinate x_3 is considered positive when measured in the downward normal direction.

The thickness of the core is $2\,\overline{h}$ which is uniform throughout. The thicknesses of the bottom and top facings are h' and h'', respectively. This implies that H ($\equiv h' + 2\,\overline{h} + h''$) is the total thickness of the structure. For the identification purposes unless otherwise noted, quantities with single primes refer to the bottom face and those with double primes refer to the top face. These primes may be placed on the right or left of the respective quantity. The geometrically nonlinear theory of doubly curved sandwich panels is based on a series of assumptions which are as follows:

The face sheets are constructed of a number of orthotropic material layers, the axes of orthotropy of the individual plies being not necessarily coincident with the geometrical axes x_α ($\alpha = 1,2$) of the structure.

1. The thickness of the core is much larger than those of the face sheets, that is, $2\overline{h} \gg h', h''$
2. The core material features orthotropic properties, the axes of orthotropy being parallel to the geometrical axes x_α
3. The cases of both the *weak* and *strong* core type sandwich structures are considered. In the weak-core case, the core is capable of carrying transverse shear

stresses only. Whereas in the strong core case, the core can carry both tangential and transverse shear stresses.

4. A perfect bonding between the face sheets and between the faces and the core exists.
5. All three layers, the facings and the core, are incompressible in the transverse direction.
6. The geometrical nonlinearities in the von Kármán sense with the geometrical imperfections are included in the sandwich structure model.
7. The principles of shallow shell theory apply.

2.3 Basic Equations

2.3.1 Displacement Field

In line with the typical plate and shell theory, a power series expansion is utilized for the displacement field. For the case of the sandwich structure, with three different layers, each layer assumes its own displacement field. The First-Order Shear Deformation Theory will be assumed for each layer which is based on the Mindlin–Reissner Theory. Each layer is treated as a separate monocoque shell. It is assumed that the normal to the mid-surface remains straight but not necessarily normal after deformation. The distance measured from each layer's respective mid-surface is measured with respect to the global mid-surface. The displacement field for a point in each layer is then given by

Bottom Face $\left(\overline{h} \leq x_3 \leq \overline{h} + h^{/}\right)$

$$\begin{aligned}
{}^{/}V_1(x_1,x_2,x_3) &= {}^{/}\overset{\circ}{V}_1(x_1,x_2) + \left(x_3 - a^{/}\right)\psi_1^{/}(x_1,x_2) \\
{}^{/}V_2(x_1,x_2,x_3) &= {}^{/}\overset{\circ}{V}_2(x_1,x_2) + \left(x_3 - a^{/}\right)\psi_2^{/}(x_1,x_2) \\
{}^{/}V_3(x_1,x_2,x_3) &= {}^{/}v_3(x_1,x_2)
\end{aligned} \tag{2.1a – c}$$

Core $\left(-\overline{h} \leq x_3 \leq \overline{h}\right)$

$$\begin{aligned}
\overline{V}_1(x_1,x_2,x_3) &= \overset{\circ}{\overline{V}}_1(x_1,x_2) + x_3\overline{\psi}_1(x_1,x_2) \\
\overline{V}_2(x_1,x_2,x_3) &= \overset{\circ}{\overline{V}}_2(x_1,x_2) + x_3\overline{\psi}_2(x_1,x_2) \\
\overline{V}_3(x_1,x_2,x_3) &= \overline{v}_3(x_1,x_2)
\end{aligned} \tag{2.2a – c}$$

Top Face $\left(-\overline{h} - h^{//} \leq x_3 \leq -\overline{h}\right)$

$$\begin{aligned} {}^{//}V_1(x_1,x_2,x_3) &= {}^{//}V_1^{\circ}(x_1,x_2) + \left(x_3 + a^{//}\right)\psi_1^{//}(x_1,x_2) \\ {}^{//}V_2(x_1,x_2,x_3) &= {}^{//}V_2^{\circ}(x_1,x_2) + \left(x_3 + a^{//}\right)\psi_2^{//}(x_1,x_2) \\ {}^{//}V_3(x_1,x_2,x_3) &= {}^{//}v_3(x_1,x_2) \end{aligned} \tag{2.3a – c}$$

where ${}^{/}V_{\alpha}^{\circ}, \overline{V}_{\alpha}, {}^{//}V_{\alpha}^{\circ}$ $(\alpha = 1,2)$ are the tangential displacements for points which lie on the mid-surface and ${}^{/}\psi_{\alpha}, \overline{\psi}_{\alpha}, {}^{//}\psi_{\alpha}$ are the shear angles for points on the mid-surface. In addition, $(a^{/} = \overline{h} + h^{/}/2)$ and $(a^{//} = \overline{h} + h^{//}/2)$ which are the distances from the global mid-surface to the mid-surface of the bottom and top faces, respectively.

At the interfaces between both facings and the core, continuity must exist regarding the displacements. For this reason, the kinematic continuity conditions must be fulfilled at these interfaces (see Librescu et al. (1997), Hause et al. (1998, 2000), Librescu (1975)). The kinematic continuity conditions can be expressed mathematically as

Bottom face sheet/core interface $\left(x_3 = \overline{h}\right)$

$$\overline{V}_1 = {}^{/}V_1 \text{ and } \overline{V}_2 = {}^{/}V_2 \tag{2.4a, b}$$

Top face sheet/core interface $\left(x_3 = -\overline{h}\right)$

$$\overline{V}_1 = {}^{//}V_1 \text{ and } \overline{V}_2 = {}^{//}V_2 \tag{2.5a, b}$$

Because the core is considered incompressible,

$${}^{/}V_3 = \overline{V}_3 = {}^{//}V_3 = v_3. \tag{2.6}$$

When Eqs. (2.4a, b)–(2.6) are used in conjunction with Eqs. (2.1a–c)–(2.3a–c), the following 3D displacement relationships result fulfilling the kinematic continuity conditions.

Bottom Face $\left(\overline{h} \le x_3 \le \overline{h} + h^{/}\right)$

$$\begin{aligned} {}^{/}V_1(x_1,x_2,x_3) &= \xi_1(x_1,x_2) + \eta_1(x_1,x_2) + \left(x_3 - a^{/}\right)\psi_1^{/}(x_1,x_2) \\ {}^{/}V_2(x_1,x_2,x_3) &= \xi_2(x_1,x_2) + \eta_2(x_1,x_2) + \left(x_3 - a^{/}\right)\psi_2^{/}(x_1,x_2) \\ {}^{/}V_3(x_1,x_2,x_3) &= v_3(x_1,x_2) \end{aligned} \tag{2.7a – c}$$

Core $\left(-\overline{h} \le x_3 \le \overline{h}\right)$

$$\overline{V}_1(x_1,x_2,x_3) = \xi_1(x_1,x_2) - \frac{1}{4}\left[{}^{/}h\psi_1^{/}(x_1,x_2) - {}^{//}h\psi_1^{//}(x_1x_2)\right]$$
$$+(x_3/\overline{h})\left\{\eta_1(x_1,x_2) - \frac{1}{4}\left[{}^{/}h\psi_1^{/}(x_1,x_2) + {}^{//}h\psi_1^{//}(x_1,x_2)\right]\right\}$$
$$\overline{V}_2(x_1,x_2,x_3) = \xi_2(x_1,x_2) - \frac{1}{4}\left[{}^{/}h\psi_2^{/}(x_1,x_2) - {}^{//}h\psi_2^{//}(x_1,x_2)\right]$$
$$+(x_3/\overline{h})\left\{\eta_2(x_1,x_2) - \frac{1}{4}\left[{}^{/}h\psi_2^{/}(x_1,x_2) + {}^{//}h\psi_2^{//}(x_1,x_2)\right]\right\}$$
$$\overline{V}_3(x_1,x_2,x_3) = v_3(x_1,x_2) \qquad (2.8\text{a}-\text{c})$$

Top Face $\left(-\overline{h} - h^{//} \leq x_3 \leq -\overline{h}\right)$

$${}^{//}V_1(x_1,x_2,x_3) = \xi_1(x_1,x_2) - \eta_1(x_1,x_2) + \left(x_3 + a^{//}\right)\psi_1^{//}(x_1,x_2)$$
$${}^{//}V_2(x_1,x_2,x_3) = \xi_2(x_1,x_2) - \eta_2(x_1,x_2) + \left(x_3 + a^{//}\right)\psi_2^{//}(x_1,x_2) \qquad (2.9\text{a}-\text{c})$$
$${}^{//}V_3(x_1,x_2,x_3) = v_3(x_1,x_2)$$

Within the above displacement equations, $\xi_1(x_1,x_2)$, $\xi_2(x_1,x_2)$, $\eta_1(x_1,x_2)$, $\eta_2(x_1, x_2)$ are defined as:

$$\xi_1(x_1,x_2) = \frac{{}^{/}V_1^{\circ} + {}^{//}V_1^{\circ}}{2}, \xi_2(x_1,x_2) = \frac{{}^{/}V_2^{\circ} + {}^{//}V_2^{\circ}}{2} \qquad (2.10\text{a, b})$$

$$\eta_1(x_1,x_2) = \frac{{}^{/}V_1^{\circ} - {}^{//}V_1^{\circ}}{2}, \eta_2(x_1,x_2) = \frac{{}^{/}V_2^{\circ} - {}^{//}V_2^{\circ}}{2} \qquad (2.11\text{a, b})$$

With these newly defined 2D tangential displacement measures in hand, the problem is reduced from 15 displacement quantities to 9 which are:

$$\xi_\alpha(x_1,x_2),\ \eta_\alpha(x_1,x_2),\ v_3(x_1,x_2),\ {}^{/}\psi_\alpha(x_1,x_2),\ {}^{//}\psi_\alpha(x_1,x_2), \qquad (\alpha = 1,\ 2)$$

2.3.2 Strain–Displacement Relationships

The strain–displacement relationships at a point in the structure are given by the Green–Lagrange strain tensor for a shallow shell. With the assumption that the transverse displacements are much larger than the in-plane or tangential displacements, in addition to the geometric imperfections, only the nonlinear terms, associated with the transverse displacements, are retained in the Green–Lagrange strain tensor (Librescu and Chang (1993)). These expressions are given by

$$e_{11} = \frac{\partial V_1}{\partial x_1} - \frac{V_3}{R_1} + \frac{1}{2}\left(\frac{\partial V_3}{\partial x_1}\right)^2 \tag{2.12}$$

$$e_{22} = \frac{\partial V_2}{\partial x_2} - \frac{V_3}{R_2} + \frac{1}{2}\left(\frac{\partial V_3}{\partial x_2}\right)^2 \tag{2.13}$$

$$\gamma_{12} = \frac{\partial V_2}{\partial x_1} + \frac{\partial V_1}{\partial x_2} + \frac{\partial V_3}{\partial x_1}\frac{\partial V_3}{\partial x_2} \tag{2.14}$$

$$\gamma_{13} = 2e_{13} = \frac{\partial V_1}{\partial x_3} + \frac{\partial V_3}{\partial x_1} + \frac{\partial V_3}{\partial x_1}\frac{\partial V_3}{\partial x_3} \tag{2.15}$$

$$\gamma_{23} = 2e_{23} = \frac{\partial V_2}{\partial x_3} + \frac{\partial V_3}{\partial x_2} + \frac{\partial V_3}{\partial x_2}\frac{\partial V_3}{\partial x_3} \tag{2.16}$$

$$e_{33} = \frac{\partial V_3}{\partial x_3} + \frac{1}{2}\left(\frac{\partial V_3}{\partial x_3}\right)^2 \tag{2.17}$$

With the introduction of a stress free initial geometric imperfection $\overset{\circ}{V}_3$, in Eqs. (2.12)–(2.17), The Green–Lagrange strain tensor for finite deformations with imperfections becomes as follows:

$$e_{11} = \frac{\partial V_1}{\partial x_1} - \frac{V_3}{R_1} + \frac{1}{2}\left(\frac{\partial V_3}{\partial x_1}\right)^2 + \frac{\partial V_3}{\partial x_1}\frac{\partial \overset{\circ}{V}_3}{\partial x_1} \tag{2.18}$$

$$e_{22} = \frac{\partial V_2}{\partial x_2} - \frac{V_3}{R_2} + \frac{1}{2}\left(\frac{\partial V_3}{\partial x_2}\right)^2 + \frac{\partial V_3}{\partial x_2}\frac{\partial \overset{\circ}{V}_3}{\partial x_2} \tag{2.19}$$

$$2e_{12} = \frac{\partial V_2}{\partial x_1} + \frac{\partial V_1}{\partial x_2} + \frac{\partial V_3}{\partial x_1}\frac{\partial V_3}{\partial x_2} + \frac{\partial V_3}{\partial x_1}\frac{\partial \overset{\circ}{V}_3}{\partial x_2} + \frac{\partial V_3}{\partial x_2}\frac{\partial \overset{\circ}{V}_3}{\partial x_1} \tag{2.20}$$

$$2e_{13} = \frac{\partial V_1}{\partial x_3} + \frac{\partial V_3}{\partial x_1} + \frac{\partial V_3}{\partial x_1}\frac{\partial V_3}{\partial x_3} + \frac{\partial V_3}{\partial x_1}\frac{\partial \overset{\circ}{V}_3}{\partial x_3} + \frac{\partial V_3}{\partial x_3}\frac{\partial \overset{\circ}{V}_3}{\partial x_1} \tag{2.21}$$

$$2e_{23} = \frac{\partial V_2}{\partial x_3} + \frac{\partial V_3}{\partial x_2} + \frac{\partial V_3}{\partial x_2}\frac{\partial V_3}{\partial x_3} + \frac{\partial V_3}{\partial x_2}\frac{\partial \overset{\circ}{V}_3}{\partial x_3} + \frac{\partial V_3}{\partial x_3}\frac{\partial \overset{\circ}{V}_3}{\partial x_2} \tag{2.22}$$

$$e_{33} = \frac{\partial V_3}{\partial x_3} + \frac{1}{2}\left(\frac{\partial V_3}{\partial x_3}\right)^2 + \frac{\partial V_3}{\partial x_3}\frac{\partial \overset{\circ}{V}_3}{\partial x_3} \tag{2.23}$$

substituting Eqs. (2.7a–c)–(2.9a–c) into Eqs. (2.18)–(2.23) one obtains the 3D strain quantities in terms of the 2D strain measures for each respective layer of the sandwich structure. These 3D strain quantities are given as

Bottom Facings $\left(\bar{h} \le x_3 \le \bar{h} + h^{/}\right)$

$$\begin{aligned}
{}^{/}e_{11} &= {}^{/}\varepsilon_{11} + \left(x_3 - a^{/}\right)\kappa_{11}^{/} \\
{}^{/}e_{22} &= {}^{/}\varepsilon_{22} + \left(x_3 - a^{/}\right)\kappa_{22}^{/} \\
2^{/}e_{12} &= {}^{/}\gamma_{12} + \left(x_3 - a^{/}\right)\kappa_{12}^{/} \\
2^{/}e_{13} &= {}^{/}\gamma_{13} \\
2^{/}e_{23} &= {}^{/}\gamma_{23} \\
{}^{/}e_{33} &= 0
\end{aligned} \tag{2.24a – f}$$

Core $\left(-\bar{h} \le x_3 \le \bar{h}\right)$

$$\begin{aligned}
\bar{e}_{11} &= \bar{\varepsilon}_{11} + x_3\bar{\kappa}_{11} \\
\bar{e}_{22} &= \bar{\varepsilon}_{22} + x_3\bar{\kappa}_{22} \\
2\bar{e}_{12} &= \bar{\varepsilon}_{12} + x_3\bar{\kappa}_{12} \\
2\bar{e}_{13} &= \bar{\gamma}_{13} \\
2\bar{e}_{23} &= \bar{\gamma}_{23} \\
\bar{e}_{33} &= 0
\end{aligned} \tag{2.25a – f}$$

Top Face $\left(-\bar{h} - h^{//} \le x_3 \le -\bar{h}\right)$

$$\begin{aligned}
{}^{//}e_{11} &= {}^{//}\varepsilon_{11} + \left(x_3 + a^{//}\right)\kappa_{11}^{//} \\
{}^{//}e_{22} &= {}^{//}\varepsilon_{22} + \left(x_3 + a^{//}\right)\kappa_{22}^{//} \\
{}^{//}2e_{12} &= {}^{//}\gamma_{12} + \left(x_3 + a^{//}\right)\kappa_{12}^{//} \\
{}^{//}2e_{13} &= {}^{//}\gamma_{13} \\
{}^{//}2e_{23} &= {}^{//}\gamma_{23} \\
{}^{//}e_{33} &= 0
\end{aligned} \tag{2.26a – f}$$

In the above equations, $\varepsilon_{ij}^{/}$, $\varepsilon_{ij}^{//}$, $\varepsilon_{i3}^{/}$, $\varepsilon_{i3}^{//}$, $\gamma_{ij}^{/}$, $\gamma_{ij}^{//}$ and $\bar{\varepsilon}_{i3}$ $(i = 1, 2, 3)$ represent the 2D tangential and transverse strain measures, respectively. While $\kappa_{ij}^{/}$ and $\kappa_{ij}^{//}$ represent the bending strains. Their expressions are given as

Bottom Face $\left(\bar{h} \leq x_3 \leq \bar{h} + h^{/}\right)$

$$ {}^{/}\varepsilon_{11} = \frac{\partial \xi_1}{\partial x_1} + \frac{\partial \eta_1}{\partial x_1} + \frac{1}{2}\left(\frac{\partial v_3}{\partial x_1}\right)^2 + \frac{\partial v_3}{\partial x_1}\frac{\partial \mathring{v}_3}{\partial x_1} - \frac{v_3}{R_1} \tag{2.27} $$

$$ {}^{/}\varepsilon_{22} = \frac{\partial \xi_2}{\partial x_2} + \frac{\partial \eta_2}{\partial x_2} + \frac{1}{2}\left(\frac{\partial v_3}{\partial x_2}\right)^2 + \frac{\partial v_3}{\partial x_2}\frac{\partial \mathring{v}_3}{\partial x_2} - \frac{v_3}{R_2} \tag{2.28} $$

$$ {}^{/}\gamma_{12} = \frac{\partial \xi_1}{\partial x_2} + \frac{\partial \xi_2}{\partial x_1} + \frac{\partial \eta_1}{\partial x_2} + \frac{\partial \eta_2}{\partial x_1} + \frac{\partial v_3}{\partial x_1}\frac{\partial v_3}{\partial x_2} + \frac{\partial \mathring{v}_3}{\partial x_1}\frac{\partial v_3}{\partial x_2} + \frac{\partial v_3}{\partial x_1}\frac{\partial \mathring{v}_3}{\partial x_2} \tag{2.29} $$

$$ {}^{/}\gamma_{13} = {}^{/}\psi_1 + \frac{\partial v_3}{\partial x_1} \tag{2.30} $$

$$ {}^{/}\gamma_{23} = {}^{/}\psi_2 + \frac{\partial v_3}{\partial x_2} \tag{2.31} $$

$$ {}^{/}\kappa_{11} = \frac{\partial {}^{/}\psi_1}{\partial x_1} \tag{2.32} $$

$$ {}^{/}\kappa_{22} = \frac{\partial {}^{/}\psi_2}{\partial x_2} \tag{2.33} $$

$$ {}^{/}\kappa_{12} = \frac{\partial {}^{/}\psi_1}{\partial x_2} + \frac{\partial {}^{/}\psi_2}{\partial x_1} \tag{2.34} $$

Core $\left(-\bar{h} \leq x_3 \leq \bar{h}\right)$

$$ \bar{\varepsilon}_{11} = \frac{\partial \xi_1}{\partial x_1} - \frac{1}{4}\left(h^{/}\frac{\partial \psi_1^{/}}{\partial x_1} - h^{//}\frac{\partial \psi_1^{//}}{\partial x_1}\right) + \frac{1}{2}\left(\frac{\partial v_3}{\partial x_1}\right)^2 + \frac{\partial \mathring{v}_3}{\partial x_1}\frac{\partial v_3}{\partial x_1} - \frac{v_3}{R_1} \tag{2.35} $$

$$ \bar{\varepsilon}_{22} = \frac{\partial \xi_2}{\partial x_2} - \frac{1}{4}\left(h^{/}\frac{\partial \psi_2^{/}}{\partial x_2} - h^{//}\frac{\partial \psi_2^{//}}{\partial x_2}\right) + \frac{1}{2}\left(\frac{\partial v_3}{\partial x_2}\right)^2 + \frac{\partial \mathring{v}_3}{\partial x_2}\frac{\partial v_3}{\partial x_2} - \frac{v_3}{R_2} \tag{2.36} $$

$$ \begin{aligned} \bar{\gamma}_{12} = {} & \frac{\partial \xi_1}{\partial x_2} + \frac{\partial \xi_2}{\partial x_1} - \frac{1}{4}\left[h^{/}\left(\frac{\partial \psi_1^{/}}{\partial x_2} + \frac{\partial \psi_2^{/}}{\partial x_1}\right) - h^{//}\left(\frac{\partial \psi_1^{//}}{\partial x_2} + \frac{\partial \psi_2^{//}}{\partial x_1}\right)\right] + \frac{\partial v_3}{\partial x_1} \\ & \times \frac{\partial v_3}{\partial x_2} + \frac{\partial \mathring{v}_3}{\partial x_1}\frac{\partial v_3}{\partial x_2} + \frac{\partial v_3}{\partial x_1}\frac{\partial \mathring{v}_3}{\partial x_2} \end{aligned} \tag{2.37} $$

$$\bar{\gamma}_{13} = \frac{1}{\bar{h}}\left\{\eta_1 - \frac{1}{4}\left(h^{/}\psi_1^{/} + h^{//}\psi_1^{//}\right)\right\} + \frac{\partial v_3}{\partial x_1} \tag{2.38}$$

$$\bar{\gamma}_{23} = \frac{1}{\bar{h}}\left\{\eta_2 - \frac{1}{4}\left(h^{/}\psi_2^{/} + h^{//}\psi_2^{//}\right)\right\} + \frac{\partial v_3}{\partial x_2} \tag{2.39}$$

$$\bar{\kappa}_{11} = \frac{1}{\bar{h}}\left\{\frac{\partial \eta_1}{\partial x_1} - \frac{1}{4}\left(h^{/}\frac{\partial \psi_1^{/}}{\partial x_1} + h^{//}\frac{\partial \psi_1^{//}}{\partial x_1}\right)\right\} \tag{2.40}$$

$$\bar{\kappa}_{22} = \frac{1}{\bar{h}}\left\{\frac{\partial \eta_2}{\partial x_2} - \frac{1}{4}\left(h^{/}\frac{\partial \psi_2^{/}}{\partial x_2} + h^{//}\frac{\partial \psi_2^{//}}{\partial x_2}\right)\right\} \tag{2.41}$$

$$\bar{\kappa}_{12} = \frac{1}{\bar{h}}\left\{\frac{\partial \eta_1}{\partial x_2} + \frac{\partial \eta_2}{\partial x_1} - \frac{1}{4}\left[h^{/}\left(\frac{\partial \psi_1^{/}}{\partial x_2} + \frac{\partial \psi_2^{/}}{\partial x_1}\right) + h^{//}\left(\frac{\partial \psi_1^{//}}{\partial x_{12}} + \frac{\partial \psi_2^{//}}{\partial x_1}\right)\right]\right\} \tag{2.42}$$

Top Face $\left(-\bar{h} - h^{//} \le x_3 \le -\bar{h}\right)$

$$^{//}\varepsilon_{11} = \frac{\partial \xi_1}{\partial x_1} - \frac{\partial \eta_1}{\partial x_1} + \frac{1}{2}\left(\frac{\partial v_3}{\partial x_1}\right)^2 + \frac{\partial v_3}{\partial x_1}\frac{\partial \mathring{v}_3}{\partial x_1} - \frac{v_3}{R_1} \tag{2.43}$$

$$^{//}\varepsilon_{22} = \frac{\partial \xi_2}{\partial x_2} - \frac{\partial \eta_2}{\partial x_2} + \frac{1}{2}\left(\frac{\partial v_3}{\partial x_2}\right)^2 + \frac{\partial v_3}{\partial x_2}\frac{\partial \mathring{v}_3}{\partial x_2} - \frac{v_3}{R_2} \tag{2.44}$$

$$^{//}\gamma_{12} = \frac{\partial \xi_1}{\partial x_2} + \frac{\partial \xi_2}{\partial x_1} - \frac{\partial \eta_1}{\partial x_2} - \frac{\partial \eta_2}{\partial x_1} + \frac{\partial v_3}{\partial x_1}\frac{\partial v_3}{\partial x_2} + \frac{\partial \mathring{v}_3}{\partial x_1}\frac{\partial v_3}{\partial x_2} + \frac{\partial v_3}{\partial x_1}\frac{\partial \mathring{v}_3}{\partial x_2} \tag{2.45}$$

$$^{//}\gamma_{13} = {}^{//}\psi_1 + \frac{\partial v_3}{\partial x_1} \tag{2.46}$$

$$^{//}\gamma_{23} = {}^{//}\psi_2 + \frac{\partial v_3}{\partial x_2} \tag{2.47}$$

$$^{//}\kappa_{11} = \frac{\partial^{//}\psi_1}{\partial x_1} \tag{2.48}$$

$$^{//}\kappa_{22} = \frac{\partial^{//}\psi_2}{\partial x_2} \tag{2.49}$$

$$^{//}\kappa_{12} = \frac{\partial^{//}\psi_1}{\partial x_2} + \frac{\partial^{//}\psi_2}{\partial x_1} \tag{2.50}$$

Up to this point it has been assumed that the facings are thick enough to contain transverse shear stresses of significant value. If the facings are regarded as being inherently thin the transverse shear stresses can be neglected. This is based on the Love–Kirchhoff assumption. By setting the transverse shear stresses to zero in Eqs. (2.30), (2.31), (2.46), and Eq. (2.47) the shear angles become

$$^{/}\psi_1 = {^{//}\psi_1} = -\frac{\partial v_3}{\partial x_1} \text{ and } ^{/}\psi_2 = {^{//}\psi_2} = -\frac{\partial v_3}{\partial x_2}. \tag{2.51a, b}$$

In addition, if the sandwich structure is globally symmetric (with respect to the global mid-surface) and both of the facings share the same thickness. This implies that

$$h^{/} = h^{//} = h \text{ and } a^{/} = a^{//} = a\left(\equiv \overline{h} + h/2\right). \tag{2.52a, b}$$

With the Love–Kirchhoff assumption and assuming local and global symmetry, the strain-displacement Eqs. (2.27)–(2.50) become

Bottom Face $\left(\overline{h} \leq x_3 \leq \overline{h} + h^{/}\right)$

$$^{/}\varepsilon_{11} = \frac{\partial \xi_1}{\partial x_1} + \frac{\partial \eta_1}{\partial x_1} + \frac{1}{2}\left(\frac{\partial v_3}{\partial x_1}\right)^2 + \frac{\partial v_3}{\partial x_1}\frac{\partial \overset{\circ}{v}_3}{\partial x_1} - \frac{v_3}{R_1} \tag{2.53}$$

$$^{/}\varepsilon_{22} = \frac{\partial \xi_2}{\partial x_2} + \frac{\partial \eta_2}{\partial x_2} + \frac{1}{2}\left(\frac{\partial v_3}{\partial x_2}\right)^2 + \frac{\partial v_3}{\partial x_2}\frac{\partial \overset{\circ}{v}_3}{\partial x_2} - \frac{v_3}{R_2} \tag{2.54}$$

$$^{/}\gamma_{12} = \frac{\partial \xi_1}{\partial x_2} + \frac{\partial \xi_2}{\partial x_1} + \frac{\partial \eta_1}{\partial x_2} + \frac{\partial \eta_2}{\partial x_1} + \frac{\partial v_3}{\partial x_1}\frac{\partial v_3}{\partial x_2} + \frac{\partial \overset{\circ}{v}_3}{\partial x_1}\frac{\partial v_3}{\partial x_2} + \frac{\partial v_3}{\partial x_1}\frac{\partial \overset{\circ}{v}_3}{\partial x_2} \tag{2.55}$$

$$^{/}\gamma_{13} = 0 \tag{2.56}$$

$$^{/}\gamma_{23} = 0 \tag{2.57}$$

$$^{/}\kappa_{11} = -\frac{\partial^2 v_3}{\partial x_1^2} \tag{2.58}$$

$$^{/}\kappa_{22} = -\frac{\partial^2 v_3}{\partial x_2^2} \tag{2.59}$$

$$^{/}\kappa_{12} = -2\frac{\partial^2 v_3}{\partial x_1 \partial x_2} \tag{2.60}$$

Core $\left(-\overline{h} \leq x_3 \leq \overline{h}\right)$

$$\bar{\varepsilon}_{11} = \frac{\partial \xi_1}{\partial x_1} + \frac{1}{2}\left(\frac{\partial v_3}{\partial x_1}\right)^2 + \frac{\partial v_3^\circ}{\partial x_1}\frac{\partial v_3}{\partial x_1} \tag{2.61}$$

$$\bar{\varepsilon}_{22} = \frac{\partial \xi_2}{\partial x_2} + \frac{1}{2}\left(\frac{\partial v_3}{\partial x_2}\right)^2 + \frac{\partial v_3^\circ}{\partial x_2}\frac{\partial v_3}{\partial x_2} \tag{2.62}$$

$$\bar{\gamma}_{12} = \frac{\partial \xi_1}{\partial x_2} + \frac{\partial \xi_2}{\partial x_1} + \frac{\partial v_3}{\partial x_1}\frac{\partial v_3}{\partial x_2} + \frac{\partial v_3^\circ}{\partial x_1}\frac{\partial v_3}{\partial x_2} + \frac{\partial v_3}{\partial x_1}\frac{\partial v_3^\circ}{\partial x_2} \tag{2.63}$$

$$\bar{\kappa}_{11} = \frac{1}{\bar{h}}\left\{\frac{\partial \eta_1}{\partial x_1} + \frac{h}{2}\frac{\partial^2 v_3}{\partial x_1^2}\right\} \tag{2.64}$$

$$\bar{\kappa}_{22} = \frac{1}{\bar{h}}\left\{\frac{\partial \eta_2}{\partial x_2} + \frac{h}{2}\frac{\partial^2 v_3}{\partial x_2^2}\right\} \tag{2.65}$$

$$\bar{\kappa}_{12} = \frac{1}{\bar{h}}\left\{\frac{\partial \eta_1}{\partial x_2} + \frac{\partial \eta_2}{\partial x_1} + h\frac{\partial^2 v_3}{\partial x_1 \partial x_2}\right\} \tag{2.66}$$

$$\bar{\gamma}_{13} = \frac{1}{\bar{h}}\left\{\eta_1 + \frac{h}{2}\frac{\partial v_3}{\partial x_1}\right\} + \frac{\partial v_3}{\partial x_1} \tag{2.67}$$

$$\bar{\gamma}_{23} = \frac{1}{\bar{h}}\left\{\eta_2 + \frac{h}{2}\frac{\partial v_3}{\partial x_2}\right\} + \frac{\partial v_3}{\partial x_2} \tag{2.68}$$

Top Face $\left(-\bar{h} - h^{//} \le x_3 \le -\bar{h}\right)$

$${}^{//}\varepsilon_{11} = \frac{\partial \xi_1}{\partial x_1} - \frac{\partial \eta_1}{\partial x_1} + \frac{1}{2}\left(\frac{\partial v_3}{\partial x_1}\right)^2 + \frac{\partial v_3}{\partial x_1}\frac{\partial \mathring{v}_3}{\partial x_1} - \frac{v_3}{R_1} \tag{2.69}$$

$${}^{//}\varepsilon_{22} = \frac{\partial \xi_2}{\partial x_2} - \frac{\partial \eta_2}{\partial x_2} + \frac{1}{2}\left(\frac{\partial v_3}{\partial x_2}\right)^2 + \frac{\partial v_3}{\partial x_2}\frac{\partial \mathring{v}_3}{\partial x_2} - \frac{v_3}{R_2} \tag{2.70}$$

$$^{//}\gamma_{12} = \frac{\partial \xi_1}{\partial x_2} + \frac{\partial \xi_2}{\partial x_1} - \frac{\partial \eta_1}{\partial x_2} - \frac{\partial \eta_2}{\partial x_1} + \frac{\partial v_3}{\partial x_1}\frac{\partial v_3}{\partial x_2} + \frac{\partial \overset{\circ}{v}_3}{\partial x_1}\frac{\partial v_3}{\partial x_2} + \frac{\partial v_3}{\partial x_1}\frac{\partial \overset{\circ}{v}_3}{\partial x_2} \tag{2.71}$$

$$^{//}\gamma_{13} = 0 \tag{2.72}$$

$$^{//}\gamma_{23} = 0 \tag{2.73}$$

$$^{//}\kappa_{11} = -\frac{\partial^2 v_3}{\partial x_1^2} \tag{2.74}$$

$$^{//}\kappa_{22} = -\frac{\partial^2 v_3}{\partial x_2^2} \tag{2.75}$$

$$^{//}\kappa_{12} = -2\frac{\partial^2 v_3}{\partial x_1 \partial x_2} \tag{2.76}$$

2.3.3 Constitutive Equations

Within the 3D geometrically nonlinear elasticity theory, the constitutive equations are described by the linear relationship between the second Piola–Kirchhoff stress and Lagrange strain tensor. As a result, for an anisotropic material featuring monoclinic symmetry Hooke's Law for the kth layer of a composite lamina can be described by (see Reddy 2004 and Jones 1999)

$$\begin{Bmatrix} \sigma_{11} \\ \sigma_{22} \\ \sigma_{12} \end{Bmatrix}_k = \begin{bmatrix} \widehat{Q}_{11} & \widehat{Q}_{12} & \widehat{Q}_{16} \\ \widehat{Q}_{12} & \widehat{Q}_{22} & \widehat{Q}_{26} \\ \widehat{Q}_{16} & \widehat{Q}_{26} & \widehat{Q}_{66} \end{bmatrix}_k \begin{Bmatrix} e_{11} \\ e_{22} \\ \gamma_{12} \end{Bmatrix}_k - \begin{Bmatrix} \widehat{\lambda}_{11} \\ \widehat{\lambda}_{22} \\ \widehat{\lambda}_{12} \end{Bmatrix}_k \Delta T - \begin{Bmatrix} \widehat{\beta}_{11} \\ \widehat{\beta}_{22} \\ \widehat{\beta}_{12} \end{Bmatrix}_k \Delta M \tag{2.77a}$$

$$\begin{Bmatrix} \sigma_{23} \\ \sigma_{13} \end{Bmatrix}_k = K^2 \begin{bmatrix} \widehat{Q}_{44} & \widehat{Q}_{45} \\ \widehat{Q}_{45} & \widehat{Q}_{55} \end{bmatrix}_k \begin{Bmatrix} 2e_{23} \\ 2e_{13} \end{Bmatrix}_k \tag{2.77b}$$

where, $\widehat{Q}_{ij}, \widehat{\lambda}_{ij}, \widehat{\mu}_{ij}$ are termed the reduced elastic moduli, the reduced thermal moduli, and the reduced moisture moduli, respectively. The expressions for these quantities are provided as

$$
\begin{aligned}
\widehat{Q}_{11}^{k} &= Q_{11}^{k}\cos^{4}\theta_k + 2\left(Q_{12}^{k} + 2Q_{66}^{k}\right)\sin^{2}\theta_k\cos^{2}\theta_k + Q_{22}^{k}\sin^{4}\theta_k \\
\widehat{Q}_{12}^{k} &= \left(Q_{11}^{k} + Q_{22}^{k} - 4Q_{66}^{k}\right)\sin^{2}\theta_k\cos^{2}\theta_k + Q_{12}^{k}\left(\sin^{4}\theta_k + \cos^{4}\theta_k\right) \\
\widehat{Q}_{22}^{k} &= Q_{11}^{k}\sin^{4}\theta_k + 2\left(Q_{12}^{k} + 2Q_{66}^{k}\right)\sin^{2}\theta_k\cos^{2}\theta_k + Q_{22}^{k}\cos^{4}\theta_k \\
\widehat{Q}_{16}^{k} &= \left(Q_{11}^{k} - Q_{12}^{k} - 2Q_{66}^{k}\right)\sin\theta_k\cos^{3}\theta_k + \left(Q_{12}^{k} - Q_{22}^{k} + 2Q_{66}^{k}\right)\sin^{3}\theta_k\cos\theta_k \\
\widehat{Q}_{26}^{k} &= \left(Q_{11}^{k} - Q_{12}^{k} - 2Q_{66}^{k}\right)\sin^{3}\theta_k\cos\theta_k + \left(Q_{12}^{k} - Q_{22}^{k} + 2Q_{66}^{k}\right)\sin\theta_k\cos^{3}\theta_k \\
\widehat{Q}_{66}^{k} &= \left(Q_{11}^{k} + Q_{22}^{k} - 2Q_{12}^{k} - 2Q_{66}^{k}\right)\sin^{2}\theta_k\cos^{2}\theta_k + Q_{66}^{k}\left(\sin^{4}\theta_k + \cos^{4}\theta_k\right) \\
\widehat{Q}_{44}^{k} &= Q_{44}^{k}\cos^{2}\theta_k + Q_{55}^{k}\sin^{2}\theta_k \\
\widehat{Q}_{45}^{k} &= \left(Q_{55}^{k} - Q_{44}^{k}\right)\cos\theta_k\sin\theta_k \\
\widehat{Q}_{55}^{k} &= Q_{55}^{k}\cos^{2}\theta_k + Q_{44}^{k}\sin^{2}\theta_k \\
\widehat{\lambda}_{11}^{k} &= \widehat{Q}_{11}^{k}\widehat{\alpha}_{11}^{k} + \widehat{Q}_{12}^{k}\widehat{\alpha}_{22}^{k} + \widehat{Q}_{16}^{k}\widehat{\alpha}_{12}^{k} \\
\widehat{\lambda}_{22}^{k} &= \widehat{Q}_{22}^{k}\widehat{\alpha}_{22}^{k} + \widehat{Q}_{12}^{k}\widehat{\alpha}_{22}^{k} + \widehat{Q}_{26}^{k}\widehat{\alpha}_{12}^{k} \\
\widehat{\lambda}_{12}^{k} &= \widehat{Q}_{16}^{k}\widehat{\alpha}_{11}^{k} + \widehat{Q}_{26}^{k}\widehat{\alpha}_{22}^{k} + \widehat{Q}_{66}^{k}\widehat{\alpha}_{12}^{k} \\
\widehat{\mu}_{11}^{k} &= \widehat{Q}_{11}^{k}\widehat{\beta}_{11}^{k} + \widehat{Q}_{12}^{k}\widehat{\beta}_{22}^{k} + \widehat{Q}_{16}^{k}\widehat{\beta}_{12}^{k} \\
\widehat{\mu}_{22}^{k} &= \widehat{Q}_{22}^{k}\widehat{\beta}_{22}^{k} + \widehat{Q}_{12}^{k}\widehat{\beta}_{11}^{k} + \widehat{Q}_{26}^{k}\widehat{\beta}_{12}^{k} \\
\widehat{\mu}_{12}^{k} &= \widehat{Q}_{16}^{k}\widehat{\beta}_{11}^{k} + \widehat{Q}_{26}^{k}\widehat{\beta}_{22}^{k} + \widehat{Q}_{66}^{k}\widehat{\beta}_{12}^{k} \\
\widehat{\alpha}_{11}^{k} &= \alpha_{11}^{k}\cos^{2}\theta_k + \alpha_{22}^{k}\sin^{2}\theta_k \\
\widehat{\alpha}_{22}^{k} &= \alpha_{22}^{k}\cos^{2}\theta_k + \alpha_{11}^{k}\sin^{2}\theta_k \\
\widehat{\alpha}_{12}^{k} &= \left(\alpha_{11}^{k} - \alpha_{22}^{k}\right)\cos\theta_k\sin\theta_k \\
\widehat{\beta}_{11}^{k} &= \beta_{11}^{k}\cos^{2}\theta_k + \beta_{22}^{k}\sin^{2}\theta_k \\
\widehat{\beta}_{22}^{k} &= \beta_{22}^{k}\cos^{2}\theta_k + \beta_{11}^{k}\sin^{2}\theta_k \\
\widehat{\beta}_{12}^{k} &= \left(\beta_{11}^{k} - \beta_{22}^{k}\right)\cos\theta_k\sin\theta_k
\end{aligned}
\tag{2.78a – u}
$$

where the material stiffnesses are

$$
Q_{11}^{k} = \frac{E_1^{k}}{1 - \nu_{12}^{k}\nu_{21}^{k}}, \quad Q_{22} = \frac{E_2^{k}}{1 - \nu_{12}^{k}\nu_{21}^{k}}, \quad Q_{12} = \frac{\nu_{21}^{k}E_1^{k}}{1 - \nu_{12}^{k}\nu_{21}^{k}} = \frac{\nu_{12}^{k}E_2^{k}}{1 - \nu_{12}^{k}\nu_{21}^{k}},
$$

$$
Q_{44}^{k} = G_{23}^{k}, \qquad Q_{55}^{k} = G_{13}^{k}, \qquad Q_{66}^{k} = G_{12}^{k}
\tag{2.79a – f}
$$

While E_i, ν_{ij}, G_{ij}, G_{i3} $(i,\ j = 1,2)$ are the Young's Modulus, Poisson's Ratio, Shear Modulus of the facings, and the core shear moduli, respectively; While β_{ij}, α_{ij} are the moisture expansion coefficient and the coefficient of thermal expansion, respectively. Assigning the constitutive equations to each layer of the sandwich panel results in

Bottom Face $\left(\bar{h} \le x_3 \le \bar{h} + h^{/}\right)$

$$\begin{Bmatrix} {}^{/}\sigma_{11} \\ {}^{/}\sigma_{22} \\ {}^{/}\sigma_{12} \end{Bmatrix}_k = \begin{bmatrix} {}^{/}\widehat{Q}_{11} & {}^{/}\widehat{Q}_{12} & {}^{/}\widehat{Q}_{16} \\ {}^{/}\widehat{Q}_{12} & {}^{/}\widehat{Q}_{22} & {}^{/}\widehat{Q}_{26} \\ {}^{/}\widehat{Q}_{16} & {}^{/}\widehat{Q}_{26} & {}^{/}\widehat{Q}_{66} \end{bmatrix}_k \begin{Bmatrix} {}^{/}e_{11} \\ {}^{/}e_{22} \\ {}^{/}\gamma_{12} \end{Bmatrix}_k - \begin{Bmatrix} {}^{/}\widehat{\lambda}_{11} \\ {}^{/}\widehat{\lambda}_{22} \\ {}^{/}\widehat{\lambda}_{12} \end{Bmatrix}_k \Delta T - \begin{Bmatrix} {}^{/}\widehat{\mu}_{11} \\ {}^{/}\widehat{\mu}_{22} \\ {}^{/}\widehat{\mu}_{12} \end{Bmatrix}_k \Delta M \tag{2.80a}$$

$$\begin{Bmatrix} {}^{/}\sigma_{23} \\ {}^{/}\sigma_{13} \end{Bmatrix}_k = {}^{/}K^2 \begin{bmatrix} {}^{/}\widehat{Q}_{44} & {}^{/}\widehat{Q}_{45} \\ {}^{/}\widehat{Q}_{45} & \widehat{Q}_{55} \end{bmatrix}_k \begin{Bmatrix} 2{}^{/}e_{23} \\ 2{}^{/}e_{13} \end{Bmatrix}_k \tag{2.80b}$$

Core $\left(-\bar{h} \le x_3 \le \bar{h}\right)$

$$\begin{Bmatrix} \bar{\sigma}_{11} \\ \bar{\sigma}_{22} \\ \bar{\sigma}_{12} \end{Bmatrix} = \begin{bmatrix} \bar{Q}_{11} & \bar{Q}_{12} & 0 \\ \bar{Q}_{12} & \bar{Q}_{22} & 0 \\ 0 & 0 & \bar{Q}_{66} \end{bmatrix} \begin{Bmatrix} \bar{e}_{11} \\ \bar{e}_{22} \\ \bar{\gamma}_{12} \end{Bmatrix} - \begin{Bmatrix} \bar{\lambda}_{11} \\ \bar{\lambda}_{22} \\ \bar{\lambda}_{12} \end{Bmatrix} \Delta T - \begin{Bmatrix} \bar{\mu}_{11} \\ \bar{\mu}_{22} \\ \bar{\mu}_{12} \end{Bmatrix} \Delta M \tag{2.81a}$$

$$\begin{Bmatrix} \bar{\sigma}_{23} \\ \bar{\sigma}_{13} \end{Bmatrix} = \bar{K}^2 \begin{bmatrix} \bar{Q}_{44} & 0 \\ 0 & \bar{Q}_{55} \end{bmatrix}_k \begin{Bmatrix} 2\bar{e}_{23} \\ 2\bar{e}_{13} \end{Bmatrix} \tag{2.81b}$$

Top Face $\left(-\bar{h} - h^{//} \le x_3 \le -\bar{h}\right)$

$$\begin{Bmatrix} {}^{//}\sigma_{11} \\ {}^{//}\sigma_{22} \\ {}^{//}\sigma_{12} \end{Bmatrix}_k = \begin{bmatrix} {}^{//}\widehat{Q}_{11} & {}^{//}\widehat{Q}_{12} & {}^{//}\widehat{Q}_{16} \\ {}^{//}\widehat{Q}_{12} & {}^{//}\widehat{Q}_{22} & {}^{//}\widehat{Q}_{26} \\ {}^{//}\widehat{Q}_{16} & {}^{//}\widehat{Q}_{26} & {}^{//}\widehat{Q}_{66} \end{bmatrix}_k \begin{Bmatrix} {}^{//}e_{11} \\ {}^{//}e_{22} \\ {}^{//}\gamma_{12} \end{Bmatrix}_k - \begin{Bmatrix} {}^{//}\widehat{\lambda}_{11} \\ {}^{//}\widehat{\lambda}_{22} \\ {}^{//}\widehat{\lambda}_{12} \end{Bmatrix}_k \Delta T - \begin{Bmatrix} {}^{//}\widehat{\mu}_{11} \\ {}^{//}\widehat{\mu}_{22} \\ {}^{//}\widehat{\mu}_{12} \end{Bmatrix}_k \Delta M \tag{2.82a}$$

$$\begin{Bmatrix} {}^{//}\sigma_{23} \\ {}^{//}\sigma_{13} \end{Bmatrix}_k = {}^{//}K^2 \begin{bmatrix} {}^{//}\widehat{Q}_{44} & {}^{//}\widehat{Q}_{45} \\ {}^{//}\widehat{Q}_{45} & {}^{//}\widehat{Q}_{55} \end{bmatrix}_k \begin{Bmatrix} 2^{//}e_{23} \\ 2^{//}e_{13} \end{Bmatrix}_k \tag{2.82b}$$

2.3.4 Stress and Moment Resultants

By definition and consistent with shallow shell theory, the stress resultants for each layer of the sandwich structure are defined below. Unless otherwise stated, $(\alpha, \beta = 1, 2)$.

Bottom Face $\left(\overline{h} \le x_3 \le \overline{h} + h^{/}\right)$

$$\left\{N^{/}_{\alpha\beta}, M^{/}_{\alpha\beta}\right\} = \sum_{k=1}^{n^{/}} \int_{(x_3)_{k-1}}^{(x_3)_k} \left({}^{/}\sigma_{\alpha\beta}\right)_k \left\{1, \left(x_3 - a^{/}\right)\right\} dx_3 \tag{2.83a}$$

$$\left\{N^{/}_{\alpha 3}\right\} = \sum_{k=1}^{n^{/}} \int_{(x_3)_{k-1}}^{(x_3)_k} \left({}^{/}\sigma_{\alpha 3}\right)_k dx_3 \tag{2.83b}$$

Core $\left(-\overline{h} \le x_3 \le \overline{h}\right)$

$$\left\{\overline{N}_{\alpha\beta}, \overline{M}_{\alpha\beta}\right\} = \int_{-\overline{h}}^{\overline{h}} \overline{\sigma}_{\alpha\beta}\{1, x_3\} dx_3 \tag{2.84a}$$

$$\overline{N}_{\alpha 3} = \int_{-\overline{h}}^{\overline{h}} \overline{\sigma}_{\alpha 3} dx_3 \tag{2.84b}$$

Top Face $\left(-\overline{h} - h^{//} \le x_3 \le -\overline{h}\right)$

$$\left\{N^{//}_{\alpha\beta}, M^{//}_{\alpha\beta}\right\} = \sum_{k=1}^{n^{//}} \int_{(x_3)_{k-1}}^{(x_3)_k} \left({}^{//}\sigma_{\alpha\beta}\right)_k \left\{1, \left(x_3 + a^{//}\right)\right\} dx_3 \tag{2.85a}$$

$$\left\{N^{//}_{\alpha 3}\right\} = \sum_{k=1}^{n^{//}} \int_{(x_3)_{k-1}}^{(x_3)_k} \left({}^{//}\sigma_{\alpha 3}\right)_k dx_3 \tag{2.85b}$$

It should be noted that $(x_3)_k$ and $(x_3)_{k-1}$ denote the distances from the global mid-surface to the upper and bottom interfaces of the kth layer, respectively. These definitions are similar to the ones defined in Librescu (1970, 1975).

Substituting Eqs. (2.80a) and (2.80b) into Eqs. (2.83a) and (2.83b) gives the stress and stress couple resultants in terms of the strain measures for the bottom face of the sandwich panel as

Bottom Face $\left(\bar{h} \le x_3 \le \bar{h} + h^{/}\right)$

$$\begin{Bmatrix} {}^{/}N_{11} \\ {}^{/}N_{22} \\ {}^{/}N_{12} \\ {}^{/}M_{11} \\ {}^{/}M_{22} \\ {}^{/}M_{12} \end{Bmatrix} = \begin{bmatrix} {}^{/}A_{11} & {}^{/}A_{12} & {}^{/}A_{16} & {}^{/}E_{11} & {}^{/}E_{12} & {}^{/}E_{16} \\ {}^{/}A_{12} & {}^{/}A_{22} & {}^{/}A_{26} & {}^{/}E_{12} & {}^{/}E_{22} & {}^{/}E_{26} \\ {}^{/}A_{16} & {}^{/}A_{26} & {}^{/}A_{66} & {}^{/}E_{16} & {}^{/}E_{26} & {}^{/}E_{66} \\ {}^{/}E_{11} & {}^{/}E_{12} & {}^{/}E_{16} & {}^{/}F_{11} & {}^{/}F_{12} & {}^{/}F_{16} \\ {}^{/}E_{12} & {}^{/}E_{22} & {}^{/}E_{26} & {}^{/}F_{12} & {}^{/}F_{22} & {}^{/}F_{26} \\ {}^{/}E_{16} & {}^{/}E_{26} & {}^{/}E_{66} & {}^{/}F_{16} & {}^{/}F_{26} & {}^{/}F_{66} \end{bmatrix} \begin{Bmatrix} {}^{/}\varepsilon_{11} \\ {}^{/}\varepsilon_{22} \\ {}^{/}\gamma_{12} \\ {}^{/}\kappa_{11} \\ {}^{/}\kappa_{22} \\ {}^{/}\kappa_{12} \end{Bmatrix} - \begin{Bmatrix} {}^{/}N^{T}_{11} \\ {}^{/}N^{T}_{22} \\ {}^{/}N^{T}_{12} \\ {}^{/}M^{T}_{11} \\ {}^{/}M^{T}_{22} \\ {}^{/}M^{T}_{12} \end{Bmatrix} - \begin{Bmatrix} {}^{/}N^{m}_{11} \\ {}^{/}N^{m}_{22} \\ {}^{/}N^{m}_{12} \\ {}^{/}M^{m}_{11} \\ {}^{/}M^{m}_{22} \\ {}^{/}M^{m}_{12} \end{Bmatrix} \tag{2.86a}$$

$$\begin{Bmatrix} {}^{/}N_{23} \\ {}^{/}N_{13} \end{Bmatrix} = {}^{/}K^2 \begin{bmatrix} {}^{/}A_{44} & {}^{/}A_{45} \\ {}^{/}A_{45} & {}^{/}A_{55} \end{bmatrix} \begin{Bmatrix} {}^{/}\gamma_{23} \\ {}^{/}\gamma_{13} \end{Bmatrix} \tag{2.86b}$$

where the stiffnesses are defined as

$$\left\{A^{/}_{\omega\rho}, B^{/}_{\omega\rho}, D^{/}_{\omega\rho}\right\} = \sum_{k=1}^{n^{/}} \int_{(x_3)_{k-1}}^{(x_3)_k} \left(\widehat{Q}^{/}_{\omega\rho}\right)_k \left(1, x_3, x_3^2\right) dx_3 \qquad (\omega, \rho = 1, 2, 6) \tag{2.87a}$$

$$A^{/}_{IJ} = \sum_{k=1}^{n^{/}} \int_{(x_3)_{k-1}}^{(x_3)_k} \left(\widehat{Q}^{/}_{IJ}\right)_k dx_3 \qquad (I, J = 4, 5) \tag{2.87b}$$

$$E'_{\omega\rho} = B'_{\omega\rho} - a' A'_{\omega\rho}, \qquad F'_{\omega\rho} = D'_{\omega\rho} - 2a' B'_{\omega\rho} + \left(a'\right)^2 A'_{\omega\rho} \tag{2.87c, d}$$

and the thermal stress and stress couple resultants are defined as

$$\begin{aligned} {}^{/}N^{T}_{\alpha\beta} &= \sum_{k=1}^{n'} \int_{(x_3)_{k-1}}^{(x_3)_k} \left({}^{/}\widehat{\lambda}_{\alpha\beta}\right)_k \Delta T dx_3 \\ {}^{/}M^{T}_{\alpha\beta} &= \sum_{k=1}^{n'} \int_{(x_3)_{k-1}}^{(x_3)_k} \left(x_3 - a'\right)\left({}^{/}\widehat{\lambda}_{\alpha\beta}\right)_k \Delta T dx_3 \end{aligned} \tag{2.88a, b}$$

$$\begin{aligned} {}^{/}N^{m}_{\alpha\beta} &= \sum_{k=1}^{n'} \int_{(x_3)_{k-1}}^{(x_3)_k} \left({}^{/}\widehat{\beta}_{\alpha\beta}\right)_k \Delta M dx_3 \\ {}^{/}M^{m}_{\alpha\beta} &= \sum_{k=1}^{n'} \int_{(x_3)_{k-1}}^{(x_3)_k} \left(x_3 - a'\right)\left({}^{/}\widehat{\beta}_{\alpha\beta}\right)_k \Delta M dx_3 \end{aligned} \tag{2.89a, b}$$

Core $\left(-\overline{h} \le x_3 \le \overline{h}\right)$

Substituting Eqs. (2.81a) and (2.81b) into Eqs. (2.84a) and (2.84b) gives the stress and stress couple resultants for the strong core layer in terms of the strain measures which are determined to be

$$\begin{Bmatrix} \overline{N}_{11} \\ \overline{N}_{22} \\ \overline{N}_{12} \end{Bmatrix} = 2\overline{h} \begin{bmatrix} \overline{Q}_{11} & \overline{Q}_{12} & 0 \\ \overline{Q}_{12} & \overline{Q}_{22} & 0 \\ 0 & 0 & \overline{Q}_{66} \end{bmatrix} \begin{Bmatrix} \overline{\varepsilon}_{11} \\ \overline{\varepsilon}_{22} \\ \overline{\gamma}_{12} \end{Bmatrix} - \begin{Bmatrix} \overline{N}^{T}_{11} \\ \overline{N}^{T}_{22} \\ 0 \end{Bmatrix} - \begin{Bmatrix} \overline{N}^{m}_{11} \\ \overline{N}^{m}_{22} \\ 0 \end{Bmatrix} \tag{2.90a}$$

$$\begin{Bmatrix} \overline{M}_{11} \\ \overline{M}_{22} \\ \overline{M}_{12} \end{Bmatrix} = \frac{2}{3}\overline{h}^{3} \begin{bmatrix} \overline{Q}_{11} & \overline{Q}_{12} & 0 \\ \overline{Q}_{12} & \overline{Q}_{22} & 0 \\ 0 & 0 & \overline{Q}_{66} \end{bmatrix} \begin{Bmatrix} \overline{\kappa}_{11} \\ \overline{\kappa}_{22} \\ \overline{\kappa}_{12} \end{Bmatrix} - \begin{Bmatrix} \overline{M}^{T}_{11} \\ \overline{M}^{T}_{22} \\ 0 \end{Bmatrix} - \begin{Bmatrix} \overline{M}^{m}_{11} \\ \overline{M}^{m}_{22} \\ 0 \end{Bmatrix} \tag{2.90b}$$

$$\begin{Bmatrix} \overline{N}_{23} \\ \overline{N}_{13} \end{Bmatrix} = 2\overline{h}\overline{K}^{2} \begin{bmatrix} \overline{Q}_{44} & 0 \\ 0 & \overline{Q}_{55} \end{bmatrix} \begin{Bmatrix} \overline{\gamma}_{23} \\ \overline{\gamma}_{13} \end{Bmatrix} \tag{2.90c}$$

where $\overline{Q}_{ij}$ $(i, j = 1, 2, 6)$ are given by Eqs. (2.79a–f). Finally, the thermal stress and stress couple resultants in Eqs. (2.90a) and (2.90b) are expressed as

$$\left(\overline{N}^{T}_{\alpha\beta}, \overline{M}^{T}_{\alpha\beta}\right) = \int_{-\overline{h}}^{\overline{h}} (1, x_3)\overline{\lambda}_{\alpha\beta}\Delta T dx_3 \tag{2.91a}$$

$$\left(\overline{N}^{m}_{\alpha\beta}, \overline{M}^{m}_{\alpha\beta}\right) = \int_{-\overline{h}}^{\overline{h}} (1, x_3)\overline{\mu}_{\alpha\beta}\Delta M dx_3 \tag{2.91b}$$

Top Face $\left(-\overline{h} - h^{//} \leq x_3 \leq -\overline{h}\right)$

Substituting Eqs. (2.82a) and (2.82b) into Eqs. (2.85a) and (2.85b) gives the stress and stress couple resultants in terms of the strain measures for the top face as

$$\begin{Bmatrix} {}^{//}N_{11} \\ {}^{//}N_{22} \\ {}^{//}N_{12} \\ {}^{//}M_{11} \\ {}^{//}M_{22} \\ {}^{//}M_{12} \end{Bmatrix} = \begin{bmatrix} {}^{//}A_{11} & {}^{//}A_{12} & {}^{//}A_{16} & {}^{//}E_{11} & {}^{//}E_{12} & {}^{//}E_{16} \\ {}^{//}A_{12} & {}^{//}A_{22} & {}^{//}A_{26} & {}^{//}E_{12} & {}^{//}E_{22} & {}^{//}E_{26} \\ {}^{//}A_{16} & {}^{//}A_{26} & {}^{//}A_{66} & {}^{//}E_{16} & {}^{//}E_{26} & {}^{//}E_{66} \\ {}^{//}E_{11} & {}^{//}E_{12} & {}^{//}E_{16} & {}^{//}F_{11} & {}^{//}F_{12} & {}^{//}F_{16} \\ {}^{//}E_{12} & {}^{//}E_{22} & {}^{//}E_{26} & {}^{//}F_{12} & {}^{//}F_{22} & {}^{//}F_{26} \\ {}^{//}E_{16} & {}^{//}E_{26} & {}^{//}E_{66} & {}^{//}F_{16} & {}^{//}F_{26} & {}^{//}F_{66} \end{bmatrix}$$

$$\times \begin{Bmatrix} {}^{//}\varepsilon_{11} \\ {}^{//}\varepsilon_{22} \\ {}^{//}\gamma_{12} \\ {}^{//}\kappa_{11} \\ {}^{//}\kappa_{22} \\ {}^{//}\kappa_{12} \end{Bmatrix} - \begin{Bmatrix} {}^{//}N^{T}_{11} \\ {}^{//}N^{T}_{22} \\ {}^{//}N^{T}_{12} \\ {}^{//}M^{T}_{11} \\ {}^{//}M^{T}_{22} \\ {}^{//}M^{T}_{12} \end{Bmatrix} - \begin{Bmatrix} {}^{//}N^{m}_{11} \\ {}^{//}N^{m}_{22} \\ {}^{//}N^{m}_{12} \\ {}^{//}M^{m}_{11} \\ {}^{//}M^{m}_{22} \\ {}^{//}M^{m}_{12} \end{Bmatrix} \tag{2.92a}$$

$$\begin{Bmatrix} {}^{//}N_{23} \\ {}^{//}N_{13} \end{Bmatrix} = {}^{//}K^2 \begin{bmatrix} {}^{//}A_{44} & {}^{//}A_{45} \\ {}^{//}A_{45} & {}^{//}A_{55} \end{bmatrix} \begin{Bmatrix} {}^{//}\gamma_{23} \\ {}^{//}\gamma_{13} \end{Bmatrix} \tag{2.92b}$$

where the stiffnesses for the top face are defined as

$$\left\{A^{//}_{\omega\rho}, B^{//}_{\omega\rho}, D^{//}_{\omega\rho}\right\} = \sum_{k=1}^{n^{//}} \int_{(x_3)_{k-1}}^{(x_3)_k} \left(\widehat{Q}^{//}_{\omega\rho}\right)_k \left(1, x_3, x_3^2\right) dx_3 \qquad (\omega, \rho = 1, 2, 6) \tag{2.93a}$$

$$A_{IJ}^{//} = \sum_{k=1}^{n^{//}} \int_{(x_3)_{k-1}}^{(x_3)_k} \left(\widehat{Q}_{IJ}^{//}\right)_k dx_3 \qquad (I, J = 4, 5) \tag{2.93b}$$

$$E_{\omega\rho}^{//} = B_{\omega\rho}^{//} + a^{//} A_{\omega\rho}^{//}, \qquad F_{\omega\rho}^{//} = D_{\omega\rho}^{//} + 2a^{//} B_{\omega\rho}^{//} + \left(a^{//}\right)^2 A_{\omega\rho}^{//} \tag{2.93c}$$

and the thermal stress and stress couple resultants are defined as

$$\begin{aligned} {}^{//}N_{\alpha\beta}^{T} &= \sum_{k=1}^{n^{//}} \int_{(x_3)_{k-1}}^{(x_3)_k} \left({}^{//}\widehat{\lambda}_{\alpha\beta}\right)_k \Delta T dx_3 \\ {}^{//}M_{\alpha\beta}^{T} &= \sum_{k=1}^{n^{//}} \int_{(x_3)_{k-1}}^{(x_3)_k} \left(x_3 + a^{//}\right)\left({}^{//}\widehat{\lambda}_{\alpha\beta}\right)_k \Delta T dx_3 \end{aligned} \tag{2.94a, b}$$

$$\begin{aligned} {}^{//}N_{\alpha\beta}^{m} &= \sum_{k=1}^{n^{//}} \int_{(x_3)_{k-1}}^{(x_3)_k} \left({}^{//}\widehat{\beta}_{\alpha\beta}\right)_k \Delta M dx_3 \\ {}^{//}M_{\alpha\beta}^{m} &= \sum_{k=1}^{n^{//}} \int_{(x_3)_{k-1}}^{(x_3)_k} \left(x_3 + a^{//}\right)\left({}^{//}\widehat{\beta}_{\alpha\beta}\right)_k \Delta M dx_3 \end{aligned} \tag{2.95c, d}$$

In the above equations, $K^{/}$, $\overline{K}$, $K^{//}$ are known as the shear correction factors.

2.4 Hamilton's Principle

2.4.1 Hamilton's Equation

To derive the equations of motion, Hamilton's variation principle will be adopted. The advantage of using this method is that, as a byproduct, the boundary conditions appear within the developments. Hamilton's equation is given by

$$\delta J = \delta \int_{t_0}^{t_1} (U - W - T)dt = 0 \tag{2.96}$$

where t_0, t_1 are two arbitrary instants of time. U denotes the strain energy within a deformable body, W represents the work due to surface tractions, the work due to edge loads, and the work due to body forces. T denotes the kinetic energy of the 3D body.

2.4.2 Strain Energy

The strain energy stored in a 3D elastic body is expressed mathematically by

$$U = \frac{1}{2}\int_{\tau} \sigma_{ij} e_{ij} d\tau \tag{2.97}$$

$d\tau$ implies a volume element ($d\tau = d\sigma dx_3$) where $d\sigma$ represents the planar area of an element. The total strain energy in a sandwich structure is a summation of the strain energies in the upper facings, the bottom facings, and the core. A variation in the total energy, considering three separate layers of the sandwich panel, is expressed as

$$\delta U = \frac{1}{2}\int_{\sigma}\left[\int_{\bar{h}}^{\bar{h}+h^{/}} \sigma^{/}_{ij}\delta e^{/}_{ij} + \int_{-\bar{h}}^{\bar{h}} \bar{\sigma}_{ij}\delta \bar{e}_{ij} + \int_{-\bar{h}-h^{//}}^{-\bar{h}} \sigma^{//}_{ij}\delta e^{//}_{ij} + \right] dx_3 d\sigma \quad (i, j = 1, 2, 3) \tag{2.98}$$

An alternative form which involves the temperature terms will become useful later on and is expressed as

$$\begin{aligned}\delta U = \frac{1}{2}\delta\int_{\sigma}\Bigg\{\int_{\bar{h}}^{\bar{h}+h^{/}} \left(\widehat{Q}^{/}_{\alpha\beta\omega\rho} e^{/}_{\alpha\beta} e^{/}_{\omega\rho} - 2\widehat{\lambda}^{/}_{\alpha\beta} T e^{/}_{\alpha\beta}\right) dx_3 + \int_{-\bar{h}-h^{/}}^{-\bar{h}} \Big(\widehat{Q}^{//}_{\alpha\beta\omega\rho} e^{//}_{\alpha\beta} e^{//}_{\omega\rho} \\ - 2\widehat{\lambda}^{//}_{\alpha\beta} T e^{//}_{\alpha\beta}\Big) dx_3 + \int_{-\bar{h}}^{\bar{h}} \left(\underline{\widehat{\bar{Q}}_{\alpha\beta\omega\rho} \bar{e}_{\alpha\beta} \bar{e}_{\omega\rho} - 2\widehat{\bar{\lambda}}_{\alpha\beta} T \bar{e}_{\alpha\beta}} + \bar{Q}_{\alpha 3\omega 3} \bar{e}_{\alpha 3} \bar{e}_{\omega 3}\right) dx_3 \Bigg\} d\sigma\end{aligned} \tag{2.99}$$

Equation (2.99) is valid for both the weak- and strong-core model. For the weak-core model, the underlined terms should be discarded. The Greek indices assume the values 1,2 and the summation convention is held over a repeated index. The terms

$$\widehat{Q}_{\alpha\beta\omega\rho}\left(\equiv Q_{\alpha\beta\omega\rho} - \frac{Q_{\alpha\beta 33} Q_{33\omega\rho}}{Q_{3333}}\right) \quad \text{and} \quad \widehat{\lambda}_{\alpha\beta}\left(\equiv \lambda_{\alpha\beta} - \frac{Q_{\alpha\beta 33}}{Q_{3333}}\lambda_{33}\right) \tag{2.100}$$

are referred to as the modified elastic moduli and the thermal compliance expansion coefficients, respectively (see Librescu, 1975). It is also assumed that Q_{ijmn} and λ_{ij} are temperature independent. Expanding Eq. (2.98) based on Einstein's summation convention gives

$$\delta U=\int_{A}\left\{\int_{\bar{h}}^{\bar{h}+h^{/}}\left(\sigma_{11}^{/}\delta e_{11}^{/}+\sigma_{22}^{/}\delta e_{22}^{/}+\sigma_{12}^{/}\delta\gamma_{12}^{/}+\sigma_{13}^{/}\delta\gamma_{13}^{/}+\sigma_{23}^{/}\delta\gamma_{23}^{/}\right)dx_{3}\right.$$
$$+\int_{-\bar{h}}^{\bar{h}}\left(\bar{\sigma}_{11}\delta\bar{e}_{11}+\bar{\sigma}_{22}\delta\bar{e}_{22}+\bar{\sigma}_{12}\delta\bar{\gamma}_{12}+\bar{\sigma}_{13}\delta\bar{\gamma}_{13}+\bar{\sigma}_{23}\delta\bar{\gamma}_{23}\right)dx_{3}+$$
$$\left.+\int_{-\bar{h}-h^{//}}^{-\bar{h}}\left(\sigma_{11}^{//}\delta e_{11}^{//}+\sigma_{22}^{//}\delta e_{22}^{//}+\sigma_{12}^{//}\delta\gamma_{12}^{//}+\sigma_{13}^{//}\delta\gamma_{13}^{//}+\sigma_{23}^{//}\delta\gamma_{23}^{//}\right)dx_{3}\right\}dA \tag{2.101}$$

where σ_{ij} are the components of the second Piola–Kirchhoff stress tensor and A is the area of the sandwich structure. Substituting Eqs. (2.24a–f)–(2.26a–f) into Eq. (2.101) results in

$$\delta U=\int_{A}\left\{\int_{\bar{h}}^{\bar{h}+h^{/}}\left\{\sigma_{11}^{/}\left(\delta\varepsilon_{11}^{/}+\left(x_{3}-a^{/}\right)\delta\kappa_{11}^{/}\right)+\sigma_{22}^{/}\left(\delta\varepsilon_{22}^{/}+\left(x_{3}-a^{/}\right)\delta\kappa_{22}^{/}\right)+\right.\right.$$
$$\sigma_{12}^{/}\left(\delta\gamma_{12}^{/}+\left(x_{3}-a^{/}\right)\delta\kappa_{12}^{/}\right)+\sigma_{13}^{/}\delta\gamma_{13}^{/}+\sigma_{23}^{/}\delta\gamma_{23}^{/}\Big\}dx_{3}+\int_{-\bar{h}}^{\bar{h}}\{\bar{\sigma}_{11}(\delta\bar{\varepsilon}_{11}+$$
$$x_{3}\delta\bar{\kappa}_{11})+\bar{\sigma}_{22}(\delta\bar{\varepsilon}_{22}+x_{3}\delta\bar{\kappa}_{22})+\bar{\sigma}_{12}(\delta\bar{\varepsilon}_{12}+x_{3}\delta\bar{\kappa}_{12})+\bar{\sigma}_{13}\delta\bar{\gamma}_{13}+$$
$$\bar{\sigma}_{13}\delta\bar{\gamma}_{13}\}dx_{3}+\int_{-\bar{h}-h^{//}}^{-\bar{h}}\sigma_{11}^{//}\left(\delta\varepsilon_{11}^{//}+\left(x_{3}+a^{//}\right)\delta\kappa_{11}^{//}\right)+\sigma_{22}^{//}\left(\delta\varepsilon_{22}^{//}+(x_{3}+\right.$$
$$\left.\left.a^{//}\right)\delta\kappa_{22}^{//}\right)+\sigma_{12}^{//}\left(\delta\gamma_{12}^{//}+\left(x_{3}+a^{//}\right)\delta\kappa_{12}^{//}\right)+\sigma_{13}^{//}\delta\gamma_{13}^{//}+\sigma_{23}^{//}\delta\gamma_{23}^{//}\Big\}dx_{3}\Big\}dA \tag{2.102}$$

Using the definitions of the local stress resultants and stress couples from Eqs. (2.83a, b)–(2.85a, b) allows the variation in the strain energy to be expressed as

$$\delta U=\int_{A}\Big[N_{11}^{/}\delta\varepsilon_{11}^{/}+M_{11}^{/}\delta\kappa_{11}^{/}+N_{22}^{/}\delta\varepsilon_{22}^{/}+M_{22}^{/}\delta\kappa_{22}^{/}+N_{12}^{/}\delta\gamma_{12}^{/}+M_{12}^{/}\delta\kappa_{12}^{/}+$$
$$N_{13}^{/}\delta\gamma_{13}^{/}+N_{23}^{/}\delta\gamma_{23}^{/}+\overline{N}_{11}\delta\overline{\varepsilon}_{11}+\overline{M}_{11}\delta\overline{\kappa}_{11}+\overline{N}_{22}\delta\overline{\varepsilon}_{22}+\overline{M}_{22}\delta\overline{\kappa}_{22}+\overline{N}_{12}\delta\overline{\varepsilon}_{12}+$$
$$\overline{M}_{12}\delta\overline{\kappa}_{12}+\overline{N}_{13}\delta\overline{\gamma}_{13}+\overline{N}_{23}\delta\overline{\gamma}_{23}+N_{11}^{//}\delta\varepsilon_{11}^{//}+M_{11}^{//}\delta\kappa_{11}^{//}+N_{22}^{//}\delta\varepsilon_{22}^{//}+M_{22}^{//}\delta\kappa_{22}^{//}+$$
$$N_{12}^{//}\delta\gamma_{12}^{//}+M_{12}^{//}\delta\kappa_{12}^{//}+N_{13}^{//}\delta\gamma_{13}^{//}+N_{23}^{//}\delta\gamma_{23}^{//}\Big]dA \tag{2.103}$$

Substituting the strain–displacement equations, Eqs. (2.53)–(2.76) into Eq. (2.103) integrating by parts, simplifying, and gathering coefficients of like virtual displacements, results in

$$
\begin{aligned}
\delta U = & \int_0^{l_2}\int_0^{l_1}\left\langle -\left(\frac{\partial N_{11}}{\partial x_1}+\frac{\partial N_{12}}{\partial x_2}\right)\delta\xi_1-\left(\frac{\partial N_{22}}{\partial x_2}+\frac{\partial N_{12}}{\partial x_1}\right)\delta\xi_2-\left(\frac{\partial L_{11}}{\partial x_1}+\frac{\partial L_{12}}{\partial x_2}-\frac{\overline{N}_{13}}{\overline{h}}\right)\delta\eta_1\right.\\
& -\left(\frac{\partial L_{22}}{\partial x_2}+\frac{\partial L_{12}}{\partial x_1}-\frac{\overline{N}_{23}}{\overline{h}}\right)\delta\eta_2+\left(\frac{\partial P_{11}}{\partial x_1}+\frac{\partial P_{12}}{\partial x_2}+N'_{13}-\frac{h'}{4\overline{h}}\overline{N}_{13}\right)\delta\psi'_1+\\
& \left(\frac{\partial P_{22}}{\partial x_2}+\frac{\partial P_{12}}{\partial x_1}+N'_{23}-\frac{h'}{4\overline{h}}\overline{N}_{23}\right)\delta\psi'_2-\left(\frac{\partial S_{11}}{\partial x_1}+\frac{\partial S_{12}}{\partial x_2}-N''_{13}+\frac{h''}{4\overline{h}}\overline{N}_{13}\right)\delta\psi''_1\\
& \left(\frac{\partial S_{22}}{\partial x_2}+\frac{\partial S_{12}}{\partial x_1}-N''_{23}+\frac{h''}{4\overline{h}}\overline{N}_{23}\right)\delta\psi''_2-\left\{\left(\frac{\partial N_{11}}{\partial x_1}+\frac{\partial N_{12}}{\partial x_2}\right)\left(\frac{\partial v_3}{\partial x_1}+\frac{\partial v_3^\circ}{\partial x_1}\right)+\right.\\
& \left(\frac{\partial N_{22}}{\partial x_2}+\frac{\partial N_{12}}{\partial x_1}\right)\left(\frac{\partial v_3}{\partial x_2}+\frac{\partial v_3^\circ}{\partial x_2}\right)+N_{11}\left(\frac{\partial^2 v_3}{\partial x_1^2}+\frac{\partial^2 v_3^\circ}{\partial x_1^2}\right)+2N_{12}\left(\frac{\partial^2 v_3}{\partial x_1\partial x_2}+\frac{\partial^2 v_3^\circ}{\partial x_1\partial x_2}\right)+\\
& \left.\left. N_{22}\left(\frac{\partial^2 v_3}{\partial x_2^2}+\frac{\partial^2 v_3^\circ}{\partial x_2^2}\right)+\frac{N_{11}}{R_1}+\frac{N_{22}}{R_2}-\frac{\partial N_{13}}{\partial x_1}-\frac{\partial N_{23}}{\partial x_2}\right\}\delta v_3\right\rangle dx_1dx_2+\\
& \int_0^{l_2}\left\langle N_{11}\delta\xi_1+N_{12}\delta\xi_2+L_{11}\delta\eta_1+L_{12}\delta\eta_2-P_{11}\delta\psi'_1-P_{12}\delta\psi'_2+S_{11}\delta\psi''_1+\right.\\
& \left.\left. S_{12}\delta\psi''_2+\left\{N_{11}\left(\frac{\partial v_3}{\partial x_1}+\frac{\partial v_3^\circ}{\partial x_1}\right)+N_{12}\left(\frac{\partial v_3}{\partial x_2}+\frac{\partial v_3^\circ}{\partial x_2}\right)+N_{13}\right\}\delta v_3\right\rangle\right|_0^{l_1}dx_2\\
& \int_0^{l_1}\left\langle N_{22}\delta\xi_2+N_{12}\delta\xi_1+L_{22}\delta\eta_2+L_{12}\delta\eta_1-P_{22}\delta\psi'_2-P_{12}\delta\psi'_1+S_{22}\delta\psi''_2+\right.\\
& \left.\left. S_{12}\delta\psi''_1+\left\{N_{22}\left(\frac{\partial v_3}{\partial x_2}+\frac{\partial v_3^\circ}{\partial x_2}\right)+N_{12}\left(\frac{\partial v_3}{\partial x_1}+\frac{\partial v_3^\circ}{\partial x_1}\right)+N_{23}\right\}\delta v_3\right\rangle\right|_0^{l_2}dx_1
\end{aligned}
\tag{2.104}
$$

where the global stress and stress couple resultants appearing in this equation are defined as

$$
\begin{aligned}
& N_{11}=N'_{11}+\overline{N}_{11}+N''_{11},\quad N_{22}=N'_{22}+\overline{N}_{22}+N''_{22},\quad N_{12}=N'_{12}+\overline{N}_{12}+N''_{12}\\
& N_{13}=N'_{13}+\overline{N}_{13}+N''_{13},\quad N_{23}=N'_{23}+\overline{N}_{23}+N''_{23}\\
& L_{11}=\left(N'_{11}-N''_{11}\right)+\overline{M}_{11}/\overline{h},\quad L_{22}=\left(N'_{22}-N''_{22}\right)+\overline{M}_{22}/\overline{h}\\
& L_{12}=\overline{h}\left(N'_{12}-N''_{12}\right)+\overline{M}_{12}/\overline{h}\\
& P_{11}=\left(h'/4\right)\left(\overline{N}_{11}+\overline{M}_{11}/\overline{h}\right)-M'_{11},\quad P_{22}=\left(h'/4\right)\left(\overline{N}_{22}+\overline{M}_{22}/\overline{h}\right)-M'_{22}\\
& P_{12}=\left(h'/4\right)\left(\overline{N}_{12}+\overline{M}_{12}/\overline{h}\right)-M'_{12}\\
& S_{11}=\left(h''/4\right)\left(\overline{N}_{11}-\overline{M}_{11}/\overline{h}\right)+M''_{11},\quad S_{22}=\left(h''/4\right)\left(\overline{N}_{22}-\overline{M}_{22}/\overline{h}\right)+M''_{22}\\
& S_{12}=\left(h''/4\right)\left(\overline{N}_{12}-\overline{M}_{12}/\overline{h}\right)+M''_{12}
\end{aligned}
\tag{2.105a – n}
$$

2.4.3 Kinetic Energy

The kinetic energy of elastic body by definition is of the form

$$T = \frac{1}{2}\int_{x_3}\int_{x_2}\int_{x_1}\rho\left[\left(\frac{\partial V_1}{\partial t}\right)^2 + \left(\frac{\partial V_2}{\partial t}\right)^2\left(\frac{\partial V_3}{\partial t}\right)^2\right]dx_1dx_2dx_3 \tag{2.106}$$

where ρ is the mass density per unit area. In general, the tangential velocities can be neglected due to the small tangential displacements as a function of time. As a result, this assumption is adopted and only the transverse inertia will be considered within the expression for the kinetic energy. Considering that the kinetic energy of a sandwich structure is the summation of the kinetic energies of each of the individual layers, the variation in the kinetic energy becomes

$$\int_{t_0}^{t_1}\delta T dt = -\int_{t_0}^{t_1}\int_{\sigma}\left[\int_{\bar{h}}^{\bar{h}+h^{/}} {}^{/}\rho\frac{\partial^2 V_3^{/}}{\partial t^2}\delta V_3^{/}dx_3 + \int_{-\bar{h}}^{\bar{h}}\bar{\rho}\frac{\partial^2\overline{V}_3}{\partial t^2}\delta\overline{V}_3 dx_3 \right.$$
$$\left. + \int_{-\bar{h}-h^{//}}^{-\bar{h}} {}^{//}\rho\frac{\partial^2 V_3^{//}}{\partial t^2}\delta V_3^{//}\right]d\sigma dt \tag{2.107}$$

Substituting in the expressions for the transverse displacements, from Eqs. (2.7c), (2.8c), and (2.9c), gives

$$\int_{t_0}^{t_1}\delta T dt = -\int_{t_0}^{t_1}\int_{\sigma}\left[\int_{\bar{h}}^{\bar{h}+h^{/}} {}^{/}\rho\frac{\partial^2 v_3}{\partial t^2}\delta v_3 + \int_{-\bar{h}}^{-\bar{h}}\bar{\rho}\frac{\partial^2 v_3}{\partial t^2}\delta v_3 \right.$$
$$\left. + \int_{-\bar{h}-h^{//}}^{-\bar{h}} \rho^{//}\frac{\partial^2 v_3}{\partial t^2}\delta v_3\right]d\sigma dt \tag{2.108}$$

Simplifying Eq. (2.108) by combining the coefficients of the transverse variational displacement gives

$$\int_{t_0}^{t_1}\delta T dt = -\int_{t_0}^{t_1}\int_{\sigma} m_0\frac{\partial^2 v_3}{\partial t^2}\delta v_3 d\sigma dt \tag{2.109}$$

where, the inertia quantity, m_0 is defined as

$$m_0 = \int_{\bar{h}}^{\bar{h}+h^{/}}\rho_{(k)}^{/}dx_3 + \int_{-\bar{h}}^{\bar{h}}\bar{\rho}dx_3 + \int_{-\bar{h}-h^{//}}^{-\bar{h}}\rho_{(k)}^{//}dx_3 \tag{2.110}$$

2.4.4 Work Done by External Loads

The total work consists of the work due to edge loads, surface tractions, and body forces. Surface tractions includes such loadings as lateral or transverse pressure loading. Body forces are forces which act through the body such as gravity, electrical forces, or magnetic forces. Edge loads are tangential, vertical, or normal loadings on the edge of the structure. The total work due to these loadings is the sum of the work applied to all three layers, the top face, the core, and the bottom face. This is expressed mathematically as

$$W_{\text{total}} = W_{\text{body forces}} + W_{\text{edge loads}} + W_{\text{surface tractions}} \tag{2.111}$$

- *Work due to body forces*

$$\delta W_b = \int_\sigma \left\{ \int_{\bar{h}}^{\bar{h}+h'} \rho' H_i' \delta V_i' dx_3 + \int_{-\bar{h}}^{\bar{h}} \bar{\rho} \bar{H}_i \delta \bar{V}_i dx_3 + \int_{-\bar{h}-h''}^{-\bar{h}} \rho'' H_i'' \delta V_i'' dx_3 \right\} d\sigma \tag{2.112}$$

where ρ is the mass density and H_i is the body force vector. The body forces are neglected such that gravitational, electrical, and magnetic forces are irrelevant.

- *Work due to surface tractions*

$$\delta W_{st} = \int_0^{l_2} \int_0^{l_1} q_3(x_1, x_2) \delta V_3 dx_1 dx_2 \tag{2.113}$$

- *Work due to edge loads*

  The work due to edge loads along the boundaries is expressed fundamentally as

$$\begin{aligned}
\delta W_{el} = & \int_0^{l_1} \left\{ \int_{\bar{h}}^{\bar{h}+h'} \left(\tilde{\sigma}_{22}' \delta V_2' + \tilde{\sigma}_{21}' \delta V_1' + \tilde{\sigma}_{23}' \delta V_3' \right) dx_3 \right. \\
& + \int_{-\bar{h}}^{\bar{h}} \left(\tilde{\bar{\sigma}}_{22} \delta \bar{V}_2 + \tilde{\bar{\sigma}}_{21} \delta \bar{V}_1 + \tilde{\bar{\sigma}}_{23} \delta \bar{V}_3 \right) dx_3 \\
& \left. + \int_{-\bar{h}-h''}^{-\bar{h}} \left(\tilde{\sigma}_{22}'' \delta V_2'' + \tilde{\sigma}_{21}'' \delta V_1'' + \tilde{\sigma}_{23}'' \delta V_3'' \right) dx_3 \right\} dx_1 \\
& + \int_0^{l_2} \left\{ \int_{\bar{h}}^{\bar{h}+h'} \left(\tilde{\sigma}_{11}' \delta V_1' + \tilde{\sigma}_{12}' \delta V_2' + \tilde{\sigma}_{13}' \delta V_3' \right) dx_3 \right. \\
& + \int_{-\bar{h}}^{\bar{h}} \left(\tilde{\bar{\sigma}}_{11} \delta \bar{V}_1 + \tilde{\bar{\sigma}}_{12} \delta \bar{V}_2 + \tilde{\bar{\sigma}}_{13} \delta \bar{V}_3 \right) dx_3 + \\
& \left. \int_{-\bar{h}-h''}^{-\bar{h}} \left(\tilde{\sigma}_{11}'' \delta V_1'' + \tilde{\sigma}_{12}'' \delta V_2'' + \tilde{\sigma}_{13}'' \delta V_3'' \right) dx_3 \right\} dx_2
\end{aligned} \tag{2.114}$$

(Note: Quantities with a tilde on top implies on the boundary.) Substituting in the expressions for the displacement quantities ${}^{/}V_i$, $\overline{V}_i$, and ${}^{//}V_i$ from Eqs. (2.7a–c)–(2.9a–c) while using the definition of stress and stress couples resultants defined earlier in Eqs. (2.83a, b)–(2.85a, b) results in

$$
\begin{aligned}
\delta W_{el} = &\int_0^{l_1} \Big[{}^{/}\widetilde{N}_{22}\delta\xi_2 + {}^{/}\widetilde{N}_{22}\delta\eta_2 + {}^{/}\widetilde{N}_{21}\delta\xi_1 + {}^{/}\widetilde{N}_{21}\delta\eta_1 + {}^{/}\widetilde{N}_{23}\delta v_3 + {}^{/}\widetilde{M}_{22}\delta\psi_2^{/} + {}^{/}\widetilde{M}_{21}\delta\psi_1^{/} \\
&+ \widetilde{\overline{N}}_{22}\delta\xi_2 + \widetilde{\overline{N}}_{21}\delta\xi_1 + \frac{\widetilde{\overline{M}}_{21}}{\overline{h}}\delta\eta_1 + \frac{\widetilde{\overline{M}}_{22}}{\overline{h}}\delta\eta_2 - \left(\widetilde{\overline{N}}_{21}\frac{h^{/}}{4} + \widetilde{\overline{M}}_{21}\frac{h^{/}}{4\overline{h}}\right)\delta\psi_1^{/} \\
&+ \left(\widetilde{\overline{N}}_{21}\frac{h^{//}}{4} - \widetilde{\overline{M}}_{21}\frac{h^{//}}{4\overline{h}}\right)\delta\psi_1^{//} - \left(\widetilde{\overline{N}}_{22}\frac{h^{/}}{4} + \widetilde{\overline{M}}_{22}\frac{h^{/}}{4\overline{h}}\right)\delta\psi_2^{/} + \left(\widetilde{\overline{N}}_{22}\frac{h^{//}}{4} - \widetilde{\overline{M}}_{22}\frac{h^{//}}{4\overline{h}}\right)\delta\psi_2^{//} \\
&+ \widetilde{\overline{N}}_{23}\delta v_3 + {}^{//}\widetilde{N}_{22}\delta\xi_2 - {}^{//}\widetilde{N}_{22}\delta\eta_2 {}^{//}\widetilde{N}_{21}\delta\xi_1 - {}^{//}\widetilde{N}_{21}\delta\eta_1 + {}^{//}\widetilde{N}_{23}\delta v_3 + {}^{//}\widetilde{M}_{22}\delta\psi_2^{//} \\
&+ {}^{//}\widetilde{M}_{21}\delta\psi_1^{//} \Big] dx_1 \\
&+ \int_0^{l_2} \Big[{}^{/}\widetilde{N}_{11}\delta\xi_1 + {}^{/}\widetilde{N}_{11}\delta\eta_1 + {}^{/}\widetilde{N}_{12}\delta\xi_2 + {}^{/}\widetilde{N}_{12}\delta\eta_2 + {}^{/}\widetilde{N}_{13}\delta v_3 \\
&+ {}^{/}\widetilde{M}_{11}\delta\psi_1^{/} + {}^{/}\widetilde{M}_{12}\delta\psi_2^{/} + \widetilde{\overline{N}}_{11}\delta\xi_1 + \widetilde{\overline{N}}_{12}\delta\xi_2 + \frac{\widetilde{\overline{M}}_{12}}{\overline{h}}\delta\eta_2 + \frac{\widetilde{\overline{M}}_{11}}{\overline{h}}\delta\eta_1 \\
&- \left(\widetilde{\overline{N}}_{12}\frac{h^{/}}{4} + \widetilde{\overline{M}}_{12}\frac{h^{/}}{4\overline{h}}\right)\delta\psi_2^{/} + \left(\widetilde{\overline{N}}_{12}\frac{h^{//}}{4} - \widetilde{\overline{M}}_{12}\frac{h^{//}}{4\overline{h}}\right)\delta\psi_2^{//} - \left(\widetilde{\overline{N}}_{11}\frac{h^{/}}{4} + \widetilde{\overline{M}}_{11}\frac{h^{/}}{4\overline{h}}\right)\delta\psi_1^{/} \\
&+ \left(\widetilde{\overline{N}}_{11}\frac{h^{//}}{4} - \widetilde{\overline{M}}_{11}\frac{h^{//}}{4\overline{h}}\right)\delta\psi_1^{//} + \widetilde{\overline{N}}_{13}\delta v_3 + {}^{//}\widetilde{N}_{11}\delta\xi_1 - {}^{//}\widetilde{N}_{11}\delta\eta_1 \\
&{}^{//}\widetilde{N}_{12}\delta\xi_2 - {}^{//}\widetilde{N}_{12}\delta\eta_2 + {}^{//}\widetilde{N}_{13}\delta v_3 + {}^{//}\widetilde{M}_{11}\delta\psi_1^{//} + {}^{//}\widetilde{M}_{12}\delta\psi_2^{//} \Big] dx_2 +
\end{aligned}
\tag{2.115}
$$

Simplifying and combining like terms gives the total variation in the work due to edge loads shown as

$$
\begin{aligned}
\delta W_{el} = &\int_0^{l_1} \Big\{ \widetilde{N}_{12}\delta\xi_1 + \widetilde{N}_{22}\delta\xi_2 + \widetilde{L}_{12}\delta\eta_1 + \widetilde{L}_{22}\delta\eta_2 - \widetilde{P}_{12}\delta\psi_1^{/} - \widetilde{P}_{22}\delta\psi_2^{/} + \widetilde{S}_{12}\delta\psi_1^{//} + \\
&\qquad \widetilde{S}_{22}\delta\psi_2^{//} + \widetilde{N}_{23}\delta v_3 \Big\} dx_1 \\
&+ \int_0^{l_2} \Big\{ \widetilde{N}_{11}\delta\xi_1 + \widetilde{N}_{12}\delta\xi_2 + \widetilde{L}_{11}\delta\eta_1 + \widetilde{L}_{12}\delta\eta_2 - \widetilde{P}_{11}\delta\psi_1^{/} - \widetilde{P}_{12}\delta\psi_2^{/} + \widetilde{S}_{11}\delta\psi_1^{//} + \\
&\qquad \widetilde{S}_{12}\delta\psi_2^{//} + \widetilde{N}_{13}\delta v_3 \Big\} dx_2
\end{aligned}
\tag{2.116}
$$

where the boundary global stress resultants and stress couples are given by

$$\widetilde{N}_{11} = \widetilde{N}_{11}^{/} + \widetilde{\overline{N}}_{11} + \widetilde{N}_{11}^{//}, \quad \widetilde{N}_{22} = \widetilde{N}_{22}^{/} + \widetilde{\overline{N}}_{22} + \widetilde{N}_{22}^{//}, \quad \widetilde{N}_{12} = \widetilde{N}_{12}^{/} + \widetilde{\overline{N}}_{12} + \widetilde{N}_{12}^{//}$$

$$\widetilde{N}_{13} = \widetilde{N}_{13}^{/} + \widetilde{\overline{N}}_{13} + \widetilde{N}_{13}^{//}, \quad \widetilde{N}_{23} = \widetilde{N}_{23}^{/} + \widetilde{\overline{N}}_{23} + \widetilde{N}_{23}^{//}$$

$$\widetilde{L}_{11} = \left(\widetilde{N}_{11}^{/} - \widetilde{N}_{11}^{//}\right) + \widetilde{\overline{M}}_{11}/\overline{h}, \quad \widetilde{L}_{22} = \left(\widetilde{N}_{22}^{/} - \widetilde{N}_{22}^{//}\right) + \widetilde{\overline{M}}_{22}/\overline{h}$$

$$\widetilde{L}_{12} = \left(\widetilde{N}_{12}^{/} - \widetilde{N}_{12}^{//}\right) + \widetilde{\overline{M}}_{12}/\overline{h}$$

$$\widetilde{P}_{11} = \left(h^{/}/4\right)\left(\widetilde{\overline{N}}_{11} + \widetilde{\overline{M}}_{11}/\overline{h}\right) - \widetilde{M}_{11}^{/}, \quad \widetilde{P}_{22} = \left(h^{/}/4\right)\left(\widetilde{\overline{N}}_{22} + \widetilde{\overline{M}}_{22}/\overline{h}\right) - \widetilde{M}_{22}^{/}$$

$$\widetilde{P}_{12} = \left(h^{/}/4\right)\left(\widetilde{\overline{N}}_{12} + \widetilde{\overline{M}}_{12}/\overline{h}\right) - \widetilde{M}_{12}^{/}$$

$$\widetilde{S}_{11} = \left(h^{//}/4\right)\left(\widetilde{\overline{N}}_{11} - \widetilde{\overline{M}}_{11}/\overline{h}\right) + \widetilde{M}_{11}^{//}, \quad \widetilde{S}_{22} = \left(h^{//}/4\right)\left(\widetilde{\overline{N}}_{22} - \widetilde{\overline{M}}_{22}/\overline{h}\right) + \widetilde{M}_{22}^{//}$$

$$\widetilde{S}_{12} = \left(h^{//}/4\right)\left(\widetilde{\overline{N}}_{12} - \widetilde{\overline{M}}_{12}/\overline{h}\right) + \widetilde{M}_{12}^{//}$$

(2.117a – n)

2.5 Equations of Motion – Nonlinear Formulation

2.5.1 The Mixed Formulation for Sandwich Plate and Shells

The mixed formulation which is expressed in terms of the transversal displacement and the stress and stress couple resultants can now easily be obtained by substituting the expressions for $\delta U,\ \delta T,\ \delta W$ from Eqs. (2.104), (2.109), (2.113), and (2.116) into Eq. (2.96) this results in Hamilton's Equation being expressed in a more useful form as

$$\int_{t_0}^{t_1}\int_0^{l_2}\int_0^{l_1}\left\{-\left(\frac{\partial N_{11}}{\partial x_1} + \frac{\partial N_{12}}{\partial x_2}\right)\delta\xi_1 - \left(\frac{\partial N_{22}}{\partial x_2} + \frac{\partial N_{12}}{\partial x_1}\right)\delta\xi_2 - \left(\frac{\partial L_{11}}{\partial x_1} + \frac{\partial L_{12}}{\partial x_2} - \frac{\overline{N}_{13}}{\overline{h}}\right)\delta\eta_1 -\right.$$

$$\left(\frac{\partial L_{22}}{\partial x_2} + \frac{\partial L_{12}}{\partial x_1} - \frac{\overline{N}_{23}}{\overline{h}}\right)\delta\eta_2 + \left(\frac{\partial P_{11}}{\partial x_1} + \frac{\partial P_{12}}{\partial x_2} + N_{13}^{/} - \frac{h^{/}}{4\overline{h}}\overline{N}_{13}\right)\delta\psi_1^{/} + \left(\frac{\partial P_{22}}{\partial x_2} + \frac{\partial P_{12}}{\partial x_1} + \right.$$

$$\left. N_{23}^{/} - \frac{h^{/}}{4\overline{h}}\overline{N}_{23}\right)\delta\psi_2^{/} - \left(\frac{\partial S_{11}}{\partial x_1} + \frac{\partial S_{12}}{\partial x_2} + N_{13}^{//} - \frac{h^{//}}{4\overline{h}}\overline{N}_{13}\right)\delta\psi_1^{//} - \left(\frac{\partial S_{22}}{\partial x_2} + \frac{\partial S_{12}}{\partial x_1} + N_{23}^{//} - \right.$$

$$\left.\frac{h^{//}}{4\overline{h}}\overline{N}_{23}\right)\delta\psi_2^{//} + \left[\left(\frac{\partial N_{11}}{\partial x_1} + \frac{\partial N_{12}}{\partial x_2}\right)\left(\frac{\partial v_3}{\partial x_1} + \frac{\partial v_3^{\circ}}{\partial x_1}\right) + + N_{11}\left(\frac{\partial^2 v_3}{\partial x_1^2} + \frac{\partial^2 v_3^{\circ}}{\partial x_1^2}\right)\right.$$

$$+2N_{12}\left(\frac{\partial^2 v_3}{\partial x_1 \partial x_2} + \frac{\partial^2 v_3^{\circ}}{\partial x_1 \partial x_2}\right) + N_{22}\left(\frac{\partial^2 v_3}{\partial x_2^2} + \frac{\partial^2 v_3^{\circ}}{\partial x_2^2}\right) + \frac{N_{11}}{R_1} + \frac{N_{22}}{R_2} - \frac{\partial N_{13}}{\partial x_1} - \frac{\partial N_{23}}{\partial x_2}$$

$$\left.\left. -m_0\frac{\partial^2 v_3}{\partial t^2} - q_3\right]\delta v_3\right\}dx_1 dx_2 dt$$

$$+\int_{t_0}^{t_1}\int_0^{l_2}\left\{(N_{11} - \widetilde{N}_{11})\delta\xi_1 + (N_{12} - \widetilde{N}_{12})\delta\xi_2 + (L_{11} - \widetilde{L}_{11})\delta\eta_1 + (L_{12} - \widetilde{L}_{12})\delta\eta_2\right.$$

$$-(P_{11}-\widetilde{P}_{11})\delta\psi_1^{/}-(P_{12}-\widetilde{P}_{12})\delta\psi_2^{/}+\left(S_{11}-\widetilde{S}_{11}\right)\delta\psi_1^{//}+\left(S_{12}-\widetilde{S}_{12}\right)\delta\psi_2^{//}$$

$$+\left[N_{11}\left(\frac{\partial v_3}{\partial x_1}+\frac{\partial v_3^{\circ}}{\partial x_1}\right)+N_{12}\left(\frac{\partial v_3}{\partial x_2}+\frac{\partial v_3^{\circ}}{\partial x_2}\right)+N_{13}-\widetilde{N}_{13}\right]\delta v_3\Bigg\}dx_2dt+$$

$$+\int_{t_0}^{t_1}\int_0^{l_1}\Bigg\{(N_{22}-\widetilde{N}_{22})\delta\xi_2+(N_{12}-\widetilde{N}_{12})\delta\xi_1+(L_{22}-\widetilde{L}_{22})\delta\eta_2+(L_{12}-\widetilde{L}_{12})\delta\eta_1$$

$$-(P_{22}-\widetilde{P}_{22})\delta\psi_2^{/}-(P_{12}-\widetilde{P}_{12})\delta\psi_1^{/}+\left(S_{22}-\widetilde{S}_{22}\right)\delta\psi_2^{//}+\left(S_{12}-\widetilde{S}_{12}\right)\delta\psi_1^{//}$$

$$+\left[N_{22}\left(\frac{\partial v_3}{\partial x_2}+\frac{\partial v_3^{\circ}}{\partial x_2}\right)+N_{12}\left(\frac{\partial v_3}{\partial x_1}+\frac{\partial v_3^{\circ}}{\partial x_1}\right)+N_{23}-\widetilde{N}_{23}\right]\delta v_3\Bigg\}dx_1dt=0 \tag{2.118}$$

The above equation can only be satisfied if each of the triple and double integrals are zero. Since the coefficients of the variational displacements are arbitrary the integrals are only satisfied if the coefficients are set equal to zero. Setting the coefficients to zero gives nine equations of motion and nine prescribed boundary conditions along each edge which are listed below as

$$\delta\xi_1: \quad \frac{\partial N_{11}}{\partial x_1}+\frac{\partial N_{12}}{\partial x_2}=0 \tag{2.119}$$

$$\delta\xi_2: \quad \frac{\partial N_{22}}{\partial x_2}+\frac{\partial N_{12}}{\partial x_1}=0 \tag{2.120}$$

$$\delta\eta_1: \quad \frac{\partial L_{11}}{\partial x_1}+\frac{\partial L_{12}}{\partial x_2}-\frac{\overline{N}_{13}}{\overline{h}}=0 \tag{2.121}$$

$$\delta\eta_2: \quad \frac{\partial L_{22}}{\partial x_2}+\frac{\partial L_{12}}{\partial x_1}-\frac{\overline{N}_{23}}{\overline{h}}=0 \tag{2.122}$$

$$\delta\psi_1^{/}: \quad \frac{\partial P_{11}}{\partial x_1}+\frac{\partial P_{12}}{\partial x_2}+N_{13}^{/}-\left(h^{/}/4\overline{h}\right)\overline{N}_{13}=0 \tag{2.123}$$

$$\delta\psi_2^{/}: \quad \frac{\partial P_{22}}{\partial x_2}+\frac{\partial P_{12}}{\partial x_1}+N_{23}^{/}-\left(h^{/}/4\overline{h}\right)\overline{N}_{23}=0 \tag{2.124}$$

$$\delta\psi_1^{//}: \quad \frac{\partial S_{11}}{\partial x_1}+\frac{\partial S_{12}}{\partial x_2}+N_{13}^{//}-\left(h^{//}/4\overline{h}\right)\overline{N}_{13}=0 \tag{2.125}$$

$$\delta\psi_2^{//}: \quad \frac{\partial S_{22}}{\partial x_2} + \frac{\partial S_{12}}{\partial x_1} + N_{23}^{//} - \left(h^{//}/4\bar{h}\right)\overline{N}_{23} = 0 \tag{2.126}$$

$$\begin{aligned} \delta v_3: \; & N_{11}\left(\frac{\partial^2 v_3}{\partial x_1^2} + \frac{\partial^2 \overset{\circ}{v}_3}{\partial x_1^2} + \frac{1}{R_1}\right) + 2N_{12}\left(\frac{\partial^2 v_3}{\partial x_1 \partial x_2} + \frac{\partial^2 \overset{\circ}{v}_3}{\partial x_1 \partial x_2}\right) \\ & + N_{22}\left(\frac{\partial^2 v_3}{\partial x_2^2} + \frac{\partial^2 \overset{\circ}{v}_3}{\partial x_2^2} + \frac{1}{R_2}\right) + \frac{\partial N_{13}}{\partial x_1} + \frac{\partial N_{23}}{\partial x_2} + q_3 - c\frac{\partial v_3}{\partial t} = m_0 \frac{\partial^2 v_3}{\partial t^2} \end{aligned} \tag{2.127}$$

(Note: In Eq. (2.127) the damping term has been manually added.) q_3 denotes the distributed transversal load.

The associated boundary conditions along the edges $x_n = \text{const}(\text{n} = 1, 2)$ become

$$\begin{aligned} & N_{nn} = \widetilde{N}_{nn} \quad \text{or} \quad \xi_n = \widetilde{\xi}_n \\ & N_{nt} = \widetilde{N}_{nt} \quad \text{or} \quad \xi_t = \widetilde{\xi}_t \\ & L_{nn} = \widetilde{L}_{nn} \quad \text{or} \quad \eta_n = \widetilde{\eta}_n \\ & L_{nt} = \widetilde{L}_{nt} \quad \text{or} \quad \eta_t = \widetilde{\eta}_t \\ & P_{nn} = \widetilde{P}_{nn} \quad \text{or} \quad \psi_n^{/} = \widetilde{\psi}_n^{/} \\ & P_{nt} = \widetilde{P}_{nt} \quad \text{or} \quad \psi_t^{/} = \widetilde{\psi}_t^{/} \\ & S_{nn} = \widetilde{S}_{nn} \quad \text{or} \quad \psi_n^{//} = \widetilde{\psi}_n^{//} \\ & S_{nt} = \widetilde{S}_{nt} \quad \text{or} \quad \psi_t^{//} = \widetilde{\psi}_t^{//} \\ & N_{nt}\left(\frac{\partial v_3}{\partial x_t} + \frac{\partial \overset{\circ}{v}_3}{\partial x_t}\right) + N_{nn}\left(\frac{\partial v_3}{\partial x_n} + \frac{\partial \overset{\circ}{v}_3}{\partial x_n}\right) + N_{n3} = \widetilde{N}_{n3} \quad \text{or} \quad v_3 = \widetilde{v}_3 \end{aligned} \tag{2.128a – i}$$

The subscripts n and t are used to imply the normal and tangential in-plane directions to an edge and hence, $n = 1$ when $t = 2$, and vice versa. There are nine boundary conditions which means the governing equations are of the eighteenth order. There are a few special cases of these equations presented next.

Special Cases

- *The discarding of the transverse shear effects in the facings*

When the facings are considered thin, the transverse shear stresses become negligible. Therefore, the Love–Kirchhoff assumption is adopted for the facings. Using the variational principle Eq. (2.96) in conjunction with Eqs. (2.51a, b) the equations of motion become

$$\delta\xi_1: \quad \frac{\partial N_{11}}{\partial x_1} + \frac{\partial N_{12}}{\partial x_2} = 0 \tag{2.129}$$

$$\delta \xi_2 : \quad \frac{\partial N_{22}}{\partial x_2} + \frac{\partial N_{12}}{\partial x_1} = 0 \tag{2.130}$$

$$\delta \eta_1 : \quad \frac{\partial L_{11}}{\partial x_1} + \frac{\partial L_{12}}{\partial x_2} - \frac{\overline{N}_{13}}{\overline{h}} = 0 \tag{2.131}$$

$$\delta \eta_2 : \quad \frac{\partial L_{22}}{\partial x_2} + \frac{\partial L_{12}}{\partial x_1} - \frac{\overline{N}_{23}}{\overline{h}} = 0 \tag{2.132}$$

$$\begin{aligned}\delta v_3 : \quad & N_{11}\left(\frac{\partial^2 v_3}{\partial x_1^2} + \frac{\partial^2 \overset{\circ}{v}_3}{\partial x_1^2} + \frac{1}{R_1})\right) + 2N_{12}\left(\frac{\partial^2 v_3}{\partial x_1 \partial x_2} + \frac{\partial^2 \overset{\circ}{v}_3}{\partial x_1 \partial x_2}\right) \\ & + N_{22}\left(\frac{\partial^2 v_3}{\partial x_2^2} + \frac{\partial^2 \overset{\circ}{v}_3}{\partial x_2^2} + \frac{1}{R_2}\right) - C_2\left(\frac{\partial^2 \bar{N}_{11}}{\partial x_1^2} + 2\frac{\partial^2 \bar{N}_{12}}{\partial x_1 \partial x_2} + \frac{\partial^2 \bar{N}_{22}}{\partial x_2^2}\right) \\ & + \left(1 + \frac{C_1}{\overline{h}}\right)\left(\frac{\partial \bar{N}_{13}}{\partial x_1} + \frac{\partial \bar{N}_{23}}{\partial x_2}\right) + \frac{\partial^2 M_{11}}{\partial x_1^2} + 2\frac{\partial^2 M_{12}}{\partial x_1 \partial x_2} + \frac{\partial^2 M_{22}}{\partial x_2^2} + q_3 \\ & - c\frac{\partial v_3}{\partial t} = m_0 \frac{\partial^2 v_3}{\partial t^2}\end{aligned} \tag{2.133}$$

The associated boundary conditions along the edges $x_n = \text{const}(n = 1, 2)$ become

$$\begin{aligned}& N_{nn} = \widetilde{N}_{nn} \quad \text{or} \quad \xi_n = \widetilde{\xi}_n \\ & N_{nt} = \widetilde{N}_{nt} \quad \text{or} \quad \xi_t = \widetilde{\xi}_t \\ & L_{nn} = \widetilde{L}_{nn} \quad \text{or} \quad \eta_n = \widetilde{\eta}_n \\ & L_{nt} = \widetilde{L}_{nt} \quad \text{or} \quad \eta_t = \widetilde{\eta}_t \\ & C_2\overline{N}_{nn} - M_{nn} = C_2\widetilde{\overline{N}}_{nn} - \widetilde{M}_{nn} \quad \text{or} \quad \frac{\partial v_3}{\partial x_n} = \frac{\partial \widetilde{v}_3}{\partial x_n} \\ & N_{nt}\left(\frac{\partial v_3}{\partial x_t} + \frac{\partial \overset{\circ}{v}_3}{\partial x_t}\right) + N_{nn}\left(\frac{\partial v_3}{\partial x_n} + \frac{\partial \overset{\circ}{v}_3}{\partial x_n}\right) + \frac{\partial M_{nn}}{\partial x_n} + 2\frac{\partial M_{nt}}{\partial x_t} \quad \text{or} \quad v_3 = \widetilde{v}_3 \\ & -C_2\left(\frac{\partial \overline{N}_{nn}}{\partial x_n} + 2\frac{\partial \overline{N}_{nt}}{\partial x_t}\right) + \left(1 + \frac{C_1}{\overline{h}}\right)\overline{N}_{n3} = -C_2\frac{\partial \widetilde{\overline{N}}_{nt}}{\partial x_t} + \frac{\partial \widetilde{M}_{nt}}{\partial x_t} + \widetilde{\overline{N}}_{n3}\end{aligned} \tag{2.134a – f}$$

For this case, the local stress and stress couple resultants, $N_{13}^{/} = N_{13}^{//} = N_{23}^{/} = N_{23}^{//} = 0$. This implies that the global stress resultants, N_{13} and N_{23} are reduced to

$$N_{13} = \overline{N}_{13} \tag{2.135a}$$

$$N_{23} = \overline{N}_{23} \tag{2.136b}$$

while the global stress couple resultants $M_{11},\ M_{22},\ M_{12}$ are defined as

$$
\begin{aligned}
M_{11} &= M_{11}^{/} + M_{11}^{//} - (C_1/\overline{h})\overline{M}_{11} \\
M_{22} &= M_{22}^{/} + M_{22}^{//} - (C_1/\overline{h})\overline{M}_{22} \\
M_{12} &= M_{12}^{/} + M_{12}^{//} - (C_1/\overline{h})\overline{M}_{12}
\end{aligned}
\tag{2.137a – c}
$$

where

$$
C_1 = \left(h^{/} + h^{//}\right)/4, \qquad C_2 = \left(h^{/} - h^{//}\right)/4 \tag{2.138a, b}
$$

For this case there are six boundary conditions required at each edge. This reduces the governing equations to the twelfth order.

- *Weak (soft) core with symmetric facings*

In this case the core is only capable of carrying the transverse shear stresses. As a result, $\overline{N}_{\alpha\beta}$ and $\overline{M}_{\alpha\beta}$ become immaterial. Because of this, the equations of motion, the stress and stress couple resultants, and the associated boundary conditions can be further simplified. Also with symmetric facings with respect to the global mid-surface (mid-surface of the core), $C_1 = h/2$, $C_2 = 0$. The equations of motion simplify to

$$
\delta\xi_1 : \qquad \frac{\partial N_{11}}{\partial x_1} + \frac{\partial N_{12}}{\partial x_2} = 0 \tag{2.139}
$$

$$
\delta\xi_2 : \qquad \frac{\partial N_{22}}{\partial x_2} + \frac{\partial N_{12}}{\partial x_1} = 0 \tag{2.140}
$$

$$
\delta\eta_1 : \qquad \frac{\partial L_{11}}{\partial x_1} + \frac{\partial L_{12}}{\partial x_2} - \frac{\overline{N}_{13}}{\overline{h}} = 0 \tag{2.141}
$$

$$
\delta\eta_2 : \qquad \frac{\partial L_{22}}{\partial x_2} + \frac{\partial L_{12}}{\partial x_1} - \frac{\overline{N}_{23}}{\overline{h}} = 0 \tag{2.142}
$$

$$
\begin{aligned}
\delta v_3 : \; & N_{11}\left(\frac{\partial^2 v_3}{\partial x_1^2} + \frac{\partial^2 \overset{\circ}{v}_3}{\partial x_1^2} + \frac{1}{R_1}\right) + 2N_{12}\left(\frac{\partial^2 v_3}{\partial x_1 \partial x_2} + \frac{\partial^2 \overset{\circ}{v}_3}{\partial x_1 \partial x_2}\right) \\
& + N_{22}\left(\frac{\partial^2 v_3}{\partial x_2^2} + \frac{\partial^2 \overset{\circ}{v}_3}{\partial x_2^2} + \frac{1}{R_2}\right) + \left(1 + \frac{h}{2\overline{h}}\right)\left(\frac{\partial \overline{N}_{13}}{\partial x_1} + \frac{\partial \overline{N}_{23}}{\partial x_2}\right) + \frac{\partial^2 M_{11}}{\partial x_1^2} \\
& + 2\frac{\partial^2 M_{12}}{\partial x_1 \partial x_2} + \frac{\partial^2 M_{22}}{\partial x_2^2} + q_3 - c\frac{\partial v_3}{\partial t} - m_0\frac{\partial^2 v_3}{\partial t^2} = 0
\end{aligned}
\tag{2.143}
$$

The associated boundary conditions along the edges $x_n = \text{const}(n = 1, 2)$ simplify to

$$
\begin{aligned}
&N_{nn} = \widetilde{N}_{nn} \quad \text{or} \quad \xi_n = \widetilde{\xi}_n \\
&N_{nt} = \widetilde{N}_{nt} \quad \text{or} \quad \xi_t = \widetilde{\xi}_t \\
&L_{nn} = \widetilde{L}_{nn} \quad \text{or} \quad \eta_n = \widetilde{\eta}_n \\
&L_{nt} = \widetilde{L}_{nt} \quad \text{or} \quad \eta_t = \widetilde{\eta}_t \\
&M_{nn} = \widetilde{M}_{nn} \quad \text{or} \quad \frac{\partial v_3}{\partial x_n} = \frac{\partial \widetilde{v}_3}{\partial x_n} \\
&N_{nt}\left(\frac{\partial v_3}{\partial x_t} + \frac{\partial \overset{\circ}{v}_3}{\partial x_t}\right) + N_{nn}\left(\frac{\partial v_3}{\partial x_n} + \frac{\partial \overset{\circ}{v}_3}{\partial x_n}\right) + \frac{\partial M_{nn}}{\partial x_n} + 2\frac{\partial M_{nt}}{\partial x_t} \quad \text{or} \quad v_3 = \widetilde{v}_3 \\
&+\left(1 + \frac{h}{2\overline{h}}\right)\overline{N}_{n3} = \frac{\partial \widetilde{M}_{nt}}{\partial x_t} + \widetilde{\overline{N}}_{n3}
\end{aligned}
\tag{2.144a – f}
$$

The global stress and stress couple resultants reduce to

$$
\begin{aligned}
&N_{11} = N_{11}^{/} + N_{11}^{//}, \quad N_{22} = N_{22}^{/} + N_{22}^{//}, \quad N_{12} = N_{12}^{/} + N_{12}^{//} \\
&N_{13} = \overline{N}_{13}, \quad N_{23} = \overline{N}_{23} \\
&L_{11} = N_{11}^{/} - N_{11}^{//}, \quad L_{22} = N_{22}^{/} - N_{22}^{//}, \quad L_{12} = N_{12}^{/} - N_{12}^{//} \\
&M_{11} = {}^{/}M_{11} + {}^{//}M_{11}, \quad M_{22} = {}^{/}M_{22} + {}^{//}M_{22}, \quad M_{12} = {}^{/}M_{12} + {}^{//}M_{12}
\end{aligned}
\tag{2.145a – k}
$$

2.5.2 *Displacement Formulation for Sandwich Plates and Shells*

- **Strong Core Formulation**

The displacement formulation can be developed by replacement of equations Eqs. (2.99), (2.109), (2.113), (2.116), (2.24a–f)–(2.26a–f), and (2.53)–(2.76) into Eq. (2.96) carrying out the integration with respect to x_3, integrating by parts wherever possible and invoking the arbitrary character of the variations $\delta\eta_1$, $\delta\eta_2$, $\delta\xi_1$, $\delta\xi_2$, and δv_3 by setting the variation of these five coefficients to zero. The result is five equations of motion and a set of boundary conditions as by-products. This governing system is valid for double curved sandwich shells with a strong core and symmetric facings neglecting the transverse shear in the facings. It is assumed that symmetry exists both locally and globally. This system of equations is provided as

$$
\begin{aligned}
\delta\xi_1 :\ & \Lambda_{11}\left\{\frac{\partial^2\xi_1}{\partial x_1^2}+\frac{\partial v_3}{\partial x_1}\frac{\partial^2 v_3}{\partial x_1^2}+\frac{\partial^2 \mathring{v}_3}{\partial x_1^2}\frac{\partial v_3}{\partial x_1}+\frac{\partial \mathring{v}_3}{\partial x_1}\frac{\partial^2 v_3}{\partial x_1^2}-\frac{1}{R_1}\frac{\partial v_3}{\partial x_1}\right\}+\\
& \Lambda_{66}\left\{\frac{\partial^2\xi_1}{\partial x_2^2}+\frac{\partial^2\xi_2}{\partial x_1\partial x_2}+\frac{\partial v_3}{\partial x_2}\frac{\partial^2 v_3}{\partial x_1\partial x_2}+\frac{\partial v_3}{\partial x_1}\frac{\partial^2 v_3}{\partial x_2^2}+\frac{\partial v_3}{\partial x_2}\frac{\partial^2 \mathring{v}_3}{\partial x_1\partial x_2}\right.\\
& \left.+\frac{\partial \mathring{v}_3}{\partial x_1}\frac{\partial^2 v_3}{\partial x_2^2}+\frac{\partial \mathring{v}_3}{\partial x_2}\frac{\partial^2 v_3}{\partial x_1\partial x_2}+\frac{\partial v_3}{\partial x_1}\frac{\partial^2 \mathring{v}_3}{\partial x_2^2}\right\}+\Lambda_{12}\left\{\frac{\partial^2\xi_2}{\partial x_1\partial x_2}+\frac{\partial v_3}{\partial x_2}\frac{\partial^2 v_3}{\partial x_1\partial x_2}\right.\\
& \left.+\frac{\partial v_3}{\partial x_2}\frac{\partial^2 \mathring{v}_3}{\partial x_1\partial x_2}+\frac{\partial \mathring{v}_3}{\partial x_2}\frac{\partial^2 v_3}{\partial x_1\partial x_2}-\frac{1}{R_2}\frac{\partial v_3}{\partial x_1}\right\}-A_{16}\left\{2\frac{\partial^2\xi_1}{\partial x_1\partial x_2}+\frac{\partial^2\xi_2}{\partial x_1^2}\right.\\
& +2\frac{\partial v_3}{\partial x_1}\frac{\partial^2 v_3}{\partial x_1\partial x_2}+2\frac{\partial v_3}{\partial x_1}\frac{\partial^2 \mathring{v}_3}{\partial x_1\partial x_2}+2\frac{\partial \mathring{v}_3}{\partial x_1}\frac{\partial^2 v_3}{\partial x_1\partial x_2}+\frac{\partial v_3}{\partial x_2}\frac{\partial^2 v_3}{\partial x_1^2}+\frac{\partial v_3}{\partial x_2}\frac{\partial^2 \mathring{v}_3}{\partial x_1^2}+\\
& \left.\frac{\partial \mathring{v}_3}{\partial x_2}\frac{\partial^2 v_3}{\partial x_1^2}-\frac{1}{R_1}\frac{\partial v_3}{\partial x_2}\right\}-A_{26}\left\{\frac{\partial^2\xi_2}{\partial x_2^2}+\frac{\partial v_3}{\partial x_2}\frac{\partial^2 v_3}{\partial x_2^2}+\frac{\partial v_3}{\partial x_2}\frac{\partial^2 \mathring{v}_3}{\partial x_2^2}+\frac{\partial \mathring{v}_3}{\partial x_2}\frac{\partial^2 v_3}{\partial x_2^2}-\frac{1}{R_2}\frac{\partial v_3}{\partial x_2}\right\}\\
& -\frac{\partial N_{11}^T}{\partial x_1}-\frac{\partial N_{12}^T}{\partial x_2}=0
\end{aligned}
\tag{2.146}
$$

$$
\begin{aligned}
\delta\xi_2 :\ & \Lambda_{22}\left\{\frac{\partial^2\xi_2}{\partial x_2^2}+\frac{\partial v_3}{\partial x_2}\frac{\partial^2 v_3}{\partial x_2^2}+\frac{\partial^2 \mathring{v}_3}{\partial x_2^2}\frac{\partial v_3}{\partial x_2}+\frac{\partial \mathring{v}_3}{\partial x_2}\frac{\partial^2 v_3}{\partial x_2^2}-\frac{1}{R_2}\frac{\partial v_3}{\partial x_2}\right\}+\Lambda_{66}\left\{\frac{\partial^2\xi_2}{\partial x_1^2}+\frac{\partial^2\xi_1}{\partial x_1\partial x_2}\right.\\
& \left.+\frac{\partial v_3}{\partial x_1}\frac{\partial^2 v_3}{\partial x_1\partial x_2}+\frac{\partial v_3}{\partial x_2}\frac{\partial^2 v_3}{\partial x_1^2}+\frac{\partial v_3}{\partial x_1}\frac{\partial^2 \mathring{v}_3}{\partial x_1\partial x_2}+\frac{\partial \mathring{v}_3}{\partial x_2}\frac{\partial^2 v_3}{\partial x_1^2}+\frac{\partial \mathring{v}_3}{\partial x_1}\frac{\partial^2 v_3}{\partial x_1\partial x_2}+\frac{\partial v_3}{\partial x_2}\frac{\partial^2 \mathring{v}_3}{\partial x_1^2}\right\}+\\
& \Lambda_{12}\left\{\frac{\partial^2\xi_1}{\partial x_1\partial x_2}+\frac{\partial v_3}{\partial x_1}\frac{\partial^2 v_3}{\partial x_1\partial x_2}+\frac{\partial v_3}{\partial x_1}\frac{\partial^2 \mathring{v}_3}{\partial x_1\partial x_2}+\frac{\partial \mathring{v}_3}{\partial x_1}\frac{\partial^2 v_3}{\partial x_1\partial x_2}-\frac{1}{R_1}\frac{\partial v_3}{\partial x_2}\right\}-A_{26}\left\{2\frac{\partial^2\xi_2}{\partial x_1\partial x_2}\right.\\
& +\frac{\partial^2\xi_1}{\partial x_2^2}+2\frac{\partial v_3}{\partial x_2}\frac{\partial^2 v_3}{\partial x_1\partial x_2}+2\frac{\partial v_3}{\partial x_2}\frac{\partial^2 \mathring{v}_3}{\partial x_1\partial x_2}+2\frac{\partial \mathring{v}_3}{\partial x_2}\frac{\partial^2 v_3}{\partial x_1\partial x_2}+\frac{\partial v_3}{\partial x_1}\frac{\partial^2 v_3}{\partial x_2^2}+\frac{\partial v_3}{\partial x_1}\frac{\partial^2 \mathring{v}_3}{\partial x_2^2}+\\
& \left.\frac{\partial \mathring{v}_3}{\partial x_1}\frac{\partial^2 v_3}{\partial x_2^2}-\frac{1}{R_2}\frac{\partial v_3}{\partial x_1}\right\}-A_{16}\left\{\frac{\partial^2\xi_1}{\partial x_1^2}+\frac{\partial v_3}{\partial x_1}\frac{\partial^2 v_3}{\partial x_1^2}+\frac{\partial v_3}{\partial x_1}\frac{\partial^2 \mathring{v}_3}{\partial x_1^2}+\frac{\partial \mathring{v}_3}{\partial x_1}\frac{\partial^2 v_3}{\partial x_1^2}-\frac{1}{R_1}\frac{\partial v_3}{\partial x_1}\right\}\\
& -\frac{\partial N_{22}^T}{\partial x_2}-\frac{\partial N_{12}^T}{\partial x_1}=0
\end{aligned}
\tag{2.147}
$$

$$
\begin{aligned}
\delta\eta_1 :\ & \widehat{\Lambda}_{11}\frac{\partial^2\eta_1}{\partial x_1^2}+\underline{\frac{h\bar{h}\overline{Q}_{11}}{3}\frac{\partial^3 v_3}{\partial x_1^3}}+\widehat{\Lambda}_{66}\left\{\frac{\partial^2\eta_1}{\partial x_2^2}+\frac{\partial^2\eta_2}{\partial x_1\partial x_2}\right\}+\underline{\frac{2h\bar{h}\overline{Q}_{66}}{3}\frac{\partial^3 v_3}{\partial x_1\partial x_2^2}}+\widehat{\Lambda}_{12}\frac{\partial^2\eta_2}{\partial x_1\partial x_2}\\
& +\underline{\frac{h\bar{h}\overline{Q}_{12}}{3}\frac{\partial^3 v_3}{\partial x_1\partial x_2^2}}+A_{16}\left\{\frac{\partial^2\eta_2}{\partial x_1^2}+2\frac{\partial^2\eta_1}{\partial x_1\partial x_2}\right\}+A_{26}\frac{\partial^2\eta_2}{\partial x_2^2}-d_1\left\{\eta_1+a\frac{\partial v_3}{\partial x_1}\right\}-\frac{1}{\bar{\bar{h}}}\frac{\partial\widehat{N}_{11}^T}{\partial x_1}\\
& -\frac{1}{\bar{\bar{h}}}\frac{\partial\widehat{N}_{12}^T}{\partial x_2}=0
\end{aligned}
\tag{2.148}
$$

$$\delta\eta_2 : \ \widehat{\Lambda}_{22}\frac{\partial^2\eta_2}{\partial x_2^2} + \underline{\frac{h\overline{hQ}_{22}}{3}\frac{\partial^3 v_3}{\partial x_2^3}} + \widehat{\Lambda}_{66}\left\{\frac{\partial^2\eta_2}{\partial x_2^2} + \frac{\partial^2\eta_1}{\partial x_1\partial x_2}\right\} + \underline{\frac{2h\overline{hQ}_{66}}{3}\frac{\partial^3 v_3}{\partial x_1^2\partial x_2}} + \widehat{\Lambda}_{12}\frac{\partial^2\eta_1}{\partial x_1\partial x_2}$$
$$+\underline{\frac{h\overline{hQ}_{12}}{3}\frac{\partial^3 v_3}{\partial x_1^2\partial x_2}} + A_{26}\left\{\frac{\partial^2\eta_1}{\partial x_2^2} + 2\frac{\partial^2\eta_2}{\partial x_1\partial x_2}\right\} + A_{16}\frac{\partial^2\eta_1}{\partial x_1^2} - d_2\left\{\eta_2 + a\frac{\partial v_3}{\partial x_2}\right\} - \frac{1}{\overline{h}}\frac{\partial\widehat{N}_{22}^T}{\partial x_2}$$
$$-\frac{1}{\overline{h}}\frac{\partial\widehat{N}_{12}^T}{\partial x_1} = 0$$
$$(2.149)$$

$$\delta v_3 : \ \Lambda_{11}\left\{\frac{\partial\xi_1}{\partial x_1}\left(\frac{\partial^2 v_3}{\partial x_1^2} + \frac{\partial^2 \mathring{v}_3}{\partial x_1^2}\right) + \frac{1}{2}\left(\frac{\partial v_3}{\partial x_1}\right)^2\left(\frac{\partial^2 v_3}{\partial x_1^2} + \frac{\partial^2 \mathring{v}_3}{\partial x_1^2}\right) + \frac{\partial v_3}{\partial x_1}\frac{\partial \mathring{v}_3}{\partial x_1}\right.$$
$$\left.\left(\frac{\partial^2 v_3}{\partial x_1^2} + \frac{\partial^2 \mathring{v}_3}{\partial x_1^2}\right) + \frac{1}{R_1}\left[\frac{\partial\xi_1}{\partial x_1} + \frac{1}{2}\left(\frac{\partial v_3}{\partial x_1}\right)^2 + \frac{\partial v_3}{\partial x_1}\frac{\partial \mathring{v}_3}{\partial x_1} - v_3\left(\frac{\partial^2 v_3}{\partial x_1^2} + \frac{\partial^2 \mathring{v}_3}{\partial x_1^2}\right) - \frac{v_3}{R_1}\right]\right\}$$
$$-\underline{\frac{h\overline{hQ}_{11}}{3}\left\{\frac{\partial^3\eta_1}{\partial x_1^3} + \frac{h}{2}\frac{\partial^4 v_3}{\partial x_1^4}\right\}} + \Lambda_{12}\left\{\frac{\partial\xi_1}{\partial x_1}\left(\frac{\partial^2 v_3}{\partial x_2^2} + \frac{\partial^2 \mathring{v}_3}{\partial x_2^2}\right) + \frac{\partial\xi_2}{\partial x_2}\left(\frac{\partial^2 v_3}{\partial x_1^2} + \frac{\partial^2 \mathring{v}_3}{\partial x_1^2}\right)\right.$$
$$+\frac{1}{2}\left(\frac{\partial v_3}{\partial x_1}\right)^2\left(\frac{\partial^2 v_3}{\partial x_2^2} + \frac{\partial^2 \mathring{v}_3}{\partial x_2^2}\right) + \frac{1}{2}\left(\frac{\partial v_3}{\partial x_2}\right)^2\left(\frac{\partial^2 v_3}{\partial x_1^2} + \frac{\partial^2 \mathring{v}_3}{\partial x_1^2}\right) + \frac{\partial v_3}{\partial x_1}\frac{\partial \mathring{v}_3}{\partial x_1}\left(\frac{\partial^2 v_3}{\partial x_2^2} + \frac{\partial^2 \mathring{v}_3}{\partial x_2^2}\right)$$
$$+\frac{\partial v_3}{\partial x_2}\frac{\partial \mathring{v}_3}{\partial x_2}\left(\frac{\partial^2 v_3}{\partial x_1^2} + \frac{\partial^2 \mathring{v}_3}{\partial x_1^2}\right) + \frac{1}{R_1}\left[\frac{\partial\xi_2}{\partial x_2} + \frac{1}{2}\left(\frac{\partial v_3}{\partial x_2}\right)^2 + \frac{\partial v_3}{\partial x_2}\frac{\partial \mathring{v}_3}{\partial x_2} - v_3\left(\frac{\partial^2 v_3}{\partial x_2^2} + \frac{\partial^2 \mathring{v}_3}{\partial x_2^2}\right) - \frac{v_3}{R_2}\right]$$
$$\left.+\frac{1}{R_2}\left[\frac{\partial\xi_1}{\partial x_1} + \frac{1}{2}\left(\frac{\partial v_3}{\partial x_1}\right)^2 + \frac{\partial v_3}{\partial x_1}\frac{\partial \mathring{v}_3}{\partial x_1} - v_3\left(\frac{\partial^2 v_3}{\partial x_1^2} + \frac{\partial^2 \mathring{v}_3}{\partial x_1^2}\right) - \frac{v_3}{R_1}\right]\right\}$$
$$-\underline{\frac{h\overline{hQ}_{12}}{3}\left\{\frac{\partial^3\eta_1}{\partial x_1\partial x_2^2} + \frac{\partial^3\eta_2}{\partial x_1^2\partial x_2} h\frac{\partial^4 v_3}{\partial x_1^2\partial x_2^2}\right\}} + \Lambda_{22}\left\{\frac{\partial\xi_2}{\partial x_2}\left(\frac{\partial^2 v_3}{\partial x_2^2} + \frac{\partial^2 \mathring{v}_3}{\partial x_2^2}\right) + \frac{1}{2}\left(\frac{\partial v_3}{\partial x_2}\right)^2\right.$$
$$\left(\frac{\partial^2 v_3}{\partial x_2^2} + \frac{\partial^2 \mathring{v}_3}{\partial x_2^2}\right) + \frac{\partial v_3}{\partial x_2}\frac{\partial \mathring{v}_3}{\partial x_2}\left(\frac{\partial^2 v_3}{\partial x_2^2} + \frac{\partial^2 \mathring{v}_3}{\partial x_2^2}\right) + \frac{1}{R_2}\left[\frac{\partial\xi_2}{\partial x_2} + \frac{1}{2}\left(\frac{\partial v_3}{\partial x_2}\right)^2 + \frac{\partial v_3}{\partial x_2}\frac{\partial \mathring{v}_3}{\partial x_2}\right.$$
$$\left.\left.-v_3\left(\frac{\partial^2 v_3}{\partial x_2^2} + \frac{\partial^2 \mathring{v}_3}{\partial x_2^2}\right) - \frac{v_3}{R_2}\right]\right\} - \underline{\frac{h\overline{hQ}_{22}}{3}\left\{\frac{\partial^3\eta_2}{\partial x_2^3} + \frac{h}{2}\frac{\partial^4 v_3}{\partial x_2^4}\right\}}$$
$$+2\Lambda_{66}\left\{\frac{\partial\xi_1}{\partial x_2}\left(\frac{\partial^2 v_3}{\partial x_1\partial x_2} + \frac{\partial^2 \mathring{v}_3}{\partial x_1\partial x_2}\right) + \frac{\partial\xi_2}{\partial x_1}\left(\frac{\partial^2 v_3}{\partial x_1\partial x_2} + \frac{\partial^2 \mathring{v}_3}{\partial x_1\partial x_2}\right) + \frac{\partial v_3}{\partial x_1}\frac{\partial v_3}{\partial x_2}\right.$$
$$\left(\frac{\partial^2 v_3}{\partial x_1\partial x_2} + \frac{\partial^2 \mathring{v}_3}{\partial x_1\partial x_2}\right) + \frac{\partial \mathring{v}_3}{\partial x_1}\frac{\partial v_3}{\partial x_2}\left(\frac{\partial^2 v_3}{\partial x_1\partial x_2} + \frac{\partial^2 \mathring{v}_3}{\partial x_1\partial x_2}\right) + \frac{\partial v_3}{\partial x_1}\frac{\partial \mathring{v}_3}{\partial x_2}\left(\frac{\partial^2 v_3}{\partial x_1\partial x_2} +\right.$$

$$\left.\left.\frac{\partial^2 \overset{\circ}{v}_3}{\partial x_1 \partial x_2}\right)\right\} - \underline{\frac{2h\overline{h}\overline{Q}_{66}}{3}\left\{\frac{\partial^3 \eta_1}{\partial x_1 \partial x_2^2} + \frac{\partial^3 \eta_2}{\partial x_1^2 \partial x_2} + h\frac{\partial^4 v_3}{\partial x_1^2 \partial x_2^2}\right\}} + A_{16}\left\{\frac{\partial \xi_1}{\partial x_2}\left(\frac{\partial^2 v_3}{\partial x_1^2} + \frac{\partial^2 \overset{\circ}{v}_3}{\partial x_1^2}\right)\right.$$

$$+\frac{\partial \xi_2}{\partial x_1}\left(\frac{\partial^2 v_3}{\partial x_1^2} + \frac{\partial^2 \overset{\circ}{v}_3}{\partial x_1^2}\right) + \frac{\partial v_3}{\partial x_1}\frac{\partial v_3}{\partial x_2}\left(\frac{\partial^2 v_3}{\partial x_1^2} + \frac{\partial^2 \overset{\circ}{v}_3}{\partial x_1^2}\right) + \frac{\partial \overset{\circ}{v}_3}{\partial x_1}\frac{\partial v_3}{\partial x_2}\left(\frac{\partial^2 v_3}{\partial x_1^2} + \frac{\partial^2 \overset{\circ}{v}_3}{\partial x_1^2}\right)$$

$$+\frac{\partial v_3}{\partial x_1}\frac{\partial \overset{\circ}{v}_3}{\partial x_2}\left(\frac{\partial^2 v_3}{\partial x_1^2} + \frac{\partial^2 \overset{\circ}{v}_3}{\partial x_1^2}\right) + 2\frac{\partial \xi_1}{\partial x_1}\left(\frac{\partial^2 v_3}{\partial x_1 \partial x_2} + \frac{\partial^2 \overset{\circ}{v}_3}{\partial x_1 \partial x_2}\right)$$

$$+\left(\frac{\partial v_3}{\partial x_1}\right)^2\left(\frac{\partial^2 v_3}{\partial x_1 \partial x_2} + \frac{\partial^2 \overset{\circ}{v}_3}{\partial x_1 \partial x_2}\right) + 2\frac{\partial v_3}{\partial x_1}\frac{\partial \overset{\circ}{v}_3}{\partial x_1}\left(\frac{\partial^2 v_3}{\partial x_1 \partial x_2} + \frac{\partial^2 \overset{\circ}{v}_3}{\partial x_1 \partial x_2}\right)$$

$$\left.+\frac{1}{R_1}\left[\frac{\partial \xi_1}{\partial x_2} + \frac{\partial \xi_2}{\partial x_1} + \frac{\partial v_3}{\partial x_1}\frac{\partial v_3}{\partial x_2} + \frac{\partial \overset{\circ}{v}_3}{\partial x_1}\frac{\partial v_3}{\partial x_2} + \frac{\partial v_3}{\partial x_1}\frac{\partial \overset{\circ}{v}_3}{\partial x_2} - 2v_3\left(\frac{\partial^2 v_3}{\partial x_1 \partial x_2} + \frac{\partial^2 \overset{\circ}{v}_3}{\partial x_1 \partial x_2}\right)\right]\right\}$$

$$+A_{26}\left\{\frac{\partial \xi_2}{\partial x_1}\left(\frac{\partial^2 v_3}{\partial x_2^2} + \frac{\partial^2 \overset{\circ}{v}_3}{\partial x_2^2}\right) + \frac{\partial \xi_1}{\partial x_2}\left(\frac{\partial^2 v_3}{\partial x_2^2} + \frac{\partial^2 \overset{\circ}{v}_3}{\partial x_2^2}\right) + \frac{\partial v_3}{\partial x_2}\frac{\partial v_3}{\partial x_1}\left(\frac{\partial^2 v_3}{\partial x_2^2} + \frac{\partial^2 \overset{\circ}{v}_3}{\partial x_2^2}\right)\right.$$

$$+\frac{\partial \overset{\circ}{v}_3}{\partial x_2}\frac{\partial v_3}{\partial x_1}\left(\frac{\partial^2 v_3}{\partial x_2^2} + \frac{\partial^2 \overset{\circ}{v}_3}{\partial x_2^2}\right) + \frac{\partial v_3}{\partial x_2}\frac{\partial \overset{\circ}{v}_3}{\partial x_1}\left(\frac{\partial^2 v_3}{\partial x_2^2} + \frac{\partial^2 \overset{\circ}{v}_3}{\partial x_2^2}\right)$$

$$+2\frac{\partial \xi_2}{\partial x_2}\left(\frac{\partial^2 v_3}{\partial x_1 \partial x_2} + \frac{\partial^2 \overset{\circ}{v}_3}{\partial x_1 \partial x_2}\right) + \left(\frac{\partial v_3}{\partial x_2}\right)^2\left(\frac{\partial^2 v_3}{\partial x_2 \partial x_1} + \frac{\partial^2 \overset{\circ}{v}_3}{\partial x_2 \partial x_1}\right)$$

$$+2\frac{\partial v_3}{\partial x_2}\frac{\partial \overset{\circ}{v}_3}{\partial x_2}\left(\frac{\partial^2 v_3}{\partial x_1 \partial x_2} + \frac{\partial^2 \overset{\circ}{v}_3}{\partial x_1 \partial x_2}\right)$$

$$\left.+\frac{1}{R_2}\left[\frac{\partial \xi_2}{\partial x_1} + \frac{\partial \xi_1}{\partial x_2} + \frac{\partial v_3}{\partial x_1}\frac{\partial v_3}{\partial x_2} + \frac{\partial \overset{\circ}{v}_3}{\partial x_2}\frac{\partial v_3}{\partial x_1} + \frac{\partial v_3}{\partial x_2}\frac{\partial \overset{\circ}{v}_3}{\partial x_1} - 2v_3\left(\frac{\partial^2 v_3}{\partial x_1 \partial x_2} + \frac{\partial^2 \overset{\circ}{v}_3}{\partial x_1 \partial x_2}\right)\right]\right\}$$

$$-F_{11}\frac{\partial^4 v_3}{\partial x_1^4} - 2(F_{12} + 2F_{66})\frac{\partial^4 v_3}{\partial x_1^2 \partial x_2^2} - F_{22}\frac{\partial^4 v_3}{\partial x_2^4} - 4F_{16}\frac{\partial^4 v_3}{\partial x_1^3 \partial x_2}$$

$$-4F_{26}\frac{\partial^4 v_3}{\partial x_1 \partial x_2^3} + 2\overline{h}\overline{K}^2\overline{G}_{13}\left(1 + \frac{h}{2\overline{h}}\right)\left\{\frac{1}{\overline{h}}\frac{\partial \eta_1}{\partial x_1} + \left(1 + \frac{C_1}{\overline{h}}\right)\frac{\partial^2 v_3}{\partial x_1^2}\right\} + 2\overline{h}\overline{K}^2\overline{G}_{23}$$

$$\left(1 + \frac{h}{2\overline{h}}\right)\left\{\frac{1}{\overline{h}}\frac{\partial \eta_2}{\partial x_2} + \left(1 + \frac{C_1}{\overline{h}}\right)\frac{\partial^2 v_3}{\partial x_2^2}\right\} - N_{11}^T\left(\frac{\partial^2 v_3}{\partial x_1^2} + \frac{\partial^2 \overset{\circ}{v}_3}{\partial x_1^2} + \frac{1}{R_1}\right)$$

$$-N_{12}^T\left(\frac{\partial^2 v_3}{\partial x_1 \partial x_2}+\frac{\partial^2 \overset{\circ}{v}_3}{\partial x_1 \partial x_2}\right)-N_{22}^T\left(\frac{\partial^2 v_3}{\partial x_2^2}+\frac{\partial^2 \overset{\circ}{v}_3}{\partial x_2^2}+\frac{1}{R_2}\right)$$

$$-N_{11}^m\left(\frac{\partial^2 v_3}{\partial x_1^2}+\frac{\partial^2 \overset{\circ}{v}_3}{\partial x_1^2}+\frac{1}{R_1}\right)-N_{12}^m\left(\frac{\partial^2 v_3}{\partial x_1 \partial x_2}+\frac{\partial^2 \overset{\circ}{v}_3}{\partial x_1 \partial x_2}\right)$$

$$-N_{22}^m\left(\frac{\partial^2 v_3}{\partial x_2^2}+\frac{\partial^2 \overset{\circ}{v}_3}{\partial x_2^2}+\frac{1}{R_2}\right)+C\frac{\partial v_3}{\partial t}+m_o\frac{\partial^2 v_3}{\partial t^2}=q_3 \tag{2.150}$$

It should be noted that the terms containing the squares of the geometrical imperfections can be discarded. The associated boundary conditions along the edges $x_n = \text{const}(n = 1, 2)$ are

$$\begin{aligned}
&N_{nn}=\widetilde{N}_{nn} \quad \text{or} \quad \xi_n=\widetilde{\xi}_n\\
&N_{nt}=\widetilde{N}_{nt} \quad \text{or} \quad \xi_t=\widetilde{\xi}_t\\
&L_{nn}=\widetilde{L}_{nn} \quad \text{or} \quad \eta_n=\widetilde{\eta}_n\\
&L_{nt}=\widetilde{L}_{nt} \quad \text{or} \quad \eta_t=\widetilde{\eta}_t\\
&M_{nn}=\widetilde{M}_{nn} \quad \text{or} \quad \frac{\partial v_3}{\partial x_n}=\frac{\partial \widetilde{v}_3}{\partial x_n}\\
&N_{nt}\left(\frac{\partial v_3}{\partial x_t}+\frac{\partial \overset{\circ}{v}_3}{\partial x_t}\right)+N_{nn}\left(\frac{\partial v_3}{\partial x_n}+\frac{\partial \overset{\circ}{v}_3}{\partial x_n}\right)+\frac{\partial M_{nn}}{\partial x_n}+2\frac{\partial M_{nt}}{\partial x_t} \quad \text{or} \quad v_3=\widetilde{v}_3\\
&+(a/\overline{h})\overline{N}_{n3}=\frac{\partial \widetilde{M}_{nt}}{\partial x_t}+\widetilde{\overline{N}}_{n3}
\end{aligned} \tag{2.151a – f}$$

In the above equations of motion, the stiffness coefficients are defined as

$$\begin{aligned}
&\Lambda_{11}=A_{11}+2\overline{h}\overline{Q}_{11}, \quad \Lambda_{22}=A_{22}+2\overline{h}\overline{Q}_{22}, \quad \Lambda_{12}=A_{12}+2\overline{h}\overline{Q}_{12}\\
&\Lambda_{66}=A_{66}+2\overline{h}\overline{Q}_{66}\\
&\widehat{\Lambda}_{11}=A_{11}+(2\overline{h}/3)\overline{Q}_{11}, \quad \widehat{\Lambda}_{22}=A_{22}+(2\overline{h}/3)\overline{Q}_{22}, \quad \widehat{\Lambda}_{12}=A_{12}+(2\overline{h}/3)\overline{Q}_{12}\\
&\widehat{\Lambda}_{66}=A_{66}+(2\overline{h}/3)\overline{Q}_{66}
\end{aligned} \tag{2.152a – h}$$

While the global mechanical stiffness measures $A_{\omega\rho}$, $F_{\omega\rho}$ are defined as

$$A_{\omega\rho}={}^{/}A_{\omega\rho}+{}^{//}A_{\omega\rho}, \quad F_{\omega\rho}={}^{/}F_{\omega\rho}+{}^{//}F_{\omega\rho} \quad (\omega,\rho=1,2,6) \tag{2.153a, b}$$

if the face sheets are symmetric with respect to both their local and global mid-surfaces then,

$$ {}^{/}A_{\omega\rho} = {}^{//}A_{\omega\rho}, \qquad {}^{/}F_{\omega\rho} = {}^{//}F_{\omega\rho} \tag{2.154c, d} $$

The global thermal stress and stress couple resultants appearing in the equations are defined as

$$ \begin{aligned} & N_{11}^{T} = {}^{/}N_{11}^{T} + \overline{N}_{11}^{T} + {}^{//}N_{11}^{T}, \quad N_{22}^{T} = {}^{/}N_{22}^{T} + \overline{N}_{22}^{T} + {}^{//}N_{22}^{T}, \quad N_{12}^{T} = {}^{/}N_{12}^{T} + {}^{//}N_{12}^{T} \\ & \widehat{N}_{11}^{T} = \overline{h}\left({}^{/}N_{11}^{T} - {}^{//}N_{11}^{T}\right), \quad \widehat{N}_{22}^{T} = \overline{h}\left({}^{/}N_{22}^{T} - {}^{//}N_{22}^{T}\right), \quad \widehat{N}_{12}^{T} = \overline{h}\left({}^{/}N_{12}^{T} - {}^{//}N_{12}^{T}\right) \\ & \widehat{M}_{11}^{T} = {}^{/}M_{11}^{T} + {}^{//}M_{11}^{T} - (h/2\overline{h})\overline{M}_{11}^{T}, \quad \widehat{M}_{22}^{T} = {}^{/}M_{22}^{T} + {}^{//}M_{22}^{T} - (h/2\overline{h})\overline{M}_{22}^{T} \\ & \widehat{M}_{12}^{T} = {}^{/}M_{12}^{T} + {}^{//}M_{12}^{T} \end{aligned} \tag{2.155a – i} $$

- **Weak/Soft-Core Formulation**

Eqs. (2.146)–(2.150) can be reduced somewhat if one considers the weak core model by discarding all of the underlined terms appearing in these equations, as well as in the stiffness coefficients. For consideration of the weak core with symmetric facings both locally and globally the governing equations of motion can be modified to appear as

$$ \begin{aligned} \delta\xi_1 : \; & A_{11}\left\{\frac{\partial^2\xi_1}{\partial x_1^2} + \frac{\partial v_3}{\partial x_1}\frac{\partial^2 v_3}{\partial x_1^2} + \frac{\partial^2 \overset{\circ}{v}_3}{\partial x_1^2}\frac{\partial v_3}{\partial x_1} + \frac{\partial \overset{\circ}{v}_3}{\partial x_1}\frac{\partial^2 v_3}{\partial x_1^2} - \frac{1}{R_1}\frac{\partial v_3}{\partial x_1}\right\} + A_{66}\left\{\frac{\partial^2\xi_1}{\partial x_2^2} + \frac{\partial^2\xi_2}{\partial x_1 \partial x_2}\right. \\ & + \frac{\partial v_3}{\partial x_2}\frac{\partial^2 v_3}{\partial x_1 \partial x_2} + \frac{\partial v_3}{\partial x_1}\frac{\partial^2 v_3}{\partial x_2^2} + \frac{\partial v_3}{\partial x_2}\frac{\partial^2 \overset{\circ}{v}_3}{\partial x_1 \partial x_2} + \frac{\partial \overset{\circ}{v}_3}{\partial x_1}\frac{\partial^2 v_3}{\partial x_2^2} + \frac{\partial \overset{\circ}{v}_3}{\partial x_2}\frac{\partial^2 v_3}{\partial x_1 \partial x_2} + \left.\frac{\partial v_3}{\partial x_1}\frac{\partial^2 \overset{\circ}{v}_3}{\partial x_2^2}\right\} + \\ & A_{12}\left\{\frac{\partial^2\xi_2}{\partial x_1 \partial x_2} + \frac{\partial v_3}{\partial x_2}\frac{\partial^2 v_3}{\partial x_1 \partial x_2} + \frac{\partial v_3}{\partial x_2}\frac{\partial^2 \overset{\circ}{v}_3}{\partial x_1 \partial x_2} + \frac{\partial \overset{\circ}{v}_3}{\partial x_2}\frac{\partial^2 v_3}{\partial x_1 \partial x_2} - \frac{1}{R_2}\frac{\partial v_3}{\partial x_1}\right\} - A_{16}\left\{2\frac{\partial^2\xi_1}{\partial x_1 \partial x_2}\right. \\ & + \frac{\partial^2\xi_2}{\partial x_1^2} + 2\frac{\partial v_3}{\partial x_1}\frac{\partial^2 v_3}{\partial x_1 \partial x_2} + 2\frac{\partial v_3}{\partial x_1}\frac{\partial^2 \overset{\circ}{v}_3}{\partial x_1 \partial x_2} + 2\frac{\partial \overset{\circ}{v}_3}{\partial x_1}\frac{\partial^2 v_3}{\partial x_1 \partial x_2} + \frac{\partial v_3}{\partial x_2}\frac{\partial^2 v_3}{\partial x_1^2} + \frac{\partial v_3}{\partial x_2}\frac{\partial^2 \overset{\circ}{v}_3}{\partial x_1^2} + \\ & \left.\frac{\partial \overset{\circ}{v}_3}{\partial x_2}\frac{\partial^2 v_3}{\partial x_1^2} - \frac{1}{R_1}\frac{\partial v_3}{\partial x_2}\right\} - A_{26}\left\{\frac{\partial^2\xi_2}{\partial x_2^2} + \frac{\partial v_3}{\partial x_2}\frac{\partial^2 v_3}{\partial x_2^2} + \frac{\partial v_3}{\partial x_2}\frac{\partial^2 \overset{\circ}{v}_3}{\partial x_2^2} + \frac{\partial \overset{\circ}{v}_3}{\partial x_2}\frac{\partial^2 v_3}{\partial x_2^2} - \frac{1}{R_2}\frac{\partial v_3}{\partial x_2}\right\} \\ & - \frac{\partial N_{11}^{T}}{\partial x_1} - \frac{\partial N_{12}^{T}}{\partial x_2} = 0 \end{aligned} \tag{2.156} $$

$$
\begin{aligned}
\delta\xi_2 :\ & A_{22}\left\{\frac{\partial^2\xi_2}{\partial x_2^2}+\frac{\partial v_3}{\partial x_2}\frac{\partial^2 v_3}{\partial x_2^2}+\frac{\partial^2\mathring{v}_3}{\partial x_2^2}\frac{\partial v_3}{\partial x_2}+\frac{\partial\mathring{v}_3}{\partial x_2}\frac{\partial^2 v_3}{\partial x_2^2}-\frac{1}{R_2}\frac{\partial v_3}{\partial x_2}\right\}+A_{66}\bigg\{\frac{\partial^2\xi_2}{\partial x_1^2}+\frac{\partial^2\xi_1}{\partial x_1\partial x_2}\\
&+\frac{\partial v_3}{\partial x_1}\frac{\partial^2 v_3}{\partial x_1\partial x_2}+\frac{\partial v_3}{\partial x_2}\frac{\partial^2 v_3}{\partial x_1^2}+\frac{\partial v_3}{\partial x_1}\frac{\partial^2\mathring{v}_3}{\partial x_1\partial x_2}+\frac{\partial\mathring{v}_3}{\partial x_2}\frac{\partial^2 v_3}{\partial x_1^2}+\frac{\partial\mathring{v}_3}{\partial x_1}\frac{\partial^2 v_3}{\partial x_1\partial x_2}+\frac{\partial v_3}{\partial x_2}\frac{\partial^2\mathring{v}_3}{\partial x_1^2}\bigg\}+\\
&A_{12}\left\{\frac{\partial^2\xi_1}{\partial x_1\partial x_2}+\frac{\partial v_3}{\partial x_1}\frac{\partial^2 v_3}{\partial x_1\partial x_2}+\frac{\partial v_3}{\partial x_1}\frac{\partial^2\mathring{v}_3}{\partial x_1\partial x_2}+\frac{\partial\mathring{v}_3}{\partial x_1}\frac{\partial^2 v_3}{\partial x_1\partial x_2}-\frac{1}{R_1}\frac{\partial v_3}{\partial x_2}\right\}-A_{26}\bigg\{2\frac{\partial^2\xi_2}{\partial x_1\partial x_2}\\
&+\frac{\partial^2\xi_1}{\partial x_2^2}+2\frac{\partial v_3}{\partial x_2}\frac{\partial^2 v_3}{\partial x_1\partial x_2}+2\frac{\partial v_3}{\partial x_2}\frac{\partial^2\mathring{v}_3}{\partial x_1\partial x_2}+2\frac{\partial\mathring{v}_3}{\partial x_2}\frac{\partial^2 v_3}{\partial x_1\partial x_2}+\frac{\partial v_3}{\partial x_1}\frac{\partial^2 v_3}{\partial x_2^2}+\frac{\partial v_3}{\partial x_1}\frac{\partial^2\mathring{v}_3}{\partial x_2^2}+\\
&\frac{\partial\mathring{v}_3}{\partial x_1}\frac{\partial^2 v_3}{\partial x_2^2}-\frac{1}{R_2}\frac{\partial v_3}{\partial x_1}\bigg\}-A_{16}\left\{\frac{\partial^2\xi_1}{\partial x_1^2}+\frac{\partial v_3}{\partial x_1}\frac{\partial^2 v_3}{\partial x_1^2}+\frac{\partial v_3}{\partial x_1}\frac{\partial^2\mathring{v}_3}{\partial x_1^2}+\frac{\partial\mathring{v}_3}{\partial x_1}\frac{\partial^2 v_3}{\partial x_1^2}-\frac{1}{R_1}\frac{\partial v_3}{\partial x_1}\right\}\\
&-\frac{\partial N_{22}^T}{\partial x_2}-\frac{\partial N_{12}^T}{\partial x_1}=0
\end{aligned}
\tag{2.157}
$$

$$
\begin{aligned}
\delta\eta_1 :\ & A_{11}\frac{\partial^2\eta_1}{\partial x_1^2}+A_{66}\left\{\frac{\partial^2\eta_1}{\partial x_2^2}+\frac{\partial^2\eta_2}{\partial x_1\partial x_2}\right\}+A_{12}\frac{\partial^2\eta_2}{\partial x_1\partial x_2}+A_{16}\left\{\frac{\partial^2\eta_2}{\partial x_1^2}+2\frac{\partial^2\eta_1}{\partial x_1\partial x_2}\right\}\\
&+A_{26}\frac{\partial^2\eta_2}{\partial x_2^2}-d_1\left\{\eta_1+a\frac{\partial v_3}{\partial x_1}\right\}-\frac{1}{\bar{h}}\frac{\partial\widehat{N}_{11}^T}{\partial x_1}-\frac{1}{\bar{h}}\frac{\partial\widehat{N}_{12}^T}{\partial x_2}=0
\end{aligned}
\tag{2.158}
$$

$$
\begin{aligned}
\delta\eta_2 :\ & A_{22}\frac{\partial^2\eta_2}{\partial x_2^2}+A_{66}\left\{\frac{\partial^2\eta_2}{\partial x_2^2}+\frac{\partial^2\eta_1}{\partial x_1\partial x_2}\right\}+A_{12}\frac{\partial^2\eta_1}{\partial x_1\partial x_2}+A_{26}\left\{\frac{\partial^2\eta_1}{\partial x_2^2}+2\frac{\partial^2\eta_2}{\partial x_1\partial x_2}\right\}\\
&+A_{16}\frac{\partial^2\eta_1}{\partial x_1^2}-d_2\left\{\eta_2+a\frac{\partial v_3}{\partial x_2}\right\}-\frac{1}{\bar{h}}\frac{\partial\widehat{N}_{22}^T}{\partial x_2}-\frac{1}{\bar{h}}\frac{\partial\widehat{N}_{12}^T}{\partial x_1}=0
\end{aligned}
\tag{2.159}
$$

$$
\begin{aligned}
\delta v_3 :\ & A_{11}\Bigg\{\frac{\partial\xi_1}{\partial x_1}\left(\frac{\partial^2 v_3}{\partial x_1^2}+\frac{\partial^2\mathring{v}_3}{\partial x_1^2}\right)+\frac{1}{2}\left(\frac{\partial v_3}{\partial x_1}\right)^2\left(\frac{\partial^2 v_3}{\partial x_1^2}+\frac{\partial^2\mathring{v}_3}{\partial x_1^2}\right)+\frac{\partial v_3}{\partial x_1}\frac{\partial\mathring{v}_3}{\partial x_1}\\
&\left(\frac{\partial^2 v_3}{\partial x_1^2}+\frac{\partial^2\mathring{v}_3}{\partial x_1^2}\right)+\frac{1}{R_1}\left[\frac{\partial\xi_1}{\partial x_1}+\frac{1}{2}\left(\frac{\partial v_3}{\partial x_1}\right)^2+\frac{\partial v_3}{\partial x_1}\frac{\partial\mathring{v}_3}{\partial x_1}-v_3\left(\frac{\partial^2 v_3}{\partial x_1^2}+\frac{\partial^2\mathring{v}_3}{\partial x_1^2}\right)-\frac{v_3}{R_1}\right]\Bigg\}\\
&+A_{12}\Bigg\{\frac{\partial\xi_1}{\partial x_1}\left(\frac{\partial^2 v_3}{\partial x_2^2}+\frac{\partial^2\mathring{v}_3}{\partial x_2^2}\right)+\frac{\partial\xi_2}{\partial x_2}\left(\frac{\partial^2 v_3}{\partial x_1^2}+\frac{\partial^2\mathring{v}_3}{\partial x_1^2}\right)+\frac{1}{2}\left(\frac{\partial v_3}{\partial x_1}\right)^2\left(\frac{\partial^2 v_3}{\partial x_2^2}+\frac{\partial^2\mathring{v}_3}{\partial x_2^2}\right)\\
&+\frac{1}{2}\left(\frac{\partial v_3}{\partial x_2}\right)^2\left(\frac{\partial^2 v_3}{\partial x_1^2}+\frac{\partial^2\mathring{v}_3}{\partial x_1^2}\right)+\frac{\partial v_3}{\partial x_1}\frac{\partial\mathring{v}_3}{\partial x_1}\left(\frac{\partial^2 v_3}{\partial x_2^2}+\frac{\partial^2\mathring{v}_3}{\partial x_2^2}\right)+\frac{\partial v_3}{\partial x_2}\frac{\partial\mathring{v}_3}{\partial x_2}\left(\frac{\partial^2 v_3}{\partial x_1^2}+\frac{\partial^2\mathring{v}_3}{\partial x_1^2}\right)
\end{aligned}
$$

$$+\frac{1}{R_1}\left[\frac{\partial\xi_2}{\partial x_2}+\frac{1}{2}\left(\frac{\partial v_3}{\partial x_2}\right)^2+\frac{\partial v_3}{\partial x_2}\frac{\partial \overset{\circ}{v}_3}{\partial x_2}-v_3\left(\frac{\partial^2 v_3}{\partial x_2^2}+\frac{\partial^2 \overset{\circ}{v}_3}{\partial x_2^2}\right)-\frac{v_3}{R_2}\right]$$

$$+\frac{1}{R_2}\left[\frac{\partial\xi_1}{\partial x_1}+\frac{1}{2}\left(\frac{\partial v_3}{\partial x_1}\right)^2+\frac{\partial v_3}{\partial x_1}\frac{\partial \overset{\circ}{v}_3}{\partial x_1}-v_3\left(\frac{\partial^2 v_3}{\partial x_1^2}+\frac{\partial^2 \overset{\circ}{v}_3}{\partial x_1^2}\right)-\frac{v_3}{R_1}\right]\Bigg\}$$

$$+A_{22}\left\{\frac{\partial\xi_2}{\partial x_2}\left(\frac{\partial^2 v_3}{\partial x_2^2}+\frac{\partial^2 \overset{\circ}{v}_3}{\partial x_2^2}\right)+\frac{1}{2}\left(\frac{\partial v_3}{\partial x_2}\right)^2\left(\frac{\partial^2 v_3}{\partial x_2^2}+\frac{\partial^2 \overset{\circ}{v}_3}{\partial x_2^2}\right)+\frac{\partial v_3}{\partial x_2}\frac{\partial \overset{\circ}{v}_3}{\partial x_2}\left(\frac{\partial^2 v_3}{\partial x_2^2}+\frac{\partial^2 \overset{\circ}{v}_3}{\partial x_2^2}\right)\right.$$

$$\left.+\frac{1}{R_2}\left[\frac{\partial\xi_2}{\partial x_2}+\frac{1}{2}\left(\frac{\partial v_3}{\partial x_2}\right)^2+\frac{\partial v_3}{\partial x_2}\frac{\partial \overset{\circ}{v}_3}{\partial x_2}-v_3\left(\frac{\partial^2 v_3}{\partial x_2^2}+\frac{\partial^2 \overset{\circ}{v}_3}{\partial x_2^2}\right)-\frac{v_3}{R_2}\right]\right\}$$

$$+2A_{66}\left\{\frac{\partial\xi_1}{\partial x_2}\left(\frac{\partial^2 v_3}{\partial x_1\partial x_2}+\frac{\partial^2 \overset{\circ}{v}_3}{\partial x_1\partial x_2}\right)+\frac{\partial\xi_2}{\partial x_1}\left(\frac{\partial^2 v_3}{\partial x_1\partial x_2}+\frac{\partial^2 \overset{\circ}{v}_3}{\partial x_1\partial x_2}\right)\right.$$

$$+\frac{\partial v_3}{\partial x_1}\frac{\partial v_3}{\partial x_2}\left(\frac{\partial^2 v_3}{\partial x_1\partial x_2}+\frac{\partial^2 \overset{\circ}{v}_3}{\partial x_1\partial x_2}\right)+\frac{\partial \overset{\circ}{v}_3}{\partial x_1}\frac{\partial v_3}{\partial x_2}\left(\frac{\partial^2 v_3}{\partial x_1\partial x_2}+\frac{\partial^2 \overset{\circ}{v}_3}{\partial x_1\partial x_2}\right)$$

$$\left.+\frac{\partial v_3}{\partial x_1}\frac{\partial \overset{\circ}{v}_3}{\partial x_2}\left(\frac{\partial^2 v_3}{\partial x_1\partial x_2}+\frac{\partial^2 \overset{\circ}{v}_3}{\partial x_1\partial x_2}\right)\right\}+A_{16}\left\{\frac{\partial\xi_1}{\partial x_2}\left(\frac{\partial^2 v_3}{\partial x_1^2}+\frac{\partial^2 \overset{\circ}{v}_3}{\partial x_1^2}\right)\right.$$

$$+\frac{\partial\xi_2}{\partial x_1}\left(\frac{\partial^2 v_3}{\partial x_1^2}+\frac{\partial^2 \overset{\circ}{v}_3}{\partial x_1^2}\right)+\frac{\partial v_3}{\partial x_1}\frac{\partial v_3}{\partial x_2}\left(\frac{\partial^2 v_3}{\partial x_1^2}+\frac{\partial^2 \overset{\circ}{v}_3}{\partial x_1^2}\right)+\frac{\partial \overset{\circ}{v}_3}{\partial x_1}\frac{\partial v_3}{\partial x_2}\left(\frac{\partial^2 v_3}{\partial x_1^2}+\frac{\partial^2 \overset{\circ}{v}_3}{\partial x_1^2}\right)$$

$$+\frac{\partial v_3}{\partial x_1}\frac{\partial \overset{\circ}{v}_3}{\partial x_2}\left(\frac{\partial^2 v_3}{\partial x_1^2}+\frac{\partial^2 \overset{\circ}{v}_3}{\partial x_1^2}\right)+2\frac{\partial\xi_1}{\partial x_1}\left(\frac{\partial^2 v_3}{\partial x_1\partial x_2}+\frac{\partial^2 \overset{\circ}{v}_3}{\partial x_1\partial x_2}\right)$$

$$+\left(\frac{\partial v_3}{\partial x_1}\right)^2\left(\frac{\partial^2 v_3}{\partial x_1\partial x_2}+\frac{\partial^2 \overset{\circ}{v}_3}{\partial x_1\partial x_2}\right)+2\frac{\partial v_3}{\partial x_1}\frac{\partial \overset{\circ}{v}_3}{\partial x_1}\left(\frac{\partial^2 v_3}{\partial x_1\partial x_2}+\frac{\partial^2 \overset{\circ}{v}_3}{\partial x_1\partial x_2}\right)$$

$$\left.+\frac{1}{R_1}\left[\frac{\partial\xi_1}{\partial x_2}+\frac{\partial\xi_2}{\partial x_1}+\frac{\partial v_3}{\partial x_1}\frac{\partial v_3}{\partial x_2}+\frac{\partial \overset{\circ}{v}_3}{\partial x_1}\frac{\partial v_3}{\partial x_2}+\frac{\partial v_3}{\partial x_1}\frac{\partial \overset{\circ}{v}_3}{\partial x_2}-2v_3\left(\frac{\partial^2 v_3}{\partial x_1\partial x_2}+\frac{\partial^2 \overset{\circ}{v}_3}{\partial x_1\partial x_2}\right)\right]\right\}$$

$$+A_{26}\left\{\frac{\partial\xi_2}{\partial x_1}\left(\frac{\partial^2 v_3}{\partial x_2^2}+\frac{\partial^2 \overset{\circ}{v}_3}{\partial x_2^2}\right)+\frac{\partial\xi_1}{\partial x_2}\left(\frac{\partial^2 v_3}{\partial x_2^2}+\frac{\partial^2 \overset{\circ}{v}_3}{\partial x_2^2}\right)+\frac{\partial v_3}{\partial x_2}\frac{\partial v_3}{\partial x_1}\left(\frac{\partial^2 v_3}{\partial x_2^2}+\frac{\partial^2 \overset{\circ}{v}_3}{\partial x_2^2}\right)\right.$$

$$+\frac{\partial \overset{\circ}{v}_3}{\partial x_2}\frac{\partial v_3}{\partial x_1}\left(\frac{\partial^2 v_3}{\partial x_2^2}+\frac{\partial^2 \overset{\circ}{v}_3}{\partial x_2^2}\right)+\frac{\partial v_3}{\partial x_2}\frac{\partial \overset{\circ}{v}_3}{\partial x_1}\left(\frac{\partial^2 v_3}{\partial x_2^2}+\frac{\partial^2 \overset{\circ}{v}_3}{\partial x_2^2}\right)+2\frac{\partial \xi_2}{\partial x_2}\left(\frac{\partial^2 v_3}{\partial x_1 \partial x_2}+\frac{\partial^2 \overset{\circ}{v}_3}{\partial x_1 \partial x_2}\right)$$

$$+\left(\frac{\partial v_3}{\partial x_2}\right)^2\left(\frac{\partial^2 v_3}{\partial x_2 \partial x_1}+\frac{\partial^2 \overset{\circ}{v}_3}{\partial x_2 \partial x_1}\right)+2\frac{\partial v_3}{\partial x_2}\frac{\partial \overset{\circ}{v}_3}{\partial x_2}\left(\frac{\partial^2 v_3}{\partial x_1 \partial x_2}+\frac{\partial^2 \overset{\circ}{v}_3}{\partial x_1 \partial x_2}\right)$$

$$+\frac{1}{R_2}\left[\frac{\partial \xi_2}{\partial x_1}+\frac{\partial \xi_1}{\partial x_2}+\frac{\partial v_3}{\partial x_1}\frac{\partial v_3}{\partial x_2}+\frac{\partial \overset{\circ}{v}_3}{\partial x_2}\frac{\partial v_3}{\partial x_1}+\frac{\partial v_3}{\partial x_2}\frac{\partial \overset{\circ}{v}_3}{\partial x_1}-2v_3\left(\frac{\partial^2 v_3}{\partial x_1 \partial x_2}+\frac{\partial^2 \overset{\circ}{v}_3}{\partial x_1 \partial x_2}\right)\right]\Bigg\}$$

$$-F_{11}\frac{\partial^4 v_3}{\partial x_1^4}-2(F_{12}+2F_{66})\frac{\partial^4 v_3}{\partial x_1^2 \partial x_2^2}-F_{22}\frac{\partial^4 v_3}{\partial x_2^4}-4F_{16}\frac{\partial^4 v_3}{\partial x_1^3 \partial x_2}-4F_{26}\frac{\partial^4 v_3}{\partial x_1 \partial x_2^3}$$

$$+2\overline{h}\overline{K}^2\overline{G}_{13}\left(1+\frac{h}{2\overline{h}}\right)\times\left\{\frac{1}{\overline{\overline{h}}}\frac{\partial \eta_1}{\partial x_1}+\left(1+\frac{C_1}{\overline{h}}\right)\frac{\partial^2 v_3}{\partial x_1^2}\right\}+2\overline{h}\overline{K}^2\overline{G}_{23}\left(1+\frac{h}{2\overline{h}}\right)$$

$$\left\{\frac{1}{\overline{\overline{h}}}\frac{\partial \eta_2}{\partial x_2}+\left(1+\frac{C_1}{\overline{h}}\right)\frac{\partial^2 v_3}{\partial x_2^2}\right\}-N_{11}^T\left(\frac{\partial^2 v_3}{\partial x_1^2}+\frac{\partial^2 \overset{\circ}{v}_3}{\partial x_1^2}+\frac{1}{R_1}\right)-N_{12}^T\left(\frac{\partial^2 v_3}{\partial x_1 \partial x_2}+\frac{\partial^2 \overset{\circ}{v}_3}{\partial x_1 \partial x_2}\right)$$

$$-N_{22}^T\left(\frac{\partial^2 v_3}{\partial x_2^2}+\frac{\partial^2 \overset{\circ}{v}_3}{\partial x_2^2}+\frac{1}{R_2}\right)-N_{11}^m\left(\frac{\partial^2 v_3}{\partial x_1^2}+\frac{\partial^2 \overset{\circ}{v}_3}{\partial x_1^2}+\frac{1}{R_1}\right)-N_{12}^m\left(\frac{\partial^2 v_3}{\partial x_1 \partial x_2}+\frac{\partial^2 \overset{\circ}{v}_3}{\partial x_1 \partial x_2}\right)$$

$$-N_{22}^m\left(\frac{\partial^2 v_3}{\partial x_2^2}+\frac{\partial^2 \overset{\circ}{v}_3}{\partial x_2^2}+\frac{1}{R_2}\right)+C\frac{\partial v_3}{\partial t}+m_o\frac{\partial^2 v_3}{\partial t^2}=q_3 \tag{2.160}$$

The associated boundary conditions along the edges $x_n = \text{const}(n = 1, 2)$ are

$$N_{nn}=\widetilde{N}_{nn} \quad \text{or} \quad \xi_n=\widetilde{\xi}_n$$
$$N_{nt}=\widetilde{N}_{nt} \quad \text{or} \quad \xi_t=\widetilde{\xi}_t$$
$$L_{nn}=\widetilde{L}_{nn} \quad \text{or} \quad \eta_n=\widetilde{\eta}_n$$
$$L_{nt}=\widetilde{L}_{nt} \quad \text{or} \quad \eta_t=\widetilde{\eta}_t$$
$$M_{nn}=\widetilde{M}_{nn} \quad \text{or} \quad \frac{\partial v_3}{\partial x_n}=\frac{\partial \widetilde{v}_3}{\partial x_n}$$
$$N_{nt}\left(\frac{\partial v_3}{\partial x_t}+\frac{\partial \overset{\circ}{v}_3}{\partial x_t}\right)+N_{nn}\left(\frac{\partial v_3}{\partial x_n}+\frac{\partial \overset{\circ}{v}_3}{\partial x_n}\right)+\frac{\partial M_{nn}}{\partial x_n}+2\frac{\partial M_{nt}}{\partial x_t} \quad \text{or} \quad v_3=\widetilde{v}_3$$
$$+(a/\overline{h})\overline{N}_{n3}=\frac{\partial \widetilde{M}_{nt}}{\partial x_t}+\widetilde{\overline{N}}_{n3}$$

$$(2.161\text{a}-\text{f})$$

2.5.3 Displacement Formulation for Sandwich Plates

Considering the weak core model and discarding the terms with curvatures in Eqs. (2.156)–(2.160) results in the nonlinear equations of motion for flat sandwich panels given as

$$
\begin{aligned}
\delta\xi_1:\ & A_{11}\left\{\frac{\partial^2\xi_1}{\partial x_1^2}+\frac{\partial v_3}{\partial x_1}\frac{\partial^2 v_3}{\partial x_1^2}+\frac{\partial^2\mathring{v}_3}{\partial x_1^2}\frac{\partial v_3}{\partial x_1}+\frac{\partial\mathring{v}_3}{\partial x_1}\frac{\partial^2 v_3}{\partial x_1^2}\right\}+A_{66}\left\{\frac{\partial^2\xi_1}{\partial x_2^2}+\frac{\partial^2\xi_2}{\partial x_1\partial x_2}+\right.\\
&\left.\frac{\partial v_3}{\partial x_2}\frac{\partial^2 v_3}{\partial x_1\partial x_2}+\frac{\partial v_3}{\partial x_1}\frac{\partial^2 v_3}{\partial x_2^2}+\frac{\partial v_3}{\partial x_2}\frac{\partial^2\mathring{v}_3}{\partial x_1\partial x_2}+\frac{\partial\mathring{v}_3}{\partial x_1}\frac{\partial^2 v_3}{\partial x_2^2}+\frac{\partial\mathring{v}_3}{\partial x_2}\frac{\partial^2 v_3}{\partial x_1\partial x_2}+\frac{\partial v_3}{\partial x_1}\frac{\partial^2\mathring{v}_3}{\partial x_2^2}\right\}+\\
& A_{12}\left\{\frac{\partial^2\xi_2}{\partial x_1\partial x_2}+\frac{\partial v_3}{\partial x_2}\frac{\partial^2 v_3}{\partial x_1\partial x_2}+\frac{\partial v_3}{\partial x_2}\frac{\partial^2\mathring{v}_3}{\partial x_1\partial x_2}+\frac{\partial\mathring{v}_3}{\partial x_2}\frac{\partial^2 v_3}{\partial x_1\partial x_2}\right\}-A_{16}\left\{2\frac{\partial^2\xi_1}{\partial x_1\partial x_2}\right.\\
&+\frac{\partial^2\xi_2}{\partial x_1^2}+2\frac{\partial v_3}{\partial x_1}\frac{\partial^2 v_3}{\partial x_1\partial x_2}+2\frac{\partial v_3}{\partial x_1}\frac{\partial^2\mathring{v}_3}{\partial x_1\partial x_2}+2\frac{\partial\mathring{v}_3}{\partial x_1}\frac{\partial^2 v_3}{\partial x_1\partial x_2}+\frac{\partial v_3}{\partial x_2}\frac{\partial^2 v_3}{\partial x_1^2}+\frac{\partial v_3}{\partial x_2}\frac{\partial^2\mathring{v}_3}{\partial x_1^2}+\\
&\left.\frac{\partial\mathring{v}_3}{\partial x_2}\frac{\partial^2 v_3}{\partial x_1^2}\right\}-A_{26}\left\{\frac{\partial^2\xi_2}{\partial x_2^2}+\frac{\partial v_3}{\partial x_2}\frac{\partial^2 v_3}{\partial x_2^2}+\frac{\partial v_3}{\partial x_2}\frac{\partial^2\mathring{v}_3}{\partial x_2^2}+\frac{\partial\mathring{v}_3}{\partial x_2}\frac{\partial^2 v_3}{\partial x_2^2}\right\}-\frac{\partial N_{11}^T}{\partial x_1}-\frac{\partial N_{12}^T}{\partial x_2}=0
\end{aligned}
\tag{2.162}
$$

$$
\begin{aligned}
\delta\xi_2:\ & A_{22}\left\{\frac{\partial^2\xi_2}{\partial x_2^2}+\frac{\partial v_3}{\partial x_2}\frac{\partial^2 v_3}{\partial x_2^2}+\frac{\partial^2\mathring{v}_3}{\partial x_2^2}\frac{\partial v_3}{\partial x_2}+\frac{\partial\mathring{v}_3}{\partial x_2}\frac{\partial^2 v_3}{\partial x_2^2}\right\}+A_{66}\left\{\frac{\partial^2\xi_2}{\partial x_1^2}+\frac{\partial^2\xi_1}{\partial x_1\partial x_2}+\right.\\
&\left.\frac{\partial v_3}{\partial x_1}\frac{\partial^2 v_3}{\partial x_1\partial x_2}+\frac{\partial v_3}{\partial x_2}\frac{\partial^2 v_3}{\partial x_1^2}+\frac{\partial v_3}{\partial x_1}\frac{\partial^2\mathring{v}_3}{\partial x_1\partial x_2}+\frac{\partial\mathring{v}_3}{\partial x_2}\frac{\partial^2 v_3}{\partial x_1^2}+\frac{\partial\mathring{v}_3}{\partial x_1}\frac{\partial^2 v_3}{\partial x_1\partial x_2}+\frac{\partial v_3}{\partial x_2}\frac{\partial^2\mathring{v}_3}{\partial x_1^2}\right\}+\\
& A_{12}\left\{\frac{\partial^2\xi_1}{\partial x_1\partial x_2}+\frac{\partial v_3}{\partial x_1}\frac{\partial^2 v_3}{\partial x_1\partial x_2}+\frac{\partial v_3}{\partial x_1}\frac{\partial^2\mathring{v}_3}{\partial x_1\partial x_2}+\frac{\partial\mathring{v}_3}{\partial x_1}\frac{\partial^2 v_3}{\partial x_1\partial x_2}\right\}-A_{26}\left\{2\frac{\partial^2\xi_2}{\partial x_1\partial x_2}\right.\\
&+\frac{\partial^2\xi_1}{\partial x_2^2}+2\frac{\partial v_3}{\partial x_2}\frac{\partial^2 v_3}{\partial x_1\partial x_2}+2\frac{\partial v_3}{\partial x_2}\frac{\partial^2\mathring{v}_3}{\partial x_1\partial x_2}+2\frac{\partial\mathring{v}_3}{\partial x_2}\frac{\partial^2 v_3}{\partial x_1\partial x_2}+\frac{\partial v_3}{\partial x_1}\frac{\partial^2 v_3}{\partial x_2^2}+\frac{\partial v_3}{\partial x_1}\frac{\partial^2\mathring{v}_3}{\partial x_2^2}+\\
&\left.\frac{\partial\mathring{v}_3}{\partial x_1}\frac{\partial^2 v_3}{\partial x_2^2}\right\}-A_{16}\left\{\frac{\partial^2\xi_1}{\partial x_1^2}+\frac{\partial v_3}{\partial x_1}\frac{\partial^2 v_3}{\partial x_1^2}+\frac{\partial v_3}{\partial x_1}\frac{\partial^2\mathring{v}_3}{\partial x_1^2}+\frac{\partial\mathring{v}_3}{\partial x_1}\frac{\partial^2 v_3}{\partial x_1^2}\right\}\\
&-\frac{\partial N_{22}^T}{\partial x_2}-\frac{\partial N_{12}^T}{\partial x_1}=0
\end{aligned}
\tag{2.163}
$$

$$\delta\eta_1: \quad A_{11}\frac{\partial^2\eta_1}{\partial x_1^2}+A_{66}\left\{\frac{\partial^2\eta_1}{\partial x_2^2}+\frac{\partial^2\eta_2}{\partial x_1\partial x_2}\right\}+A_{12}\frac{\partial^2\eta_2}{\partial x_1\partial x_2}+A_{16}\left\{\frac{\partial^2\eta_2}{\partial x_1^2}+2\frac{\partial^2\eta_1}{\partial x_1\partial x_2}\right\}$$
$$+A_{26}\frac{\partial^2\eta_2}{\partial x_2^2}-d_1\left\{\eta_1+a\frac{\partial v_3}{\partial x_1}\right\}-\frac{1}{\overline{\overline{h}}}\frac{\partial\widehat{N}_{11}^T}{\partial x_1}-\frac{1}{\overline{\overline{h}}}\frac{\partial\widehat{N}_{12}^T}{\partial x_2}=0 \tag{2.164}$$

$$\delta\eta_2: \quad A_{22}\frac{\partial^2\eta_2}{\partial x_2^2}+A_{66}\left\{\frac{\partial^2\eta_2}{\partial x_2^2}+\frac{\partial^2\eta_1}{\partial x_1\partial x_2}\right\}+A_{12}\frac{\partial^2\eta_1}{\partial x_1\partial x_2}+A_{26}\left\{\frac{\partial^2\eta_1}{\partial x_2^2}+2\frac{\partial^2\eta_2}{\partial x_1\partial x_2}\right\}$$
$$+A_{16}\frac{\partial^2\eta_1}{\partial x_1^2}-d_2\left\{\eta_2+a\frac{\partial v_3}{\partial x_2}\right\}-\frac{1}{\overline{\overline{h}}}\frac{\partial\widehat{N}_{22}^T}{\partial x_2}-\frac{1}{\overline{\overline{h}}}\frac{\partial\widehat{N}_{12}^T}{\partial x_1}=0 \tag{2.165}$$

$$\delta v_3: \quad A_{11}\left\{\frac{\partial\xi_1}{\partial x_1}\left(\frac{\partial^2 v_3}{\partial x_1^2}+\frac{\partial^2 v_3^\circ}{\partial x_1^2}\right)+\frac{1}{2}\left(\frac{\partial v_3}{\partial x_1}\right)^2\left(\frac{\partial^2 v_3}{\partial x_1^2}+\frac{\partial^2 v_3^\circ}{\partial x_1^2}\right)+\frac{\partial v_3}{\partial x_1}\frac{\partial v_3^\circ}{\partial x_1}\right.$$
$$\left.\left(\frac{\partial^2 v_3}{\partial x_1^2}+\frac{\partial^2 v_3^\circ}{\partial x_1^2}\right)\right\}+A_{12}\left\{\frac{\partial\xi_1}{\partial x_1}\left(\frac{\partial^2 v_3}{\partial x_2^2}+\frac{\partial^2 v_3^\circ}{\partial x_2^2}\right)+\frac{\partial\xi_2}{\partial x_2}\left(\frac{\partial^2 v_3}{\partial x_1^2}+\frac{\partial^2 v_3^\circ}{\partial x_1^2}\right)\right.$$
$$+\frac{1}{2}\left(\frac{\partial v_3}{\partial x_1}\right)^2\left(\frac{\partial^2 v_3}{\partial x_2^2}+\frac{\partial^2 v_3^\circ}{\partial x_2^2}\right)+\frac{1}{2}\left(\frac{\partial v_3}{\partial x_2}\right)^2\left(\frac{\partial^2 v_3}{\partial x_1^2}+\frac{\partial^2 v_3^\circ}{\partial x_1^2}\right)+\frac{\partial v_3}{\partial x_1}\frac{\partial v_3^\circ}{\partial x_1}\left(\frac{\partial^2 v_3}{\partial x_2^2}+\frac{\partial^2 v_3^\circ}{\partial x_2^2}\right)$$
$$\left.+\frac{\partial v_3}{\partial x_2}\frac{\partial v_3^\circ}{\partial x_2}\left(\frac{\partial^2 v_3}{\partial x_1^2}+\frac{\partial^2 v_3^\circ}{\partial x_1^2}\right)\right\}+A_{22}\left\{\frac{\partial\xi_2}{\partial x_2}\left(\frac{\partial^2 v_3}{\partial x_2^2}+\frac{\partial^2 v_3^\circ}{\partial x_2^2}\right)+\frac{1}{2}\left(\frac{\partial v_3}{\partial x_2}\right)^2\left(\frac{\partial^2 v_3}{\partial x_2^2}+\frac{\partial^2 v_3^\circ}{\partial x_2^2}\right)\right.$$
$$\left.+\frac{\partial v_3}{\partial x_2}\frac{\partial v_3^\circ}{\partial x_2}\left(\frac{\partial^2 v_3}{\partial x_2^2}+\frac{\partial^2 v_3^\circ}{\partial x_2^2}\right)\right\}+2A_{66}\left\{\frac{\partial\xi_1}{\partial x_2}\left(\frac{\partial^2 v_3}{\partial x_1\partial x_2}+\frac{\partial^2 v_3^\circ}{\partial x_1\partial x_2}\right)+\frac{\partial\xi_2}{\partial x_1}\right.$$
$$\left(\frac{\partial^2 v_3}{\partial x_1\partial x_2}+\frac{\partial^2 v_3^\circ}{\partial x_1\partial x_2}\right)+\frac{\partial v_3}{\partial x_1}\frac{\partial v_3}{\partial x_2}\left(\frac{\partial^2 v_3}{\partial x_1\partial x_2}+\frac{\partial^2 v_3^\circ}{\partial x_1\partial x_2}\right)+\frac{\partial v_3^\circ}{\partial x_1}\frac{\partial v_3}{\partial x_2}\left(\frac{\partial^2 v_3}{\partial x_1\partial x_2}+\frac{\partial^2 v_3^\circ}{\partial x_1\partial x_2}\right)$$
$$\left.+\frac{\partial v_3}{\partial x_1}\frac{\partial v_3^\circ}{\partial x_2}\left(\frac{\partial^2 v_3}{\partial x_1\partial x_2}+\frac{\partial^2 v_3^\circ}{\partial x_1\partial x_2}\right)\right\}+A_{16}\left\{\frac{\partial\xi_1}{\partial x_2}\left(\frac{\partial^2 v_3}{\partial x_1^2}+\frac{\partial^2 v_3^\circ}{\partial x_1^2}\right)+\frac{\partial\xi_2}{\partial x_1}\left(\frac{\partial^2 v_3}{\partial x_1^2}+\frac{\partial^2 v_3^\circ}{\partial x_1^2}\right)\right.$$
$$+\frac{\partial v_3}{\partial x_1}\frac{\partial v_3}{\partial x_2}\left(\frac{\partial^2 v_3}{\partial x_1^2}+\frac{\partial^2 v_3^\circ}{\partial x_1^2}\right)+\frac{\partial v_3^\circ}{\partial x_1}\frac{\partial v_3}{\partial x_2}\left(\frac{\partial^2 v_3}{\partial x_1^2}+\frac{\partial^2 v_3^\circ}{\partial x_1^2}\right)+\frac{\partial v_3}{\partial x_1}\frac{\partial v_3^\circ}{\partial x_2}\left(\frac{\partial^2 v_3}{\partial x_1^2}+\frac{\partial^2 v_3^\circ}{\partial x_1^2}\right)$$
$$+2\frac{\partial\xi_1}{\partial x_1}\left(\frac{\partial^2 v_3}{\partial x_1\partial x_2}+\frac{\partial^2 v_3^\circ}{\partial x_1\partial x_2}\right)+\left(\frac{\partial v_3}{\partial x_1}\right)^2\left(\frac{\partial^2 v_3}{\partial x_1\partial x_2}+\frac{\partial^2 v_3^\circ}{\partial x_1\partial x_2}\right)$$
$$\left.+2\frac{\partial v_3}{\partial x_1}\frac{\partial v_3^\circ}{\partial x_1}\left(\frac{\partial^2 v_3}{\partial x_1\partial x_2}+\frac{\partial^2 v_3^\circ}{\partial x_1\partial x_2}\right)\right\}+A_{26}\left\{\frac{\partial\xi_2}{\partial x_1}\left(\frac{\partial^2 v_3}{\partial x_2^2}+\frac{\partial^2 v_3^\circ}{\partial x_2^2}\right)\right.$$

$$+\frac{\partial \xi_1}{\partial x_2}\left(\frac{\partial^2 v_3}{\partial x_2^2}+\frac{\partial^2 \overset{\circ}{v}_3}{\partial x_2^2}\right)+\frac{\partial v_3}{\partial x_2}\frac{\partial v_3}{\partial x_1}\left(\frac{\partial^2 v_3}{\partial x_2^2}+\frac{\partial^2 \overset{\circ}{v}_3}{\partial x_2^2}\right)+\frac{\partial \overset{\circ}{v}_3}{\partial x_2}\frac{\partial v_3}{\partial x_1}\left(\frac{\partial^2 v_3}{\partial x_2^2}+\frac{\partial^2 \overset{\circ}{v}_3}{\partial x_2^2}\right)$$

$$+\frac{\partial v_3}{\partial x_2}\frac{\partial \overset{\circ}{v}_3}{\partial x_1}\left(\frac{\partial^2 v_3}{\partial x_2^2}+\frac{\partial^2 \overset{\circ}{v}_3}{\partial x_2^2}\right)+2\frac{\partial \xi_2}{\partial x_2}\left(\frac{\partial^2 v_3}{\partial x_1 \partial x_2}+\frac{\partial^2 \overset{\circ}{v}_3}{\partial x_1 \partial x_2}\right)+\left(\frac{\partial v_3}{\partial x_2}\right)^2$$

$$\left(\frac{\partial^2 v_3}{\partial x_2 \partial x_1}+\frac{\partial^2 \overset{\circ}{v}_3}{\partial x_2 \partial x_1}\right)+2\frac{\partial v_3}{\partial x_2}\frac{\partial \overset{\circ}{v}_3}{\partial x_2}\left(\frac{\partial^2 v_3}{\partial x_1 \partial x_2}+\frac{\partial^2 \overset{\circ}{v}_3}{\partial x_1 \partial x_2}\right)\Bigg\}-F_{11}\frac{\partial^4 v_3}{\partial x_1^4}$$

$$-2(F_{12}+2F_{66})\frac{\partial^4 v_3}{\partial x_1^2 \partial x_2^2}-F_{22}\frac{\partial^4 v_3}{\partial x_2^4}-4F_{16}\frac{\partial^4 v_3}{\partial x_1^3 \partial x_2}-4F_{26}\frac{\partial^4 v_3}{\partial x_1 \partial x_2^3}$$

$$+2\overline{h}\overline{K}^2\overline{G}_{13}\left(1+\frac{h}{2\overline{h}}\right)\left\{\frac{1}{\overline{h}}\frac{\partial \eta_1}{\partial x_1}+\left(1+\frac{C_1}{\overline{h}}\right)\frac{\partial^2 v_3}{\partial x_1^2}\right\}+2\overline{h}\overline{K}^2\overline{G}_{23}\left(1+\frac{h}{2\overline{h}}\right)\left\{\frac{1}{\overline{h}}\frac{\partial \eta_2}{\partial x_2}\right.$$

$$\left.+\left(1+\frac{C_1}{\overline{h}}\right)\frac{\partial^2 v_3}{\partial x_2^2}\right\}-N_{11}^T\left(\frac{\partial^2 v_3}{\partial x_1^2}+\frac{\partial^2 \overset{\circ}{v}_3}{\partial x_1^2}\right)-N_{12}^T\left(\frac{\partial^2 v_3}{\partial x_1 \partial x_2}+\frac{\partial^2 \overset{\circ}{v}_3}{\partial x_1 \partial x_2}\right)$$

$$-N_{22}^T\left(\frac{\partial^2 v_3}{\partial x_2^2}+\frac{\partial^2 \overset{\circ}{v}_3}{\partial x_2^2}\right)-N_{11}^m\left(\frac{\partial^2 v_3}{\partial x_1^2}+\frac{\partial^2 \overset{\circ}{v}_3}{\partial x_1^2}\right)-N_{12}^m\left(\frac{\partial^2 v_3}{\partial x_1 \partial x_2}+\frac{\partial^2 \overset{\circ}{v}_3}{\partial x_1 \partial x_2}\right)$$

$$-N_{22}^m\left(\frac{\partial^2 v_3}{\partial x_2^2}+\frac{\partial^2 \overset{\circ}{v}_3}{\partial x_2^2}\right)+C\frac{\partial v_3}{\partial t}+m_o\frac{\partial^2 v_3}{\partial t^2}=q_3 \tag{2.166}$$

The associated boundary conditions along the edges $x_n = \text{const}(n = 1, 2)$ are

$$\begin{aligned}
&N_{nn} = \widetilde{N}_{nn} \quad &&\text{or} \quad \xi_n = \widetilde{\xi}_n \\
&N_{nt} = \widetilde{N}_{nt} \quad &&\text{or} \quad \xi_t = \widetilde{\xi}_t \\
&L_{nn} = \widetilde{L}_{nn} \quad &&\text{or} \quad \eta_n = \widetilde{\eta}_n \\
&L_{nt} = \widetilde{L}_{nt} \quad &&\text{or} \quad \eta_t = \widetilde{\eta}_t \\
&M_{nn} = \widetilde{M}_{nn} \quad &&\text{or} \quad \frac{\partial v_3}{\partial x_n} = \frac{\partial \widetilde{v}_3}{\partial x_n} \\
&N_{nt}\left(\frac{\partial v_3}{\partial x_t}+\frac{\partial \overset{\circ}{v}_3}{\partial x_t}\right)+N_{nn}\left(\frac{\partial v_3}{\partial x_n}+\frac{\partial \overset{\circ}{v}_3}{\partial x_n}\right)+\frac{\partial M_{nn}}{\partial x_n}+2\frac{\partial M_{nt}}{\partial x_t} && \text{or} \quad v_3 = \widetilde{v}_3 \\
&\quad +(a/\overline{h})\overline{N}_{n3} = \frac{\partial \widetilde{M}_{nt}}{\partial x_t}+\widetilde{\overline{N}}_{n3}
\end{aligned}$$

(2.167a – f)

2.6 Equations of Motion – Linear Formulation

2.6.1 Displacement Formulation for Sandwich Shells

Considering Eqs. (2.156)–(2.160), the following linear equations of motion for sandwich shells can be obtained by discarding all of the nonlinear terms and neglecting the thermal and moisture terms. This results in

$$\begin{aligned}\delta\xi_1:\quad & A_{11}\left(\frac{\partial^2\xi_1}{\partial x_1^2}-\frac{1}{R_1}\frac{\partial v_3}{\partial x_1}\right)+A_{66}\left(\frac{\partial^2\xi_1}{\partial x_2^2}+\frac{\partial^2\xi_2}{\partial x_1\partial x_2}\right)+A_{12}\left(\frac{\partial^2\xi_2}{\partial x_1\partial x_2}-\frac{1}{R_2}\frac{\partial v_3}{\partial x_2}\right)-\\ & A_{16}\left(2\frac{\partial^2\xi_1}{\partial x_1\partial x_2}+\frac{\partial^2\xi_2}{\partial x_1^2}-\frac{1}{R_1}\frac{\partial v_3}{\partial x_1}\right)-A_{26}\left(\frac{\partial^2\xi_2}{\partial x_2^2}-\frac{1}{R_2}\frac{\partial v_3}{\partial x_2}\right)=0\end{aligned} \tag{2.168}$$

$$\begin{aligned}\delta\xi_2:\quad & A_{22}\left(\frac{\partial^2\xi_2}{\partial x_2^2}-\frac{1}{R_2}\frac{\partial v_3}{\partial x_2}\right)+A_{66}\left(\frac{\partial^2\xi_2}{\partial x_1^2}+\frac{\partial^2\xi_1}{\partial x_1\partial x_2}\right)+A_{12}\left(\frac{\partial^2\xi_1}{\partial x_1\partial x_2}-\frac{1}{R_1}\frac{\partial v_3}{\partial x_1}\right)-\\ & A_{26}\left(2\frac{\partial^2\xi_2}{\partial x_1\partial x_2}+\frac{\partial^2\xi_1}{\partial x_2^2}-\frac{1}{R_2}\frac{\partial v_3}{\partial x_2}\right)-A_{16}\left(\frac{\partial^2\xi_1}{\partial x_1^2}-\frac{1}{R_1}\frac{\partial v_3}{\partial x_1}\right)=0\end{aligned} \tag{2.169}$$

$$\begin{aligned}\delta\eta_1:\quad & A_{11}\frac{\partial^2\eta_1}{\partial x_1^2}+A_{66}\left(\frac{\partial^2\eta_1}{\partial x_2^2}+\frac{\partial^2\eta_2}{\partial x_1\partial x_2}\right)+A_{12}\frac{\partial^2\eta_2}{\partial x_1\partial x_2}\\ & +A_{16}\left(\frac{\partial^2\eta_2}{\partial x_1^2}+2\frac{\partial^2\eta_1}{\partial x_1\partial x_2}\right)+A_{26}\frac{\partial^2\eta_2}{\partial x_2^2}-d_1\left(\eta_1+a\frac{\partial v_3}{\partial x_1}\right)=0\end{aligned} \tag{2.170}$$

$$\begin{aligned}\delta\eta_2:\quad & A_{22}\frac{\partial^2\eta_2}{\partial x_2^2}+A_{66}\left(\frac{\partial^2\eta_2}{\partial x_2^2}+\frac{\partial^2\eta_1}{\partial x_1\partial x_2}\right)+A_{12}\frac{\partial^2\eta_1}{\partial x_1\partial x_2}\\ & +A_{26}\left(\frac{\partial^2\eta_1}{\partial x_2^2}+2\frac{\partial^2\eta_2}{\partial x_1\partial x_2}\right)+A_{16}\frac{\partial^2\eta_1}{\partial x_1^2}-d_2\left(\eta_2+a\frac{\partial v_3}{\partial x_2}\right)=0\end{aligned} \tag{2.171}$$

$$\begin{aligned}\delta v_3:\quad & -\left(\frac{A_{11}}{R_1}+\frac{A_{12}}{R_2}\right)\frac{\partial\xi_1}{\partial x_1}-\left(\frac{A_{16}}{R_1}+\frac{A_{26}}{R_2}\right)\frac{\partial\xi_1}{\partial x_2}-\left(\frac{A_{22}}{R_2}+\frac{A_{12}}{R_1}\right)\frac{\partial\xi_2}{\partial x_2}-\left(\frac{A_{26}}{R_2}+\frac{A_{16}}{R_1}\right)\frac{\partial\xi_2}{\partial x_1}\\ & -\left(\frac{A_{11}}{R_1^2}+\frac{A_{22}}{R_2^2}+\frac{2A_{12}}{R_1R_2}\right)v_3-d_1a\left(\frac{\partial\eta_1}{\partial x_1}+a\frac{\partial^2v_3}{\partial x_1^2}\right)-d_2a\left(\frac{\partial\eta_2}{\partial x_2}+a\frac{\partial^2v_3}{\partial x_2^2}\right)+F_{11}\frac{\partial^4v_3}{\partial x_1^4}\\ & +F_{22}\frac{\partial^4v_3}{\partial x_2^4}+2(F_{12}+2F_{66})\frac{\partial^4v_3}{\partial x_1^2\partial x_2^2}+4F_{16}\frac{\partial^4v_3}{\partial x_1^3\partial x_2}+4F_{26}\frac{\partial^4v_3}{\partial x_1\partial x_2^3}+N_{11}^0\frac{\partial^2v_3}{\partial x_1^2}+\\ & +2N_{12}^0\frac{\partial^2v_3}{\partial x_1\partial x_2}+N_{22}^0\frac{\partial^2v_3}{\partial x_2^2}-c\frac{\partial v_3}{\partial t}-m_0\frac{\partial^2v_3}{\partial t^2}=-q_3\end{aligned} \tag{2.172}$$

where N_{11}^0, N_{22}^0, and N_{12}^0 are the prescribed edge loads. The associated boundary conditions are along the edges $x_n = \text{const}(n = 1, 2)$

$$
\begin{aligned}
&N_{nn} = \widetilde{N}_{nn} \quad \text{or} \quad \xi_n = \widetilde{\xi}_n \\
&N_{nt} = \widetilde{N}_{nt} \quad \text{or} \quad \xi_t = \widetilde{\xi}_t \\
&L_{nn} = \widetilde{L}_{nn} \quad \text{or} \quad \eta_n = \widetilde{\eta}_n \\
&L_{nt} = \widetilde{L}_{nt} \quad \text{or} \quad \eta_t = \widetilde{\eta}_t \\
&M_{nn} = \widetilde{M}_{nn} \quad \text{or} \quad \frac{\partial v_3}{\partial x_n} = \frac{\partial \widetilde{v}_3}{\partial x_n} \\
&N_{nt}\left(\frac{\partial v_3}{\partial x_t} + \frac{\partial \overset{\circ}{v}_3}{\partial x_t}\right) + N_{nn}\left(\frac{\partial v_3}{\partial x_n} + \frac{\partial \overset{\circ}{v}_3}{\partial x_n}\right) + \frac{\partial M_{nn}}{\partial x_n} + 2\frac{\partial M_{nt}}{\partial x_t} \quad \text{or} \quad v_3 = \widetilde{v}_3 \\
&\quad + (a/\overline{h})\overline{N}_{n3} = \frac{\partial \widetilde{M}_{nt}}{\partial x_t} + \widetilde{\overline{N}}_{n3}
\end{aligned}
\tag{2.173a – f}
$$

2.6.2 *Displacement Formulation for Sandwich Plates*

The governing equations of motion for flat sandwich plates can be obtained by setting the curvatures to zero in Eqs. (2.168)–(2.172). This results in decoupling occurring between stretching and bending, leaving the first two equations of motion Eqs. (2.168) and (2.169) decoupled from the last three. As a result, the system of equations, Eqs. (2.168)–(2.172) are reduced to

$$
\begin{aligned}
\delta\eta_1 : \quad & A_{11}\frac{\partial^2 \eta_1}{\partial x_1^2} + A_{66}\left(\frac{\partial^2 \eta_1}{\partial x_2^2} + \frac{\partial^2 \eta_2}{\partial x_1 \partial x_2}\right) + A_{12}\frac{\partial^2 \eta_2}{\partial x_1 \partial x_2} \\
& + A_{16}\left(\frac{\partial^2 \eta_2}{\partial x_1^2} + 2\frac{\partial^2 \eta_1}{\partial x_1 \partial x_2}\right) + A_{26}\frac{\partial^2 \eta_2}{\partial x_2^2} - d_1\left(\eta_1 + a\frac{\partial v_3}{\partial x_1}\right) = 0
\end{aligned}
\tag{2.174}
$$

$$
\begin{aligned}
\delta\eta_2 : \quad & A_{22}\frac{\partial^2 \eta_2}{\partial x_2^2} + A_{66}\left(\frac{\partial^2 \eta_2}{\partial x_2^2} + \frac{\partial^2 \eta_1}{\partial x_1 \partial x_2}\right) + A_{12}\frac{\partial^2 \eta_1}{\partial x_1 \partial x_2} \\
& + A_{26}\left(\frac{\partial^2 \eta_1}{\partial x_2^2} + 2\frac{\partial^2 \eta_2}{\partial x_1 \partial x_2}\right) + A_{16}\frac{\partial^2 \eta_1}{\partial x_1^2} - d_2\left(\eta_2 + a\frac{\partial v_3}{\partial x_2}\right) = 0
\end{aligned}
\tag{2.175}
$$

$$\delta v_3: \quad -d_1 a\left(\frac{\partial \eta_1}{\partial x_1}+a\frac{\partial^2 v_3}{\partial x_1^2}\right)-d_2 a\left(\frac{\partial \eta_2}{\partial x_2}+a\frac{\partial^2 v_3}{\partial x_2^2}\right)+F_{11}\frac{\partial^4 v_3}{\partial x_1^4}+F_{22}\frac{\partial^4 v_3}{\partial x_2^4}$$
$$+2(F_{12}+2F_{66})\frac{\partial^4 v_3}{\partial x_1^2\partial x_2^2}+4F_{16}\frac{\partial^4 v_3}{\partial x_1^3\partial x_2}+4F_{26}\frac{\partial^4 v_3}{\partial x_1\partial x_2^3}+N_{11}^0\frac{\partial^2 v_3}{\partial x_1^2}+$$
$$+2N_{12}^0\frac{\partial^2 v_3}{\partial x_1\partial x_2}+N_{22}^0\frac{\partial^2 v_3}{\partial x_2^2}-c\frac{\partial v_3}{\partial t}-m_0\frac{\partial^2 v_3}{\partial t^2}=-q_3 \tag{2.176}$$

Because of the decoupling, the associated boundary conditions are reduced to the following four boundary conditions along the edges $x_n = \text{const}(n = 1, 2)$

$$L_{nn}=\widetilde{L}_{nn} \quad \text{or} \quad \eta_n=\widetilde{\eta}_n$$
$$L_{nt}=\widetilde{L}_{nt} \quad \text{or} \quad \eta_t=\widetilde{\eta}_t$$
$$M_{nn}=\widetilde{M}_{nn} \quad \text{or} \quad \frac{\partial v_3}{\partial x_n}=\frac{\partial \widetilde{v}_3}{\partial x_n}$$
$$N_{nt}\left(\frac{\partial v_3}{\partial x_t}+\frac{\partial \overset{\circ}{v}_3}{\partial x_t}\right)+N_{nn}\left(\frac{\partial v_3}{\partial x_n}+\frac{\partial \overset{\circ}{v}_3}{\partial x_n}\right)+\frac{\partial M_{nn}}{\partial x_n}+2\frac{\partial M_{nt}}{\partial x_t} \quad \text{or} \quad v_3=\widetilde{v}_3$$
$$+(a/\overline{h})\overline{N}_{n3}=\frac{\partial \widetilde{M}_{nt}}{\partial x_t}+\widetilde{\overline{N}}_{n3} \tag{2.177a – d}$$

2.7 Boundary Conditions

Two types of boundary conditions that will be considered here are simply supported and clamped. Referring to the boundary conditions in Eqs. (2.151a–f) and (2.161a–f), the following determination is made. For **Simply Supported** edge conditions of the sandwich shell there are two cases.

Case I: The edges $x_n = 0$, L_n is *loaded in compression and freely movable*. In this case, along these edges, the following conditions need to be fulfilled.

$$N_{nn}=-\widetilde{N}_{nn},\ N_{nt}=0,\ \eta_n=0,\ \eta_t=0,\ M_{nn}=0,\ v_3=0 \tag{2.178}$$

Case II: The edges x_n = constant is *unloaded and immovable*. For this case, the following conditions have to be fulfilled.

$$\xi_n=0,\ N_{nt}=0,\ \eta_n=0,\ \eta_t=0,\ M_{nn}=0,\ v_3=0 \tag{2.179}$$

For **Clamped** boundary conditions, along the edges x_n= 0, L_n,

$$\xi_n = 0, \ \ \xi_t = 0, \ \ \eta_n = 0, \ \ \eta_t = 0, \ \ \frac{\partial v_3}{\partial x_n}, \ \ v_3 = 0. \tag{2.180}$$

$n =$ normal. The subscripts n and t are used to designate the normal and tangential in-plane directions to an edge and hence, $n = 1$ when $t = 2$ and vice versa. For the case expressing the immovability of the edges $\xi_n = 0$, it is implied that this is fulfilled in an average sense which is expressed mathematically as

$$\int_0^{l_2} \int_0^{l_1} \frac{\partial \xi_n}{\partial x_n} dx_n dx_t = 0 \tag{2.181}$$

Analogously, the static boundary conditions, Eqs. (2.178a, b) and (2.179b) are also fulfilled in an average sense expressed as

$$\int_0^{L_t} N_{nn} dx_t = -\widetilde{N}_{nn} L_n \tag{2.182}$$

$$\int_0^{L_t} N_{nt} dx_t = 0, \qquad \begin{pmatrix} n = 1, 2 \\ t = 2, 1 \end{pmatrix} \ \not\sum_{n,t} \tag{2.183}$$

where $\widetilde{N}_{nn}$ denotes the compressive edge load on the edges $x_n = 0, \ L_n$. The sign $\not\sum$ means no summation on n, t.

2.8 Summary

In this chapter, the governing equations for the case of the core being incompressible have been derived. First the derivations started out with the most general case considering the strong core with the transverse stresses in the facings. Hamilton's principle was then leveraged for the theoretical developments. The equations of motion were derived for both the mixed formulation and the displacement formulation. Later on, special cases were considered which were the discarding of the transverse shear stresses in the facings, adopting the Love–Kirchhoff assumptions, and the case of the weak core neglecting the in-plane stresses. Only the transverse stresses were considered in the weak core case. Finally, various types of boundary conditions were discussed. These governing equations will be utilized in later chapters on buckling, post-buckling, and the dynamic response where several solution methodologies will be applied to these equations. Several references to the equations in this chapter and the next few chapters will be forthcoming.

References

Hause, T., Librescu, L., & Johnson, T. F. (1998). Thermomechanical load-carrying capacity of sandwich flat panels. *Journal of Thermal Stresses, 21*(6), 627–653.

Hause, T., Johnson, T. F., & Librescu, L. (2000). Effect of face-sheet anisotropy on buckling and post buckling of flat sandwich panels. *Journal of Spacecraft and Rockets, 37*(3), 331–341.

Jones, R. M. (1999). *Mechanics of composite materials* (2nd ed.). New York/London: Taylor and Francis.

Librescu, L. (1970). On a geometrically non-linear theory of elastic anisotropic sandwich type plates. *Revue Roumaine des Sciences techniques-Mecanique Appliquee, 15*(2), 323–339.

Librescu, L. (1975). *Elastostatics and kinetics of anisotropic and heterogeneous shell-type structures*. Leyden: Noordhoff International Publishing.

Librescu, L., & Chang, M. J. (1993). Effects of geometric imperfections on vibration of compressed shear deformable composite curved panels. *Acta Mechanica, 96*, 203–224.

Librescu, L., Hause, T., & Camarda, C. J. (1997). Geometrically nonlinear theory of initially imperfect sandwich plates and shells incorporating non-classical effects. *AIAA Journal, 35*(8), 1393–1403.

Reddy, J. N. (2004). *Mechanics of laminated composite plates and shells-theory and analysis* (2nd ed.). Boca Raton: CRC Press.

Chapter 3
Buckling of Sandwich Plates and Shells

Abstract Application of the governing linear equations of equilibrium and the associated boundary conditions as applied to the stability of sandwich plates and shells with anisotropic laminated composite facings and a weak core is addressed while the facings are considered symmetric with respect to both their local and global mid-surfaces. Both sandwich plates and shells are considered with implications of various structural configurations. Such structural configurations consider the structural tailoring of the face sheets, the panel face thickness, the aspect ratio, etc. on the buckling load. Finally, validations are made with results found within the existing literature.

3.1 Introduction

This chapter is concerned with the static or stability behavior of anisotropic laminated composite sandwich plates and shells. Both flat and curved sandwich panels are discussed and treated separately. This study concerns symmetric laminated facings both locally and globally in conjunction with a weak or soft core. In addition, the facings are assumed thin so that the Kirchhoff–Love assumptions apply. Simply supported boundary conditions with all four edges freely movable are also considered. The chosen solution methodology is devoted to the extended Galerkin method, while the influence of a number of kinematical and physical parameters, on the load-carrying capacity of the structure, are investigated. Finally, validations are presented to highlight the accuracy of the theory.

3.2 Preliminaries and Basic Assumptions

The governing linear theory for both flat and doubly curved sandwich panels is based on a number of previously mentioned assumptions in Chap. 2 but are repeated here for reference. These assumptions are relisted as

T. J. Hause, *Sandwich Structures: Theory and Responses*,
https://doi.org/10.1007/978-3-030-71895-4_3

1. The face sheets are constructed of a number of orthotropic material layers, the axes of orthotropy of the individual plies being not necessarily coincident with the geometrical axes x_α (α =1, 2) of the structure.
2. The thickness of the core is much larger than those of the face sheets, that is, $2\bar{h} \gg h^{/},\ h^{//}$.
3. The core material features orthotropic properties, the axes of orthotropy being parallel to the geometrical axes x_α.
4. The cases of the *weak*-core type are considered.
5. A perfect bonding between the face sheets and between the faces and the core exists.
6. All three layers, the facings and the core, are incompressible in the transverse direction.
7. The global middle surface of the structure is selected to coincide with that of the core layer which is referred to a curvilinear and orthogonal coordinate system, $x_\alpha(\alpha = 1,\ 2)$. The transverse normal coordinate x_3 is considered positive when measured in the direction of the downward normal.
8. The principles of shallow shell theory apply in regard to doubly curved sandwich panels.

3.3 Buckling of Flat Sandwich Panels

3.3.1 Governing System

This section is concerned with the stability of sandwich plates at the critical load. The solution methodology begins with the linear governing system of partial differential equations for sandwich plates considering a weak core and the transverse shear effects neglected in the facings. The pertinent equations are a subset of the governing equations for shells. These equations were developed in Chap. 2 as Eqs. (2.174)–(2.176) with the reduced set of boundary conditions, Eqs. (2.177a–d). These governing equations are

- *Equations of Equilibrium*:

$$A_{11}\frac{\partial^2\eta_1}{\partial x_1^2}+A_{66}\left(\frac{\partial^2\eta_1}{\partial x_2^2}+\frac{\partial^2\eta_2}{\partial x_1\partial x_2}\right)+A_{12}\frac{\partial^2\eta_2}{\partial x_1\partial x_2}+A_{16}\left(\frac{\partial^2\eta_2}{\partial x_1^2}+2\frac{\partial^2\eta_1}{\partial x_1\partial x_2}\right)$$
$$+A_{26}\frac{\partial^2\eta_2}{\partial x_2^2}-d_1\left(\eta_1+a\frac{\partial v_3}{\partial x_1}\right)=0 \tag{3.1}$$

$$A_{22}\frac{\partial^2\eta_2}{\partial x_2^2}+A_{66}\left(\frac{\partial^2\eta_2}{\partial x_2^2}+\frac{\partial^2\eta_1}{\partial x_1\partial x_2}\right)+A_{12}\frac{\partial^2\eta_1}{\partial x_1\partial x_2}+A_{26}\left(\frac{\partial^2\eta_1}{\partial x_2^2}+2\frac{\partial^2\eta_2}{\partial x_1\partial x_2}\right)$$
$$+A_{16}\frac{\partial^2\eta_1}{\partial x_1^2}-d_2\left(\eta_2+a\frac{\partial v_3}{\partial x_2}\right)=0 \tag{3.2}$$

$$-d_1a\left(\frac{\partial\eta_1}{\partial x_1}+a\frac{\partial^2 v_3}{\partial x_1^2}\right)-d_2a\left(\frac{\partial\eta_2}{\partial x_2}+a\frac{\partial^2 v_3}{\partial x_2^2}\right)+F_{11}\frac{\partial^4 v_3}{\partial x_1^4}+F_{22}\frac{\partial^4 v_3}{\partial x_2^4}$$
$$+2(F_{12}+2F_{66})\frac{\partial^4 v_3}{\partial x_1^2\partial x_2^2}+4F_{16}\frac{\partial^4 v_3}{\partial x_1^3\partial x_2}+4F_{26}\frac{\partial^4 v_3}{\partial x_1\partial x_2^3}+N_{11}^0\frac{\partial^2 v_3}{\partial x_1^2} \tag{3.3}$$
$$+2N_{12}^0\frac{\partial^2 v_3}{\partial x_1\partial x_2}+N_{22}^0\frac{\partial^2 v_3}{\partial x_2^2}=0$$

where N_{11}^0, N_{22}^0, and N_{12}^0 are the prescribed edge loadings in both the x_1 and x_2 directions. It should be noted that the damping, inertia, and transversal load terms have been neglected. Because of the decoupling between stretching and bending, the associated boundary conditions are reduced to the following four boundary conditions along the edges $x_n = 0,\ L_n$ which are

- *Boundary conditions*:

$$L_{nn}=\widetilde{L}_{nn} \quad \text{or} \quad \eta_n=\widetilde{\eta}_n \tag{3.4}$$

$$L_{nt}=\widetilde{L}_{nt} \quad \text{or} \quad \eta_t=\widetilde{\eta}_t \tag{3.5}$$

$$M_{nn}=\widetilde{M}_{nn} \quad \text{or} \quad \frac{\partial v_3}{\partial x_n}=\frac{\partial\widetilde{v}_3}{\partial x_n} \tag{3.6}$$

$$N_{nt}\frac{\partial v_3}{\partial x_t}+N_{nn}\frac{\partial v_3}{\partial x_n}+\frac{\partial M_{nn}}{\partial x_n}+2\frac{\partial M_{nt}}{\partial x_t} \quad \text{or} \quad v_3=\widetilde{v}_3$$
$$+(a/\overline{h})\overline{N}_{n3}=\frac{\partial\widetilde{M}_{nt}}{\partial x_t}+\widetilde{\overline{N}}_{n3} \tag{3.7}$$

considering simply supported boundary conditions, the following conditions apply
Along the edges $x_1 = 0,\ L_1$

$$\eta_1=0,\ \ \eta_2=0,\ \ M_{11}=0,\ \ v_3=0 \tag{3.8a – d}$$

Along the edges $x_2 = 0,\ L_2$

$$\eta_2=0,\ \ \eta_1=0,\ \ M_{22}=0,\ \ v_3=0 \tag{3.9a – d}$$

In terms of displacements, the third boundary conditions are expressed as

$$M_{11} = F_{11}\frac{\partial^2 v_3}{\partial x_1^2} + F_{12}\frac{\partial^2 v_3}{\partial x_2^2} + 2F_{16}\frac{\partial^2 v_3}{\partial x_1 \partial x_2} = 0 \tag{3.10a}$$

$$M_{22} = F_{22}\frac{\partial^2 v_3}{\partial x_2^2} + F_{12}\frac{\partial^2 v_3}{\partial x_1^2} + 2F_{26}\frac{\partial^2 v_3}{\partial x_1 \partial x_2} = 0 \tag{3.10b}$$

These expressions will become useful later on when the displacement functions have been determined and thus the fulfillment of these particular boundary conditions will become readily apparent.

3.3.2 Solution Methodology

The solution methodology chosen here is the extended Galerkin method (EGM). The advantage of using this method relies on the fact that the unfulfilled boundary conditions are satisfied in an average sense. As a starting point, the transverse displacement can be represented as

$$v_3(x_1, x_2, t) = w_{mn} \sin \lambda_m x_1 \sin \mu_n x_2 \tag{3.11}$$

where $\lambda_m = m\pi/L_1$ and $\mu_n = n\pi/L_2$. This completely satisfies the condition that $v_3 = 0$ along all four edges. But does not satisfy the conditions for $M_{11} = 0,\ M_{22} = 0$ along the edges $x_1 = 0,\ L_1,\ x_2 = 0,\ L_2$, respectively. These will be satisfied later on, in an average sense, through the use of the EGM. The first two equations of equilibrium Eqs. (3.1) and (3.2) can be satisfied by assuming η_1 and η_2 as follows

$$\begin{Bmatrix} \eta_1(x_1, x_2) \\ \eta_2(x_1, x_2) \end{Bmatrix} = \begin{Bmatrix} H_{mn}^{(1)} \\ I_{mn}^{(1)} \end{Bmatrix} \cos \lambda_m x_1 \sin \mu_n x_2 + \begin{Bmatrix} H_{mn}^{(2)} \\ I_{mn}^{(2)} \end{Bmatrix} \sin \lambda_m x_1 \cos \mu_n x_2 \tag{3.12a, b}$$

where $H_{mn}^{(1)}, H_{mn}^{(2)}, I_{mn}^{(1)},$ and $I_{mn}^{(2)}$ are undetermined coefficients. Substituting the expressions for η_1 and η_2 back into Eqs. (3.1) and (3.2) and comparing coefficients of like trigonometric functions gives the following matrix equation in terms of the unknown coefficients as

$$\begin{bmatrix} J_{11}^{(m,n)} & J_{12}^{(m,n)} & J_{13}^{(m,n)} & J_{14}^{(m,n)} \\ & J_{11}^{(m,n)} & J_{14}^{(m,n)} & J_{13}^{(m,n)} \\ & & J_{33}^{(m,n)} & J_{34}^{(m,n)} \\ \text{Symm.} & & & J_{33}^{(m,n)} \end{bmatrix} \begin{Bmatrix} H_{mn}^{(1)} \\ H_{mn}^{(2)} \\ I_{mn}^{(1)} \\ I_{mn}^{(2)} \end{Bmatrix} = \begin{Bmatrix} K_{mn}^{(1)} \\ K_{mn}^{(2)} \\ K_{mn}^{(3)} \\ K_{mn}^{(4)} \end{Bmatrix} \tag{3.13}$$

where

$$\begin{aligned} &J_{11}^{(m,n)} = -\left(\lambda_m^2 A_{11} + \mu_n^2 A_{66} + d_1\right), \quad J_{12}^{(m,n)} = -2\lambda_m \mu_n A_{16}, \quad J_{13}^{(m,n)} = -\lambda_m^2 A_{16} - \mu_n^2 A_{26} \\ &J_{14}^{(m,n)} = -\lambda_m \mu_n (A_{12} + A_{66}), \ J_{33}^{(m,n)} = -\left(\mu_n^2 A_{22} + \lambda_m^2 A_{66} + d_2\right), \ J_{34}^{(m,n)} = -2\lambda_m \mu_n A_{26} \\ &K_{mn}^{(1)} = d_1 a \lambda_m w_{mn}, \quad K_{mn}^{(2)} = 0, \quad K_{mn}^{(3)} = 0, \quad K_{mn}^{(3)} = d_2 a \mu_n w_{mn} \end{aligned} \tag{3.14a – j}$$

From Eq. (3.13), using Cramer's rule, the coefficients $H_{mn}^{(\alpha)}$, $I_{mn}^{(\alpha)}$ $(\alpha = 1, 2)$ can be expressed as

$$\left(H_{mn}^{(1)}, H_{mn}^{(2)}, I_{mn}^{(1)}, I_{mn}^{(2)}\right) = \left(\widetilde{H}_{mn}^{(1)}, \widetilde{H}_{mn}^{(2)}, \widetilde{I}_{mn}^{(1)}, \widetilde{I}_{mn}^{(2)}\right) w_{mn}, \tag{3.15}$$

where

$$\widetilde{H}_{mn}^{(1)} = \frac{\det(J_1)}{\det(J)}, \quad \widetilde{H}_{mn}^{(2)} = \frac{\det(J_2)}{\det(J)}, \quad \widetilde{I}_{mn}^{(1)} = \frac{\det(J_3)}{\det(J)}, \quad \widetilde{I}_{mn}^{(2)} = \frac{\det(J_4)}{\det(J)} \tag{3.16a – d}$$

While,

$$J_1 = \begin{pmatrix} K_{mn}^{(1)} & J_{12}^{(m.n)} & J_{13}^{(m,n)} & J_{14}^{(m,n)} \\ K_{mn}^{(2)} & J_{11}^{(m,n)} & J_{14}^{(m,n)} & J_{13}^{(m,n)} \\ K_{mn}^{(3)} & J_{14}^{(m,n)} & J_{33}^{(m,n)} & J_{34}^{(m,n)} \\ K_{mn}^{(4)} & J_{13}^{(m,n)} & J_{34}^{(m,n)} & J_{33}^{(m,n)} \end{pmatrix}, \quad J_2 = \begin{pmatrix} J_{11}^{(m,n)} & K_{mn}^{(1)} & J_{13}^{(m,n)} & J_{14}^{(m,n)} \\ J_{12}^{(m.n)} & K_{mn}^{(2)} & J_{14}^{(m,n)} & J_{13}^{(m,n)} \\ J_{13}^{(m,n)} & K_{mn}^{(3)} & J_{33}^{(m,n)} & J_{34}^{(m,n)} \\ J_{14}^{(m,n)} & K_{mn}^{(4)} & J_{34}^{(m,n)} & J_{33}^{(m,n)} \end{pmatrix} \tag{3.17a, b}$$

$$J_3 = \begin{pmatrix} J_{11}^{(m,n)} & J_{12}^{(m.n)} & K_{mn}^{(1)} & J_{14}^{(m,n)} \\ J_{12}^{(m.n)} & J_{11}^{(m,n)} & K_{mn}^{(2)} & J_{13}^{(m,n)} \\ J_{13}^{(m,n)} & J_{14}^{(m,n)} & K_{mn}^{(3)} & J_{34}^{(m,n)} \\ J_{14}^{(m,n)} & J_{13}^{(m,n)} & K_{mn}^{(4)} & J_{33}^{(m,n)} \end{pmatrix}, \quad J_4 = \begin{pmatrix} J_{11}^{(m,n)} & J_{12}^{(m.n)} & J_{13}^{(m,n)} & K_{mn}^{(1)} \\ J_{12}^{(m.n)} & J_{11}^{(m,n)} & J_{14}^{(m,n)} & K_{mn}^{(2)} \\ J_{13}^{(m,n)} & J_{14}^{(m,n)} & J_{33}^{(m,n)} & K_{mn}^{(3)} \\ J_{14}^{(m,n)} & J_{13}^{(m,n)} & J_{34}^{(m,n)} & K_{mn}^{(4)} \end{pmatrix} \tag{3.17c, d}$$

$$J = \begin{pmatrix} J_{11}^{(m,n)} & J_{12}^{(m,n)} & J_{13}^{(m,n)} & J_{14}^{(m,n)} \\ & J_{11}^{(m,n)} & J_{14}^{(m,n)} & J_{13}^{(m,n)} \\ & & J_{33}^{(m,n)} & J_{34}^{(m,n)} \\ \text{Symm.} & & & J_{33}^{(m,n)} \end{pmatrix} \tag{3.17e}$$

Although the first two equations of equilibrium are satisfied by the assumed functional forms for η_1 and η_2, the boundary conditions for η_n and η_t along the edges $x_n = 0,\ L_n$ are not satisfied. Again, these will be satisfied in an average sense through the use of the EGM. At this juncture, all of the displacement quantities $v_3,\ \eta_1,\ \eta_2$ are known. In addition, there remains only the fifth equation of equilibrium unfulfilled along with the first, second, and third boundary conditions. To solve for the critical load, the unfulfilled fifth equation of motion and the unfulfilled boundary conditions will be retained in the energy functional with the appropriate integrations carried out. In the end, this will supply the critical load for the stability problem in terms of the geometrical and material parameters. Retaining these unfulfilled expressions in the Hamilton's energy functional results in

$$\begin{aligned} &\int_{t_0}^{t_1}\left\langle\int_0^{l_2}\int_0^{l_1}\left\{-d_1 a\left(\frac{\partial\eta_1}{\partial x_1}+a\frac{\partial^2 v_3}{\partial x_1^2}\right)-d_2 a\left(\frac{\partial\eta_2}{\partial x_2}+a\frac{\partial^2 v_3}{\partial x_2^2}\right)+F_{11}\frac{\partial^4 v_3}{\partial x_1^4}+F_{22}\frac{\partial^4 v_3}{\partial x_2^4}\right.\right. \\ &\quad +2(F_{12}+2F_{66})\frac{\partial^4 v_3}{\partial x_1^2\partial x_2^2}+4F_{16}\frac{\partial^4 v_3}{\partial x_1^3\partial x_2}+4F_{26}\frac{\partial^4 v_3}{\partial x_1\partial x_2^3}+N_{11}^0\frac{\partial^2 v_3}{\partial x_1^2}+ \\ &\quad \left.\left.+2N_{12}^0\frac{\partial^2 v_3}{\partial x_1\partial x_2}+N_{22}^0\frac{\partial^2 v_3}{\partial x_2^2}\right\}\delta v_3 dx_1 dx_2\right\rangle dt+\int_{t_0}^{t_1}\left\langle\int_0^{l_2}\left[L_{11}\delta\eta_1+L_{12}\delta\eta_2+\right.\right. \\ &\left.\left. M_{11}\delta\left(\frac{\partial v_3}{\partial x_1}\right)\right]\Big|_0^{l_1}dx_2\right\rangle dt+\int_{t_0}^{t_1}\left\langle\int_0^{l_1}\left[L_{22}\delta\eta_2+L_{12}\delta\eta_1+M_{22}\delta\left(\frac{\partial v_3}{\partial x_2}\right)\right]\Big|_0^{l_2}dx_1\right\rangle dt \\ &=0 \end{aligned} \tag{3.18}$$

Substituting in the expressions for v_3, η_1, η_2 and carrying out the indicated integrations and solving for the prescribed edge loads results in the buckling solution for angle-ply laminated sandwich structures which results in a linear algebraic equation as

$$N_{11}^0\lambda_m^2 + N_{22}^0\mu_n^2 = \lambda_m a d_1\left(\widetilde{H}_{mn}^{(1)} + a\lambda_m\right) + \mu_n a d_2\left(\widetilde{I}_{mn}^{(2)} + a\mu_n\right) + F_{11}\lambda_m^2 + F_{22}\mu_n^2 + 4F_{66}\lambda_m^2\mu_n^2 + 2F_{12}\lambda_m^2\mu_n^2 \tag{3.19}$$

Where the shear loading term, N_{12}^0 has been discarded. Nondimensionalizing Eq. (3.19) results in

$$K_x\left(m^2\pi^2 + L_R n^2\phi^2\right) = m^4 + \frac{n^4\phi^4 F_{22}}{F_{11}} + \frac{2m^2n^2\phi^2(F_{12}+2F_{66})}{F_{11}} + \frac{a^2L_1^2}{\pi^2 F_{11}}\left(m^2 d_1 + n^2\phi^2 d_2\right) + \frac{aL_1^3}{\pi^3 F_{11}}\left[m d_1\widetilde{H}_{mn}^{(1)} + n\phi d_2\widetilde{I}_{mn}^{(2)}\right] \tag{3.20}$$

Where the nondimensional parameters are defined as

$$K_x = \frac{L_1^2 N_{11}^0}{\pi^4 F_{11}}, \quad L_R = \frac{N_{22}^0}{N_{11}^0}, \quad \phi = \frac{L_1}{L_2} \tag{3.21}$$

It should be mentioned that the coefficients $\widetilde{H}_{mn}^{(1)}$ and $\widetilde{I}_{mn}^{(2)}$ should also be nondimensionalized which is not shown here.

3.3.3 Validation of the Theoretical Results

Before addressing the present results some validations are displayed which compare the results from the present theory with the experimental counterparts (Alexandrov et al. 1960). Table 3.1 lists the geometrical and material properties, in addition to the buckling response, for a three-layered flat sandwich panel with isotropic facings (Dura Aluminum, $\nu = 0.3$, $E = 6.96 \times 10^5$ kg/cm^2) and a transversely isotropic core (Penoplast). Comparing the present theory with the experimental results reasonable agreement is seen keeping in mind that with the inclusion of geometric imperfections, within the theory, would provide more exact agreement.

Results in Table 3.2 display the material and geometrical properties and the buckling response for a three-layered flat sandwich panel with isotropic facings (Dura Aluminum) and an orthotropic core. In both cases, simply supported boundary conditions are assumed. As with the results in Table 3.1, the results in Table 3.2

Table 3.1 Comparisons of theoretical and buckling predictions for a flat sandwich panel with a transversely isotropic core with isotropic faces (Validation No. 1)

Case	L_1(cm)	L_2(cm)	h (cm)	$\bar{h}$ (cm)	$\overline{G}$ (kg/cm^2)	$\widetilde{N}_{11}L_2 \times 10^3$ kg		
						Present	A. et al.	% Error
1	60	40	0.05	0.425	99.4	3.79	3.60	+5.28
2	60	40	0.10	0.650	149.6	9.03	8.25	+9.45
3	40	60	0.10	0.700	117.1	11.30	12.30	-8.13
4	40	60	0.10	1.400	96.5	17.40	16.00	+8.75
5	80	60	0.05	0.450	73.5	4.21	4.00	+5.25
6	80	60	0.05	0.450	74.1	4.24	4.10	+3.41

A. et al. → Alexandrov et al. (1960)

Table 3.2 Comparisons of theoretical and buckling predictions for a flat sandwich panel with an orthotropic core and isotropic faces (Validation No. 2)

Case	L_1(cm)	L_2(cm)	h_f (cm)	$\bar{h}$ (cm)	$\widetilde{N}_{11}L_2 \times 10^3$ kg				
					$\overline{G}_{13}^a$	$\overline{G}_{23}^a$	Present	A. et al.	% Error
1	60	40	0.05	0.45	140.4	100.8	5.29	5.85	-9.57
2	60	40	0.25	1.25	390.0	103.0	47.22	46.57	+1.11
3	60	40	0.25	1.15	337.0	97.0	38.20	36.50	+4.66
4	80	60	0.10	0.95	138.1	78.6	17.34	15.25	+13.7

[a]Units are in (kg/cm^2)
A. et al. → Alexandrov et al. (1960)

Table 3.3 Face sheet material properties

Type	Material	E_1(N/mm^2)	E_2(N/mm^2)	G_{12} (N/mm)	ν_{12}	α_1 (1/K)	α_2 (1/K)
F1	HS Graph. Ep.	1.8375	0.105	0.0735	0.28	11.34	36.9
F2	IM7/977-2	0.0812	0.0763	0.0098	0.06	1.62	1.674
F3	SCS-6/Ti-15-3	0.19404	0.12663	0.05705	0.3	–	–
F4	CFRP	2.324	0.13552	0.05327	0.32	–	–

Note: Multiply E_1, E_2, $G_{12} \times 10^5$

reveal a slight over agreement which again is most likely due to the imperfections within the test specimen.

3.3.4 *Results and Discussion*

For the following numerical results, unless specified otherwise, the material properties for the face sheets and the core are listed in Tables 3.3 and 3.4 by type.

Some additional validations are made in Fig. 3.1 where the effect of the panel face thickness on the critical buckling load for a flat sandwich panel of a fixed stacking sequence in the facings is shown for specific values of $\underline{a}$ (the distance between the

Table 3.4 Core material properties

Type	Core type	$\overline{G}_{13}$ (N/mm^2)	$\overline{G}_{23}$ (N/mm^2)
C1	Titanium honeycomb	0.0145×10	0.0066×10^5
C2	Aluminum honeycomb	148.239	91.7847
C3	Aluminum honeycomb	99.4	60.2

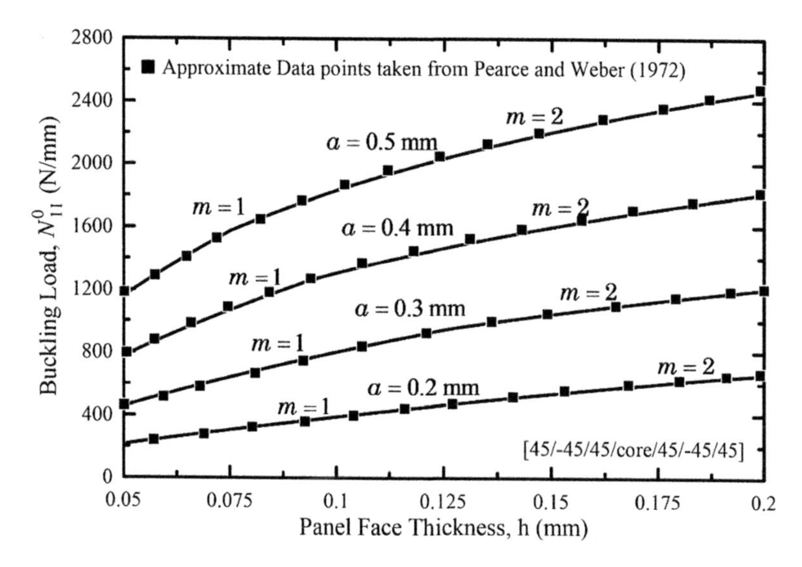

Fig. 3.1 The compressive buckling load vs. panel face thickness for fixed values of the distance between the mid-surface of the core and the mid-surface of the upper and local face sheets

mid-surface of the core and the mid-surface of the facings). The results in Fig. 3.1 are of the F4:C1 type. It can be seen that as the face thickness increases, the critical buckling load increases. Overlaid on the plot are data points from Pearce and Webber (1972) who considered the same results. Almost perfect agreement is seen. In Fig. 3.2, the effect of the ply angle on the buckling load for various aspect ratios of a four-layered flat sandwich panel are depicted for a F3:C1 type. All of the trends appear to be more flat than curved. For lower aspect ratios, the critical buckling load is higher across the entire ply angle spectrum. For higher aspect ratios, the critical buckling loads are lower. In this case, data taken from Ko and Jackson (1991, 1993) for the same configuration are overlaid on the present results and perfect agreement is seen. Figures 3.3 and 3.4 which are of the F1:C1 type display the effect of the directional material properties on the critical buckling load for a single-layered and a three-layered flat sandwich panel for the given layup. *The buckling loads in both cases appear to peak around 45 degrees then drop off up until 90 degrees.* The three-layered sandwich panel can sustain larger buckling loads

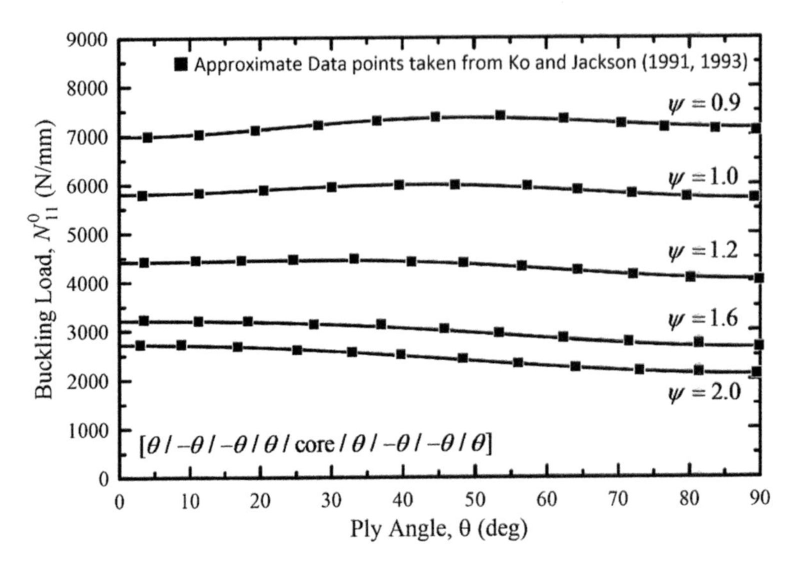

Fig. 3.2 The effect of the fiber orientation of the face sheet lamina on the buckling load of the flat sandwich panel

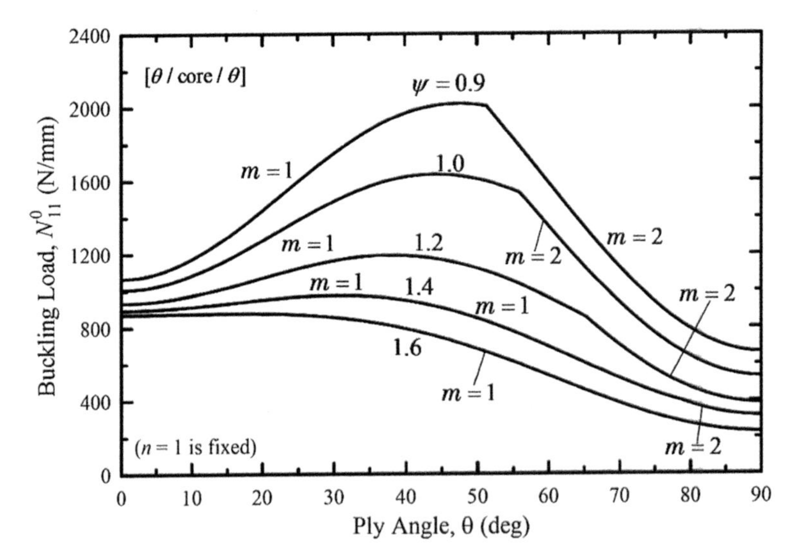

Fig. 3.3 The effect of the fiber orientation of the face sheet lamina on the buckling load of single-layered flat sandwich panel

as compared to its single-layered counterpart. The peak critical loads at their determined ply angle are recorded in Table 3.5 for later reference in Chap. 5 concerning free vibration where it is found that at the peak buckling load the eigenfrequencies vanish.

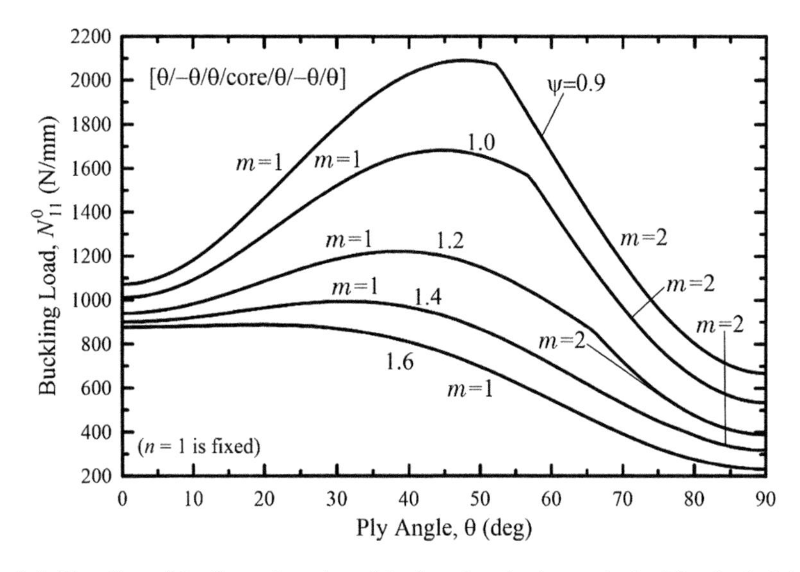

Fig. 3.4 The effect of the fiber orientation of the face sheet lamina on the buckling load of three-layered flat sandwich panel

Table 3.5 Maximum buckling loads and associated ply angles for two different layups of the face sheets for a flat sandwich panel

	ψ^a				
Layup	0.9	1.0	1.2	1.4	1.6
$[\theta/c/\theta]$	(2.021) [47.7]	(1.634) [44.1]	(1.193) [37.8]	(0.975) [31.5]	(0.878) [18.9]
$[\theta/-\theta/\theta/c]_s$	(2.089) [47.7]	(1.682) [44.1]	(1.222) [38.6]	(0.994) [31.5]	(0.889) [19.8]

[a]Sequence ( ) [ ] $\equiv \left(\widetilde{N}_{11} \times 10^3 \text{ N/mm}\right)$ $[\theta\text{deg}]$

3.4 Buckling of Doubly Curved Sandwich Panels

3.4.1 Governing System

This section is concerned with the stability of doubly curved sandwich shells. In contrast to the previous section covering flat sandwich plates, the governing equations are coupled between stretching and bending. There are two additional equations of equilibrium which exhibits this coupling along with two additional boundary conditions prescribed along each edge. The governing equations which apply to the case of doubly curved sandwich shells are Eqs. (2.168)–(2.172) in conjunction with the boundary conditions, Eqs. (2.173a–f). These are presented as

- *Equations of Equilibrium*:

$$A_{11}\left(\frac{\partial^2\xi_1}{\partial x_1^2}-\frac{1}{R_1}\frac{\partial v_3}{\partial x_1}\right)+A_{66}\left(\frac{\partial^2\xi_1}{\partial x_2^2}+\frac{\partial^2\xi_2}{\partial x_1\partial x_2}\right)+A_{12}\left(\frac{\partial^2\xi_2}{\partial x_1\partial x_2}-\frac{1}{R_2}\frac{\partial v_3}{\partial x_1}\right)$$
$$-A_{16}\left(2\frac{\partial^2\xi_1}{\partial x_1\partial x_2}+\frac{\partial^2\xi_2}{\partial x_1^2}-\frac{1}{R_1}\frac{\partial v_3}{\partial x_2}\right)-A_{26}\left(\frac{\partial^2\xi_2}{\partial x_2^2}-\frac{1}{R_2}\frac{\partial v_3}{\partial x_2}\right)=0 \tag{3.22}$$

$$A_{22}\left(\frac{\partial^2\xi_2}{\partial x_2^2}-\frac{1}{R_2}\frac{\partial v_3}{\partial x_2}\right)+A_{66}\left(\frac{\partial^2\xi_2}{\partial x_1^2}+\frac{\partial^2\xi_1}{\partial x_1\partial x_2}\right)+A_{12}\left(\frac{\partial^2\xi_1}{\partial x_1\partial x_2}-\frac{1}{R_1}\frac{\partial v_3}{\partial x_2}\right)$$
$$-A_{26}\left(2\frac{\partial^2\xi_2}{\partial x_1\partial x_2}+\frac{\partial^2\xi_1}{\partial x_2^2}-\frac{1}{R_2}\frac{\partial v_3}{\partial x_1}\right)-A_{16}\left(\frac{\partial^2\xi_1}{\partial x_1^2}-\frac{1}{R_1}\frac{\partial v_3}{\partial x_1}\right)=0 \tag{3.23}$$

$$A_{11}\frac{\partial^2\eta_1}{\partial x_1^2}+A_{66}\left(\frac{\partial^2\eta_1}{\partial x_2^2}+\frac{\partial^2\eta_2}{\partial x_1\partial x_2}\right)+A_{12}\frac{\partial^2\eta_2}{\partial x_1\partial x_2}+A_{16}\left(\frac{\partial^2\eta_2}{\partial x_1^2}+2\frac{\partial^2\eta_1}{\partial x_1\partial x_2}\right)$$
$$+A_{26}\frac{\partial^2\eta_2}{\partial x_2^2}-d_1\left(\eta_1+a\frac{\partial v_3}{\partial x_1}\right)=0 \tag{3.24}$$

$$A_{22}\frac{\partial^2\eta_2}{\partial x_2^2}+A_{66}\left(\frac{\partial^2\eta_2}{\partial x_2^2}+\frac{\partial^2\eta_1}{\partial x_1\partial x_2}\right)+A_{12}\frac{\partial^2\eta_1}{\partial x_1\partial x_2}+A_{26}\left(\frac{\partial^2\eta_1}{\partial x_2^2}+2\frac{\partial^2\eta_2}{\partial x_1\partial x_2}\right)$$
$$+A_{16}\frac{\partial^2\eta_1}{\partial x_1^2}-d_2\left(\eta_2+a\frac{\partial v_3}{\partial x_2}\right)=0 \tag{3.25}$$

$$-\left(\frac{A_{11}}{R_1}+\frac{A_{12}}{R_2}\right)\frac{\partial\xi_1}{\partial x_1}-\left(\frac{A_{16}}{R_1}+\frac{A_{26}}{R_2}\right)\frac{\partial\xi_1}{\partial x_2}-\left(\frac{A_{22}}{R_2}+\frac{A_{12}}{R_1}\right)\frac{\partial\xi_2}{\partial x_2}-\left(\frac{A_{26}}{R_2}+\frac{A_{16}}{R_1}\right)\frac{\partial\xi_2}{\partial x_1}-$$
$$-\left(\frac{A_{11}}{R_1^2}+\frac{A_{22}}{R_2^2}+\frac{2A_{12}}{R_1R_2}\right)v_3-d_1a\left(\frac{\partial\eta_1}{\partial x_1}+a\frac{\partial^2v_3}{\partial x_1^2}\right)-d_2a\left(\frac{\partial\eta_2}{\partial x_2}+a\frac{\partial^2v_3}{\partial x_2^2}\right)+F_{11}\frac{\partial^4v_3}{\partial x_1^4}+$$
$$F_{22}\frac{\partial^4v_3}{\partial x_2^4}+2(F_{12}+2F_{66})\frac{\partial^4v_3}{\partial x_1^2\partial x_2^2}+4F_{16}\frac{\partial^4v_3}{\partial x_1^3\partial x_2}+4F_{26}\frac{\partial^4v_3}{\partial x_1\partial x_2^3}+N_{11}^0\frac{\partial^2v_3}{\partial x_1^2}+$$
$$2N_{12}^0\frac{\partial^2v_3}{\partial x_1\partial x_2}+N_{22}^0\frac{\partial^2v_3}{\partial x_2^2}=0 \tag{3.26}$$

- *Boundary conditions*:

The associated boundary conditions along the edges $x_n = 0,\ L_n$ are

$$N_{nn} = \widetilde{N}_{nn} \quad \text{or} \quad \xi_n = \widetilde{\xi}_n \tag{3.27}$$

$$N_{nt} = \widetilde{N}_{nt} \quad \text{or} \quad \xi_t = \widetilde{\xi}_t \tag{3.28}$$

$$L_{nn} = \widetilde{L}_{nn} \quad \text{or} \quad \eta_n = \widetilde{\eta}_n \tag{3.29}$$

$$L_{nt} = \widetilde{L}_{nt} \quad \text{or} \quad \eta_t = \widetilde{\eta}_t \tag{3.30}$$

$$M_{nn} = \widetilde{M}_{nn} \quad \text{or} \quad \frac{\partial v_3}{\partial x_n} = \frac{\partial \widetilde{v}_3}{\partial x_n} \tag{3.31}$$

$$\begin{aligned} &N_{nt}\frac{\partial v_3}{\partial x_t} + N_{nn}\frac{\partial v_3}{\partial x_n} + \frac{\partial M_{nn}}{\partial x_n} + 2\frac{\partial M_{nt}}{\partial x_t} \qquad \text{or} \qquad v_3 = \widetilde{v}_3 \\ &+(a/\overline{h})\overline{N}_{n3} = \frac{\partial \widetilde{M}_{nt}}{\partial x_t} + \widetilde{\overline{N}}_{n3} \end{aligned} \tag{3.32}$$

In the case of simply supported boundary conditions freely movable on all four edges

at $x_1 = 0,\ L_1$

$$N_{11} = N_{12} = \eta_1 = \eta_2 = M_{11} = v_3 = 0 \tag{3.33a – f}$$

at $x_2 = 0,\ L_2$

$$N_{22} = N_{12} = \eta_1 = \eta_2 = M_{22} = v_3 = 0 \tag{3.34a – f}$$

In terms of displacements, the first, second, and fifth boundary conditions from Eqs. (3.33) and (3.34) can be written as

$$\begin{aligned} N_{11} &= A_{11}\frac{\partial \xi_1}{\partial x_1} + A_{12}\frac{\partial \xi_2}{\partial x_2} + A_{16}\left(\frac{\partial \xi_2}{\partial x_1} + \frac{\partial \xi_1}{\partial x_2}\right) - \left(\frac{A_{11}}{R_1} + \frac{A_{12}}{R_2}\right)v_3 \\ &= 0, \qquad (1 \rightleftarrows 2) \end{aligned} \tag{3.35a}$$

$$N_{12} = A_{66}\left(\frac{\partial \xi_2}{\partial x_1} + \frac{\partial \xi_1}{\partial x_2}\right) + A_{26}\frac{\partial \xi_2}{\partial x_2} + A_{16}\frac{\partial \xi_1}{\partial x_1} - \left(\frac{A_{16}}{R_1} + \frac{A_{26}}{R_2}\right)v_3 = 0 \tag{3.35b}$$

$$M_{11} = F_{11}\frac{\partial^2 v_3}{\partial x_1^2} + F_{12}\frac{\partial^2 v_3}{\partial x_2^2} + 2F_{16}\frac{\partial^2 v_3}{\partial x_1 \partial x_2} = 0 \tag{3.36a}$$

$$M_{22} = F_{22}\frac{\partial^2 v_3}{\partial x_2^2} + F_{12}\frac{\partial^2 v_3}{\partial x_1^2} + 2F_{26}\frac{\partial^2 v_3}{\partial x_1 \partial x_2} = 0 \tag{3.36b}$$

3.4.2 Solution Methodology

Following the same procedure as was carried out for flat sandwich plates, $v_3(x_1, x_2, t)$ can be assumed in the following form

$$v_3(x_1, x_2, t) = w_{mn} \sin \lambda_m x_1 \sin \mu_n x_2 \tag{3.37}$$

$v_3(x_1, x_2, t)$ identically fulfills the sixth boundary conditions provided in Eqs. (3.33f) and (3.34f). The next step is to fulfill the first two equations of equilibrium, Eqs. (3.22) and (3.23). To achieve this, ξ_1 and ξ_2 can be assumed in the following form

$$\begin{Bmatrix} \xi_1(x_1, x_2, t) \\ \xi_2(x_1, x_2, t) \end{Bmatrix} = \begin{Bmatrix} F_{mn}^{(1)} \\ G_{mn}^{(1)} \end{Bmatrix} \cos \lambda_m x_1 \sin \mu_n x_2 + \begin{Bmatrix} F_{mn}^{(2)} \\ G_{mn}^{(2)} \end{Bmatrix} \sin \lambda_m x_1 \cos \mu_n x_2 \tag{3.38a, b}$$

where $F_{mn}^{(1)}$, $F_{mn}^{(2)}$, $G_{mn}^{(1)}$, $G_{mn}^{(2)}$ are constants to be determined. Substituting the expressions for v_3, ξ_1, ξ_2 into Eqs. (3.22) and (3.23) and comparing coefficients of the same trigonometric functions gives the following matrix equation in terms of the unknown constants $F_{mn}^{(1)}$, $F_{mn}^{(2)}$, $G_{mn}^{(1)}$, $G_{mn}^{(2)}$ which is expressed as

$$\begin{bmatrix} L_{11}^{(m,n)} & L_{12}^{(m,n)} & L_{13}^{(m,n)} & L_{14}^{(m,n)} \\ & L_{11}^{(m,n)} & L_{14}^{(m,n)} & L_{13}^{(m,n)} \\ & & L_{33}^{(m,n)} & L_{34}^{(m,n)} \\ \text{Symm.} & & & L_{33}^{(m,n)} \end{bmatrix} \begin{Bmatrix} F_{mn}^{(1)} \\ F_{mn}^{(2)} \\ G_{mn}^{(1)} \\ G_{mn}^{(2)} \end{Bmatrix} = \begin{Bmatrix} M_{mn}^{(1)} \\ M_{mn}^{(2)} \\ M_{mn}^{(3)} \\ M_{mn}^{(4)} \end{Bmatrix} \tag{3.39}$$

where

$$\begin{gathered} L_{11}^{(m,n)} = -(\lambda_m^2 A_{11} + \mu_n^2 A_{66}), \quad L_{12}^{(m,n)} = 2\lambda_m \mu_n A_{16}, \quad L_{13}^{(m,n)} = \lambda_m^2 A_{16} + \mu_n^2 A_{26} \\ L_{14}^{(m,n)} = -\lambda_m \mu_n (A_{12} + A_{66}), \quad L_{33}^{(m,n)} = -(\mu_n^2 A_{22} + \lambda_m^2 A_{66}), \quad L_{34}^{(m,n)} = 2\lambda_m \mu_n A_{26} \\ M_{mn}^{(1)} = -\mu_n \left(\frac{A_{16}}{R_1} + \frac{A_{26}}{R_2} \right) w_{mn}, \quad M_{mn}^{(2)} = \lambda_m \left(\frac{A_{11}}{R_1} + \frac{A_{12}}{R_2} \right) w_{mn} \\ M_{mn}^{(3)} = \mu_n \left(\frac{A_{12}}{R_1} + \frac{A_{22}}{R_2} \right) w_{mn}, \quad M_{mn}^{(4)} = -\lambda_m \left(\frac{A_{16}}{R_1} + \frac{A_{26}}{R_2} \right) w_{mn} \end{gathered} \tag{3.40a – j}$$

From Eq. (3.39), using Cramer's rule, the coefficients $F_{mn}^{(\alpha)}$, $G_{mn}^{(\alpha)}$ $(\alpha = 1, 2)$ can be expressed as

$$\left(F_{mn}^{(1)},\ F_{mn}^{(2)},\ G_{mn}^{(1)},\ G_{mn}^{(2)}\right)=\left(\widetilde{F}_{mn}^{(1)},\ \widetilde{F}_{mn}^{(2)},\ \widetilde{G}_{mn}^{(1)},\ \widetilde{G}_{mn}^{(2)}\right)w_{mn} \tag{3.41}$$

where

$$\widetilde{F}_{mn}^{(1)}=\frac{\det(L_1)}{\det(L)},\quad \widetilde{F}_{mn}^{(2)}=\frac{\det(L_2)}{\det(L)},\quad \widetilde{G}_{mn}^{(1)}=\frac{\det(L_3)}{\det(L)},\quad \widetilde{G}_{mn}^{(2)}=\frac{\det(L_4)}{\det(L)} \tag{3.42a – d}$$

while

$$L_1=\begin{pmatrix} F_{mn}^{(1)} & L_{12}^{(m,n)} & L_{13}^{(m,n)} & L_{14}^{(m,n)} \\ F_{mn}^{(2)} & L_{11}^{(m,n)} & L_{14}^{(m,n)} & L_{13}^{(m,n)} \\ G_{mn}^{(1)} & L_{14}^{(m,n)} & L_{33}^{(m,n)} & L_{34}^{(m,n)} \\ G_{mn}^{(2)} & L_{13}^{(m,n)} & L_{34}^{(m,n)} & L_{33}^{(m,n)} \end{pmatrix},\quad L_2=\begin{pmatrix} L_{11}^{(m,n)} & F_{mn}^{(1)} & L_{13}^{(m,n)} & L_{14}^{(m,n)} \\ L_{12}^{(m,n)} & F_{mn}^{(2)} & L_{14}^{(m,n)} & L_{13}^{(m,n)} \\ L_{13}^{(m,n)} & G_{mn}^{(1)} & L_{33}^{(m,n)} & L_{34}^{(m,n)} \\ L_{14}^{(m,n)} & G_{mn}^{(2)} & L_{34}^{(m,n)} & L_{33}^{(m,n)} \end{pmatrix}$$

$$L_3=\begin{pmatrix} L_{11}^{(m,n)} & L_{12}^{(m,n)} & F_{mn}^{(1)} & L_{14}^{(m,n)} \\ L_{12}^{(m,n)} & L_{11}^{(m,n)} & F_{mn}^{(2)} & L_{13}^{(m,n)} \\ L_{13}^{(m,n)} & L_{14}^{(m,n)} & G_{mn}^{(1)} & L_{34}^{(m,n)} \\ L_{14}^{(m,n)} & L_{13}^{(m,n)} & G_{mn}^{(2)} & L_{33}^{(m,n)} \end{pmatrix},\quad L_4=\begin{pmatrix} L_{11}^{(m,n)} & L_{12}^{(m,n)} & L_{13}^{(m,n)} & F_{mn}^{(1)} \\ L_{12}^{(m,n)} & L_{11}^{(m,n)} & L_{14}^{(m,n)} & F_{mn}^{(2)} \\ L_{13}^{(m,n)} & L_{14}^{(m,n)} & L_{33}^{(m,n)} & G_{mn}^{(1)} \\ L_{14}^{(m,n)} & L_{13}^{(m,n)} & L_{34}^{(m,n)} & G_{mn}^{(2)} \end{pmatrix}$$

$$L=\begin{pmatrix} L_{11}^{(m,n)} & L_{12}^{(m,n)} & L_{13}^{(m,n)} & L_{14}^{(m,n)} \\ & L_{11}^{(m,n)} & L_{14}^{(m,n)} & L_{13}^{(m,n)} \\ & & L_{33}^{(m,n)} & L_{34}^{(m,n)} \\ \text{Symm.} & & & L_{33}^{(m,n)} \end{pmatrix} \tag{3.43a – e}$$

Following in a similar manner, as is customary, the third and fourth equations of equilibrium can be satisfied by assuming η_1 and η_2 in the usual form as

$$\begin{Bmatrix} \eta_1(x_1,x_2) \\ \eta_2(x_1,x_2) \end{Bmatrix}=\begin{Bmatrix} H_{mn}^{(1)} \\ I_{mn}^{(1)} \end{Bmatrix}\cos\lambda_m x_1 \sin\mu_n x_2 + \begin{Bmatrix} H_{mn}^{(2)} \\ I_{mn}^{(2)} \end{Bmatrix}\sin\lambda_m x_1 \cos\mu_n x_2 \tag{3.44a, b}$$

Substituting the expressions for η_1 and η_2 back into Eqs. (3.24) and (3.25) and comparing coefficients of like trigonometric functions gives the same result as was found for the buckling of flat sandwich panels. Therefore, no further development is necessary. The first four equations of equilibrium, Eqs. (3.22)–(3.25) and the boundary conditions, Eqs. (3.33f) and Eqs. (3.34f) are identically fulfilled. There remains the fifth equation of equilibrium and the remaining unfulfilled boundary conditions, Eqs. (3.26), (3.33a–e), and (3.34a–e). The identical procedure that was carried out for flat plates will be duplicated here for the case of doubly curved sandwich shells through the use of the extended Galerkin method. The unfulfilled quantities are retained in the energy functional and thus by performing the necessary operations will result in fulfilling the last equation of motion and the remaining boundary conditions in an average sense. Inserting these unfilled expressions back into Hamilton's equation gives

$$\begin{aligned}
&\int_{t_0}^{t_1}\left\langle \int_0^{l_2}\int_0^{l_1}\left\{ -\left(\frac{A_{11}}{R_1}+\frac{A_{12}}{R_2}\right)\frac{\partial \xi_1}{\partial x_1}-\left(\frac{A_{16}}{R_1}+\frac{A_{26}}{R_2}\right)\frac{\partial \xi_1}{\partial x_2}+\left(\frac{A_{22}}{R_2}+\frac{A_{12}}{R_1}\right)\frac{\partial \xi_2}{\partial x_2}-\right.\right.\\
&\left(\frac{A_{26}}{R_2}+\frac{A_{16}}{R_1}\right)\frac{\partial \xi_2}{\partial x_1}-\left(\frac{A_{11}}{R_1^2}+\frac{A_{22}}{R_2^2}+\frac{2A_{12}}{R_1R_2}\right)v_3-d_1a\left(\frac{\partial \eta_1}{\partial x_1}+a\frac{\partial^2 v_3}{\partial x_1^2}\right)-d_2a\left(\frac{\partial \eta_2}{\partial x_2}+\right.\\
&\left.a\frac{\partial^2 v_3}{\partial x_2^2}\right)+F_{11}\frac{\partial^4 v_3}{\partial x_1^4}+F_{22}\frac{\partial^4 v_3}{\partial x_2^4}+2\left(F_{12}+2F_{66}\right)\frac{\partial^4 v_3}{\partial x_1^2\partial x_2^2}+4F_{16}\frac{\partial^4 v_3}{\partial x_1^3\partial x_2}+\\
&\left.\left.4F_{26}\frac{\partial^4 v_3}{\partial x_1\partial x_2^3}+N_{11}^0\frac{\partial^2 v_3}{\partial x_1^2}+2N_{12}^0\frac{\partial^2 v_3}{\partial x_1\partial x_2}+N_{22}^0\frac{\partial^2 v_3}{\partial x_2^2}\right\}\delta v_3 dx_1 dx_2\right\rangle dt+\\
&\int_{t_0}^{t_1}\left\langle \int_0^{l_2}\left\{N_{11}\delta\xi_1+N_{12}\delta\xi_2+L_{11}\delta\eta_1+L_{12}\delta\eta_2+M_{11}\delta\left(\frac{\partial v_3}{\partial x_1}\right)\right\}\Bigg|_0^{l_1}dx_2\right\rangle dt+\\
&\int_{t_0}^{t_1}\left\langle \int_0^{l_1}\left\{N_{12}\delta\xi_1+N_{22}\delta\xi_2+L_{12}\delta\eta_1+L_{22}\delta\eta_2+M_{22}\delta\left(\frac{\partial v_3}{\partial x_2}\right)\right\}\Bigg|_0^{l_2}dx_2\right\rangle dt=0
\end{aligned} \tag{3.45}$$

Substituting in the expressions for $\eta_1,\ \eta_2,\ \xi_1,\ \xi_2,\ v_3$ into Eq. (3.45) and carrying out the indicated operations results in an algebraic equation which governs the stability of doubly curved sandwich structures with symmetric anisotropic laminated face sheets. This algebraic equation is presented as

$$\begin{aligned}
&\lambda_m^2N_{11}^0+\mu_n^2N_{22}^0=\lambda_m^4F_{11}+2\lambda_m^2\mu_n^2(F_{12}+2F_{66})+\mu_n^4F_{22}+\lambda_m ad_1\left(H_{mn}^{(1)}+a\lambda_m\right)\\
&+\mu_n ad_2\left(I_{mn}^{(2)}+a\mu_n\right)+\lambda_m\left(\frac{A_{11}}{R_1}+\frac{A_{12}}{R_2}\right)F_{mn}^{(2)}+\mu_n\left(\frac{A_{12}}{R_1}+\frac{A_{22}}{R_2}\right)G_{mn}^{(1)}\\
&+\mu_n\left(\frac{A_{16}}{R_1}+\frac{A_{26}}{R_2}\right)F_{mn}^{(1)}+\lambda_m\left(\frac{A_{16}}{R_1}+\frac{A_{26}}{R_2}\right)G_{mn}^{(2)}+\left(\frac{A_{11}}{R_1^2}+2\frac{A_{12}}{R_1R_2}+\frac{A_{22}}{R_2^2}\right)
\end{aligned} \tag{3.46}$$

where the shear load, N_{12}^{0} has been discarded. Eq. (3.46) can be nondimensionalized as

$$
\begin{aligned}
&K_x\left(m^2\psi_1^2+L_R n^2\phi^2\psi_2^2\right)=m^4+\frac{n^4\phi^4F_{22}}{F_{11}}+\frac{2m^2n^2\phi^2(F_{12}+2F_{66})}{F_{11}}\\
&+\frac{a^2L_1^2}{\pi^2F_{11}}(m^2d_1+n^2\phi^2d_2)+\frac{aL_1^3}{\pi^3F_{11}}\left[md_1\widetilde{H}_{mn}^{(1)}+n\phi d_2\widetilde{I}_{mn}^{(2)}\right]+\frac{L_1^2}{\pi^3F_{11}}\Big[m(\psi_1A_{11}+\phi\psi_2A_{12})\widetilde{F}_{mn}^{(1)}+\\
&n\phi(\psi_1A_{12}+\phi\psi_2A_{22})\widetilde{G}_{mn}^{(2)}+\left(n\phi\widetilde{F}_{mn}^{(2)}+m\widetilde{G}_{mn}^{(1)}\right)(\psi_1A_{16}+\phi\psi_2A_{26})\Big]\\
&+\frac{L_1^2}{\pi^4F_{11}}(\psi_1^2A_{11}+\psi_2^2\phi^2A_{22}+2A_{12}\psi_1\psi_2\phi)
\end{aligned}
\tag{3.47}
$$

With Eq. (3.47) in hand, the critical load can be determined for various geometrical and material parameters to study their effects on the stability of the doubly curved sandwich panel. Also, the effect of the structural tailoring can be determined as to its beneficial structural behavior.

3.4.3 Validation of the Theoretical Results

As was presented in the section for flat plates, validations are made first for a circular cylindrical panel composed of isotropic facings (Dura Aluminum, $\nu = 0.3$, $E = 6.96 \times 10^5$ kg/cm^2) and a transversely isotropic core of penoplast (see Karavanov 1960) as shown in Table 3.6. The panel geometrical properties are $L_1 = 60$ cm, $L_2 = 40$ cm, $R_2 = 100$ cm, $h_f = 0.1$ cm (facing thickness). Herein, the results again reveal overpredictions which are most likely due to imperfections

Table 3.6 Comparisons of theoretical and experimental buckling predictions for a circular cylindrical panel with a transversely isotropic core of penoplast

				$\widetilde{N}_{11}L_2 \times 10^3$kg		
Case	$\bar{h}$ (cm)	$\bar{G}$ (kg/cm^2)	$(N_{11})_{cr}$	Present	K.	% Error
1	0.750	81.3	0.365	9.114	8.2	+11.4
2	0.750	84	0.373	9.3	7.8	+19.23
3	0.475	150	0.977	10.49	8.9	+17.865
4	0.200	127	2.726	6.643	6.28	+5.78
5	0.200	566	6.708	16.35	14.6	+11.99
6	0.225	92	2.023	5.965	5.0	+19.3
7	0.500	32.6	0.431	5.087	4.4	+15.61
8	0.475	40	0.5	5.369	4.62	+16.212
9	0.500	141	0.879	10.36	9.25	+12.00
10	0.700	104	0.468	10.26	8.55	+20.00

K.→ Karavanov (1960)

within the test specimens and other miscellaneous considerations. A concept known as the knockdown factor which utilizes a numerical factor to knockdown the theoretical results to the experimental ones within a reasonable amount would most likely bring the results within agreement. The knockdown factor in practice should be no greater than 20%.

3.4.4 Results and Discussion

For the following numerical results, the material properties for the core and face sheets are displayed in Tables 3.7 and 3.8 by type below. Additionally, unless specified otherwise, $L_1 = 609.6$ mm, $\overline{h} = 12.7$ mm. h_f (Face thickness) is given in the figure caption. Herein, results are presented which show the effect of certain geometrical properties on the buckling strength of curved sandwich panels. Figure 3.5 depicts the effect of the ply angle for single-layered facings on the buckling strength of a cylindrical sandwich panel for various aspect ratios. The material characteristics for this case are of the F1:C1 type. The results show that as the aspect ratio of the panel becomes smaller the critical buckling load becomes larger and peaks around 45 degrees and drops off afterwards.

Cylindrical sandwich panels with larger aspect ratios have a much lower critical buckling load. In Fig. 3.6 considering the F1:C1 type highlights the effect of the ply angle for various aspect ratios on the critical buckling load. It is apparent that at smaller aspect ratios the panel can carry larger compressive edge loading which peaks around 45 degrees. At the larger aspect ratios, the load carrying capacity of the panel is diminished in comparison. Figure 3.7 depicts the effect of the panel aspect ratio on the critical buckling load of a cylindrical sandwich panel for various fiber orientation angles in the face sheets. These configurations are of the material type F2: C1. It is seen that as the aspect ratio becomes smaller the critical buckling load increases. As the aspect ratio increases the critical load drops significantly up to a point at which the trend seems to flatten out. As it can be seen that the 45-degree fiber

Table 3.7 Face sheet material properties

Type	Material	E_1(N/mm^2)	E_2(N/mm^2)	G_{12}(N/mm^2)	ν_{12}	α_1 (1/K)	α_2 (1/K)
F1	HS Graph. Ep	1.8375	0.105	0.0735	0.28	11.34	36.9
F2	IM7/977-2	0.0812	0.0763	0.0098	0.06	1.62	1.674

Note: Multiply E_1, E_2, $G_{12} \times 10^5$

Table 3.8 Core material properties

Type	Core type	$\overline{G}_{13}$ (N/mm^2)	$\overline{G}_{23}$ (N/mm^2)
C1	Titanium Honeycomb	0.0145×10^5	0.0066×10^5

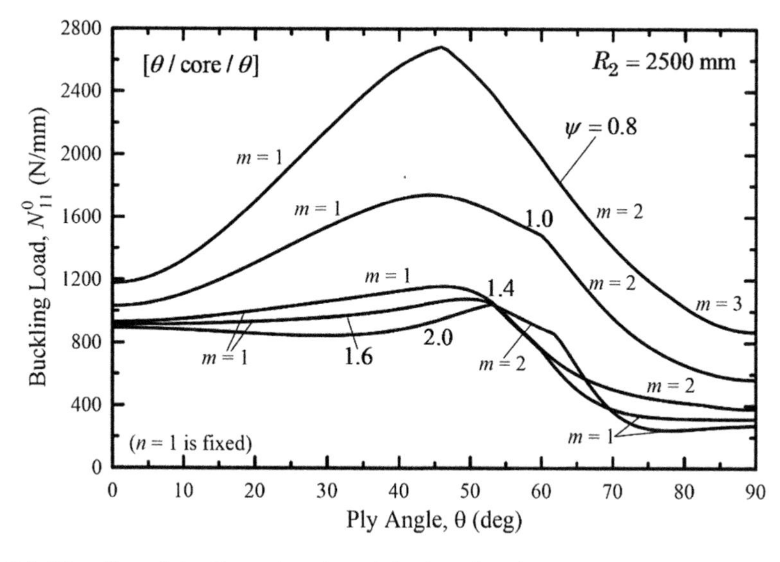

Fig. 3.5 The effect of the fiber orientation of the face sheet lamina on the buckling load for a cylindrical sandwich panel with single-layered facings ($h_f = 0.508$ mm)

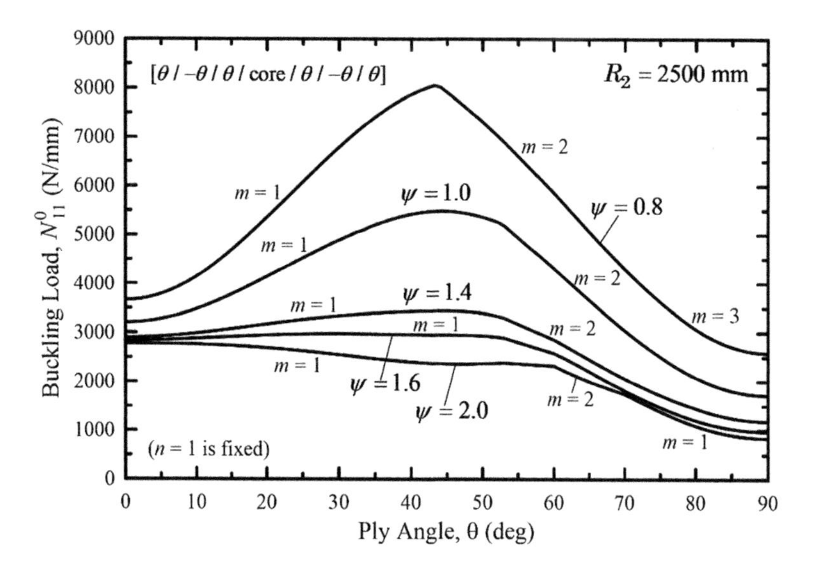

Fig. 3.6 The effect of the fiber orientation of the face sheet lamina on the buckling load for a cylindrical sandwich panel with single-layered facings ($h_f = 1.524$ mm)

orientation has the larger load carrying capacity of the panel. There also seems to be no advantage between the 0-degree and the 90-degree fiber orientation from a load-carrying capacity standpoint.

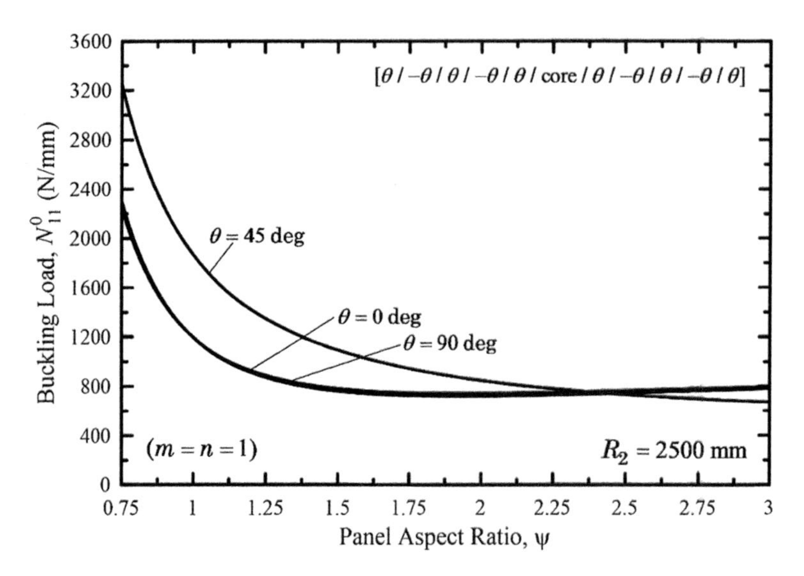

Fig. 3.7 The effect of the panel aspect ratio on the buckling load for a cylindrical sandwich panel with various ply angles in the facings ($h_f = 0.635$ mm)

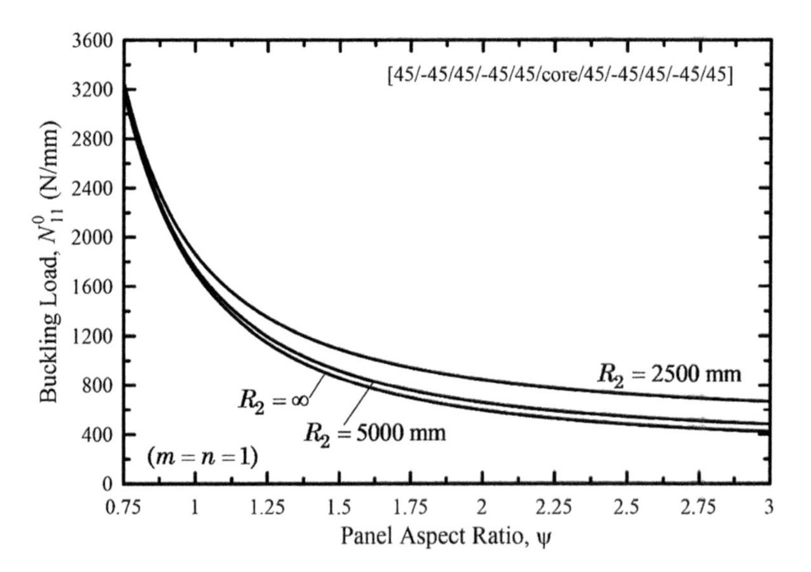

Fig. 3.8 The effect of the panel aspect ratio on the buckling load for a cylindrical sandwich panel with various curvatures ($h_f = 0.635$ mm)

Figure 3.8 which depicts results for the material type F2:C1 reveals that for the chosen stacking sequence in the face sheets that at appreciable small aspect ratios the critical load seems to be the same for all three degrees of curvature. As the aspect ratio increases, the larger load-carrying capacity of the structure appears to occur in the panel with the larger curvature.

3.5 Stress/Strain Theory

The stress and strain relationships were introduced in Chap. 2 without any discussion on their use. Now that the buckling bifurcation load can be determined it is important to know what stresses or strains are produced in the structure from this loading. The first step is to quantify the strains. Once these are known the next step is determining the stress at any location within the structure. It will be assumed that we are only interested in the mechanical stresses and that the thermal and moisture stresses will not be considered. Also, full symmetry is assumed for both the top and bottom facings. The facings are also assumed thin and the Kirchhoff–Love assumptions hold. With this in mind, considering the bottom facing, Eq. (2.86a) is simplified and decoupled to account for the full symmetry in the structure as

$$\left\{N^{/}\right\} = \left[A^{/}\right]\left\{\varepsilon^{/}\right\} \tag{3.48a}$$

$$\left\{M^{/}\right\} = \left[F^{/}\right]\left\{\kappa^{/}\right\} \tag{3.48b}$$

Because of symmetry, $[E^{/}] = 0$. Let

$$\left[{}^{/}A^{*}\right] = \left[A^{/}\right]^{-1} \quad \text{and} \quad \left[{}^{/}F^{*}\right] = \left[F^{/}\right]^{-1} \tag{3.49a, b}$$

With some simple algebraic matrix manipulation, Eqs. (3.48a, b) can be expressed as

$$\left\{\varepsilon^{/}\right\} = \left[{}^{/}A^{*}\right]\left\{N^{/}\right\} \tag{3.50}$$

and

$$\left\{\kappa^{/}\right\} = \left[{}^{/}F^{*}\right]\left\{M^{/}\right\} \tag{3.51}$$

In terms of the geometrical coordinate system, the strain relationships are given by Eq. (2.24a–f) as

$$\left\{e^{/}\right\} = \left\{\varepsilon^{/}\right\} + \left(x_3 - a^{/}\right)\left\{\kappa^{/}\right\} \tag{3.52}$$

Substituting Eqs. (3.50) and (3.51) into the above expression gives

$$\left\{e^{/}\right\} = \left[{}^{/}A^{*}\right]\left\{N^{/}\right\} + \left(x_3 - a^{/}\right)\left[{}^{/}F^{*}\right]\left\{M^{/}\right\} \tag{3.53}$$

With edge loading and applied moments on the structure,

$$\left\{N^{/}\right\} = \begin{Bmatrix} \widetilde{N}_{11} \\ \widetilde{N}_{22} \\ \widetilde{N}_{12} \end{Bmatrix} \quad \text{and} \quad \left\{M^{/}\right\} = \begin{Bmatrix} \widetilde{M}_{11} \\ \widetilde{M}_{22} \\ \widetilde{M}_{12} \end{Bmatrix} \tag{3.54a, b}$$

where $\widetilde{N}_{ij},\ \widetilde{M}_{ij}\quad (i = 1, 2)$ are the applied edge and couple loadings on the structure. If just uniaxial compressive edge loading is applied in the x_1 direction then $\widetilde{N}_{22} = \widetilde{N}_{12} = 0,$ and $M_{ij} = 0$. Eq. (3.53) can be written as

$$\begin{Bmatrix} e_{11}^{/} \\ e_{22}^{/} \\ 2\gamma_{12}^{/} \end{Bmatrix}_G = \left[{}^{/}A^{*}\right]\begin{Bmatrix} \widetilde{N}_{11} \\ \widetilde{N}_{22} \\ \widetilde{N}_{12} \end{Bmatrix} + \left(x_3 - a^{/}\right)\left[{}^{/}F^{*}\right]\begin{Bmatrix} \widetilde{M}_{11} \\ \widetilde{M}_{22} \\ \widetilde{M}_{12} \end{Bmatrix} \tag{3.55}$$

The above expressions for the strain relationships as a function of the edge loading is only valid for the geometrical coordinate system. To obtain the strains in the material coordinate system, a coordinate transformation needs to be applied. It can be shown, the details of which are not provided here, that the strains in the material coordinate system can be expressed by

$$\begin{Bmatrix} e_{11}^{/} \\ e_{22}^{/} \\ 2\gamma_{12}^{/} \end{Bmatrix}_M = [T(\theta)]^{-T}\begin{Bmatrix} e_{11}^{/} \\ e_{22}^{/} \\ 2\gamma_{12}^{/} \end{Bmatrix}_G \tag{3.56}$$

where

$$T(\theta) = \begin{bmatrix} \cos^2(\theta) & \sin^2(\theta) & 2\cos(\theta)\sin(\theta) \\ \sin^2(\theta) & \cos^2(\theta) & -2\cos(\theta)\sin(\theta) \\ -\sin(\theta)\cos(\theta) & \sin(\theta)\cos(\theta) & \cos^2(\theta) - \sin^2(\theta) \end{bmatrix} \tag{3.57}$$

where θ is the fiber orientation angle of each ply. The next step is to determine the stresses within the structure. For the bottom face, in the geometrical coordinate system, the stresses can be obtained from Eq. (2.80a) as

$$\begin{Bmatrix} {}^{/}\sigma_{11} \\ {}^{/}\sigma_{22} \\ {}^{/}\sigma_{12} \end{Bmatrix}_G = \begin{bmatrix} {}^{/}\widehat{Q}_{11} & {}^{/}\widehat{Q}_{12} & {}^{/}\widehat{Q}_{16} \\ {}^{/}\widehat{Q}_{12} & {}^{/}\widehat{Q}_{22} & {}^{/}\widehat{Q}_{26} \\ {}^{/}\widehat{Q}_{16} & {}^{/}\widehat{Q}_{26} & {}^{/}\widehat{Q}_{66} \end{bmatrix} \begin{Bmatrix} {}^{/}e_{11} \\ {}^{/}e_{22} \\ {}^{/}\gamma_{12} \end{Bmatrix}_G \tag{3.58}$$

Substituting Eq. (3.55) into Eq. (3.58) gives

$$\begin{Bmatrix} {}^{/}\sigma_{11} \\ {}^{/}\sigma_{22} \\ {}^{/}\sigma_{12} \end{Bmatrix}_G = \left[\widehat{Q}^{/}\right]\left[{}^{/}A^{*}\right] \begin{Bmatrix} \widetilde{N}_{11} \\ 0 \\ 0 \end{Bmatrix} \tag{3.59}$$

Within this expression the noncontributing elements and or terms have been dropped. In terms of the material coordinate system, the stresses can be obtained from the following expression:

$$\begin{Bmatrix} {}^{/}\sigma_{11} \\ {}^{/}\sigma_{22} \\ {}^{/}\sigma_{12} \end{Bmatrix}_M = T(\theta) \begin{Bmatrix} {}^{/}\sigma_{11} \\ {}^{/}\sigma_{22} \\ {}^{/}\sigma_{12} \end{Bmatrix}_G \tag{3.60}$$

The same approach can be applied for the strains and stresses in the top face. A quicker approach would be to just replace $a^{/}$ with $a^{//}$ and single primes with double primes in the above expressions.

3.6 Summary

The theory governing the buckling response of both flat and curved sandwich panels with a weak (soft) core have been presented and discussed in detail. The theory accounted for the directional properties of the face sheets, the orthotropy of the core all within a geometrical linear context, while the principles of shallow shell theory were adhered to. As an addendum a section on the deformation of sandwich panels was presented covering the stress–strain behavior of sandwich panels. Following the theory, results were presented for both flat and curved sandwich panels separately. The effects of the ply angle, aspect ratio, the panel face thickness, and the effects of the curvature on the critical load were discussed. Several validations were made for both flat and curved sandwich panels where excellent agreement was seen. These results play an important role in the design of aerospace vehicles and structures where lightweight materials and strength are a necessity.

References

Alexandrov, A. I., Briuker, L. E., Kurshin, L. M., & Prusakov, A. P. (1960). *Research on three layered panels* (in Russian). Moskow: Oboronghiz.

Karavanov, V. F. (1960). Stability of three-layer shallow cylindrical panels with soft core. *Izvestia Vishish Ukebnich Zavedenia-Aviatsionaia Technika, 2*, 50–60. (in Russian).

Ko, W. L., & Jackson, R. H. (1991). Combined load buckling behavior of metal matrix composite sandwich panels under different thermal environments. *NASA TM-4321.*

Ko, W. L., & Jackson, R. H. (1993). Compressive and shear buckling analysis of metal matrix composite sandwich panels under different thermal environments. *Composite Structures, 25*, 227–239.

Pearce, T. R. A., & Webber, J. P. H. (1972). Buckling of sandwich panels with laminated face plates. *Aeronautical Quarterly, 23*, 148–160.

Chapter 4
Post-Buckling

Abstract Sandwich structures are capable of sustaining loading beyond the buckling bifurcation point. With this in mind, a comprehensive geometrically nonlinear theoretical treatment of sandwich plates and shells is considered to provide insight into their structural performance within this loading region. Within this comprehensive theoretical treatment, there are few structural and material conditions considered. The face sheets are considered symmetric with respect to their local and global mid-surfaces while constructed from anisotropic laminated composites. The facings are assumed imperfect and thin thereby neglecting the transverse shear stresses in the facings. The core is assumed to be of the weak-type construction. With these theoretical aspects, two cases of laminated facings are considered, cross-ply and angle-ply. Two formulations, the mixed formulation and the displacement formulation, are presented within the theoretical developments. To arrive at the post-buckling solution, the extended Galerkin method and Newton's method are adopted. Finally, a number of kinematical and physical parameters are considered in regard to the load-carrying capacity of these sandwich panels. Several results are presented which provides sufficient insight to the behavior of these structures.

4.1 Introduction

This chapter is concerned with the post-buckling behavior of anisotropic laminated composite sandwich plates and shells. The governing nonlinear equations developed in Chap. 2 are applied to three cases for post-buckling solution. The first case considers cross-ply laminated facings which lend themselves to the stress potential solution technique. The second case considers angle-ply laminated sandwich plates. While the third case considers angle-ply laminated sandwich shells. The theory considers that the facings are symmetrically laminated both locally and globally in conjunction with a weak (soft) core. In addition, the facings are assumed thin so that the Kirchhoff–Love assumptions apply where the transverse shear stress can be neglected. Simply supported boundary conditions are considered with all four edges freely movable. The solution methodologies involve both the extended Galerkin method and Newton's method for nonlinear polynomials. Finally, the influence of a

T. J. Hause, *Sandwich Structures: Theory and Responses*,
https://doi.org/10.1007/978-3-030-71895-4_4

number of kinematical and physical parameters, as well as the tangential boundary conditions on the load-carrying capacity of the structure are investigated with several results presented.

4.2 Preliminaries and Basic Assumptions

The governing nonlinear theory for the postcritical loading of doubly curved sandwich panels is based on a number of previously mentioned assumptions which were presented in Chaps. 2 and 3. The reader is referred to those chapters for an overview. Those preliminaries and basic assumptions apply here as well.

4.3 Cross-Ply Laminated Sandwich Shells

4.3.1 Governing System

This section is concerned with the stability of sandwich plates and shells both at the critical load and above. This solution methodology begins with the nonlinear governing system of partial differential equations for sandwich shells for the case of a weak core and the transverse shear effects neglected in the facings. The applicable equations for this case are given by Eqs. (2.139)–(2.143) along with the boundary conditions Eqs. (2.144a–f). The governing equations are

- *Equations of Equilibrium*

$$\frac{\partial N_{11}}{\partial x_1}+\frac{\partial N_{12}}{\partial x_2}=0 \tag{4.1a}$$

$$\frac{\partial N_{22}}{\partial x_2}+\frac{\partial N_{12}}{\partial x_1}=0 \tag{4.1b}$$

$$\frac{\partial L_{11}}{\partial x_1}+\frac{\partial L_{12}}{\partial x_2}-\frac{\overline{N}_{13}}{\overline{h}}=0 \tag{4.1c}$$

$$\frac{\partial L_{22}}{\partial x_2}+\frac{\partial L_{12}}{\partial x_1}-\frac{\overline{N}_{23}}{\overline{h}}=0 \tag{4.1d}$$

$$\begin{aligned}&N_{11}\left(\frac{\partial^2 v_3}{\partial x_1^2}+\frac{\partial^2 \overset{\circ}{v}_3}{\partial x_1^2}+\frac{1}{R_1}\right)+2N_{12}\left(\frac{\partial^2 v_3}{\partial x_1\partial x_2}+\frac{\partial^2 \overset{\circ}{v}_3}{\partial x_1\partial x_2}\right)+N_{22}\left(\frac{\partial^2 v_3}{\partial x_2^2}+\frac{\partial^2 \overset{\circ}{v}_3}{\partial x_2^2}+\frac{1}{R_2}\right)\\&+\left(1+\frac{h}{2\overline{h}}\right)\left(\frac{\partial \overline{N}_{13}}{\partial x_1}+\frac{\partial \overline{N}_{23}}{\partial x_2}\right)+\frac{\partial^2 M_{11}}{\partial x_1^2}+2\frac{\partial^2 M_{12}}{\partial x_1\partial x_2}+\frac{\partial^2 M_{22}}{\partial x_2^2}=q_3\end{aligned} \tag{4.1e}$$

- *Boundary Conditions*

The associated boundary conditions along the edges $x_n = \text{const}(\text{n} = 1, 2)$ are

$$
\begin{aligned}
&N_{nn} = \widetilde{N}_{nn} \quad \text{or} \quad \xi_n = \widetilde{\xi}_n \\
&N_{nt} = \widetilde{N}_{nt} \quad \text{or} \quad \xi_t = \widetilde{\xi}_t \\
&L_{nn} = \widetilde{L}_{nn} \quad \text{or} \quad \eta_n = \widetilde{\eta}_n \\
&L_{nt} = \widetilde{L}_{nt} \quad \text{or} \quad \eta_t = \widetilde{\eta}_t \\
&M_{nn} = \widetilde{M}_{nn} \quad \text{or} \quad \frac{\partial v_3}{\partial x_n} = \frac{\partial \widetilde{v}_3}{\partial x_n} \\
&N_{nt}\left(\frac{\partial v_3}{\partial x_t} + \frac{\partial \overset{\circ}{v}_3}{\partial x_t}\right) + N_{nn}\left(\frac{\partial v_3}{\partial x_n} + \frac{\partial \overset{\circ}{v}_3}{\partial x_n}\right) + \frac{\partial M_{nn}}{\partial x_n} + 2\frac{\partial M_{nt}}{\partial x_t} \quad \text{or} \quad v_3 = \widetilde{v}_3 \\
&+\left(1 + \frac{h}{2\overline{h}}\right)\overline{N}_{n3} = \frac{\partial \widetilde{M}_{nt}}{\partial x_t} + \widetilde{\overline{N}}_{n3}
\end{aligned}
\tag{4.2a – f}
$$

In Eq. (4.1e), the transverse inertia and damping terms have been eliminated due to the fact that they are not part of the post stability problem. Generally, it is common to consider two types of simply supported boundary conditions, Type A and Type B. In this text only Type A will be considered. For Type A all four edges are freely movable where the following conditions require fulfillment.

Along the edges $x_n = 0,\ L_n$

$$N_{nn} = -\widetilde{N}_{nn},\ N_{nt} = 0,\ \eta_n = 0,\ \eta_t = 0,\ M_{nn} = 0,\ v_3 = 0 \tag{4.3a – f}$$

For Type B, the unloaded edges are immovable where the following conditions apply.

Along the edges $x_n = 0,\ L_n$

$$\xi_n = N_{nt} = \eta_n = \eta_t = M_{nn} = v_3 = 0 \tag{4.4a – f}$$

The boundary condition (4.4a) is fulfilled in an average sense through enforcing the following condition.

$$\int_0^{L_n}\int_0^{L_t}\left(\frac{\partial \xi_n}{\partial x_n}\right)dx_n dx_t = 0 \tag{4.5}$$

$x_t = 0,\ L_t$ denote the edges parallel to the direction of the uniaxial compressive edge load $\widetilde{N}_{nn}$. From this condition, the fictious edge load $\widetilde{N}_{nn}$ rendering the edges immovable can be determined. Once the fictitious edge load $\widetilde{N}_{nn}$ is determined it needs to be incorporated into the post-buckling equation. Also, boundary conditions (4.3a, b) and (4.3b) are fulfilled in an average sense through the following conditions.

$$\int_0^{L_t} N_{nn} dx_t = -\widetilde{N}_{nn} L_n, \qquad \int_0^{L_t} N_{nt} dx_t = 0 \quad \sum_{n,t} \begin{pmatrix} n = 1, & 2 \\ t = 2, & 1 \end{pmatrix} \tag{4.6a, b}$$

Herein, $\widetilde{N}_{nn}$ are the compressive edge loads along the edges $x_n = 0,\ L_n$.

4.3.2 Solution Methodology

The solution methodology chosen here will use a stress potential approach which will require a compatibility equation. To fulfill the complete governing system of equations, focus will begin with the first two equations of equilibrium, Eqs. (4.1a and 4.1b). These first two equations of equilibrium can be satisfied with a stress potential representation for the stress resultants $N_{11},\ N_{22}$ and N_{12} as

$$N_{11} = \frac{\partial^2 \varphi}{\partial x_2^2}, \qquad N_{22} = \frac{\partial^2 \varphi}{\partial x_1^2}, \qquad N_{12} = -\frac{\partial^2 \varphi}{\partial x_1 \partial x_2} \tag{4.7a – c}$$

Equations (4.1a and 4.1b) are now identically fulfilled in terms of the Airy's potential function. Next attention will be given to the third and fourth equations of equilibrium. In Eqs. (4.1c and 4.1d), $\overline{h}$ can be factored out of both equations. As a result, Eqs. (4.1c and 4.1d) become

$$\frac{\partial L_{11}}{\partial x_1} + \frac{\partial L_{12}}{\partial x_2} - \overline{N}_{13} = 0 \tag{4.8a}$$

$$\frac{\partial L_{22}}{\partial x_2} + \frac{\partial L_{12}}{\partial x_1} - \overline{N}_{23} = 0 \tag{4.8b}$$

where

$$L_{11} = \overline{h}\left(N_{11}^{/} - N_{11}^{//}\right),\ L_{22} = \overline{h}\left(N_{22}^{/} - N_{22}^{//}\right),\ L_{12} = \overline{h}\left(N_{12}^{/} - N_{12}^{//}\right) \tag{4.9a – c}$$

Expressing Eqs. (4.1c and 4.1d) in terms of displacements and setting $A_{16} = A_{26} = 0$ results in

$$A_{11}\frac{\partial^2 \eta_1}{\partial x_1^2} + A_{66}\left\{\frac{\partial^2 \eta_1}{\partial x_2^2} + \frac{\partial^2 \eta_2}{\partial x_1 \partial x_2}\right\} + A_{12}\frac{\partial^2 \eta_2}{\partial x_1 \partial x_2} - d_1\left\{\eta_1 + a\frac{\partial v_3}{\partial x_1}\right\} = 0 \tag{4.10a}$$

$$A_{22}\frac{\partial^2 \eta_2}{\partial x_2^2} + A_{66}\left\{\frac{\partial^2 \eta_2}{\partial x_2^2} + \frac{\partial^2 \eta_1}{\partial x_1 \partial x_2}\right\} + A_{12}\frac{\partial^2 \eta_1}{\partial x_1 \partial x_2} - d_2\left\{\eta_2 + a\frac{\partial v_3}{\partial x_2}\right\} = 0 \tag{4.10b}$$

The above equations can be satisfied by assuming η_1 and η_2 as follows

$$\eta_1(x_1, x_2) = B_{mn}^{(1)} \cos\lambda_m x_1 \sin\mu_n x_2 \tag{4.11a}$$

$$\eta_2(x_1, x_2) = C_{mn}^{(1)} \sin\lambda_m x_1 \cos\mu_n x_2 \tag{4.11b}$$

Substituting the expressions for η_1 and η_2 into Eqs. (4.10a and 4.10b) and comparing coefficients of like trigonometric functions gives the expressions for $B_{mn}^{(1)}$ and $C_{mn}^{(1)}$ as

$$B_{mn}^{(1)} = \widetilde{B}_{mn}^{(1)} w_{mn}, \qquad C_{mn}^{(1)} = \widetilde{C}_{mn}^{(1)} w_{mn} \tag{4.12a}$$

where

$$\widetilde{B}_{mn}^{(1)} = \frac{a\{[(A_{12}+A_{66})d_2 - A_{22}d_1]\lambda_m\mu_n^2 - d_1A_{66}\lambda_m^3 - d_1d_2\lambda_m\}}{\widehat{\Delta}_{mn}} \tag{4.12b}$$

$$\widetilde{C}_{mn}^{(1)} = \frac{a\{[(A_{12}+A_{66})d_1 - A_{11}d_2]\lambda_m^2\mu_n - d_2A_{66}\mu_n^3 - d_1d_2\mu_n\}}{\widehat{\Delta}_{mn}} \tag{4.12c}$$

and

$$\begin{aligned}\widehat{\Delta}_{mn} =& A_{11}A_{66}\lambda_m^4 + (d_2A_{11} + d_1A_{66})\lambda_m^2 + \left(A_{11}A_{22} - 2A_{12}A_{66} - A_{12}^2\right)\lambda_m^2\mu_n^2 + \\ & (d_1A_{22} + d_2A_{66})\mu_n^2 + A_{22}A_{66}\mu_n^4 + d_1d_2\end{aligned} \tag{4.12d}$$

Up to this point the first four equations of equilibrium are satisfied. The fifth equation of equilibrium will now be addressed which can be expressed in the following form as

$$\begin{aligned}&\frac{\partial^2\phi}{\partial x_2^2}\left(\frac{\partial^2 v_3}{\partial x_1^2} + \frac{\partial^2 v_3^\circ}{\partial x_1^2} + \frac{1}{R_1}\right) + \frac{\partial^2\phi}{\partial x_1^2}\left(\frac{\partial^2 v_3}{\partial x_2^2} + \frac{\partial^2 v_3^\circ}{\partial x_2^2} + \frac{1}{R_2}\right) \\ &-2\frac{\partial^2\phi}{\partial x_1\partial x_2}\left(\frac{\partial^2 v_3}{\partial x_1\partial x_2} + \frac{\partial^2 v_3^\circ}{\partial x_1\partial x_2}\right) - F_{11}\frac{\partial^4 v_3}{\partial x_1^4} - F_{22}\frac{\partial^4 v_3}{\partial x_2^4} - 4F_{16}\frac{\partial^4 v_3}{\partial x_1^3\partial x_2} \\ &-4F_{26}\frac{\partial^4 v_3}{\partial x\partial x_2^3} - 2(F_{12}+2F_{66})\frac{\partial^4 v_3}{\partial x_1^2\partial x_2^2} + \left(\frac{2a\overline{K}^2}{\overline{h}}\right) \\ &\left\{\overline{G}_{13}\left(\frac{\partial\eta_1}{\partial x_1} + a\frac{\partial^2 v_3}{\partial x_1^2}\right) + \overline{G}_{23}\left(\frac{\partial\eta_2}{\partial x_2} + a\frac{\partial^2 v_3}{\partial x_2^2}\right)\right\} = 0\end{aligned} \tag{4.13}$$

The above equation has three unknowns ϕ, v_3 and v_3°. Based on Seide (1974) and Simitses (1986), v_3° can be represented as

$$v_3^\circ = w_{mn}^\circ \sin \lambda_m x_1 \sin \mu_n x_2 \tag{4.14}$$

where w_{mn}° are the modal amplitudes of the initial geometric imperfection shape. Eq. (4.13) is in terms of two unknowns at this point. These quantities are ϕ and v_3. To ensure single-valued displacements, one more partial differential equation in terms of these two unknown quantities is needed. This will come from a compatibility equation. With the use of Eqs. (2.53)–(2.55) and (2.69)–(2.71), a compatibility equation can be derived by eliminating the in-displacements which is determined to be

$$\begin{aligned}
&\frac{\partial^2 \varepsilon_{11}}{\partial x_2^2} + \frac{\partial^2 \varepsilon_{22}}{\partial x_1^2} - \frac{\partial^2 \gamma_{12}}{\partial x_1 \partial x_2} + \frac{2}{R_1}\frac{\partial^2 v_3}{\partial x_2^2} + \frac{2}{R_2}\frac{\partial^2 v_3}{\partial x_1^2} - 2\left(\frac{\partial^2 v_3}{\partial x_1 \partial x_2}\right)^2 + 2\frac{\partial^2 v_3}{\partial x_1^2}\frac{\partial^2 v_3}{\partial x_2^2} \\
&+2\frac{\partial^2 v_3^\circ}{\partial x_1^2}\frac{\partial^2 v_3}{\partial x_2^2} - 4\frac{\partial^2 v_3}{\partial x_1 \partial x_2}\frac{\partial^2 v_3^\circ}{\partial x_1 \partial x_2} + 2\frac{\partial^2 v_3}{\partial x_1^2}\frac{\partial^2 v_3^\circ}{\partial x_2^2} = 0
\end{aligned} \tag{4.15}$$

where

$$\varepsilon_{11} = \varepsilon_{11}^{/} + \varepsilon_{11}^{//}, \quad \varepsilon_{22} = \varepsilon_{22}^{/} + \varepsilon_{22}^{//}, \quad \gamma_{12} = \gamma_{12}^{/} + \gamma_{12}^{//} \tag{4.16}$$

To express ε_{11}, ε_{22}, and γ_{12} in terms of ϕ, the constitutive equations will be utilized. With the use of Eqs. (2.86a) and (2.92a) and neglecting the thermal and moisture terms, the following matrix relationship can be established assuming identical symmetric laminated facings.

$$\begin{Bmatrix} N \\ M \end{Bmatrix} = \begin{bmatrix} A & E \\ E & F \end{bmatrix} \begin{bmatrix} \varepsilon \\ \kappa \end{bmatrix} \tag{4.17}$$

where

$$N = \{\mathrm{N}_{11}, \mathrm{N}_{22}, \mathrm{N}_{12}\}^{\mathrm{T}}$$

$$M = \{\mathrm{M}_{11}, \mathrm{M}_{22}, \mathrm{M}_{12}\}^{\mathrm{T}}$$

$$A = \begin{bmatrix} A_{11} & A_{12} & A_{16} \\ & A_{22} & A_{26} \\ \text{Symm.} & & A_{66} \end{bmatrix}, \quad E = \begin{bmatrix} E_{11} & E_{12} & E_{16} \\ & E_{22} & E_{26} \\ \text{Symm.} & & E_{66} \end{bmatrix},$$

$$F = \begin{bmatrix} F_{11} & F_{12} & F_{16} \\ & F_{22} & F_{26} \\ \text{Symm.} & & F_{66} \end{bmatrix}$$

$$\varepsilon = \{\varepsilon_{11}, \varepsilon_{22}, \gamma_{12}\}^T$$

$$\kappa = \{\kappa_{11}, \kappa_{22}, \kappa_{12}\}^T$$

Performing a partial inversion of this matrix equation provides

$$\begin{Bmatrix} \varepsilon \\ M \end{Bmatrix} = \begin{bmatrix} A^* & E^* \\ -(E^*)^T & F^* \end{bmatrix} \begin{Bmatrix} N \\ \kappa \end{Bmatrix} \tag{4.18}$$

where

$$A^* = A^{-1}, \quad E^* = -A^{-1}E, \quad -(E^*)^{\mathrm{T}} = EA^{-1}, \quad F^* = F - EA^{-1}E \tag{4.19}$$

For symmetric laminated facings $[E] = 0$. With the use of Eqs. (4.7a–c), Eq. (4.18) can be expressed in terms of ϕ as

$$\begin{aligned} &\varepsilon_{11} = A_{11}^* \frac{\partial^2 \varphi}{\partial x_2^2} + A_{12}^* \frac{\partial^2 \varphi}{\partial x_1^2}, \qquad \varepsilon_{22} = A_{12}^* \frac{\partial^2 \varphi}{\partial x_2^2} + A_{22}^* \frac{\partial^2 \varphi}{\partial x_1^2}, \\ &\gamma_{12} = -A_{66}^* \frac{\partial^2 \varphi}{\partial x_1 \partial x_2} \end{aligned} \tag{4.20}$$

Substituting these expressions into Eq. (4.15) gives the compatibility equation in the required form as

$$\begin{aligned} &A_{22}^* \frac{\partial^4 \phi}{\partial x_1^4} + A_{11}^* \frac{\partial^4 \phi}{\partial x_2^4} - 2A_{16}^* \frac{\partial^2 \phi}{\partial x_1 \partial x_2^3} - 2A_{26}^* \frac{\partial^2 \phi}{\partial x_1^3 \partial x_2} + \left(A_{66}^* + 2A_{12}^*\right) \frac{\partial^2 \phi}{\partial x_1^2 \partial x_2^2} \\ &+ \frac{2}{R_1} \frac{\partial^2 v_3}{\partial x_2^2} + \frac{2}{R_2} \frac{\partial^2 v_3}{\partial x_1^2} - 2\left(\frac{\partial^2 v_3}{\partial x_1 \partial x_2}\right)^2 + 2\frac{\partial^2 v_3}{\partial x_1^2} \frac{\partial^2 v_3}{\partial x_2^2} + 2\frac{\partial^2 v_3^\circ}{\partial x_1^2} \frac{\partial^2 v_3}{\partial x_2^2} \\ &-4 \frac{\partial^2 v_3}{\partial x_1 \partial x_2} \frac{\partial^2 v_3^\circ}{\partial x_1 \partial x_2} + 2\frac{\partial^2 v_3}{\partial x_1^2} \frac{\partial^2 v_3^\circ}{\partial x_2^2} = 0 \end{aligned} \tag{4.21}$$

The governing system of equations for the problem at hand are Eqs. (4.13) and (4.21) along with the appropriate boundary conditions. For simply supported boundary conditions of Type A, v_3 can be assumed as

$$v_3 = w_{mn} \sin \lambda_m x_1 \sin \mu_n x_2 \tag{4.22}$$

This representation satisfies both the fifth and sixth boundary conditions ($M_{nn} = v_3 = 0$) leaving the remaining ones unfulfilled. The only unknown at this point is ϕ. This quantity is assumed in the following form (see Librescu 1965, 1975; Librescu and Chang 1992)

$$\phi(x_1, x_2) = \phi_1(x_1, x_2) - \frac{1}{2}\left(\widetilde{N}_{11} x_2^2 + \widetilde{N}_{22} x_1^2 - 2\widetilde{N}_{12} x_1 x_2\right) \tag{4.23}$$

Here $\widetilde{N}_{11}$, $\widetilde{N}_{22}$, and $\widetilde{N}_{12}$ represent the average compressive and shear edge loadings, while ϕ_1 is the particular solution of Eq. (4.21). Substituting Eqs. (4.14), (4.22), and (4.23) into Eq. (4.21) and comparing coefficients of like trigonometric functions provides ϕ_1 in the form

$$\phi_1(x_1, x_2) = A_{mn}^{(1)} \cos 2\lambda_m x_1 + A_{mn}^{(2)} \cos 2\mu_n x_2 + A_{mn}^{(3)} \sin \lambda_m x_1 \sin \mu_n x_2 \tag{4.24}$$

Herein,

$$A_{mn}^{(1)} = \widetilde{A}_{mn}^{(1)}\left(w_{mn}^2 + 2w_{mn} \overset{\circ}{w}_{mn}\right), \quad A_{mn}^{(2)} = \widetilde{A}_{mn}^{(2)}\left(w_{mn}^2 + 2w_{mn} \overset{\circ}{w}_{mn}\right), \quad A_{mn}^{(3)} = \widetilde{A}_{mn}^{(3)} w_{mn} \tag{4.25a}$$

where

$$\widetilde{A}_{mn}^{(1)} = \frac{\left(A_{11}A_{22} - A_{12}^2\right)\mu_n^2}{32 A_{11} \lambda_m^2} \tag{4.25b}$$

$$\widetilde{A}_{mn}^{(2)} = \frac{\left(A_{11}A_{22} - A_{12}^2\right)\lambda_m^2}{32 A_{22} \mu_n^2} \tag{4.25c}$$

$$\widetilde{A}_{mn}^{(3)} = \frac{\left(\frac{\mu_n^2}{R_1} + \frac{\lambda_m^2}{R_2}\right)\left(A_{11}A_{22}A_{66} - A_{66}A_{12}^2\right)}{\widetilde{\Delta}_{mn}} \tag{4.25d}$$

and

$$\widetilde{\Delta}_{mn} = A_{11}A_{66}\lambda_m^4 + \left(A_{11}A_{22} - 2A_{12}A_{66} - A_{12}^2\right)\lambda_m^2\mu_n^2 + A_{22}A_{66}\mu_n^4 \tag{4.25e}$$

The particular solution ϕ_1 satisfies the following conditions.

$$\begin{gathered}\int_0^{L_2} \frac{\partial^2\phi_1}{\partial x_2^2}\Bigg|_0^{L_1} dx_1 = 0, \quad \int_0^{L_1} \frac{\partial^2\phi_1}{\partial x_1^2}\Bigg|_0^{L_2} dx_1 = 0, \\ \int_0^{L_2} \frac{\partial^2\phi_1}{\partial x_1 \partial x_2}\Bigg|_0^{L_1} dx_2 = 0, \quad \int_0^{L_1} \frac{\partial^2\phi_1}{\partial x_1 \partial x_2}\Bigg|_0^{L_2} dx_1 = 0\end{gathered} \tag{4.26a – d}$$

This reveals that $\widetilde{N}_{11}$, $\widetilde{N}_{22}$ acquire the meaning of average in-plane compressive edge loads expressed by

$$\int_0^{L_2} \frac{\partial^2\phi}{\partial x_2^2}\Bigg|_{x_1 = 0, L_1} dx_2 = -\widetilde{N}_{11}L_2, \qquad \int_0^{L_1} \frac{\partial^2\phi}{\partial x_1^2}\Bigg|_{x_2 = 0, L_2} dx_1 = -\widetilde{N}_{22}L_1 \tag{4.26e, f}$$

Finally, ξ_1 and ξ_2 can be determined from the expressions of stress resultants N_{11} and N_{22} expressed in terms of displacements which have the following form

$$\begin{aligned}N_{11} = \frac{\partial^2\phi}{\partial x_2^2} &= A_{11}\left\{\frac{\partial\xi_1}{\partial x_1} + \frac{1}{2}\left(\frac{\partial v_3}{\partial x_1}\right)^2 + \frac{\partial v_3^\circ}{\partial x_1}\frac{\partial v_3}{\partial x_1} - \frac{v_3}{R_1}\right\} \\ &+ A_{12}\left\{\frac{\partial\xi_2}{\partial x_2} + \frac{1}{2}\left(\frac{\partial v_3}{\partial x_2}\right)^2 + \frac{\partial v_3^\circ}{\partial x_2}\frac{\partial v_3}{\partial x_2} - \frac{v_3}{R_2}\right\}\end{aligned} \tag{4.27a}$$

$$\begin{aligned}N_{22} = \frac{\partial^2\phi}{\partial x_1^2} &= A_{22}\left\{\frac{\partial\xi_2}{\partial x_2} + \frac{1}{2}\left(\frac{\partial v_3}{\partial x_2}\right)^2 + \frac{\partial v_3^\circ}{\partial x_2}\frac{\partial v_3}{\partial x_2} - \frac{v_3}{R_2}\right\} \\ &+ A_{12}\left\{\frac{\partial\xi_1}{\partial x_1} + \frac{1}{2}\left(\frac{\partial v_3}{\partial x_1}\right)^2 + \frac{\partial v_3^\circ}{\partial x_1}\frac{\partial v_3}{\partial x_1} - \frac{v_3}{R_1}\right\}\end{aligned} \tag{4.27b}$$

$$N_{12} = -\frac{\partial^2\phi}{\partial x_1 \partial x_2} = A_{66}\left\{\frac{\partial\xi_1}{\partial x_2} + \frac{\partial\xi_2}{\partial x_1} + \frac{\partial v_3}{\partial x_1}\frac{\partial v_3}{\partial x_2} + \frac{\partial v_3^\circ}{\partial x_1}\frac{\partial v_3}{\partial x_2} + \frac{\partial v_3}{\partial x_1}\frac{\partial v_3^\circ}{\partial x_2}\right\} \tag{4.27c}$$

Equations (4.27a, 4.27b, and 4.27c) considered in conjunction with Eqs. (4.14), (4.22), (4.23), and (4.24) allow ξ_1 and ξ_2 to be determined as

$$\xi_1(x_1,x_2) = D^{(1)}_{mn}x_1 + D^{(2)}_{mn}\sin 2\lambda_m x_1 + D^{(3)}_{mn}\sin 2\lambda_m x_1 \cos 2\mu_n x_2 + D^{(4)}_{mn}\cos\lambda_m x_1 \sin\mu_n x_2 + \left(D^{(5)}_{mn}\widetilde{N}_{11} + D^{(6)}_{mn}\widetilde{N}_{22}\right)x_1 \tag{4.28}$$

$$\xi_2(x_1,x_2) = E^{(1)}_{mn}x_2 + E^{(2)}_{mn}\sin 2\mu_n x_2 + E^{(3)}_{mn}\cos 2\lambda_m x_1 \sin 2\mu_n x_2 + E^{(4)}_{mn}\sin\lambda_m x_1 \cos\mu_n x_2 + \left(E^{(5)}_{mn}\widetilde{N}_{11} + E^{(6)}_{mn}\widetilde{N}_{22}\right)x_2 \tag{4.3-29}$$

where the constants $D^{(1)}_{mn} - D^{(6)}_{mn}$ and $E^{(1)}_{mn} - E^{(6)}_{mn}$ are expressed as

$$\left(D^{(i)}_{mn}, E^{(i)}_{mn}\right) = \left(\widetilde{D}^{(i)}_{mn}, \widetilde{E}^{(i)}_{mn}\right)\left(w^2_{mn} + 2w_{mn}w^\circ_{mn}\right), \qquad (i = 1,2,3)$$
$$D^{(4)}_{mn} = \widetilde{D}^{(4)}_{mn}w_{mn}, \qquad E^{(4)}_{mn} = \widetilde{E}^{(4)}_{mn}w_{mn} \tag{4.30a – c}$$

while

$$\widetilde{D}^{(1)}_{mn} = -\frac{\lambda^2_m}{8}, \quad \widetilde{D}^{(2)}_{mn} = \frac{A_{12}\mu^2_n - A_{11}\lambda^2_m}{16A_{11}\lambda_m}, \quad \widetilde{D}^{(3)}_{mn} = \frac{\lambda_m}{16}$$
$$\widetilde{D}^{(4)}_{mn} = \frac{\left(A_{12}A_{66} + A^2_{12} - A_{11}A_{22}\right)\lambda_m\mu^2_n - A_{11}A_{66}\lambda^3_m}{R_1\widehat{\Delta}_{mn}} + \frac{A_{22}A_{66}\lambda_m\mu^2_n - A_{12}A_{66}\lambda^3_m}{R_2\widehat{\Delta}_{mn}}$$
$$D^{(5)}_{mn} = \frac{-A_{22}}{A_{11}A_{22} - A^2_{12}}, \quad D^{(6)}_{mn} = \frac{A_{12}}{A_{11}A_{22} - A^2_{12}}$$
$$\widetilde{E}^{(1)}_{mn} = -\frac{\mu^2_n}{8}, \quad \widetilde{E}^{(2)}_{mn} = \frac{A_{12}\lambda^2_m - A_{22}\mu^2_n}{16A_{22}\mu_n}, \quad \widetilde{E}^{(3)}_{mn} = \frac{\mu_n}{16}$$
$$\widetilde{E}^{(4)}_{mn} = \frac{\left(A_{12}A_{66} + A^2_{12} - A_{11}A_{22}\right)\lambda^2_m\mu_n - A_{22}A_{66}\mu^3_n}{R_2\widehat{\Delta}_{mn}} + \frac{A_{11}A_{66}\lambda^2_m\mu_n - A_{12}A_{66}\mu^3_n}{R_1\widehat{\Delta}_{mn}}$$
$$E^{(5)}_{mn} = \frac{A_{12}}{A_{11}A_{22} - A^2_{12}}, \quad E^{(6)}_{mn} = \frac{-A_{11}}{A_{11}A_{22} - A^2_{12}} \tag{4.31a – l}$$

and

$$\widehat{\Delta}_{mn} = A_{11}A_{66}\lambda^4_m + \left(A_{11}A_{22} - 2A_{12}A_{66} - A^2_{12}\right)\lambda^2_m\mu^2_n + A_{22}A_{66}\mu^4_n \tag{4.32}$$

At this juncture, all of the displacement quantities v_3, v°_3, ξ_1, ξ_2, η_1, η_2, ϕ, ϕ_1 are known. In addition, there remains only the fifth equation of equilibrium unfulfilled and the unfulfilled boundary conditions. To solve for the postcritical solution, the EGM will be leveraged (see Fulton 1961). The unfulfilled fifth equation of motion and the unfulfilled boundary conditions will be retained in the energy functional with the appropriate integrations carried out. In the end, this will supply the postcritical solution in terms of the geometrical and material parameters. Retaining these unfulfilled expressions in the Hamilton's energy functional results in

$$\begin{aligned}&\int_{t_0}^{t_1}\left\langle\int_0^{l_2}\int_0^{l_1}\left\{\frac{\partial^2\phi}{\partial x_2^2}\left(\frac{\partial^2 v_3}{\partial x_1^2}+\frac{\partial^2 v_3^\circ}{\partial x_1^2}+\frac{1}{R_1}\right)+\frac{\partial^2\phi}{\partial x_1^2}\left(\frac{\partial^2 v_3}{\partial x_2^2}+\frac{\partial^2 v_3^\circ}{\partial x_2^2}+\frac{1}{R_2}\right)\right.\right.\\&-2\frac{\partial^2\phi}{\partial x_1\partial x_2}\left(\frac{\partial^2 v_3}{\partial x_1\partial x_2}+\frac{\partial^2 v_3^\circ}{\partial x_1\partial x_2}\right)-F_{11}\frac{\partial^4 v_3}{\partial x_1^4}-F_{22}\frac{\partial^4 v_3}{\partial x_2^4}-2(F_{12}+\ 2F_{66})\frac{\partial^4 v_3}{\partial x_1^2\partial x_2^2}\\&\left.\left.+\left(\frac{2a\overline{K}^2}{\overline{h}}\right)\left\{\overline{G}_{13}\left(\frac{\partial\eta_1}{\partial x_1}+a\frac{\partial^2 v_3}{\partial x_1^2}\right)+\overline{G}_{23}\left(\frac{\partial\eta_2}{\partial x_2}+a\frac{\partial^2 v_3}{\partial x_2^2}\right)\right\}\right\}\delta v_3 dx_1 dx_2\right\rangle dt+\\&\int_{t_0}^{t_1}\left\langle\int_0^{l_2}\left[\left(\frac{\partial^2\phi}{\partial x_2^2}+\widetilde{N}_{11}\right)\delta\xi_1-\frac{\partial^2\varphi}{\partial x_1\partial x_2}\ \delta\xi_2+L_{11}\delta\eta_1+L_{12}\delta\eta_2\right]\Bigg|_0^{l_1}dx_2\right\rangle dt+\\&\int_{t_0}^{t_1}\left\langle\int_0^{l_1}\left[\left(\frac{\partial^2\phi}{\partial x_1^2}+\widetilde{N}_{22}\right)\delta\xi_2-\frac{\partial^2\varphi}{\partial x_1\partial x_2}\ \delta\xi_1+L_{22}\delta\eta_2+L_{12}\delta\eta_1\right]\Bigg|_0^{l_2}dx_1\right\rangle dt\\&=0\end{aligned}\tag{4.33}$$

Substituting in the expressions for $v_3,\ v_3^\circ,\ \xi_1,\ \xi_2,\ \eta_1,\ \eta_2,\ \phi,\ \phi_1$ and carrying out the indicated integrations results in the governing post-buckling solution for cross-ply laminated sandwich structures resulting in a nonlinear algebraic equation expressed in terms of the modal amplitudes as

$$\begin{aligned}&P_{mn}^{(5)}\left(w_{mn}^2+2w_{mn}w_{mn}^\circ\right)\left(w_{mn}+w_{mn}^\circ\right)+P_{mn}^{(4)}\left(w_{mn}+w_{mn}^\circ\right)w_{mn}\\&+P_{mn}^{(3)}\left(w_{mn}^2+2w_{mn}w_{mn}^\circ\right)+P_{mn}^{(2)}\left(w_{mn}+w_{mn}^\circ\right)+P_{mn}^{(1)}w_{mn}+P_{mn}^{(0)}+q_{mn}=0\end{aligned}\tag{4.34}$$

Where

$$P_{mn}^{(5)}=2\lambda_m^2\mu_n^2\left(\widetilde{A}_{mn}^{(1)}+\widetilde{A}_{mn}^{(2)}\right)\tag{4.35}$$

$$P_{mn}^{(4)}=-\frac{16\Delta_n^m}{3L_1L_2}\left[-\mu_n\widetilde{A}_{mn}^{(2)}\widetilde{D}_{mn}^{(4)}-\lambda_m\widetilde{A}_{mn}^{(1)}\widetilde{E}_{mn}^{(4)}+\frac{\mu_n\widetilde{A}_{mn}^{(2)}}{\lambda_m R_1}+\frac{\lambda_m\widetilde{A}_{mn}^{(1)}}{\mu_n R_2}\right]\tag{4.36}$$

$$P_{mn}^{(3)}=\left[-\frac{8\Delta_n^m\lambda_m\mu_n}{3L_1L_2}-\frac{16\Delta_n^m\lambda_m}{3L_1L_2}\left(\widetilde{D}_{mn}^{(2)}+\widetilde{D}_{mn}^{(3)}+\widetilde{E}_{mn}^{(2)}+\widetilde{E}_{mn}^{(3)}\right)\right]\widetilde{A}_{mn}^{(3)}\tag{4.37}$$

$$P_{mn}^{(2)}=-\left(\lambda_m^2N_{11}+\mu_n^2N_{22}\right)\tag{4.38}$$

$$P_{mn}^{(1)} = \Big[\big(\mu_n^2/R_1 + \lambda_m^2/R_2\big)\widetilde{A}_{mn}^{(3)} + F_{11}\lambda_m^4 + F_{22}\mu_n^4 + 2(F_{12} + 2F_{66})\lambda_m^2\mu_n^2 + d_1 a\lambda_m\Big(\widetilde{B}_{mn}^{(1)} + a\lambda_m\Big) + d_2 a\mu_n\Big(\widetilde{C}_{mn}^{(1)} + a\mu_n\Big)\Big] \tag{4.39}$$

$$P_{mn}^{(0)} = \frac{4\Delta_n^m}{\lambda_m\mu_n L_1 L_2}\left(\frac{\widetilde{N}_{11}}{R_1} + \frac{\widetilde{N}_{22}}{R_2}\right) \tag{4.40}$$

where

$$\Delta_n^m = [(-1)^m - 1][(-1)^n - 1] \tag{4.41}$$

4.3.3 Numerical Results and Discussion

In what follows are a few post-buckling results which are presented for the case of cross-ply laminated face sheets. Applicable to these results are the material properties for the face sheets and the core which are given in Tables 4.1 and 4.2. The geometrical properties are provided as, $L_1 = L_2 = 609.6$ m, $h_f = 0.635$ mm, $\overline{h} = 12.7$ mm.

Figure 4.1 shows the effect of curvature on the post-buckling response of a cylindrical sandwich panel with the given layup in the face sheets. It is apparent that as the curvature increases there is more of a tendency for snap through–type behavior beyond the buckling bifurcation point. For the case of a flat plate, the equilibrium paths are given by the standard Euler behavior.

In Fig. 4.2, the effect of the geometric imperfections on the post-buckling behavior of a cylindrical sandwich panel for the given layup is shown. The case where the imperfection is zero is considered the ideal post-buckling behavior; where the existence of an geometric imperfection is considered the real behavior that the load–deflection interaction would follow. It can be seen that the larger the geometric imperfections are, the farther the ideal load–deflection interaction is from the ideal

Table 4.1 Face sheet material properties

Type	Material	E_1(N/mm^2)	E_2(N/mm^2)	G_{12} (N/mm^2)	ν_{12}	α_1 (1/K)	α_2 (1/K)
F1	HS Graph. Ep.	1.8375	0.105	0.0735	0.28	11.34	36.9

Note: Multiply E_1, E_2, $G_{12} \times 10^5$

Table 4.2 Core material properties

Type	Core type	$\overline{G}_{13}$ (N/mm^2)	$\overline{G}_{23}$ (N/mm^2)
C1	Titanium honeycomb	0.0145×10^5	0.0066×10^5

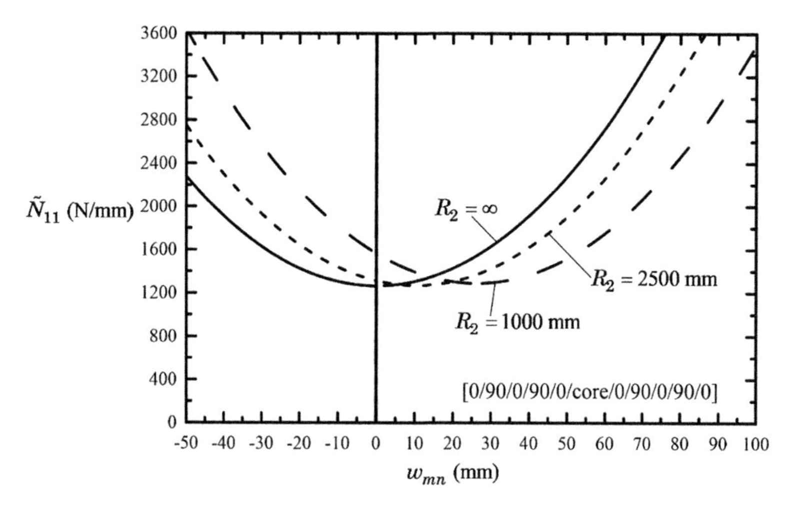

Fig. 4.1 The effect of the curvature of a cylindrical sandwich panel under uniaxial compressive edge loading for the depicted cross-ply stacking sequence

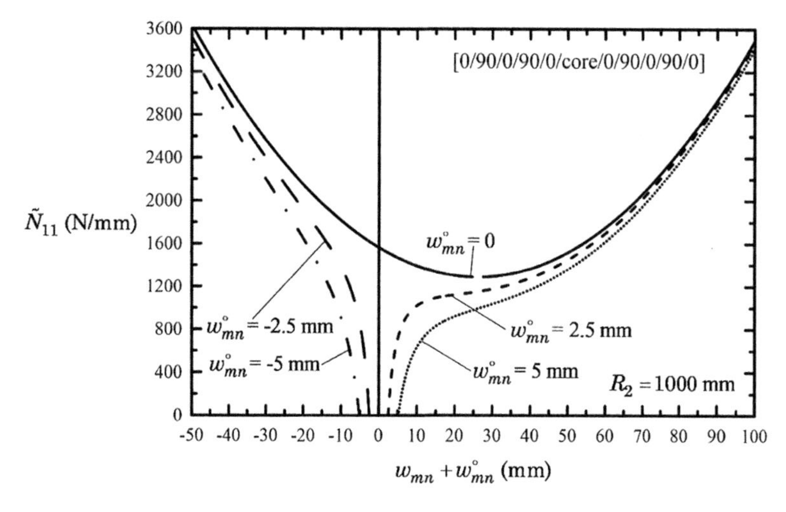

Fig. 4.2 The effect of imperfections on a cylindrical sandwich panel under uniaxial compressive edge loading for the displayed cross-ply stacking sequence

Euler behavior. Also, for the given fixed curvature, a minor snap through–type behavior is seen beyond the buckling bifurcation point for the ideal case. Figure 4.3 shows the effect of the combination of transverse pressure loading and biaxial edge loading on the post-buckling response of a cylindrical sandwich panel for the given layup and curvature. The case where the transverse pressure and biaxial edge loading is nonexistent, the load–deflection interact is the seen to be the standard Euler-type post-buckling behavior for a cylindrical panel.

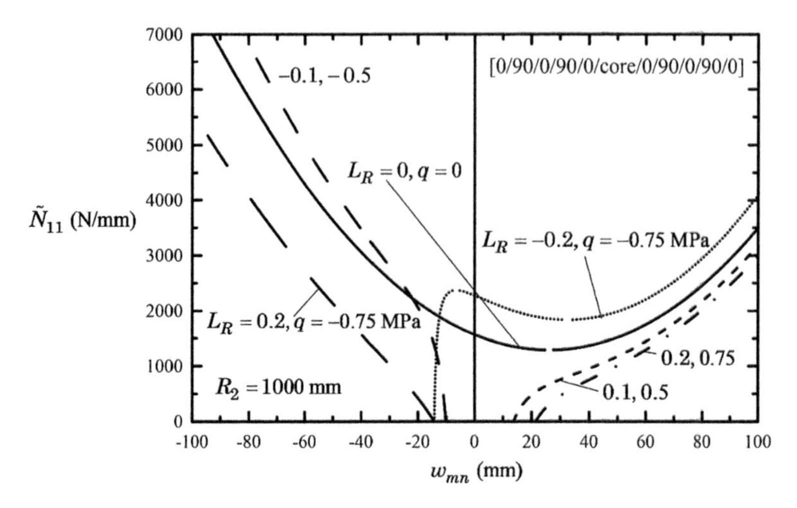

Fig. 4.3 The effect of biaxial edge loading and transverse pressure on a cylindrical sandwich panel under uniaxial compressive edge loading of cross-ply face-sheets

With the inclusion of the transverse pressure, in addition to the biaxial edge loading, the behavior is similar to the behavior seen with the presence of geometric imperfections. The transverse pressure behaves like a geometric imperfection while the biaxial edge loading determines if the imperfection type behavior due to the transverse pressure is negative or positive. Various combinations of the effect of transverse pressure and biaxial edge loading are displayed for both positive and negative loading scenarios. It is seen in general that the combination of a positive transverse pressure and a compressive biaxial edge loading (positive) that the behavior resembles a positive geometric imperfection hugging the ideal load–deflection curve. For a combination of negative transverse pressure and negative biaxial edge loading (tension), it behaves more like a negative geometric imperfection. In Fig. 4.4, the effect of the face sheet thickness on the post-buckling response is seen. The trends appear to be identical for the entire range of face sheet thicknesses. With this in mind, the higher face sheet thickness provides a higher buckling bifurcation point with a minor increase in the snap through–type behavior.

4.4 Angle-Ply Laminated Sandwich Plates

4.4.1 Governing System

In the previous section, the Airy's stress potential method was limited to sandwich panels with cross-ply laminated facings because of the global stiffness terms, A_{16}, A_{26}. When these terms are not zero, the stress potential method is not feasible and

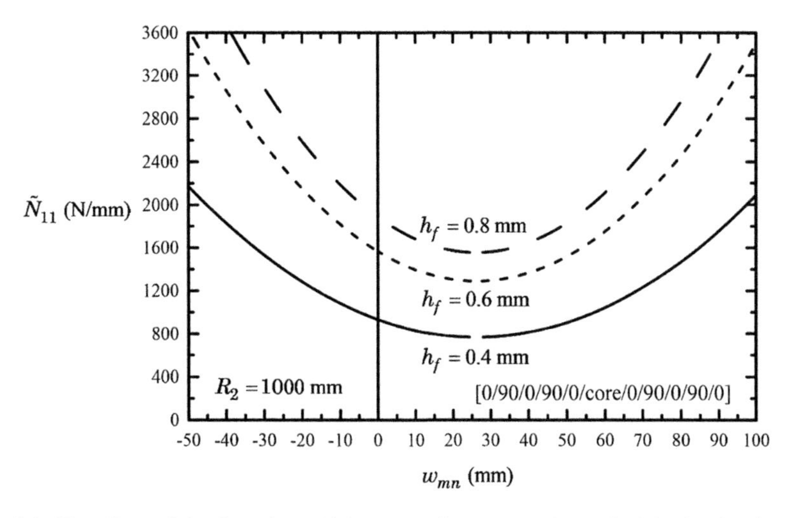

Fig. 4.4 The effect of the face sheet thickness on the compressive uniaxial edge loading of a cylindrical sandwich panel for a cross-ply stacking sequence

resorting to the displacement formulation is the most favorable approach. This will be the approach for the formulation governing angle-ply laminated flat sandwich panels in this section and for sandwich shells in Sect. 4.5. The governing nonlinear equations for sandwich plates can be obtained from Eqs. (2.162)–(2.166) as

- *Equations of Motion*

$$
\begin{aligned}
& A_{11}\left\{\frac{\partial^2\xi_1}{\partial x_1^2}+\frac{\partial v_3}{\partial x_1}\frac{\partial^2 v_3}{\partial x_1^2}+\frac{\partial^2 \overset{\circ}{v}_3}{\partial x_1^2}\frac{\partial v_3}{\partial x_1}+\frac{\partial \overset{\circ}{v}_3}{\partial x_1}\frac{\partial^2 v_3}{\partial x_1^2}\right\}+A_{66}\left\{\frac{\partial^2\xi_1}{\partial x_2^2}+\frac{\partial^2\xi_2}{\partial x_1\partial x_2}+\right.\\
& \left.\frac{\partial v_3}{\partial x_2}\frac{\partial^2 v_3}{\partial x_1\partial x_2}+\frac{\partial v_3}{\partial x_1}\frac{\partial^2 v_3}{\partial x_2^2}+\frac{\partial v_3}{\partial x_2}\frac{\partial^2 \overset{\circ}{v}_3}{\partial x_1\partial x_2}+\frac{\partial \overset{\circ}{v}_3}{\partial x_1}\frac{\partial^2 v_3}{\partial x_2^2}+\frac{\partial \overset{\circ}{v}_3}{\partial x_2}\frac{\partial^2 v_3}{\partial x_1\partial x_2}+\frac{\partial v_3}{\partial x_1}\frac{\partial^2 \overset{\circ}{v}_3}{\partial x_2^2}\right\}+\\
& A_{12}\left\{\frac{\partial^2\xi_2}{\partial x_1\partial x_2}+\frac{\partial v_3}{\partial x_2}\frac{\partial^2 v_3}{\partial x_1\partial x_2}+\frac{\partial v_3}{\partial x_2}\frac{\partial^2 \overset{\circ}{v}_3}{\partial x_1\partial x_2}+\frac{\partial \overset{\circ}{v}_3}{\partial x_2}\frac{\partial^2 v_3}{\partial x_1\partial x_2}\right\}-A_{16}\left\{2\frac{\partial^2\xi_1}{\partial x_1\partial x_2}+\right.\\
& \frac{\partial^2\xi_2}{\partial x_1^2}+2\frac{\partial v_3}{\partial x_1}\frac{\partial^2 v_3}{\partial x_1\partial x_2}+2\frac{\partial v_3}{\partial x_1}\frac{\partial^2 \overset{\circ}{v}_3}{\partial x_1\partial x_2}+2\frac{\partial \overset{\circ}{v}_3}{\partial x_1}\frac{\partial^2 v_3}{\partial x_1\partial x_2}+\frac{\partial v_3}{\partial x_2}\frac{\partial^2 v_3}{\partial x_1^2}+\frac{\partial v_3}{\partial x_2}\frac{\partial^2 \overset{\circ}{v}_3}{\partial x_1^2}+\\
& \left.\frac{\partial \overset{\circ}{v}_3}{\partial x_2}\frac{\partial^2 v_3}{\partial x_1^2}\right\}-A_{26}\left\{\frac{\partial^2\xi_2}{\partial x_2^2}+\frac{\partial v_3}{\partial x_2}\frac{\partial^2 v_3}{\partial x_2^2}+\frac{\partial v_3}{\partial x_2}\frac{\partial^2 \overset{\circ}{v}_3}{\partial x_2^2}+\frac{\partial \overset{\circ}{v}_3}{\partial x_2}\frac{\partial^2 v_3}{\partial x_2^2}\right\}=0
\end{aligned}
\tag{4.42a}
$$

$$A_{22}\left\{\frac{\partial^2\xi_2}{\partial x_2^2}+\frac{\partial v_3}{\partial x_2}\frac{\partial^2 v_3}{\partial x_2^2}+\frac{\partial^2 \overset{\circ}{v}_3}{\partial x_2^2}\frac{\partial v_3}{\partial x_2}+\frac{\partial \overset{\circ}{v}_3}{\partial x_2}\frac{\partial^2 v_3}{\partial x_2^2}\right\}+A_{66}\left\{\frac{\partial^2\xi_2}{\partial x_1^2}+\frac{\partial^2\xi_1}{\partial x_1\partial x_2}+\right.$$

$$\left.\frac{\partial v_3}{\partial x_1}\frac{\partial^2 v_3}{\partial x_1\partial x_2}+\frac{\partial v_3}{\partial x_2}\frac{\partial^2 v_3}{\partial x_1^2}+\frac{\partial v_3}{\partial x_1}\frac{\partial^2 \overset{\circ}{v}_3}{\partial x_1\partial x_2}+\frac{\partial \overset{\circ}{v}_3}{\partial x_2}\frac{\partial^2 v_3}{\partial x_1^2}+\frac{\partial \overset{\circ}{v}_3}{\partial x_1}\frac{\partial^2 v_3}{\partial x_1\partial x_2}+\frac{\partial v_3}{\partial x_2}\frac{\partial^2 \overset{\circ}{v}_3}{\partial x_1^2}\right\}+$$

$$A_{12}\left\{\frac{\partial^2\xi_1}{\partial x_1\partial x_2}+\frac{\partial v_3}{\partial x_1}\frac{\partial^2 v_3}{\partial x_1\partial x_2}+\frac{\partial v_3}{\partial x_1}\frac{\partial^2 \overset{\circ}{v}_3}{\partial x_1\partial x_2}+\frac{\partial \overset{\circ}{v}_3}{\partial x_1}\frac{\partial^2 v_3}{\partial x_1\partial x_2}\right\}-A_{26}\left\{2\frac{\partial^2\xi_2}{\partial x_1\partial x_2}\right.$$

$$+\frac{\partial^2\xi_1}{\partial x_2^2}+2\frac{\partial v_3}{\partial x_2}\frac{\partial^2 v_3}{\partial x_1\partial x_2}+2\frac{\partial v_3}{\partial x_2}\frac{\partial^2 \overset{\circ}{v}_3}{\partial x_1\partial x_2}+2\frac{\partial \overset{\circ}{v}_3}{\partial x_2}\frac{\partial^2 v_3}{\partial x_1\partial x_2}+\frac{\partial v_3}{\partial x_1}\frac{\partial^2 v_3}{\partial x_2^2}+\frac{\partial v_3}{\partial x_1}\frac{\partial^2 \overset{\circ}{v}_3}{\partial x_2^2}+$$

$$\left.\frac{\partial \overset{\circ}{v}_3}{\partial x_1}\frac{\partial^2 v_3}{\partial x_2^2}\right\}-A_{16}\left\{\frac{\partial^2\xi_1}{\partial x_1^2}+\frac{\partial v_3}{\partial x_1}\frac{\partial^2 v_3}{\partial x_1^2}+\frac{\partial v_3}{\partial x_1}\frac{\partial^2 \overset{\circ}{v}_3}{\partial x_1^2}+\frac{\partial \overset{\circ}{v}_3}{\partial x_1}\frac{\partial^2 v_3}{\partial x_1^2}\right\}=0 \tag{4.42b}$$

$$A_{11}\frac{\partial^2\eta_1}{\partial x_1^2}+A_{66}\left\{\frac{\partial^2\eta_1}{\partial x_2^2}+\frac{\partial^2\eta_2}{\partial x_1\partial x_2}\right\}+A_{12}\frac{\partial^2\eta_2}{\partial x_1\partial x_2}+A_{16}\left\{\frac{\partial^2\eta_2}{\partial x_1^2}+2\frac{\partial^2\eta_1}{\partial x_1\partial x_2}\right\}$$

$$+A_{26}\frac{\partial^2\eta_2}{\partial x_2^2}-d_1\left\{\eta_1+a\frac{\partial v_3}{\partial x_1}\right\}=0 \tag{4.42c}$$

$$A_{22}\frac{\partial^2\eta_2}{\partial x_2^2}+A_{66}\left\{\frac{\partial^2\eta_2}{\partial x_2^2}+\frac{\partial^2\eta_1}{\partial x_1\partial x_2}\right\}+A_{12}\frac{\partial^2\eta_1}{\partial x_1\partial x_2}+A_{26}\left\{\frac{\partial^2\eta_1}{\partial x_2^2}+2\frac{\partial^2\eta_2}{\partial x_1\partial x_2}\right\}$$

$$+A_{16}\frac{\partial^2\eta_1}{\partial x_1^2}-d_2\left\{\eta_2+a\frac{\partial v_3}{\partial x_2}\right\}=0 \tag{4.42d}$$

$$A_{11}\left\{\frac{\partial\xi_1}{\partial x_1}\left(\frac{\partial^2 v_3}{\partial x_1^2}+\frac{\partial^2 \overset{\circ}{v}_3}{\partial x_1^2}\right)+\frac{1}{2}\left(\frac{\partial v_3}{\partial x_1}\right)^2\left(\frac{\partial^2 v_3}{\partial x_1^2}+\frac{\partial^2 \overset{\circ}{v}_3}{\partial x_1^2}\right)+\frac{\partial v_3}{\partial x_1}\frac{\partial \overset{\circ}{v}_3}{\partial x_1}\left(\frac{\partial^2 v_3}{\partial x_1^2}+\frac{\partial^2 \overset{\circ}{v}_3}{\partial x_1^2}\right)\right\}$$

$$+A_{12}\left\{\frac{\partial\xi_1}{\partial x_1}\left(\frac{\partial^2 v_3}{\partial x_2^2}+\frac{\partial^2 \overset{\circ}{v}_3}{\partial x_2^2}\right)+\frac{\partial\xi_2}{\partial x_2}\left(\frac{\partial^2 v_3}{\partial x_1^2}+\frac{\partial^2 \overset{\circ}{v}_3}{\partial x_1^2}\right)+\frac{1}{2}\left(\frac{\partial v_3}{\partial x_1}\right)^2\left(\frac{\partial^2 v_3}{\partial x_2^2}+\frac{\partial^2 \overset{\circ}{v}_3}{\partial x_2^2}\right)\right.$$

$$\left.+\frac{1}{2}\left(\frac{\partial v_3}{\partial x_2}\right)^2\left(\frac{\partial^2 v_3}{\partial x_1^2}+\frac{\partial^2 \overset{\circ}{v}_3}{\partial x_1^2}\right)+\frac{\partial v_3}{\partial x_1}\frac{\partial \overset{\circ}{v}_3}{\partial x_1}\left(\frac{\partial^2 v_3}{\partial x_2^2}+\frac{\partial^2 \overset{\circ}{v}_3}{\partial x_2^2}\right)+\frac{\partial v_3}{\partial x_2}\frac{\partial \overset{\circ}{v}_3}{\partial x_2}\left(\frac{\partial^2 v_3}{\partial x_1^2}+\frac{\partial^2 \overset{\circ}{v}_3}{\partial x_1^2}\right)\right\}$$

$$+A_{22}\left\{\frac{\partial\xi_2}{\partial x_2}\left(\frac{\partial^2 v_3}{\partial x_2^2}+\frac{\partial^2 \overset{\circ}{v}_3}{\partial x_2^2}\right)+\frac{1}{2}\left(\frac{\partial v_3}{\partial x_2}\right)^2\left(\frac{\partial^2 v_3}{\partial x_2^2}+\frac{\partial^2 \overset{\circ}{v}_3}{\partial x_2^2}\right)+\frac{\partial v_3}{\partial x_2}\frac{\partial \overset{\circ}{v}_3}{\partial x_2}\left(\frac{\partial^2 v_3}{\partial x_2^2}+\frac{\partial^2 \overset{\circ}{v}_3}{\partial x_2^2}\right)\right\}$$

$$+2A_{66}\left\{\frac{\partial\xi_1}{\partial x_2}\left(\frac{\partial^2 v_3}{\partial x_1\partial x_2}+\frac{\partial^2 \overset{\circ}{v}_3}{\partial x_1\partial x_2}\right)+\frac{\partial\xi_2}{\partial x_1}\left(\frac{\partial^2 v_3}{\partial x_1\partial x_2}+\frac{\partial^2 \overset{\circ}{v}_3}{\partial x_1\partial x_2}\right)+\frac{\partial v_3}{\partial x_1}\frac{\partial v_3}{\partial x_2}\left(\frac{\partial^2 v_3}{\partial x_1\partial x_2}\right.\right.$$

$$\left.\left.+\frac{\partial^2 \overset{\circ}{v}_3}{\partial x_1\partial x_2}\right)+\frac{\partial \overset{\circ}{v}_3}{\partial x_1}\frac{\partial v_3}{\partial x_2}\left(\frac{\partial^2 v_3}{\partial x_1\partial x_2}+\frac{\partial^2 \overset{\circ}{v}_3}{\partial x_1\partial x_2}\right)+\frac{\partial v_3}{\partial x_1}\frac{\partial \overset{\circ}{v}_3}{\partial x_2}\left(\frac{\partial^2 v_3}{\partial x_1\partial x_2}+\frac{\partial^2 \overset{\circ}{v}_3}{\partial x_1\partial x_2}\right)\right\}$$

$$+A_{16}\left\{\frac{\partial\xi_1}{\partial x_2}\left(\frac{\partial^2 v_3}{\partial x_1^2}+\frac{\partial^2 \overset{\circ}{v}_3}{\partial x_1^2}\right)+\frac{\partial\xi_2}{\partial x_1}\left(\frac{\partial^2 v_3}{\partial x_1^2}+\frac{\partial^2 \overset{\circ}{v}_3}{\partial x_1^2}\right)+\frac{\partial v_3}{\partial x_1}\frac{\partial v_3}{\partial x_2}\left(\frac{\partial^2 v_3}{\partial x_1^2}+\frac{\partial^2 \overset{\circ}{v}_3}{\partial x_1^2}\right)\right.$$

$$+\frac{\partial \overset{\circ}{v}_3}{\partial x_1}\frac{\partial v_3}{\partial x_2}\left(\frac{\partial^2 v_3}{\partial x_1^2}+\frac{\partial^2 \overset{\circ}{v}_3}{\partial x_1^2}\right)+\frac{\partial v_3}{\partial x_1}\frac{\partial \overset{\circ}{v}_3}{\partial x_2}\left(\frac{\partial^2 v_3}{\partial x_1^2}+\frac{\partial^2 \overset{\circ}{v}_3}{\partial x_1^2}\right)+2\frac{\partial\xi_1}{\partial x_1}\left(\frac{\partial^2 v_3}{\partial x_1\partial x_2}+\frac{\partial^2 \overset{\circ}{v}_3}{\partial x_1\partial x_2}\right)$$

$$\left.+\left(\frac{\partial v_3}{\partial x_1}\right)^2\left(\frac{\partial^2 v_3}{\partial x_1\partial x_2}+\frac{\partial^2 \overset{\circ}{v}_3}{\partial x_1\partial x_2}\right)+2\frac{\partial v_3}{\partial x_1}\frac{\partial \overset{\circ}{v}_3}{\partial x_1}\left(\frac{\partial^2 v_3}{\partial x_1\partial x_2}+\frac{\partial^2 \overset{\circ}{v}_3}{\partial x_1\partial x_2}\right)\right\}$$

$$+A_{26}\left\{\frac{\partial\xi_2}{\partial x_1}\left(\frac{\partial^2 v_3}{\partial x_2^2}+\frac{\partial^2 \overset{\circ}{v}_3}{\partial x_2^2}\right)+\frac{\partial\xi_1}{\partial x_2}\left(\frac{\partial^2 v_3}{\partial x_2^2}+\frac{\partial^2 \overset{\circ}{v}_3}{\partial x_2^2}\right)+\frac{\partial v_3}{\partial x_2}\frac{\partial v_3}{\partial x_1}\left(\frac{\partial^2 v_3}{\partial x_2^2}+\frac{\partial^2 \overset{\circ}{v}_3}{\partial x_2^2}\right)\right.$$

$$+\frac{\partial \overset{\circ}{v}_3}{\partial x_2}\frac{\partial v_3}{\partial x_1}\left(\frac{\partial^2 v_3}{\partial x_2^2}+\frac{\partial^2 \overset{\circ}{v}_3}{\partial x_2^2}\right)+\frac{\partial v_3}{\partial x_2}\frac{\partial \overset{\circ}{v}_3}{\partial x_1}\left(\frac{\partial^2 v_3}{\partial x_2^2}+\frac{\partial^2 \overset{\circ}{v}_3}{\partial x_2^2}\right)+2\frac{\partial\xi_2}{\partial x_2}\left(\frac{\partial^2 v_3}{\partial x_1\partial x_2}+\frac{\partial^2 \overset{\circ}{v}_3}{\partial x_1\partial x_2}\right)$$

$$\left.+\left(\frac{\partial v_3}{\partial x_2}\right)^2\left(\frac{\partial^2 v_3}{\partial x_2\partial x_1}+\frac{\partial^2 \overset{\circ}{v}_3}{\partial x_2\partial x_1}\right)+2\frac{\partial v_3}{\partial x_2}\frac{\partial \overset{\circ}{v}_3}{\partial x_2}\left(\frac{\partial^2 v_3}{\partial x_1\partial x_2}+\frac{\partial^2 \overset{\circ}{v}_3}{\partial x_1\partial x_2}\right)\right\}-F_{11}\frac{\partial^4 v_3}{\partial x_1^4}$$

$$-2(F_{12}+2F_{66})\frac{\partial^4 v_3}{\partial x_1^2\partial x_2^2}-F_{22}\frac{\partial^4 v_3}{\partial x_2^4}-4F_{16}\frac{\partial^4 v_3}{\partial x_1^3\partial x_2}-4F_{26}\frac{\partial^4 v_3}{\partial x_1\partial x_2^3}$$

$$+2\overline{h}\overline{K}^2\overline{G}_{13}\left(1+\frac{h}{2\overline{h}}\right)\left\{\frac{1}{\overline{h}}\frac{\partial\eta_1}{\partial x_1}+\left(1+\frac{C_1}{\overline{h}}\right)\frac{\partial^2 v_3}{\partial x_1^2}\right\}$$

$$+2\overline{h}\overline{K}^2\overline{G}_{23}\left(1+\frac{h}{2\overline{h}}\right)\left\{\frac{1}{\overline{h}}\frac{\partial\eta_2}{\partial x_2}+\left(1+\frac{C_1}{\overline{h}}\right)\frac{\partial^2 v_3}{\partial x_2^2}\right\}-q_3=0 \tag{4.42e}$$

- *Boundary Conditions*

and the associated boundary conditions along the edges $x_n = \text{const}(\text{n} = 1, 2)$ are

$$
\begin{aligned}
&N_{nn} = \widetilde{N}_{nn} \quad \text{or} \quad \xi_n = \widetilde{\xi}_n \\
&N_{nt} = \widetilde{N}_{nt} \quad \text{or} \quad \xi_t = \widetilde{\xi}_t \\
&L_{nn} = \widetilde{L}_{nn} \quad \text{or} \quad \eta_n = \widetilde{\eta}_n \\
&L_{nt} = \widetilde{L}_{nt} \quad \text{or} \quad \eta_t = \widetilde{\eta}_t \\
&M_{nn} = \widetilde{M}_{nn} \quad \text{or} \quad \frac{\partial v_3}{\partial x_n} = \frac{\partial \widetilde{v}_3}{\partial x_n} \\
&N_{nt}\left(\frac{\partial v_3}{\partial x_t} + \frac{\partial \overset{\circ}{v}_3}{\partial x_t}\right) + N_{nn}\left(\frac{\partial v_3}{\partial x_n} + \frac{\partial \overset{\circ}{v}_3}{\partial x_n}\right) + \frac{\partial M_{nn}}{\partial x_n} + 2\frac{\partial M_{nt}}{\partial x_t} \quad \text{or} \quad v_3 = \widetilde{v}_3 \\
&+(a/\overline{h})\overline{N}_{n3} = \frac{\partial \widetilde{M}_{nt}}{\partial x_t} + \widetilde{\overline{N}}_{n3}
\end{aligned}
\tag{4.43a – f}
$$

Considering simply supported boundary conditions, of Type A, they are assumed as

Along the edges $x_n = 0,\ L_n$

$$
N_{nn} = -\widetilde{N}_{nn}, \quad N_{nt} = 0, \quad \eta_n = 0, \quad \eta_t = 0, \quad M_{nn} = 0, \quad v_3 = 0 \tag{4.44a – f}
$$

4.4.2 Solution Methodology

Equations (4.42a–e) and (4.44a–f) constitutes the governing system of equations for the post-buckling solution for flat sandwich panels with angle-ply laminated facings and a weak incompressible core. The method of approach will be similar as in the previous section but without the stress potential to aid in the solution. It should also be mentioned that the inertia and thermal terms have been discarded within the equations of motion which are now referred to as the equations of equilibrium. The transversal deflection and the geometric imperfection representations are the same as in the previous section for the same reasons. The boundary conditions remain the same and the assumptions for the transversal displacement and the peak geometric imperfection remain unchanged. Again, these are represented by the following expressions

$$
\begin{Bmatrix} v_3 \\ \overset{\circ}{v}_3 \end{Bmatrix} = \begin{Bmatrix} w_{mn} \\ \overset{\circ}{w}_{mn} \end{Bmatrix} \sin \lambda_m x_1 \sin \mu_n x_2 \tag{4.45a, b}
$$

Unlike the linear theory the first two equations of equilibrium are coupled with the last three. Therefore, there are five equations of motion to contend with. All five equations of equilibrium need to be fulfilled as well as the boundary conditions.

Keeping the EGM in mind and knowing whatever is not identically fulfilled is fulfilled in an average sense allows ξ_1 and ξ_2 to be assumed in the following form

$$\begin{Bmatrix} \xi_1(x_1,x_2) \\ \xi_2(x_1,x_2) \end{Bmatrix} = \begin{Bmatrix} F_{mn}^{(1)} \\ G_{mn}^{(1)} \end{Bmatrix} \sin 2\lambda_m x_1 + \begin{Bmatrix} F_{mn}^{(2)} \\ G_{mn}^{(2)} \end{Bmatrix} \sin 2\mu_n x_2 + \begin{Bmatrix} F_{mn}^{(3)} \\ G_{mn}^{(3)} \end{Bmatrix} \sin \lambda_m x_1 \cos \mu_n x_2 + \begin{Bmatrix} F_{mn}^{(4)} \\ G_{mn}^{(4)} \end{Bmatrix} \cos \lambda_m x_1 \sin \mu_n x_2 + \begin{Bmatrix} F_{mn}^{(5)} \\ G_{mn}^{(5)} \end{Bmatrix} \sin 2\lambda_m x_1 \cos 2\mu_n x_2 + \begin{Bmatrix} F_{mn}^{(6)} \\ G_{mn}^{(6)} \end{Bmatrix} \cos 2\lambda_m x_1 \sin 2\mu_n x_2 + \begin{Bmatrix} F_{mn}^{(7)} \\ G_{mn}^{(7)} \end{Bmatrix} x_1 + \begin{Bmatrix} F_{mn}^{(8)} \\ G_{mn}^{(8)} \end{Bmatrix} x_2 \tag{4.46a, b}$$

Substituting the expressions for v_3, v_3°, ξ_1, and ξ_2, from Eqs. (4.45a, b) and (4.46a, b), into Eqs. (4.42a and 4.42b) and comparing coefficients of like trigonometric functions provides $F_{mn}^{(1)} - F_{mn}^{(6)}$ and $G_{mn}^{(1)} - G_{mn}^{(6)}$ as

$$\left(F_{mn}^{(i)}, G_{mn}^{(i)}, F_{mn}^{(5)}, G_{mn}^{(6)}\right) = \left(\widetilde{F}_{mn}^{(i)}, \widetilde{G}_{mn}^{(i)}, \widetilde{F}_{mn}^{(5)}, \widetilde{G}_{mn}^{(6)}\right)\left(w_{mn}^2 + 2w_{mn}w_{mn}^\circ\right), (i = 1,\ 2) \tag{4.47a}$$

$$F_{mn}^{(j)} = G_{mn}^{(j)} = F_{mn}^{(6)} = G_{mn}^{(5)} = 0, \quad (j = 3,\ 4) \tag{4.47c}$$

where,

$$\widetilde{F}_{mn}^{(1)} = \frac{(A_{16}^2 - A_{11}A_{66})\lambda_m^2 + (A_{12}A_{66} - A_{16}A_{26})\mu_n^2}{16\lambda_m(A_{11}A_{66} - A_{16}^2)}, \quad \widetilde{F}_{mn}^{(2)} = \frac{(A_{16}A_{22} - A_{26}A_{12})\lambda_m^2}{16\mu_n(A_{22}A_{66} - A_{26}^2)}$$
$$\widetilde{G}_{mn}^{(1)} = \frac{(A_{26}A_{11} - A_{16}A_{12})\mu_n^2}{16\lambda_m(A_{11}A_{66} - A_{16}^2)}, \quad \widetilde{G}_{mn}^{(2)} = \frac{(A_{26}^2 - A_{22}A_{66})\mu_n^2 + (A_{12}A_{66} - A_{16}A_{26})\lambda_m^2}{16\mu_n(A_{22}A_{66} - A_{26}^2)}$$
$$\widetilde{F}_{mn}^{(5)} = \frac{\lambda_m}{16}, \quad \widetilde{G}_{mn}^{(6)} = \frac{\mu_n}{16} \tag{4.48a – f}$$

The constants, $F_{mn}^{(7)}$, $F_{mn}^{(8)}$, $G_{mn}^{(7)}$, $G_{mn}^{(8)}$ are four constants which remain undetermined. These are determined from enforcing the in-plane static boundary conditions which are expressed mathematically as

$$\int_0^{L_t} N_{nn}dx_t = -\widetilde{N}_{nn}L_t \quad \begin{pmatrix} n = 1,2 \\ t = 2,1 \end{pmatrix} \quad \sum\!\!\!\!/_{n,t} \tag{4.49}$$

$$\int_0^{L_t} N_{nt} dx_t = 0 \tag{4.50}$$

These conditions fulfill the boundary conditions, Eqs. (4.44a, b) in an average sense. In order to utilize these conditions the stress resultants N_{11}, N_{22}, and N_{12} need to be expressed in terms of displacements. It should be recalled that for the case of a weak core

$$N_{11} = N_{11}^{/} + N_{11}^{//}, \quad N_{22} = N_{22}^{/} + N_{22}^{//}, \quad N_{12} = N_{12}^{/} + N_{12}^{//} \tag{4.51a – c}$$

Making use of Eqs. (2.86a) and (2.92a) coupled with the strain–displacement relationships, Eqs. (2.53)–(2.55) and Eqs. (2.69)–(2.71), while discarding the terms with the curvatures gives the global stress resultants in terms of displacements ξ_1, ξ_2, v_3, and v_3° as

$$N_{11} = A_{11}\left\{\frac{\partial \xi_1}{\partial x_1} + \frac{1}{2}\left(\frac{\partial v_3}{\partial x_1}\right)^2 + \frac{\partial v_3^{\circ}}{\partial x_1}\frac{\partial v_3}{\partial x_1}\right\} + A_{12}\left\{\frac{\partial \xi_2}{\partial x_2} + \frac{1}{2}\left(\frac{\partial v_3}{\partial x_2}\right)^2 + \frac{\partial v_3^{\circ}}{\partial x_2}\frac{\partial v_3}{\partial x_2} - \right\}$$
$$+ A_{16}\left\{\frac{\partial \xi_1}{\partial x_2} + \frac{\partial \xi_2}{\partial x_1} + \frac{\partial v_3}{\partial x_1}\frac{\partial v_3}{\partial x_2} + \frac{\partial v_3^{\circ}}{\partial x_1}\frac{\partial v_3}{\partial x_2} + \frac{\partial v_3}{\partial x_1}\frac{\partial v_3^{\circ}}{\partial x_2}\right\} \tag{4.52a}$$

$$N_{22} = A_{22}\left\{\frac{\partial \xi_2}{\partial x_2} + \frac{1}{2}\left(\frac{\partial v_3}{\partial x_2}\right)^2 + \frac{\partial v_3^{\circ}}{\partial x_2}\frac{\partial v_3}{\partial x_2}\right\} + A_{12}\left\{\frac{\partial \xi_1}{\partial x_1} + \frac{1}{2}\left(\frac{\partial v_3}{\partial x_1}\right)^2 + \frac{\partial v_3^{\circ}}{\partial x_1}\frac{\partial v_3}{\partial x_1}\right\}$$
$$+ A_{26}\left\{\frac{\partial \xi_2}{\partial x_1} + \frac{\partial \xi_1}{\partial x_2} + \frac{\partial v_3}{\partial x_2}\frac{\partial v_3}{\partial x_1} + \frac{\partial v_3^{\circ}}{\partial x_2}\frac{\partial v_3}{\partial x_1} + \frac{\partial v_3}{\partial x_2}\frac{\partial v_3^{\circ}}{\partial x_1}\right\} \tag{4.52b}$$

$$N_{12} = A_{16}\left\{\frac{\partial \xi_1}{\partial x_1} + \frac{1}{2}\left(\frac{\partial v_3}{\partial x_1}\right)^2 + \frac{\partial v_3^{\circ}}{\partial x_1}\frac{\partial v_3}{\partial x_1}\right\} + A_{26}\left\{\frac{\partial \xi_2}{\partial x_2} + \frac{1}{2}\left(\frac{\partial v_3}{\partial x_2}\right)^2 + \frac{\partial v_3^{\circ}}{\partial x_2}\frac{\partial v_3}{\partial x_2}\right\}$$
$$+ A_{66}\left\{\frac{\partial \xi_1}{\partial x_2} + \frac{\partial \xi_2}{\partial x_1} + \frac{\partial v_3}{\partial x_1}\frac{\partial v_3}{\partial x_2} + \frac{\partial v_3^{\circ}}{\partial x_1}\frac{\partial v_3}{\partial x_2} + \frac{\partial v_3}{\partial x_1}\frac{\partial v_3^{\circ}}{\partial x_2}\right\} \tag{4.52c}$$

Substituting the expressions for v_3, v_3°, ξ_1, and ξ_2 into Eqs. (4.52a, 4.52b, and 4.52c) then substituting the result into the requirements for the in-plane static boundary conditions, Eqs. (4.49) and (4.50) and carrying out the indicated operations provides the constants $F_{mn}^{(7)}, F_{mn}^{(8)}, G_{mn}^{(7)}, G_{mn}^{(8)}$ as

$$F_{mn}^{(7)} = -\frac{\lambda_m^2}{8}\left(w_{mn}^2 + 2w_{mn}w_{mn}^{\circ}\right) + \frac{A_{26}^2 - A_{22}A_{66}}{\Omega}\widetilde{N}_{11} + \frac{A_{12}A_{66} - A_{16}A_{26}}{\Omega}\widetilde{N}_{22} \quad (4.53a)$$

$$G_{mn}^{(8)} = -\frac{\mu_n^2}{8}\left(w_{mn}^2 + 2w_{mn}w_{mn}^{\circ}\right) + \frac{A_{12}A_{66} - A_{16}A_{26}}{\Omega}\widetilde{N}_{11} + \frac{A_{16}^2 - A_{11}A_{66}}{\Omega}\widetilde{N}_{22} \quad (4.53b)$$

where

$$\Omega = A_{11}A_{22}A_{66} - A_{12}^2A_{66} + 2A_{12}A_{16}A_{26} - A_{11}A_{26}^2 - A_{22}A_{16}^2 \quad (4.53c)$$

$F_{mn}^{(8)}$ and $G_{mn}^{(7)}$ are arbitrary and have no effect on the final post-buckling solution. With the first two equilibrium equations fulfilled, in addition to the boundary condition Eq. (4.44f), attention is now given to the third and fourth equilibrium Eqs. (4.42c and 4.42d). It can be seen that these two governing equations have no curvature terms in them, as a result, the curvature plays no part on the fulfillment of these two equations. Except for the consideration of cross-ply laminated sandwich plates and shells where the A_{16} and A_{26} terms are zero the expressions for η_1 and η_2 will remain the same for other pre- and postcritical stability problems. The expressions for these displacement functions have already been determined in Chap. 3 (see Eqs. 3.12a, b) – (3.17a–e). Therefore, the first four equations of equilibrium are fulfilled with the boundary condition (4.44f). The fifth equation of equilibrium and the unfulfilled boundary conditions will be satisfied in an average sense through the application of the EGM. Retaining the fifth equilibrium equation along with the unfulfilled boundary conditions within the energy functional gives

$$\int_{t_0}^{t_1}\left\langle\int_0^{l_1}\int_0^{l_2}\left[A_{11}\left\{\frac{\partial\xi_1}{\partial x_1}\left(\frac{\partial^2 v_3}{\partial x_1^2}+\frac{\partial^2 \overset{\circ}{v}_3}{\partial x_1^2}\right)+\frac{1}{2}\left(\frac{\partial v_3}{\partial x_1}\right)^2\left(\frac{\partial^2 v_3}{\partial x_1^2}+\frac{\partial^2 \overset{\circ}{v}_3}{\partial x_1^2}\right)\right.\right.\right.$$

$$\left.+\frac{\partial v_3}{\partial x_1}\frac{\partial \overset{\circ}{v}_3}{\partial x_1}\left(\frac{\partial^2 v_3}{\partial x_1^2}+\frac{\partial^2 \overset{\circ}{v}_3}{\partial x_1^2}\right)\right\}+A_{12}\left\{\frac{\partial\xi_1}{\partial x_1}\left(\frac{\partial^2 v_3}{\partial x_2^2}+\frac{\partial^2 \overset{\circ}{v}_3}{\partial x_2^2}\right)+\frac{\partial\xi_2}{\partial x_2}\left(\frac{\partial^2 v_3}{\partial x_1^2}+\frac{\partial^2 \overset{\circ}{v}_3}{\partial x_1^2}\right)\right.$$

$$+\frac{1}{2}\left(\frac{\partial v_3}{\partial x_1}\right)^2\left(\frac{\partial^2 v_3}{\partial x_2^2}+\frac{\partial^2 \overset{\circ}{v}_3}{\partial x_2^2}\right)+\frac{1}{2}\left(\frac{\partial v_3}{\partial x_2}\right)^2\left(\frac{\partial^2 v_3}{\partial x_1^2}+\frac{\partial^2 \overset{\circ}{v}_3}{\partial x_1^2}\right)$$

$$\left.+\frac{\partial v_3}{\partial x_1}\frac{\partial \overset{\circ}{v}_3}{\partial x_1}\left(\frac{\partial^2 v_3}{\partial x_2^2}+\frac{\partial^2 \overset{\circ}{v}_3}{\partial x_2^2}\right)+\frac{\partial v_3}{\partial x_2}\frac{\partial \overset{\circ}{v}_3}{\partial x_2}\left(\frac{\partial^2 v_3}{\partial x_1^2}+\frac{\partial^2 \overset{\circ}{v}_3}{\partial x_1^2}\right)\right\}$$

$$+A_{22}\left\{\frac{\partial\xi_2}{\partial x_2}\left(\frac{\partial^2 v_3}{\partial x_2^2}+\frac{\partial^2 \overset{\circ}{v}_3}{\partial x_2^2}\right)+\frac{1}{2}\left(\frac{\partial v_3}{\partial x_2}\right)^2\left(\frac{\partial^2 v_3}{\partial x_2^2}+\frac{\partial^2 \overset{\circ}{v}_3}{\partial x_2^2}\right)+\frac{\partial v_3}{\partial x_2}\frac{\partial \overset{\circ}{v}_3}{\partial x_2}\left(\frac{\partial^2 v_3}{\partial x_2^2}+\frac{\partial^2 \overset{\circ}{v}_3}{\partial x_2^2}\right)\right\}$$

$$+2A_{66}\left\{\frac{\partial\xi_1}{\partial x_2}\left(\frac{\partial^2 v_3}{\partial x_1\partial x_2}+\frac{\partial^2 \overset{\circ}{v}_3}{\partial x_1\partial x_2}\right)+\frac{\partial\xi_2}{\partial x_1}\left(\frac{\partial^2 v_3}{\partial x_1\partial x_2}+\frac{\partial^2 \overset{\circ}{v}_3}{\partial x_1\partial x_2}\right)+\frac{\partial v_3}{\partial x_1}\frac{\partial v_3}{\partial x_2}\right.$$

$$\left(\frac{\partial^2 v_3}{\partial x_1\partial x_2}+\frac{\partial^2 \overset{\circ}{v}_3}{\partial x_1\partial x_2}\right)+\frac{\partial \overset{\circ}{v}_3}{\partial x_1}\frac{\partial v_3}{\partial x_2}\left(\frac{\partial^2 v_3}{\partial x_1\partial x_2}+\frac{\partial^2 \overset{\circ}{v}_3}{\partial x_1\partial x_2}\right)+\frac{\partial v_3}{\partial x_1}\frac{\partial \overset{\circ}{v}_3}{\partial x_2}$$

$$\left.\left(\frac{\partial^2 v_3}{\partial x_1\partial x_2}+\frac{\partial^2 \overset{\circ}{v}_3}{\partial x_1\partial x_2}\right)\right\}+A_{16}\left\{\frac{\partial\xi_1}{\partial x_2}\left(\frac{\partial^2 v_3}{\partial x_1^2}+\frac{\partial^2 \overset{\circ}{v}_3}{\partial x_1^2}\right)+\frac{\partial\xi_2}{\partial x_1}\left(\frac{\partial^2 v_3}{\partial x_1^2}+\frac{\partial^2 \overset{\circ}{v}_3}{\partial x_1^2}\right)\right.$$

$$+\frac{\partial v_3}{\partial x_1}\frac{\partial v_3}{\partial x_2}\left(\frac{\partial^2 v_3}{\partial x_1^2}+\frac{\partial^2 \overset{\circ}{v}_3}{\partial x_1^2}\right)+\frac{\partial \overset{\circ}{v}_3}{\partial x_1}\frac{\partial v_3}{\partial x_2}\left(\frac{\partial^2 v_3}{\partial x_1^2}+\frac{\partial^2 \overset{\circ}{v}_3}{\partial x_1^2}\right)+\frac{\partial v_3}{\partial x_1}\frac{\partial \overset{\circ}{v}_3}{\partial x_2}$$

$$\left(\frac{\partial^2 v_3}{\partial x_1^2}+\frac{\partial^2 \overset{\circ}{v}_3}{\partial x_1^2}\right)+2\frac{\partial\xi_1}{\partial x_1}\left(\frac{\partial^2 v_3}{\partial x_1\partial x_2}+\frac{\partial^2 \overset{\circ}{v}_3}{\partial x_1\partial x_2}\right)+\left(\frac{\partial v_3}{\partial x_1}\right)^2\left(\frac{\partial^2 v_3}{\partial x_1\partial x_2}+\frac{\partial^2 \overset{\circ}{v}_3}{\partial x_1\partial x_2}\right)$$

$$\left.+2\frac{\partial v_3}{\partial x_1}\frac{\partial \overset{\circ}{v}_3}{\partial x_1}\left(\frac{\partial^2 v_3}{\partial x_1\partial x_2}+\frac{\partial^2 \overset{\circ}{v}_3}{\partial x_1\partial x_2}\right)\right\}+A_{26}\left\{\frac{\partial\xi_2}{\partial x_1}\left(\frac{\partial^2 v_3}{\partial x_2^2}+\frac{\partial^2 \overset{\circ}{v}_3}{\partial x_2^2}\right)\right.$$

$$+\frac{\partial\xi_1}{\partial x_2}\left(\frac{\partial^2 v_3}{\partial x_2^2}+\frac{\partial^2 \overset{\circ}{v}_3}{\partial x_2^2}\right)+\frac{\partial v_3}{\partial x_2}\frac{\partial v_3}{\partial x_1}\left(\frac{\partial^2 v_3}{\partial x_2^2}+\frac{\partial^2 \overset{\circ}{v}_3}{\partial x_2^2}\right)+\frac{\partial \overset{\circ}{v}_3}{\partial x_2}\frac{\partial v_3}{\partial x_1}\left(\frac{\partial^2 v_3}{\partial x_2^2}+\frac{\partial^2 \overset{\circ}{v}_3}{\partial x_2^2}\right)$$

$$+\frac{\partial v_3}{\partial x_2}\frac{\partial \overset{\circ}{v}_3}{\partial x_1}\left(\frac{\partial^2 v_3}{\partial x_2^2}+\frac{\partial^2 \overset{\circ}{v}_3}{\partial x_2^2}\right)+2\frac{\partial\xi_2}{\partial x_2}\left(\frac{\partial^2 v_3}{\partial x_1\partial x_2}+\frac{\partial^2 \overset{\circ}{v}_3}{\partial x_1\partial x_2}\right)$$

$$\left.+\left(\frac{\partial v_3}{\partial x_2}\right)^2\left(\frac{\partial^2 v_3}{\partial x_2\partial x_1}+\frac{\partial^2 \overset{\circ}{v}_3}{\partial x_2\partial x_1}\right)+2\frac{\partial v_3}{\partial x_2}\frac{\partial \overset{\circ}{v}_3}{\partial x_2}\left(\frac{\partial^2 v_3}{\partial x_1\partial x_2}+\frac{\partial^2 \overset{\circ}{v}_3}{\partial x_1\partial x_2}\right)\right\}-F_{11}\frac{\partial^4 v_3}{\partial x_1^4}$$

$$-2(F_{12}+2F_{66})\frac{\partial^4 v_3}{\partial x_1^2\partial x_2^2}-F_{22}\frac{\partial^4 v_3}{\partial x_2^4}-4F_{16}\frac{\partial^4 v_3}{\partial x_1^3\partial x_2}-4F_{26}\frac{\partial^4 v_3}{\partial x_1\partial x_2^3}$$

$$+2\overline{h}\overline{K}^2\overline{G}_{13}\left(1+\frac{h}{2\overline{h}}\right)\left\{\frac{1}{\overline{h}}\frac{\partial\eta_1}{\partial x_1}+\left(1+\frac{C_1}{\overline{h}}\right)\frac{\partial^2 v_3}{\partial x_1^2}\right\}$$

$$\left.\left.+2\overline{h}\overline{K}^2\overline{G}_{23}\left(1+\frac{h}{2\overline{h}}\right)\left\{\frac{1}{\overline{h}}\frac{\partial\eta_2}{\partial x_2}+\left(1+\frac{C_1}{\overline{h}}\right)\frac{\partial^2 v_3}{\partial x_2^2}\right\}-q_3=0\right]\delta v_3 dx_1 dx_2\right\rangle dt+$$

$$+\int_{t_0}^{t_1}\left\langle\int_0^{l_2}\left[(N_{11}+\widetilde{N}_{11})\delta\xi_1+N_{12}\delta\xi_2+L_{11}\delta\eta_1+L_{12}\delta\eta_2+M_{11}\delta\left(\frac{\partial v_3}{\partial x_1}\right)\right]\Bigg|_0^{l_1}dx_2\right\rangle dt$$

$$+\int_{t_0}^{t_1}\left\langle\int_0^{l_1}\left[(N_{22}+\widetilde{N}_{22})\delta\xi_2+N_{12}\delta\xi_1+L_{22}\delta\eta_2+L_{12}\delta\eta_1+M_{22}\delta\left(\frac{\partial v_3}{\partial x_2}\right)\right]\Bigg|_0^{l_2}dx_1\right\rangle dt$$

$$=0 \tag{4.54}$$

Substituting the expressions for v_3, v_3°, ξ_1, ξ_2, η_1, η_2 into Eq. (4.54) and carrying out the indicated integrations results in the governing post-buckling solution for angle-ply laminated flat sandwich panels which results in a nonlinear algebraic equation expressed in terms of the modal amplitudes as

$$P_{mn}^{(5)}\left(w_{mn}^2 + 2w_{mn}w_{mn}^\circ\right)\left(w_{mn} + w_{mn}^\circ\right) + P_{mn}^{(2)}\left(w_{mn} + w_{mn}^\circ\right) + P_{mn}^{(1)}w_{mn} + q_{mn} = 0 \tag{4.55}$$

Where

$$P_{mn}^{(5)} = -\left(\lambda_m^3 A_{11} + \lambda_m\mu_n^2 A_{12}\right)\widetilde{F}_{mn}^{(1)} - \left(\mu_n^3 A_{22} + \lambda_m^2\mu_n A_{12}\right)\widetilde{G}_{mn}^{(2)} - \left(\lambda_m^3 A_{16} + \lambda_m\mu_n^2 A_{26}\right)\widetilde{G}_{mn}^{(1)} - \left(\mu_n^3 A_{26} + \lambda_m^2\mu_n A_{16}\right)F_{mn}^{(2)} \tag{4.56}$$

$$P_{mn}^{(2)} = -\left(\lambda_m^2\widetilde{N}_{11} + \mu_n^2\widetilde{N}_{22}\right) \tag{4.57}$$

$$P_{mn}^{(1)} = F_{11}\lambda_m^4 + 2(F_{12} + 2F_{66})\lambda_m^2\mu_n^2 + F_{22}\mu_n^4 + d_1 a\lambda_m\left(\widetilde{H}_{mn}^{(1)} + a\lambda_m\right) + d_2 a\mu_n\left(\widetilde{I}_{mn}^{(2)} + a\mu_n\right) \tag{4.58}$$

Equation (4.55) is put in a more conducive form for generating results as

$$P_{mn}^{(5)}w_{mn}^3 + \left(3P_{mn}^{(5)}w_{mn}^\circ\right)w_{mn}^2 + \left(2P_{mn}^{(5)}\left(w_{mn}^\circ\right)^2 + P_{mn}^{(2)} + P_{mn}^{(1)}\right)w_{mn} + P_{mn}^{(2)}w_{mn}^\circ + q_{mn} = 0 \tag{4.59}$$

which can be further expressed as

$$L_1 w_{mn}^3 + L_2 w_{mn}^2 + L_3 w_{mn} + L_4 = 0 \tag{4.60}$$

where

$$L_1 = P_{mn}^{(5)} \tag{4.61a}$$

$$L_2 = 3P_{mn}^{(5)}w_{mn}^\circ \tag{4.61b}$$

$$L_3 = 2P_{mn}^{(5)}\left(w_{mn}^\circ\right)^2 + P_{mn}^{(2)} + P_{mn}^{(1)} \tag{4.61c}$$

$$L_4 = P_{mn}^{(2)}w_{mn}^\circ + q_{mn} \tag{4.61d}$$

Equation (4.60) is easily solved via Newton's method for the various equilibrium configurations of the structure. Usually, a more convenient form is to introduce the following Parameter δ as

$$\delta = w_{mn} + w^{\circ}_{mn} \quad \Rightarrow w_{mn} = \delta - w^{\circ}_{mn} \tag{4.62}$$

Substituting Eq. (4.62) into Eq. (4.60) gives

$$L_1\left(\delta - w^{\circ}_{mn}\right)^3 + L_2\left(\delta - w^{\circ}_{mn}\right)^2 + L_3\left(\delta - w^{\circ}_{mn}\right) + L_4 = 0 \tag{4.63}$$

Expanding, combining like terms, and simplifying gives

$$L_1\delta^3_{mn} + \left(L_2 - 3L_1w^{\circ}_{mn}\right)\delta^2_{mn} + \left(L_3 - 2L_2w^{\circ}_{mn} - L_1\left(w^{\circ}_{mn}\right)^2\right)\delta_{mn} + L_4 - L_1\left(w^{\circ}_{mn}\right)^3 + L_2\left(w^{\circ}_{mn}\right)^2 - L_3w^{\circ}_{mn} = 0 \tag{4.64}$$

Solving Eq. (4.64) via Newton's method gives the compressive uniaxial load vs δ_{mn} (amplitude of deflection plus amplitude of imperfection).

4.4.3 Numerical Results and Discussion

A few results of the applied theory of sandwich plates as it pertains to post-buckling are presented next. For the following numerical illustrations which follow, Tables 4.3 and 4.4 contain the material and geometrical properties for both the face sheets and the core.

Figure 4.5 depicts the effect of the geometrical imperfections on the uniaxial compressive edge loading of a flat sandwich panel versus amplitude of deflection for a fixed layup of the facing. The entire map of the stability paths is shown. For the case of zero imperfection, there is a bifurcation point which shows several directions

Table 4.3 The geometrical and material properties for the face sheets

t^a_k (mm)	L_1 (mm)	L_2 (mm)	E_1 (GPa)	E_2 (GPa)	G_{12} (GPa)	ν_{12}
0.005	609.6 (24in)	609.6 (24in)	180.987	10.342	7.239	0.28

t^a_k implies the ply thickness

Table 4.4 The geometrical material properties for the core

$\bar{h}$ in (mm)	$\bar{G}_{13}$(MPa)	$\bar{G}_{23}$(MPa)
12.7 (0.5in)	1.437	0.651

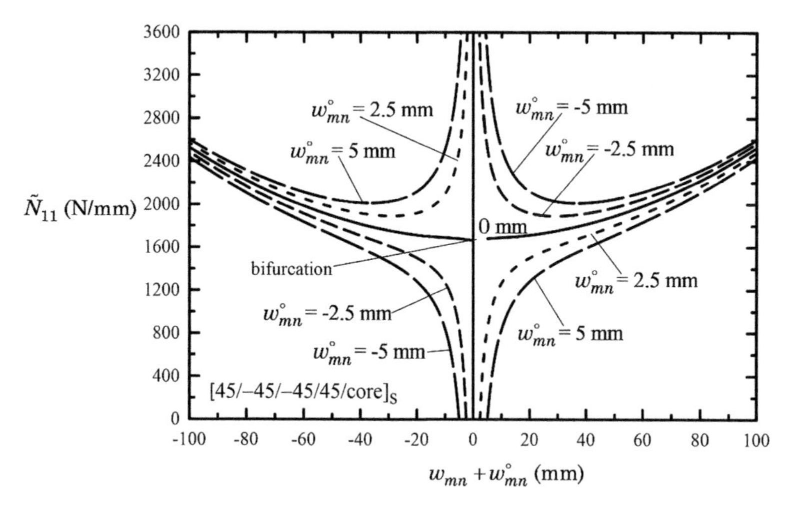

Fig. 4.5 The effect of geometric imperfections on the compressive uniaxial edge loading vs. amplitude of deflection for a flat sandwich panel

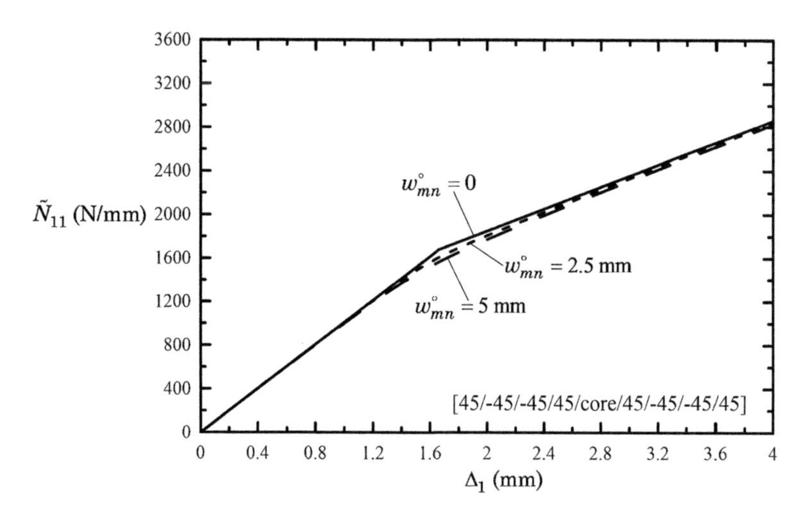

Fig. 4.6 The counterpart of Fig. 4.5 in the compressive edge load-end shortening plane for various geometric imperfections of a flat sandwich panel

in which the loading can occur. This bifurcation point is the point at which the panel starts to buckle. The results also reveal that with an imperfection present, the bifurcation disappears and there is an initial deflection with the start of compressive loading. As the imperfection increases, the initial deflection increases, and the direction of loading appears to proceed farther away from the bifurcation point. Figure 4.6 is the

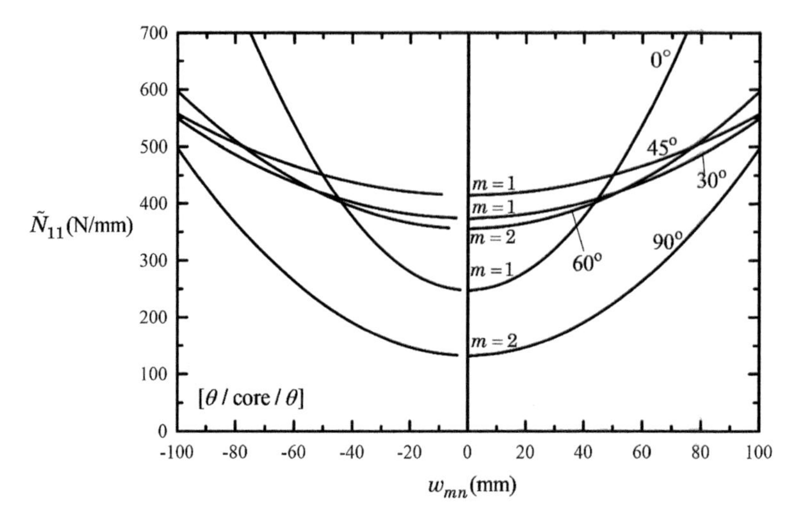

Fig. 4.7 The effect of the ply angle on the compressive edge loading for a single-layered flat sandwich panel

counterpart of Fig. 4.5 revealing the end-shorting of panel. As the compressive loading increases, the edges displace inward. This amount of displacement is depicted for various amounts of imperfections inherent within the structure. With no imperfection there is a discontinuity within the trend line. This discontinuity is the buckling bifurcation point for the panel which matches with the buckling point in Fig. 4.5. Beyond the bifurcation point, the panel continues to displace transversely as the compressive loading increases until the panel fails. Also, as the imperfection increases, the end-shortening increases for a fixed compressive edge loading.

Figures 4.7 and 4.8 highlight the influence played by the various ply angles on the uniaxial compressive strength of a flat sandwich panel for two different layups. In each case, it appears that the ply angle of 45 degrees contains the higher buckling bifurcation point whereas at a ply angle of zero degrees the panel has a larger load-carrying capacity from a load–deflection standpoint. In Fig. 4.9, the effect of biaxial edge loading is depicted. When both compression and tension are present simultaneously the effect seems to increase the point at which the structure buckles beyond which the loading increases until the panel fails. In comparison, when both edges are loaded in compression the effect seems to diminish the point at which the structure buckles.

Figure 4.10 illustrates the effect of the ply angle for the given layup of a flat sandwich panel on the transverse pressure deflection interaction. It clearly reveals that the structure has a larger load-carrying capacity at a ply angle of 45 degrees as compared with the other ply angle configurations. It can also be seen that 0- and 90-degree ply angle configurations have the least load-carrying capacity.

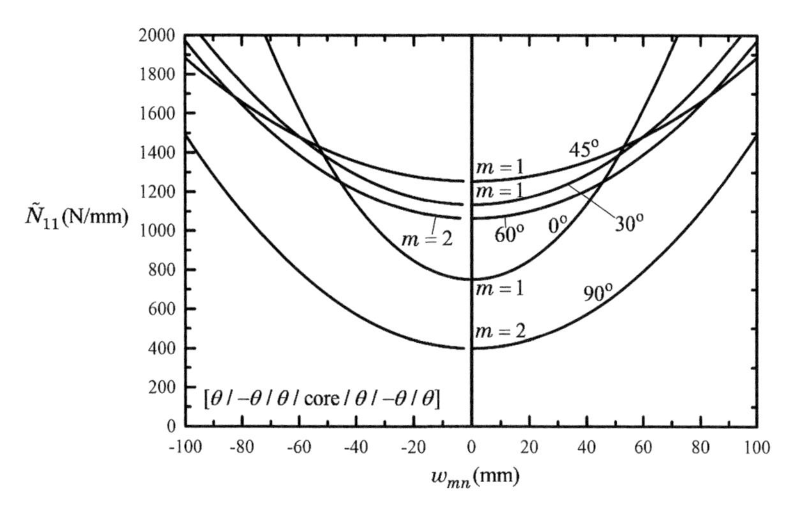

Fig. 4.8 The effect of the ply angle on the compressive edge loading for the given layup in the facings

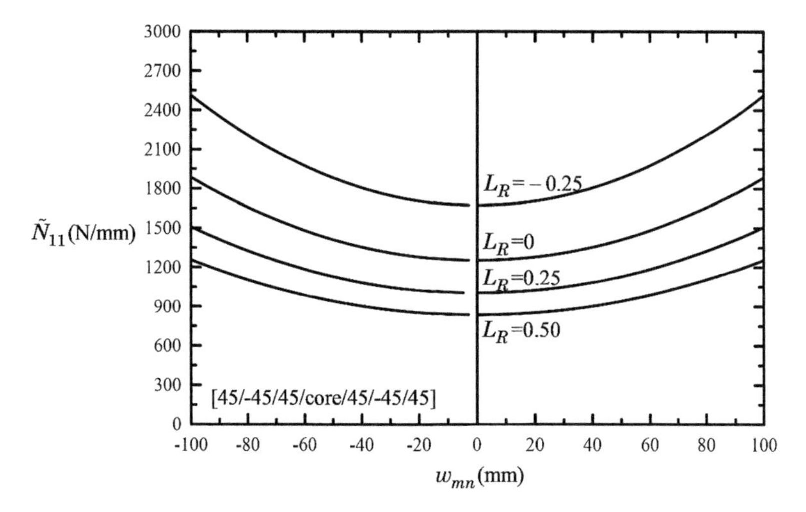

Fig. 4.9 The effect of the biaxial edge loading, for the given fixed layup, on the uniaxial compressive edge load

4.5 Angle-Ply Laminated Sandwich Shells

4.5.1 Governing System

In contrast to the previous section, this section considers the curvature terms included in the governing equations. The solution methodology is exactly the

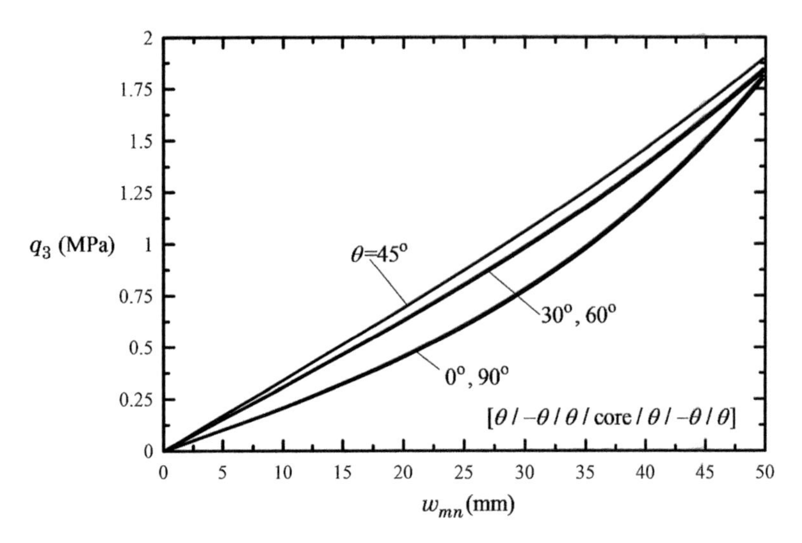

Fig. 4.10 The effect of the transverse pressure on the deflection of a flat sandwich panel for various ply angles

same as in the previous section with only an additional level of computational effort. The governing equations can be obtained from Eqs. (2.156)–(2.160) and (2.161a–f) which are given as

- *Equations of Equilibrium*

$$
\begin{aligned}
&A_{11}\left\{\frac{\partial^2\xi_1}{\partial x_1^2}+\frac{\partial v_3}{\partial x_1}\frac{\partial^2 v_3}{\partial x_1^2}+\frac{\partial^2 \overset{\circ}{v}_3}{\partial x_1^2}\frac{\partial v_3}{\partial x_1}+\frac{\partial \overset{\circ}{v}_3}{\partial x_1}\frac{\partial^2 v_3}{\partial x_1^2}-\frac{1}{R_1}\frac{\partial v_3}{\partial x_1}\right\}+A_{66}\left\{\frac{\partial^2\xi_1}{\partial x_2^2}+\frac{\partial^2\xi_2}{\partial x_1\partial x_2}\right.\\
&\left.+\frac{\partial v_3}{\partial x_2}\frac{\partial^2 v_3}{\partial x_1\partial x_2}+\frac{\partial v_3}{\partial x_1}\frac{\partial^2 v_3}{\partial x_2^2}+\frac{\partial v_3}{\partial x_2}\frac{\partial^2 \overset{\circ}{v}_3}{\partial x_1\partial x_2}+\frac{\partial \overset{\circ}{v}_3}{\partial x_1}\frac{\partial^2 v_3}{\partial x_2^2}+\frac{\partial \overset{\circ}{v}_3}{\partial x_2}\frac{\partial^2 v_3}{\partial x_1\partial x_2}+\frac{\partial v_3}{\partial x_1}\frac{\partial^2 \overset{\circ}{v}_3}{\partial x_2^2}\right\}\\
&+A_{12}\left\{\frac{\partial^2\xi_2}{\partial x_1\partial x_2}+\frac{\partial v_3}{\partial x_2}\frac{\partial^2 v_3}{\partial x_1\partial x_2}+\frac{\partial v_3}{\partial x_2}\frac{\partial^2 \overset{\circ}{v}_3}{\partial x_1\partial x_2}+\frac{\partial \overset{\circ}{v}_3}{\partial x_2}\frac{\partial^2 v_3}{\partial x_1\partial x_2}-\frac{1}{R_2}\frac{\partial v_3}{\partial x_2}\right\}\\
&-A_{16}\left\{2\frac{\partial^2\xi_1}{\partial x_1\partial x_2}+\frac{\partial^2\xi_2}{\partial x_1^2}+2\frac{\partial v_3}{\partial x_1}\frac{\partial^2 v_3}{\partial x_1\partial x_2}+2\frac{\partial v_3}{\partial x_1}\frac{\partial^2 \overset{\circ}{v}_3}{\partial x_1\partial x_2}+2\frac{\partial \overset{\circ}{v}_3}{\partial x_1}\frac{\partial^2 v_3}{\partial x_1\partial x_2}\right.\\
&\left.+\frac{\partial v_3}{\partial x_2}\frac{\partial^2 v_3}{\partial x_1^2}+\frac{\partial v_3}{\partial x_2}\frac{\partial^2 \overset{\circ}{v}_3}{\partial x_1^2}+\frac{\partial^2 v_3}{\partial x_1^2}-\frac{1}{R_1}\frac{\partial v_3}{\partial x_1}\right\}\\
&-A_{26}\left\{\frac{\partial^2\xi_2}{\partial x_2^2}+\frac{\partial v_3}{\partial x_2}\frac{\partial^2 v_3}{\partial x_2^2}+\frac{\partial v_3}{\partial x_2}\frac{\partial^2 \overset{\circ}{v}_3}{\partial x_2^2}+\frac{\partial \overset{\circ}{v}_3}{\partial x_2}\frac{\partial^2 v_3}{\partial x_2^2}-\frac{1}{R_2}\frac{\partial v_3}{\partial x_2}\right\}=0
\end{aligned}
\tag{4.65a}
$$

$$A_{22}\left\{\frac{\partial^2\xi_2}{\partial x_2^2}+\frac{\partial v_3}{\partial x_2}\frac{\partial^2 v_3}{\partial x_2^2}+\frac{\partial^2 \overset{\circ}{v}_3}{\partial x_2^2}\frac{\partial v_3}{\partial x_2}+\frac{\partial \overset{\circ}{v}_3}{\partial x_2}\frac{\partial^2 v_3}{\partial x_2^2}-\frac{1}{R_2}\frac{\partial v_3}{\partial x_2}\right\}+A_{66}\left\{\frac{\partial^2\xi_2}{\partial x_1^2}+\frac{\partial^2\xi_1}{\partial x_1\partial x_2}+\right.$$
$$\left.\frac{\partial v_3}{\partial x_1}\frac{\partial^2 v_3}{\partial x_1\partial x_2}+\frac{\partial v_3}{\partial x_2}\frac{\partial^2 v_3}{\partial x_1^2}+\frac{\partial v_3}{\partial x_1}\frac{\partial^2 \overset{\circ}{v}_3}{\partial x_1\partial x_2}+\frac{\partial \overset{\circ}{v}_3}{\partial x_2}\frac{\partial^2 v_3}{\partial x_1^2}+\frac{\partial \overset{\circ}{v}_3}{\partial x_1}\frac{\partial^2 v_3}{\partial x_1\partial x_2}+\frac{\partial v_3}{\partial x_2}\frac{\partial^2 \overset{\circ}{v}_3}{\partial x_1^2}\right\}+$$
$$A_{12}\left\{\frac{\partial^2\xi_1}{\partial x_1\partial x_2}+\frac{\partial v_3}{\partial x_1}\frac{\partial^2 v_3}{\partial x_1\partial x_2}+\frac{\partial v_3}{\partial x_1}\frac{\partial^2 \overset{\circ}{v}_3}{\partial x_1\partial x_2}+\frac{\partial \overset{\circ}{v}_3}{\partial x_1}\frac{\partial^2 v_3}{\partial x_1\partial x_2}-\frac{1}{R_1}\frac{\partial v_3}{\partial x_2}\right\}-A_{26}\left\{2\frac{\partial^2\xi_2}{\partial x_1\partial x_2}\right.$$
$$+\frac{\partial^2\xi_1}{\partial x_2^2}+2\frac{\partial v_3}{\partial x_2}\frac{\partial^2 v_3}{\partial x_1\partial x_2}+2\frac{\partial v_3}{\partial x_2}\frac{\partial^2 \overset{\circ}{v}_3}{\partial x_1\partial x_2}+2\frac{\partial \overset{\circ}{v}_3}{\partial x_2}\frac{\partial^2 v_3}{\partial x_1\partial x_2}+\frac{\partial v_3}{\partial x_1}\frac{\partial^2 v}{\partial x_2^2}+\frac{\partial v_3}{\partial x_1}\frac{\partial^2 \overset{\circ}{v}_3}{\partial x_2^2}+$$
$$\left.\frac{\partial \overset{\circ}{v}_3}{\partial x_1}\frac{\partial^2 v_3}{\partial x_2^2}-\frac{1}{R_2}\frac{\partial v_3}{\partial x_2}\right\}-A_{16}\left\{\frac{\partial^2\xi_1}{\partial x_1^2}+\frac{\partial v_3}{\partial x_1}\frac{\partial^2 v_3}{\partial x_1^2}+\frac{\partial v_3}{\partial x_1}\frac{\partial^2 \overset{\circ}{v}_3}{\partial x_1^2}+\frac{\partial \overset{\circ}{v}_3}{\partial x_1}\frac{\partial^2 v_3}{\partial x_1^2}-\frac{1}{R_1}\frac{\partial v_3}{\partial x_1}\right\}=0$$

(4.65b)

$$A_{11}\frac{\partial^2\eta_1}{\partial x_1^2}+A_{66}\left\{\frac{\partial^2\eta_1}{\partial x_2^2}+\frac{\partial^2\eta_2}{\partial x_1\partial x_2}\right\}+A_{12}\frac{\partial^2\eta_2}{\partial x_1\partial x_2}+A_{16}\left\{\frac{\partial^2\eta_2}{\partial x_1^2}+2\frac{\partial^2\eta_1}{\partial x_1\partial x_2}\right\}$$
$$+A_{26}\frac{\partial^2\eta_2}{\partial x_2^2}-d_1\left\{\eta_1+a\frac{\partial v_3}{\partial x_1}\right\}=0$$

(4.65c)

$$A_{22}\frac{\partial^2\eta_2}{\partial x_2^2}+A_{66}\left\{\frac{\partial^2\eta_2}{\partial x_2^2}+\frac{\partial^2\eta_1}{\partial x_1\partial x_2}\right\}+A_{12}\frac{\partial^2\eta_1}{\partial x_1\partial x_2}+A_{26}\left\{\frac{\partial^2\eta_1}{\partial x_2^2}+2\frac{\partial^2\eta_2}{\partial x_1\partial x_2}\right\}$$
$$+A_{16}\frac{\partial^2\eta_1}{\partial x_1^2}-d_2\left\{\eta_2+a\frac{\partial v_3}{\partial x_2}\right\}=0$$

(4.65d)

$$A_{11}\left\{\frac{\partial\xi_1}{\partial x_1}\left(\frac{\partial^2 v_3}{\partial x_1^2}+\frac{\partial^2 \overset{\circ}{v}_3}{\partial x_1^2}\right)+\frac{1}{2}\left(\frac{\partial v_3}{\partial x_1}\right)^2\left(\frac{\partial^2 v_3}{\partial x_1^2}+\frac{\partial^2 \overset{\circ}{v}_3}{\partial x_1^2}\right)+\frac{\partial v_3}{\partial x_1}\frac{\partial \overset{\circ}{v}_3}{\partial x_1}\left(\frac{\partial^2 v_3}{\partial x_1^2}+\frac{\partial^2 \overset{\circ}{v}_3}{\partial x_1^2}\right)\right.$$
$$\left.+\frac{1}{R_1}\left[\frac{\partial\xi_1}{\partial x_1}+\frac{1}{2}\left(\frac{\partial v_3}{\partial x_1}\right)^2+\frac{\partial v_3}{\partial x_1}\frac{\partial \overset{\circ}{v}_3}{\partial x_1}-v_3\left(\frac{\partial^2 v_3}{\partial x_1^2}+\frac{\partial^2 \overset{\circ}{v}_3}{\partial x_1^2}\right)-\frac{v_3}{R_1}\right]\right\}+A_{12}\left\{\frac{\partial\xi_1}{\partial x_1}\left(\frac{\partial^2 v_3}{\partial x_2^2}\right.\right.$$
$$\left.+\frac{\partial^2 \overset{\circ}{v}_3}{\partial x_2^2}\right)+\frac{\partial\xi_2}{\partial x_2}\left(\frac{\partial^2 v_3}{\partial x_1^2}+\frac{\partial^2 \overset{\circ}{v}_3}{\partial x_1^2}\right)+\frac{1}{2}\left(\frac{\partial v_3}{\partial x_1}\right)^2\left(\frac{\partial^2 v_3}{\partial x_2^2}+\frac{\partial^2 \overset{\circ}{v}_3}{\partial x_2^2}\right)+\frac{1}{2}\left(\frac{\partial v_3}{\partial x_2}\right)^2$$
$$\left(\frac{\partial^2 v_3}{\partial x_1^2}+\frac{\partial^2 \overset{\circ}{v}_3}{\partial x_1^2}\right)+\frac{\partial v_3}{\partial x_1}\frac{\partial \overset{\circ}{v}_3}{\partial x_1}\left(\frac{\partial^2 v_3}{\partial x_2^2}+\frac{\partial^2 \overset{\circ}{v}_3}{\partial x_2^2}\right)+\frac{\partial v_3}{\partial x_2}\frac{\partial \overset{\circ}{v}_3}{\partial x_2}\left(\frac{\partial^2 v_3}{\partial x_1^2}+\frac{\partial^2 \overset{\circ}{v}_3}{\partial x_1^2}\right)$$

$$+\frac{1}{R_1}\left[\frac{\partial \xi_2}{\partial x_2}+\frac{1}{2}\left(\frac{\partial v_3}{\partial x_2}\right)^2+\frac{\partial v_3}{\partial x_2}\frac{\partial \mathring{v}_3}{\partial x_2}-v_3\left(\frac{\partial^2 v_3}{\partial x_2^2}+\frac{\partial^2 \mathring{v}_3}{\partial x_2^2}\right)-\frac{v_3}{R_2}\right]$$

$$+\frac{1}{R_2}\left[\frac{\partial \xi_1}{\partial x_1}+\frac{1}{2}\left(\frac{\partial v_3}{\partial x_1}\right)^2+\frac{\partial v_3}{\partial x_1}\frac{\partial \mathring{v}_3}{\partial x_1}-v_3\left(\frac{\partial^2 v_3}{\partial x_1^2}+\frac{\partial^2 \mathring{v}_3}{\partial x_1^2}\right)-\frac{v_3}{R_1}\right]\Bigg\}$$

$$+A_{22}\left\{\frac{\partial \xi_2}{\partial x_2}\left(\frac{\partial^2 v_3}{\partial x_2^2}+\frac{\partial^2 \mathring{v}_3}{\partial x_2^2}\right)+\frac{1}{2}\left(\frac{\partial v_3}{\partial x_2}\right)^2\left(\frac{\partial^2 v_3}{\partial x_2^2}+\frac{\partial^2 \mathring{v}_3}{\partial x_2^2}\right)+\frac{\partial v_3}{\partial x_2}\frac{\partial \mathring{v}_3}{\partial x_2}\left(\frac{\partial^2 v_3}{\partial x_2^2}+\frac{\partial^2 \mathring{v}_3}{\partial x_2^2}\right)\right.$$

$$\left.+\frac{1}{R_2}\left[\frac{\partial \xi_2}{\partial x_2}+\frac{1}{2}\left(\frac{\partial v_3}{\partial x_2}\right)^2+\frac{\partial v_3}{\partial x_2}\frac{\partial \mathring{v}_3}{\partial x_2}-v_3\left(\frac{\partial^2 v_3}{\partial x_2^2}+\frac{\partial^2 \mathring{v}_3}{\partial x_2^2}\right)-\frac{v_3}{R_2}\right]\right\}$$

$$+2A_{66}\left\{\frac{\partial \xi_1}{\partial x_2}\left(\frac{\partial^2 v_3}{\partial x_1\partial x_2}+\frac{\partial^2 \mathring{v}_3}{\partial x_1\partial x_2}\right)+\frac{\partial \xi_2}{\partial x_1}\left(\frac{\partial^2 v_3}{\partial x_1\partial x_2}+\frac{\partial^2 \mathring{v}_3}{\partial x_1\partial x_2}\right)+\frac{\partial v_3}{\partial x_1}\frac{\partial v_3}{\partial x_2}\right.$$

$$\left(\frac{\partial^2 v_3}{\partial x_1\partial x_2}+\frac{\partial^2 \mathring{v}_3}{\partial x_1\partial x_2}\right)+\frac{\partial \mathring{v}_3}{\partial x_1}\frac{\partial v_3}{\partial x_2}\left(\frac{\partial^2 v_3}{\partial x_1\partial x_2}+\frac{\partial^2 \mathring{v}_3}{\partial x_1\partial x_2}\right)+\frac{\partial v_3}{\partial x_1}\frac{\partial \mathring{v}_3}{\partial x_2}\left(\frac{\partial^2 v_3}{\partial x_1\partial x_2}+\right.$$

$$\left.\left.\frac{\partial^2 \mathring{v}_3}{\partial x_1\partial x_2}\right)\right\}+A_{16}\left\{\frac{\partial \xi_1}{\partial x_2}\left(\frac{\partial^2 v_3}{\partial x_1^2}+\frac{\partial^2 \mathring{v}_3}{\partial x_1^2}\right)+\frac{\partial \xi_2}{\partial x_1}\left(\frac{\partial^2 v_3}{\partial x_1^2}+\frac{\partial^2 \mathring{v}_3}{\partial x_1^2}\right)+\frac{\partial v_3}{\partial x_1}\frac{\partial v_3}{\partial x_2}\right.$$

$$\left(\frac{\partial^2 v_3}{\partial x_1^2}+\frac{\partial^2 \mathring{v}_3}{\partial x_1^2}\right)+\frac{\partial \mathring{v}_3}{\partial x_1}\frac{\partial v_3}{\partial x_2}\left(\frac{\partial^2 v_3}{\partial x_1^2}+\frac{\partial^2 \mathring{v}_3}{\partial x_1^2}\right)+\frac{\partial v_3}{\partial x_1}\frac{\partial \mathring{v}_3}{\partial x_2}\left(\frac{\partial^2 v_3}{\partial x_1^2}+\frac{\partial^2 \mathring{v}_3}{\partial x_1^2}\right)$$

$$+2\frac{\partial \xi_1}{\partial x_1}\left(\frac{\partial^2 v_3}{\partial x_1\partial x_2}+\frac{\partial^2 \mathring{v}_3}{\partial x_1\partial x_2}\right)+\left(\frac{\partial v_3}{\partial x_1}\right)^2\left(\frac{\partial^2 v_3}{\partial x_1\partial x_2}+\frac{\partial^2 \mathring{v}_3}{\partial x_1\partial x_2}\right)+2\frac{\partial v_3}{\partial x_1}\frac{\partial \mathring{v}_3}{\partial x_1}$$

$$\left(\frac{\partial^2 v_3}{\partial x_1\partial x_2}+\frac{\partial^2 \mathring{v}_3}{\partial x_1\partial x_2}\right)+\frac{1}{R_1}\left[\frac{\partial \xi_1}{\partial x_2}+\frac{\partial \xi_2}{\partial x_1}+\frac{\partial v_3}{\partial x_1}\frac{\partial v_3}{\partial x_2}+\frac{\partial \mathring{v}_3}{\partial x_1}\frac{\partial v_3}{\partial x_2}+\frac{\partial v_3}{\partial x_1}\frac{\partial \mathring{v}_3}{\partial x_2}\right.$$

$$\left.\left.-2v_3\left(\frac{\partial^2 v_3}{\partial x_1\partial x_2}+\frac{\partial^2 \mathring{v}_3}{\partial x_1\partial x_2}\right)\right]\right\}+A_{26}\left\{\frac{\partial \xi_2}{\partial x_1}\left(\frac{\partial^2 v_3}{\partial x_2^2}+\frac{\partial^2 \mathring{v}_3}{\partial x_2^2}\right)+\frac{\partial \xi_1}{\partial x_2}\right.$$

$$\left(\frac{\partial^2 v_3}{\partial x_2^2}+\frac{\partial^2 \mathring{v}_3}{\partial x_2^2}\right)+\frac{\partial v_3}{\partial x_2}\frac{\partial v_3}{\partial x_1}\left(\frac{\partial^2 v_3}{\partial x_2^2}+\frac{\partial^2 \mathring{v}_3}{\partial x_2^2}\right)+\frac{\partial \mathring{v}_3}{\partial x_2}\frac{\partial v_3}{\partial x_1}\left(\frac{\partial^2 v_3}{\partial x_2^2}+\frac{\partial^2 \mathring{v}_3}{\partial x_2^2}\right)$$

$$+\frac{\partial v_3}{\partial x_2}\frac{\partial \mathring{v}_3}{\partial x_1}\left(\frac{\partial^2 v_3}{\partial x_2^2}+\frac{\partial^2 \mathring{v}_3}{\partial x_2^2}\right)+2\frac{\partial \xi_2}{\partial x_2}\left(\frac{\partial^2 v_3}{\partial x_1\partial x_2}+\frac{\partial^2 \mathring{v}_3}{\partial x_1\partial x_2}\right)+$$

$$\left(\frac{\partial v_3}{\partial x_2}\right)^2\left(\frac{\partial^2 v_3}{\partial x_2 \partial x_1}+\frac{\partial^2 \overset{\circ}{v}_3}{\partial x_2 \partial x_1}\right)+2\frac{\partial v_3}{\partial x_2}\frac{\partial \overset{\circ}{v}_3}{\partial x_2}\left(\frac{\partial^2 v_3}{\partial x_1 \partial x_2}+\frac{\partial^2 \overset{\circ}{v}_3}{\partial x_1 \partial x_2}\right)$$

$$+\frac{1}{R_2}\left[\frac{\partial \xi_2}{\partial x_1}+\frac{\partial \xi_1}{\partial x_2}+\frac{\partial v_3}{\partial x_1}\frac{\partial v_3}{\partial x_2}+\frac{\partial \overset{\circ}{v}_3}{\partial x_2}\frac{\partial v_3}{\partial x_1}+\frac{\partial v_3}{\partial x_2}\frac{\partial \overset{\circ}{v}_3}{\partial x_1}-2v_3\left(\frac{\partial^2 v_3}{\partial x_1 \partial x_2}+\frac{\partial^2 \overset{\circ}{v}_3}{\partial x_1 \partial x_2}\right)\right]\Bigg\}$$

$$-F_{11}\frac{\partial^4 v_3}{\partial x_1^4}-2(F_{12}+2F_{66})\frac{\partial^4 v_3}{\partial x_1^2 \partial x_2^2}-F_{22}\frac{\partial^4 v_3}{\partial x_2^4}-4F_{16}\frac{\partial^4 v_3}{\partial x_1^3 \partial x_2}-4F_{26}\frac{\partial^4 v_3}{\partial x_1 \partial x_2^3}$$

$$+2\overline{h}\overline{K}^2\overline{G}_{13}\left(1+\frac{h}{2\overline{h}}\right)\times\left\{\frac{1}{\overline{h}}\frac{\partial \eta_1}{\partial x_1}+\left(1+\frac{C_1}{\overline{h}}\right)\frac{\partial^2 v_3}{\partial x_1^2}\right\}+2\overline{h}\overline{K}^2\overline{G}_{23}\left(1+\frac{h}{2\overline{h}}\right)$$

$$\left\{\frac{1}{\overline{h}}\frac{\partial \eta_2}{\partial x_2}+\left(1+\frac{C_1}{\overline{h}}\right)\frac{\partial^2 v_3}{\partial x_2^2}\right\}-N_{11}^T\left(\frac{\partial^2 v_3}{\partial x_1^2}+\frac{\partial^2 \overset{\circ}{v}_3}{\partial x_1^2}+\frac{1}{R_1}\right)$$

$$-N_{12}^T\left(\frac{\partial^2 v_3}{\partial x_1 \partial x_2}+\frac{\partial^2 \overset{\circ}{v}_3}{\partial x_1 \partial x_2}\right)-N_{22}^T\left(\frac{\partial^2 v_3}{\partial x_2^2}+\frac{\partial^2 \overset{\circ}{v}_3}{\partial x_2^2}+\frac{1}{R_2}\right)$$

$$-N_{11}^m\left(\frac{\partial^2 v_3}{\partial x_1^2}+\frac{\partial^2 \overset{\circ}{v}_3}{\partial x_1^2}+\frac{1}{R_1}\right)-N_{12}^m\left(\frac{\partial^2 v_3}{\partial x_1 \partial x_2}+\frac{\partial^2 \overset{\circ}{v}_3}{\partial x_1 \partial x_2}\right)$$

$$-N_{22}^m\left(\frac{\partial^2 v_3}{\partial x_2^2}+\frac{\partial^2 \overset{\circ}{v}_3}{\partial x_2^2}+\frac{1}{R_2}\right)-q_3=0 \tag{4.65e}$$

- *Boundary Conditions*

  and the associated boundary conditions along the edges $x_n = \text{const}(\text{n} = 1, 2)$ are

$$\begin{aligned}
&N_{nn}=\widetilde{N}_{nn} \quad \text{or} \quad \xi_n=\widetilde{\xi}_n\\
&N_{nt}=\widetilde{N}_{nt} \quad \text{or} \quad \xi_t=\widetilde{\xi}_t\\
&L_{nn}=\widetilde{L}_{nn} \quad \text{or} \quad \eta_n=\widetilde{\eta}_n\\
&L_{nt}=\widetilde{L}_{nt} \quad \text{or} \quad \eta_t=\widetilde{\eta}_t\\
&M_{nn}=\widetilde{M}_{nn} \quad \text{or} \quad \frac{\partial v_3}{\partial x_n}=\frac{\partial \widetilde{v}_3}{\partial x_n}\\
&N_{nt}\left(\frac{\partial v_3}{\partial x_t}+\frac{\partial \overset{\circ}{v}_3}{\partial x_t}\right)+N_{nn}\left(\frac{\partial v_3}{\partial x_n}+\frac{\partial \overset{\circ}{v}_3}{\partial x_n}\right)+\frac{\partial M_{nn}}{\partial x_n}+2\frac{\partial M_{nt}}{\partial x_t} \quad \text{or} \quad v_3=\widetilde{v}_3\\
&+(a/\overline{h})\overline{N}_{n3}=\frac{\partial \widetilde{M}_{nt}}{\partial x_t}+\widetilde{\overline{N}}_{n3}
\end{aligned} \tag{4.66a – f}$$

As was assumed previously, simply supported boundary conditions of Type A are assumed and restated as:

Along the edges $x_n = 0,\ L_n$

$$N_{nn} = -\widetilde{N}_{nn}, \quad N_{nt} = 0, \quad \eta_n = 0, \quad \eta_t = 0, \quad M_{nn} = 0, \quad v_3 = 0 \tag{4.67a – f}$$

4.5.2 Solution Methodology

Equations (4.65a–e) and (4.67a–f) constitute the governing system of equations concerning the post-buckling problem for doubly curved sandwich panels with angle-ply laminated facings and a weak incompressible core. The method of approach will be the same as in the previous section. As previously the inertia and thermal terms have been discarded. The assumed forms for transversal deflection and the geometric imperfection representations are represented by the following expressions

$$\begin{Bmatrix} v_3 \\ v_3^\circ \end{Bmatrix} = \begin{Bmatrix} w_{mn} \\ w_{mn}^\circ \end{Bmatrix} \sin \lambda_m x_1 \sin \mu_n x_2 \tag{4.68a, b}$$

additionally, the first two equations of equilibrium, Eqs. (4.65a and 4.65b) can be fulfilled by assuming ξ_1 and ξ_2 in the following form which is identical to the previously assumed form in the previous section. The difference will appear in the expressions for the coefficients.

$$\begin{aligned}\begin{Bmatrix} \xi_1(x_1,x_2) \\ \xi_2(x_1,x_2) \end{Bmatrix} = &\begin{Bmatrix} F_{mn}^{(1)} \\ G_{mn}^{(1)} \end{Bmatrix} \sin 2\lambda_m x_1 + \begin{Bmatrix} F_{mn}^{(2)} \\ G_{mn}^{(2)} \end{Bmatrix} \sin 2\mu_n x_2 + \begin{Bmatrix} F_{mn}^{(3)} \\ G_{mn}^{(3)} \end{Bmatrix} \sin \lambda_m x_1 \cos \mu_n x_2 + \\ &\begin{Bmatrix} F_{mn}^{(4)} \\ G_{mn}^{(4)} \end{Bmatrix} \cos \lambda_m x_1 \sin \mu_n x_2 + \begin{Bmatrix} F_{mn}^{(5)} \\ G_{mn}^{(5)} \end{Bmatrix} \sin 2\lambda_m x_1 \cos 2\mu_n x_2 + \\ &\begin{Bmatrix} F_{mn}^{(6)} \\ G_{mn}^{(6)} \end{Bmatrix} \cos 2\lambda_m x_1 \sin 2\mu_n x_2 + \begin{Bmatrix} F_{mn}^{(7)} \\ G_{mn}^{(7)} \end{Bmatrix} x_1 + \begin{Bmatrix} F_{mn}^{(8)} \\ G_{mn}^{(8)} \end{Bmatrix} x_2 \end{aligned} \tag{4.69a, b}$$

Substituting the expressions for $v_3,\ v_3^\circ,\ \xi_1,$ and ξ_2, from Eqs. (4.68a, b) and (4.69a, b), into Eqs. (4.65a and 4.65b) and comparing coefficients of like trigonometric functions provides $F_{mn}^{(1)} - F_{mn}^{(6)}$ and $G_{mn}^{(1)} - G_{mn}^{(6)}$ as

$$\left(F_{mn}^{(i)}, G_{mn}^{(i)}, F_{mn}^{(5)}, G_{mn}^{(6)}\right) = \left(\widetilde{F}_{mn}^{(i)}, \widetilde{G}_{mn}^{(i)}, \widetilde{F}_{mn}^{(5)}, \widetilde{G}_{mn}^{(6)}\right)\left(w_{mn}^2 + 2w_{mn}w_{mn}^{\circ}\right), \quad (i = 1, 2) \tag{4.70a}$$

$$\left(F_{mn}^{(j)}, G_{mn}^{(j)}\right) = \left(\widetilde{F}_{mn}^{(j)}, \widetilde{G}_{mn}^{(j)}\right)w_{mn}, \quad (j = 3, 4) \tag{4.70b}$$

$$F_{mn}^{(6)} = G_{mn}^{(5)} = 0 \tag{4.70c}$$

where

$$\widetilde{F}_{mn}^{(1)} = \frac{\left(A_{16}^2 - A_{11}A_{66}\right)\lambda_m^2 + \left(A_{12}A_{66} - A_{16}A_{26}\right)\mu_n^2}{16\lambda_m\left(A_{11}A_{66} - A_{16}^2\right)}, \quad \widetilde{F}_{mn}^{(2)} = \frac{\left(A_{16}A_{22} - A_{26}A_{12}\right)\lambda_m^2}{16\mu_n\left(A_{22}A_{66} - A_{26}^2\right)}$$
$$\widetilde{G}_{mn}^{(1)} = \frac{\left(A_{26}A_{11} - A_{16}A_{12}\right)\mu_n^2}{16\lambda_m\left(A_{11}A_{66} - A_{16}^2\right)}, \quad \widetilde{G}_{mn}^{(2)} = \frac{\left(A_{26}^2 - A_{22}A_{66}\right)\mu_n^2 + \left(A_{12}A_{66} - A_{16}A_{26}\right)\lambda_m^2}{16\mu_n\left(A_{22}A_{66} - A_{26}^2\right)}$$
$$\widetilde{F}_{mn}^{(5)} = \frac{\lambda_m}{16}, \quad \widetilde{G}_{mn}^{(6)} = \frac{\mu_n}{16} \tag{4.71a – f}$$

The constants $F_{mn}^{(3)}$, $F_{mn}^{(4)}$, $G_{mn}^{(3)}$, $G_{mn}^{(4)}$ are determined from the following matrix equation.

$$\begin{bmatrix} R_{11}^{(m,n)} & R_{12}^{(m,n)} & R_{13}^{(m,n)} & R_{14}^{(m,n)} \\ & R_{11}^{(m,n)} & R_{14}^{(m,n)} & R_{13}^{(m,n)} \\ & & R_{33}^{(m,n)} & R_{34}^{(m,n)} \\ \text{Symm.} & & & R_{33}^{(m,n)} \end{bmatrix} \begin{Bmatrix} F_{mn}^{(3)} \\ F_{mn}^{(4)} \\ G_{mn}^{(3)} \\ G_{mn}^{(4)} \end{Bmatrix} = \begin{Bmatrix} S_{mn}^{(1)} \\ S_{mn}^{(2)} \\ S_{mn}^{(3)} \\ S_{mn}^{(4)} \end{Bmatrix} \tag{4.72}$$

where

$$R_{11}^{(m,n)} = -\left(\lambda_m^2 A_{11} + \mu_n^2 A_{66}\right), \quad R_{12}^{(m,n)} = 2\lambda_m\mu_n A_{16}, \quad R_{13}^{(m,n)} = \lambda_m^2 A_{16} + \mu_n^2 A_{26}$$
$$R_{14}^{(m,n)} = -\lambda_m\mu_n\left(A_{12} + A_{66}\right), \quad R_{33}^{(m,n)} = -\left(\mu_n^2 A_{22} + \lambda_m^2 A_{66}\right), \quad R_{34}^{(m,n)} = 2\lambda_m\mu_n A_{26}$$
$$S_{mn}^{(1)} = -\mu_n\left(\frac{A_{16}}{R_1} + \frac{A_{26}}{R_2}\right)w_{mn}, \quad S_{mn}^{(2)} = \lambda_m\left(\frac{A_{11}}{R_1} + \frac{A_{12}}{R_2}\right)w_{mn}$$
$$S_{mn}^{(3)} = \mu_n\left(\frac{A_{12}}{R_1} + \frac{A_{22}}{R_2}\right)w_{mn}, \quad S_{mn}^{(2)} = -\lambda_m\left(\frac{A_{16}}{R_1} + \frac{A_{26}}{R_2}\right)w_{mn} \tag{4.73a – j}$$

From Eq. (4.72), Using Cramer's rule, the coefficients $F_{mn}^{(j)}$, $G_{mn}^{(j)}$ $(j = 3, 4)$ can be expressed as

$$\left(F_{mn}^{(3)},\ F_{mn}^{(4)},\ G_{mn}^{(3)},\ G_{mn}^{(4)}\right)=\left(\widetilde{F}_{mn}^{(3)},\ \widetilde{F}_{mn}^{(4)},\ \widetilde{G}_{mn}^{(3)},\ \widetilde{G}_{mn}^{(4)}\right)w_{mn} \tag{4.74}$$

where

$$\widetilde{\mathrm{F}}_{\mathrm{mn}}^{(3)}=\frac{\det(R_1)}{\det(R)},\quad \widetilde{\mathrm{F}}_{\mathrm{mn}}^{(4)}=\frac{\det(R_2)}{\det(R)},\quad \widetilde{\mathrm{G}}_{\mathrm{mn}}^{(3)}=\frac{\det(R_3)}{\det(R)},\quad \widetilde{\mathrm{G}}_{\mathrm{mn}}^{(4)}=\frac{\det(R_4)}{\det(R)} \tag{4.75a – d}$$

and

$$R_1=\begin{pmatrix} F_{mn}^{(3)} & L_{12}^{(m.n)} & L_{13}^{(m,n)} & L_{14}^{(m,n)} \\ F_{mn}^{(4)} & L_{11}^{(m,n)} & L_{14}^{(m,n)} & L_{13}^{(m,n)} \\ G_{mn}^{(3)} & L_{14}^{(m,n)} & L_{33}^{(m,n)} & L_{34}^{(m,n)} \\ G_{mn}^{(4)} & L_{13}^{(m,n)} & L_{34}^{(m,n)} & L_{33}^{(m,n)} \end{pmatrix},\quad R_2=\begin{pmatrix} L_{11}^{(m,n)} & F_{mn}^{(3)} & L_{13}^{(m,n)} & L_{14}^{(m,n)} \\ L_{12}^{(m.n)} & F_{mn}^{(4)} & L_{14}^{(m,n)} & L_{13}^{(m,n)} \\ L_{13}^{(m,n)} & G_{mn}^{(3)} & L_{33}^{(m,n)} & L_{34}^{(m,n)} \\ L_{14}^{(m,n)} & G_{mn}^{(4)} & L_{34}^{(m,n)} & L_{33}^{(m,n)} \end{pmatrix} \tag{4.76a, b}$$

$$R_3=\begin{pmatrix} L_{11}^{(m,n)} & L_{12}^{(m.n)} & F_{mn}^{(3)} & L_{14}^{(m,n)} \\ L_{12}^{(m.n)} & L_{11}^{(m,n)} & F_{mn}^{(4)} & L_{13}^{(m,n)} \\ L_{13}^{(m,n)} & L_{14}^{(m,n)} & G_{mn}^{(3)} & L_{34}^{(m,n)} \\ L_{14}^{(m,n)} & L_{13}^{(m,n)} & G_{mn}^{(4)} & L_{33}^{(m,n)} \end{pmatrix},\quad R_4=\begin{pmatrix} L_{11}^{(m,n)} & L_{12}^{(m.n)} & L_{13}^{(m,n)} & F_{mn}^{(3)} \\ L_{12}^{(m.n)} & L_{11}^{(m,n)} & L_{14}^{(m,n)} & F_{mn}^{(4)} \\ L_{13}^{(m,n)} & L_{14}^{(m,n)} & L_{33}^{(m,n)} & G_{mn}^{(3)} \\ L_{14}^{(m,n)} & L_{13}^{(m,n)} & L_{34}^{(m,n)} & G_{mn}^{(4)} \end{pmatrix} \tag{4.76c, d}$$

$$R=\begin{pmatrix} L_{11}^{(m,n)} & L_{12}^{(m.n)} & L_{13}^{(m,n)} & L_{14}^{(m,n)} \\ & L_{11}^{(m,n)} & L_{14}^{(m,n)} & L_{13}^{(m,n)} \\ & & L_{33}^{(m,n)} & L_{34}^{(m,n)} \\ \text{Symm.} & & & L_{33}^{(m,n)} \end{pmatrix} \tag{4.76e}$$

Finally, $F_{mn}^{(7)},\ F_{mn}^{(8)},\ G_{mn}^{(7)},\ G_{mn}^{(8)}$ are four constants which remain undetermined. These are determined as before from enforcing the in-plane static boundary conditions which are expressed mathematically as

$$\int_0^{L_t} N_{nn}dx_t=-\widetilde{N}_{nn}L_t \quad \begin{pmatrix} n=1,2 \\ t=2,1 \end{pmatrix} \quad \not\sum_{n,t} \tag{4.77}$$

$$\int_0^{L_t} N_{nt} dx_t = 0 \tag{4.78}$$

These conditions fulfill the boundary conditions, Eqs. (4.67a, b) in an average sense. To apply these conditions, N_{11}, N_{22}, and N_{12} are expressed in terms of displacements by Making use of Eqs. (2.86a) and (2.92a) coupled with the strain–displacement relationships, Eqs. (2.53)–(2.55) and Eqs. (2.69)–(2.71) gives the global stress resultants in terms of displacements as

$$\begin{aligned} N_{11} = A_{11}&\left\{\frac{\partial \xi_1}{\partial x_1} + \frac{1}{2}\left(\frac{\partial v_3}{\partial x_1}\right)^2 + \frac{\partial v_3^\circ}{\partial x_1}\frac{\partial v_3}{\partial x_1} - \frac{v_3}{R_1}\right\} \\ + A_{12}&\left\{\frac{\partial \xi_2}{\partial x_2} + \frac{1}{2}\left(\frac{\partial v_3}{\partial x_2}\right)^2 + \frac{\partial v_3^\circ}{\partial x_2}\frac{\partial v_3}{\partial x_2} - \frac{v_3}{R_2}\right\} \\ + A_{16}&\left\{\frac{\partial \xi_1}{\partial x_2} + \frac{\partial \xi_2}{\partial x_1} + \frac{\partial v_3}{\partial x_1}\frac{\partial v_3}{\partial x_2} + \frac{\partial v_3^\circ}{\partial x_1}\frac{\partial v_3}{\partial x_2} + \frac{\partial v_3}{\partial x_1}\frac{\partial v_3^\circ}{\partial x_2}\right\} \end{aligned} \tag{4.79a}$$

$$\begin{aligned} N_{22} = A_{22}&\left\{\frac{\partial \xi_2}{\partial x_2} + \frac{1}{2}\left(\frac{\partial v_3}{\partial x_2}\right)^2 + \frac{\partial v_3^\circ}{\partial x_2}\frac{\partial v_3}{\partial x_2} - \frac{v_3}{R_2}\right\} \\ + A_{12}&\left\{\frac{\partial \xi_1}{\partial x_1} + \frac{1}{2}\left(\frac{\partial v_3}{\partial x_1}\right)^2 + \frac{\partial v_3^\circ}{\partial x_1}\frac{\partial v_3}{\partial x_1} - \frac{v_3}{R_1}\right\} \\ + A_{26}&\left\{\frac{\partial \xi_2}{\partial x_1} + \frac{\partial \xi_1}{\partial x_2} + \frac{\partial v_3}{\partial x_2}\frac{\partial v_3}{\partial x_1} + \frac{\partial v_3^\circ}{\partial x_2}\frac{\partial v_3}{\partial x_1} + \frac{\partial v_3}{\partial x_2}\frac{\partial v_3^\circ}{\partial x_1}\right\} \end{aligned} \tag{4.79b}$$

$$\begin{aligned} N_{12} = A_{16}&\left\{\frac{\partial \xi_1}{\partial x_1} + \frac{1}{2}\left(\frac{\partial v_3}{\partial x_1}\right)^2 + \frac{\partial v_3^\circ}{\partial x_1}\frac{\partial v_3}{\partial x_1} - \frac{v_3}{R_1}\right\} \\ + A_{26}&\left\{\frac{\partial \xi_2}{\partial x_2} + \frac{1}{2}\left(\frac{\partial v_3}{\partial x_2}\right)^2 + \frac{\partial v_3^\circ}{\partial x_2}\frac{\partial v_3}{\partial x_2} - \frac{v_3}{R_2}\right\} \\ + A_{66}&\left\{\frac{\partial \xi_1}{\partial x_2} + \frac{\partial \xi_2}{\partial x_1} + \frac{\partial v_3}{\partial x_1}\frac{\partial v_3}{\partial x_2} + \frac{\partial v_3^\circ}{\partial x_1}\frac{\partial v_3}{\partial x_2} + \frac{\partial v_3}{\partial x_1}\frac{\partial v_3^\circ}{\partial x_2}\right\} \end{aligned} \tag{4.79c}$$

Substituting the expressions for v_3, v_3°, ξ_1, and ξ_2 into Eqs. (4.79a, 4.79b, and 4.79c) then substituting the result into the requirements for the in-plane static boundary conditions, Eqs. (4.77) and (4.78) and carrying out the indicated operations provides the relationships for the constants $F_{mn}^{(7)}, F_{mn}^{(8)}, G_{mn}^{(7)}, G_{mn}^{(8)}$ as

$$F_{mn}^{(7)} = -\frac{\lambda_m^2}{8}\left(w_{mn}^2 + 2w_{mn}\overset{\circ}{w}_{mn}\right) + \frac{A_{26}^2 - A_{22}A_{66}}{\Omega}\widetilde{N}_{11} + \frac{A_{12}A_{66} - A_{16}A_{26}}{\Omega}\widetilde{N}_{22} \tag{4.80a}$$

$$G_{mn}^{(8)} = -\frac{\mu_n^2}{8}\left(w_{mn}^2 + 2w_{mn}\overset{\circ}{w}_{mn}\right) + \frac{A_{12}A_{66} - A_{16}A_{26}}{\Omega}\widetilde{N}_{11} + \frac{A_{16}^2 - A_{11}A_{66}}{\Omega}\widetilde{N}_{22} \tag{4.80b}$$

where

$$\Omega = A_{11}A_{22}A_{66} - A_{12}^2A_{66} + 2A_{12}A_{16}A_{26} - A_{11}A_{26}^2 - A_{22}A_{16}^2 \tag{4.80c}$$

$F_{mn}^{(8)}$ and $G_{mn}^{(7)}$ are arbitrary and have no effect on the final post-buckling solution. These expressions are identical in the case of sandwich plates. With the first two equilibrium equations fulfilled, the third and fourth equations of equilibrium will be addressed in a similar manner to determine the expressions for η_1 and η_2 which were addressed in Chap. 3 (see Chap. 3, Eqs. (3.12a, b)-(3.17a–e)). Continuing in this manner with the first four equations of equilibrium and the sixth boundary condition fulfilled, the remaining unfulfilled equilibrium equation and boundary conditions will be retained in Hamilton's energy functional resulting in

$$\int_{t_0}^{t_1}\left\langle\int_0^{l_1}\int_0^{l_2}\left[A_{11}\left\{\frac{\partial\xi_1}{\partial x_1}\left(\frac{\partial^2 v_3}{\partial x_1^2} + \frac{\partial^2 \overset{\circ}{v}_3}{\partial x_1^2}\right) + \frac{1}{2}\left(\frac{\partial v_3}{\partial x_1}\right)^2\left(\frac{\partial^2 v_3}{\partial x_1^2} + \frac{\partial^2 \overset{\circ}{v}_3}{\partial x_1^2}\right) + \frac{\partial v_3}{\partial x_1}\frac{\partial \overset{\circ}{v}_3}{\partial x_1}\right.\right.\right.$$

$$\left.\left(\frac{\partial^2 v_3}{\partial x_1^2} + \frac{\partial^2 \overset{\circ}{v}_3}{\partial x_1^2}\right) + \frac{1}{R_1}\left[\frac{\partial\xi_1}{\partial x_1} + \frac{1}{2}\left(\frac{\partial v_3}{\partial x_1}\right)^2 + \frac{\partial v_3}{\partial x_1}\frac{\partial \overset{\circ}{v}_3}{\partial x_1} - v_3\left(\frac{\partial^2 v_3}{\partial x_1^2} + \frac{\partial^2 \overset{\circ}{v}_3}{\partial x_1^2}\right) - \frac{v_3}{R_1}\right]\right\}$$

$$+ A_{12}\left\{\frac{\partial\xi_1}{\partial x_1}\left(\frac{\partial^2 v_3}{\partial x_2^2} + \frac{\partial^2 \overset{\circ}{v}_3}{\partial x_2^2}\right) + \frac{\partial\xi_2}{\partial x_2}\left(\frac{\partial^2 v_3}{\partial x_1^2} + \frac{\partial^2 \overset{\circ}{v}_3}{\partial x_1^2}\right) + \frac{1}{2}\left(\frac{\partial v_3}{\partial x_1}\right)^2\left(\frac{\partial^2 v_3}{\partial x_2^2} + \frac{\partial^2 \overset{\circ}{v}_3}{\partial x_2^2}\right)\right.$$

$$+ \frac{1}{2}\left(\frac{\partial v_3}{\partial x_2}\right)^2\left(\frac{\partial^2 v_3}{\partial x_1^2} + \frac{\partial^2 \overset{\circ}{v}_3}{\partial x_1^2}\right) + \frac{\partial v_3}{\partial x_1}\frac{\partial \overset{\circ}{v}_3}{\partial x_1}\left(\frac{\partial^2 v_3}{\partial x_2^2} + \frac{\partial^2 \overset{\circ}{v}_3}{\partial x_2^2}\right) + \frac{\partial v_3}{\partial x_2}\frac{\partial \overset{\circ}{v}_3}{\partial x_2}\left(\frac{\partial^2 v_3}{\partial x_1^2} + \frac{\partial^2 \overset{\circ}{v}_3}{\partial x_1^2}\right)$$

$$+ \frac{1}{R_1}\left[\frac{\partial\xi_2}{\partial x_2} + \frac{1}{2}\left(\frac{\partial v_3}{\partial x_2}\right)^2 + \frac{\partial v_3}{\partial x_2}\frac{\partial \overset{\circ}{v}_3}{\partial x_2} - v_3\left(\frac{\partial^2 v_3}{\partial x_2^2} + \frac{\partial^2 \overset{\circ}{v}_3}{\partial x_2^2}\right) - \frac{v_3}{R_2}\right]$$

$$\left. + \frac{1}{R_2}\left[\frac{\partial\xi_1}{\partial x_1} + \frac{1}{2}\left(\frac{\partial v_3}{\partial x_1}\right)^2 + \frac{\partial v_3}{\partial x_1}\frac{\partial \overset{\circ}{v}_3}{\partial x_1} - v_3\left(\frac{\partial^2 v_3}{\partial x_1^2} + \frac{\partial^2 \overset{\circ}{v}_3}{\partial x_1^2}\right) - \frac{v_3}{R_1}\right]\right\}$$

$$+ A_{22}\left\{\frac{\partial\xi_2}{\partial x_2}\left(\frac{\partial^2 v_3}{\partial x_2^2} + \frac{\partial^2 \overset{\circ}{v}_3}{\partial x_2^2}\right) + \frac{1}{2}\left(\frac{\partial v_3}{\partial x_2}\right)^2\left(\frac{\partial^2 v_3}{\partial x_2^2} + \frac{\partial^2 \overset{\circ}{v}_3}{\partial x_2^2}\right) + \frac{\partial v_3}{\partial x_2}\frac{\partial \overset{\circ}{v}_3}{\partial x_2}\left(\frac{\partial^2 v_3}{\partial x_2^2} + \frac{\partial^2 \overset{\circ}{v}_3}{\partial x_2^2}\right)\right.$$

$$+\frac{1}{R_2}\left[\frac{\partial \xi_2}{\partial x_2}+\frac{1}{2}\left(\frac{\partial v_3}{\partial x_2}\right)^2+\frac{\partial v_3}{\partial x_2}\frac{\partial \overset{\circ}{v}_3}{\partial x_2}-v_3\left(\frac{\partial^2 v_3}{\partial x_2^2}+\frac{\partial^2 \overset{\circ}{v}_3}{\partial x_2^2}\right)-\frac{v_3}{R_2}\right]\Bigg\}$$

$$+2A_{66}\Bigg\{\frac{\partial \xi_1}{\partial x_2}\left(\frac{\partial^2 v_3}{\partial x_1\partial x_2}+\frac{\partial^2 \overset{\circ}{v}_3}{\partial x_1\partial x_2}\right)+\frac{\partial \xi_2}{\partial x_1}\left(\frac{\partial^2 v_3}{\partial x_1\partial x_2}+\frac{\partial^2 \overset{\circ}{v}_3}{\partial x_1\partial x_2}\right)+\frac{\partial v_3}{\partial x_1}\frac{\partial v_3}{\partial x_2}$$

$$\left(\frac{\partial^2 v_3}{\partial x_1\partial x_2}+\frac{\partial^2 \overset{\circ}{v}_3}{\partial x_1\partial x_2}\right)+\frac{\partial \overset{\circ}{v}_3}{\partial x_1}\frac{\partial v_3}{\partial x_2}\left(\frac{\partial^2 v_3}{\partial x_1\partial x_2}+\frac{\partial^2 \overset{\circ}{v}_3}{\partial x_1\partial x_2}\right)+\frac{\partial v_3}{\partial x_1}\frac{\partial \overset{\circ}{v}_3}{\partial x_2}$$

$$\left(\frac{\partial^2 v_3}{\partial x_1\partial x_2}+\frac{\partial^2 \overset{\circ}{v}_3}{\partial x_1\partial x_2}\right)\Bigg\}+A_{16}\Bigg\{\frac{\partial \xi_1}{\partial x_2}\left(\frac{\partial^2 v_3}{\partial x_1^2}+\frac{\partial^2 \overset{\circ}{v}_3}{\partial x_1^2}\right)+\frac{\partial \xi_2}{\partial x_1}\left(\frac{\partial^2 v_3}{\partial x_1^2}+\frac{\partial^2 \overset{\circ}{v}_3}{\partial x_1^2}\right)$$

$$+\frac{\partial v_3}{\partial x_1}\frac{\partial v_3}{\partial x_2}\left(\frac{\partial^2 v_3}{\partial x_1^2}+\frac{\partial^2 \overset{\circ}{v}_3}{\partial x_1^2}\right)+\frac{\partial \overset{\circ}{v}_3}{\partial x_1}\frac{\partial v_3}{\partial x_2}\left(\frac{\partial^2 v_3}{\partial x_1^2}+\frac{\partial^2 \overset{\circ}{v}_3}{\partial x_1^2}\right)+\frac{\partial v_3}{\partial x_1}\frac{\partial \overset{\circ}{v}_3}{\partial x_2}\left(\frac{\partial^2 v_3}{\partial x_1^2}+\frac{\partial^2 \overset{\circ}{v}_3}{\partial x_1^2}\right)$$

$$+2\frac{\partial \xi_1}{\partial x_1}\left(\frac{\partial^2 v_3}{\partial x_1\partial x_2}+\frac{\partial^2 \overset{\circ}{v}_3}{\partial x_1\partial x_2}\right)+\left(\frac{\partial v_3}{\partial x_1}\right)^2\left(\frac{\partial^2 v_3}{\partial x_1\partial x_2}+\frac{\partial^2 \overset{\circ}{v}_3}{\partial x_1\partial x_2}\right)+2\frac{\partial v_3}{\partial x_1}\frac{\partial \overset{\circ}{v}_3}{\partial x_1}$$

$$\left(\frac{\partial^2 v_3}{\partial x_1\partial x_2}+\frac{\partial^2 \overset{\circ}{v}_3}{\partial x_1\partial x_2}\right)+\frac{1}{R_1}\left[\frac{\partial \xi_1}{\partial x_2}+\frac{\partial \xi_2}{\partial x_1}+\frac{\partial v_3}{\partial x_1}\frac{\partial v_3}{\partial x_2}+\frac{\partial \overset{\circ}{v}_3}{\partial x_1}\frac{\partial v_3}{\partial x_2}+\frac{\partial v_3}{\partial x_1}\frac{\partial \overset{\circ}{v}_3}{\partial x_2}\right.$$

$$\left.-2v_3\left(\frac{\partial^2 v_3}{\partial x_1\partial x_2}+\frac{\partial^2 \overset{\circ}{v}_3}{\partial x_1\partial x_2}\right)\right]\Bigg\}+A_{26}\Bigg\{\frac{\partial \xi_2}{\partial x_1}\left(\frac{\partial^2 v_3}{\partial x_2^2}+\frac{\partial^2 \overset{\circ}{v}_3}{\partial x_2^2}\right)+\frac{\partial \xi_1}{\partial x_2}\left(\frac{\partial^2 v_3}{\partial x_2^2}+\frac{\partial^2 \overset{\circ}{v}_3}{\partial x_2^2}\right)$$

$$+\frac{\partial v_3}{\partial x_2}\frac{\partial v_3}{\partial x_1}\left(\frac{\partial^2 v_3}{\partial x_2^2}+\frac{\partial^2 \overset{\circ}{v}_3}{\partial x_2^2}\right)\frac{\partial \overset{\circ}{v}_3}{\partial x_2}\frac{\partial v_3}{\partial x_1}\left(\frac{\partial^2 v_3}{\partial x_2^2}+\frac{\partial^2 \overset{\circ}{v}_3}{\partial x_2^2}\right)+\frac{\partial v_3}{\partial x_2}\frac{\partial \overset{\circ}{v}_3}{\partial x_1}\left(\frac{\partial^2 v_3}{\partial x_2^2}+\frac{\partial^2 \overset{\circ}{v}_3}{\partial x_2^2}\right)$$

$$+2\frac{\partial \xi_2}{\partial x_2}\left(\frac{\partial^2 v_3}{\partial x_1\partial x_2}+\frac{\partial^2 \overset{\circ}{v}_3}{\partial x_1\partial x_2}\right)+\left(\frac{\partial v_3}{\partial x_2}\right)^2\left(\frac{\partial^2 v_3}{\partial x_2\partial x_1}+\frac{\partial^2 \overset{\circ}{v}_3}{\partial x_2\partial x_1}\right)+2\frac{\partial v_3}{\partial x_2}\frac{\partial \overset{\circ}{v}_3}{\partial x_2}$$

$$\left(\frac{\partial^2 v_3}{\partial x_1\partial x_2}+\frac{\partial^2 \overset{\circ}{v}_3}{\partial x_1\partial x_2}\right)+\frac{1}{R_2}\left[\frac{\partial \xi_2}{\partial x_1}+\frac{\partial \xi_1}{\partial x_2}+\frac{\partial v_3}{\partial x_1}\frac{\partial v_3}{\partial x_2}+\frac{\partial \overset{\circ}{v}_3}{\partial x_2}\frac{\partial v_3}{\partial x_1}+\frac{\partial v_3}{\partial x_2}\frac{\partial \overset{\circ}{v}_3}{\partial x_1}\right.$$

$$\left.-2v_3\left(\frac{\partial^2 v_3}{\partial x_1\partial x_2}+\frac{\partial^2 \overset{\circ}{v}_3}{\partial x_1\partial x_2}\right)\right]\Bigg\}-F_{11}\frac{\partial^4 v_3}{\partial x_1^4}-2(F_{12}+2F_{66})\frac{\partial^4 v_3}{\partial x_1^2\partial x_2^2}-F_{22}\frac{\partial^4 v_3}{\partial x_2^4}$$

$$-4F_{16}\frac{\partial^4 v_3}{\partial x_1^3\partial x_2}-4F_{26}\frac{\partial^4 v_3}{\partial x_1\partial x_2^3}+2\overline{h}\overline{K}^2\overline{G}_{13}\left(1+\frac{h}{2\overline{h}}\right)\left\{\frac{1}{\overline{h}}\frac{\partial \eta_1}{\partial x_1}+\left(1+\frac{C_1}{\overline{h}}\right)\frac{\partial^2 v_3}{\partial x_1^2}\right\}$$

$$+2\overline{h}\overline{K}^2\overline{G}_{23}\left(1+\frac{h}{2\overline{h}}\right)\left\{\frac{1}{\overline{h}}\frac{\partial \eta_2}{\partial x_2}+\left(1+\frac{C_1}{\overline{h}}\right)\frac{\partial^2 v_3}{\partial x_2^2}\right\}-q_3\Bigg]\delta v_3 dx_1 dx_2\Bigg\rangle dt$$

$$+\int_{t_0}^{t_1}\left\langle\int_0^{l_2}\left[(N_{11}+\widetilde{N}_{11})\delta\xi_1+N_{12}\delta\xi_2+L_{11}\delta\eta_1+L_{12}\delta\eta_2+M_{11}\delta\left(\frac{\partial v_3}{\partial x_1}\right)\right]\Bigg|_0^{l_1}dx_2\right\rangle dt$$

$$+\int_{t_0}^{t_1}\left\langle\int_0^{l_1}\left[(N_{22}+\widetilde{N}_{22})\delta\xi_2+N_{12}\delta\xi_1+L_{22}\delta\eta_2+L_{12}\delta\eta_1+M_{22}\delta\left(\frac{\partial v_3}{\partial x_2}\right)\right]\Bigg|_0^{l_2}dx_1\right\rangle dt$$

$$=0 \tag{4.81}$$

Substituting the expressions for v_3, v_3°, ξ_1, ξ_2, η_1, η_2, into Eq. (4.81) and carrying out the indicated integrations results in the governing post-buckling solution for angle-ply laminated doubly curved sandwich structures resulting in a nonlinear algebraic equation expressed in terms of the modal amplitudes as

$$P_{mn}^{(5)}\left(w_{mn}^2+2w_{mn}w_{mn}^\circ\right)\left(w_{mn}+w_{mn}^\circ\right)+P_{mn}^{(4)}\left(w_{mn}+w_{mn}^\circ\right)w_{mn}+P_{mn}^{(3)}\left(w_{mn}^2+2w_{mn}w_{mn}^\circ\right)+$$
$$P_{mn}^{(2)}\left(w_{mn}+w_{mn}^\circ\right)+P_{mn}^{(1)}w_{mn}+P_{mn}^{(0)}+q_{mn}=0 \tag{4.82}$$

where

$$P_{mn}=-\left(N_{51}\widetilde{F}_{mn}^{(1)}+N_{52}\widetilde{F}_{mn}^{(2)}+N_{53}\widetilde{G}_{mn}^{(1)}+N_{54}\widetilde{G}_{mn}^{(2)}\right) \tag{4.83}$$

$$\begin{aligned}P_{mn}^{(4)}=&\frac{16\Delta_n^m}{9\lambda_m\mu_nL_1L_2}\left(N_{40}+N_{41}\widetilde{F}_{mn}^{(3)}+N_{42}\widetilde{F}_{mn}^{(4)}+N_{43}\widetilde{G}_{mn}^{(3)}+N_{44}\widetilde{G}_{mn}^{(4)}\right)+\\&+\frac{16\Delta_n^m}{3\lambda_m\mu_nL_1L_2}\left(N_{45}\widetilde{F}_{mn}^{(2)}+N_{46}\widetilde{G}_{mn}^{(1)}+N_{47}\left(\widetilde{F}_{mn}^{(1)}+\widetilde{F}_{mn}^{(5)}\right)+N_{48}\left(\widetilde{G}_{mn}^{(2)}+\widetilde{G}_{mn}^{(6)}\right)\right)\end{aligned} \tag{4.84}$$

$$\begin{aligned}P_{mn}^{(3)}=&\frac{-8\Delta_n^m}{3\lambda_m\mu_nL_1L_2}\left(N_{31}\widetilde{F}_{mn}^{(1)}+N_{32}\widetilde{F}_{mn}^{(2)}+N_{33}\widetilde{G}_{mn}^{(1)}+N_{34}\widetilde{G}_{mn}^{(2)}\right)\\&+\frac{4\Delta_n^m}{\lambda_m\mu_nL_1L_2}\left(N_{35}\widetilde{F}_{mn}^{(3)}+N_{36}\widetilde{F}_{mn}^{(4)}+N_{37}\widetilde{G}_{mn}^{(3)}+N_{38}\widetilde{G}_{mn}^{(4)}\right)\end{aligned} \tag{4.85}$$

$$P_{mn}^{(2)}=-\left(\lambda_m^2N_{11}+\mu_n^2N_{22}\right) \tag{4.86}$$

$$\begin{aligned}P_{mn}^{(1)}=&\lambda_m\left(\frac{A_{11}}{R_1}+\frac{A_{12}}{R_2}\right)\widetilde{F}_{mn}^{(4)}+\mu_n\left(\frac{A_{12}}{R_1}+\frac{A_{22}}{R_2}\right)\widetilde{G}_{mn}^{(3)}+\mu_n\left(\frac{A_{16}}{R_1}+\frac{A_{26}}{R_2}\right)\widetilde{F}_{mn}^{(3)}\\&+\lambda_m\left(\frac{A_{16}}{R_1}+\frac{A_{26}}{R_2}\right)\widetilde{G}_{mn}^{(4)}+\left(\frac{A_{11}}{R_1^2}+\frac{A_{22}}{R_2^2}+\frac{2A_{12}}{R_1R_2}\right)+F_{11}\lambda_m^4+2(F_{12}+2F_{66})\lambda_m^2\mu_n^2\\&+F_{22}\mu_n^4+d_1a\lambda_m\left(\widetilde{H}_{mn}^{(1)}+a\lambda_m\right)+d_2a\mu_n\left(\widetilde{I}_{mn}^{(2)}+a\mu_n\right)\end{aligned} \tag{4.87}$$

$$P_{mn}^{(0)}=-\frac{4\Delta_n^m}{\lambda_m\mu_nL_1L_2}\left(\widetilde{P}_{11}\widetilde{N}_{11}+\widetilde{P}_{22}\widetilde{N}_{22}\right) \tag{4.88}$$

$$\widetilde{P}_{11} = \left(\frac{A_{11}}{R_1} + \frac{A_{12}}{R_2}\right)\left(\frac{A_{26}^2 - A_{22}A_{66}}{\Omega}\right) + \left(\frac{A_{12}}{R_1} + \frac{A_{22}}{R_2}\right)\left(\frac{A_{12}A_{66} - A_{16}A_{66}}{\Omega}\right) -$$
$$\lambda_m\left(1 + \left(\frac{A_{11}A_{26}^2 - A_{11}A_{22}A_{66}}{\Omega}\right) + \left(\frac{A_{12}^2A_{66}^2 - A_{12}A_{16}A_{26}}{\Omega}\right)\right)\widetilde{F}_{mn}^{(4)} +$$
$$\mu_n\left(\left(\frac{A_{22}A_{12}A_{66} - A_{22}A_{16}A_{26}}{\Omega}\right) + \left(\frac{A_{12}A_{26}^2 - A_{12}A_{22}A_{66}}{\Omega}\right)\right)\widetilde{G}_{mn}^{(3)} +$$
$$\left(\lambda_m\widetilde{G}_{mn}^{(4)} + \mu_n\widetilde{F}_{mn}^{(3)}\right)\left(\left(\frac{A_{16}A_{26}^2 - A_{16}A_{22}A_{66}}{\Omega}\right) + \left(\frac{A_{26}A_{12}A_{66} - A_{16}A_{26}^2}{\Omega}\right)\right) \tag{4.89}$$

$$\widetilde{P}_{22} = \left(\frac{A_{11}}{R_1} + \frac{A_{12}}{R_2}\right)\left(\frac{A_{12}A_{66} - A_{16}A_{26}}{\Omega}\right) + \left(\frac{A_{12}}{R_1} + \frac{A_{22}}{R_2}\right)\left(\frac{A_{16}^2 - A_{11}A_{66}}{\Omega}\right) -$$
$$\lambda_m\left(\left(\frac{A_{11}A_{12}A_{66} - A_{11}A_{16}A_{26}}{\Omega}\right) + \left(\frac{A_{12}A_{16}^2 - A_{12}A_{11}A_{66}}{\Omega}\right)\right)\widetilde{F}_{mn}^{(4)} +$$
$$\mu_n\left(1 + \left(\frac{A_{22}A_{16}^2 - A_{22}A_{11}A_{66}}{\Omega}\right) + \left(\frac{A_{12}^2A_{66} - A_{12}A_{16}A_{26}}{\Omega}\right)\right)\widetilde{G}_{mn}^{(3)} +$$
$$\left(\lambda_m\widetilde{G}_{mn}^{(4)} + \mu_n\widetilde{F}_{mn}^{(3)}\right)\left(\left(\frac{A_{16}A_{12}A_{66} - A_{16}^2A_{26}}{\Omega}\right) + \left(\frac{A_{16}^2A_{26} - A_{11}A_{26}A_{66}}{\Omega}\right)\right) \tag{4.90}$$

and

$$N_{51} = \lambda_m^3A_{11} + \lambda_m\mu_n^2A_{12}, \quad N_{52} = \lambda_m^2\mu_nA_{16} + \mu_n^3A_{26}, \quad N_{53} = \lambda_m^3A_{16} + \lambda_m\mu_n^2A_{26}$$
$$N_{54} = \mu_n^3A_{22} + \lambda_m^2\mu_nA_{12}$$
$$N_{47} = N_{48} = \lambda_m^2A_{16}\widetilde{F}_{mn}^{(3)} + \lambda_m\mu_nA_{26}\widetilde{G}_{mn}^{(4)} + \left(\lambda_m\mu_n\widetilde{F}_{mn}^{(4)} + \lambda_m^2\widetilde{G}_{mn}^{(3)}\right)A_{66}$$
$$N_{46} = \lambda_m\mu_nA_{22}\widetilde{G}_{mn}^{(4)} + \lambda_m^2A_{12}\widetilde{F}_{mn}^{(3)} + \left(\lambda_m\mu_n\widetilde{F}_{mn}^{(4)} + \lambda_m^2\widetilde{G}_{mn}^{(3)}\right)A_{26}$$
$$N_{45} = \lambda_m^2A_{11}\widetilde{F}_{mn}^{(3)} + \lambda_m\mu_nA_{12}\widetilde{G}_{mn}^{(4)} + \left(\lambda_m\mu_n\widetilde{F}_{mn}^{(4)} + \lambda_m^2\widetilde{G}_{mn}^{(3)}\right)A_{16}$$
$$N_{44} = \lambda_m^3A_{16} + \frac{3}{2}\lambda_m\mu_n^2A_{26}, \quad N_{43} = \mu_n^3A_{22} + \lambda_m^2\mu_nA_{12} + \frac{1}{2}\lambda_m^2\mu_nA_{66}$$
$$N_{42} = \lambda_m^3A_{11} + \lambda_m\mu_n^2A_{12} + \frac{1}{2}\lambda_m\mu_n^2A_{66}, \quad N_{41} = \mu_n^3A_{26} + \frac{3}{2}\lambda_m^2\mu_nA_{16}$$
$$N_{40} = \lambda_m^2\left(\frac{A_{11}}{R_1} + \frac{A_{12}}{R_2}\right) + \mu_n^2\left(\frac{A_{22}}{R_2} + \frac{A_{12}}{R_1}\right)$$
$$N_{38} = \lambda_m\left\{\frac{1}{3}\left(2\lambda_mA_{16}\widetilde{F}_{mn}^{(5)} + 2\mu_nA_{26}\left(\widetilde{G}_{mn}^{(2)} + \widetilde{G}_{mn}^{(6)}\right) + 2\mu_nA_{66}\widetilde{F}_{mn}^{(2)} - \lambda_m^2A_{16}\right)\right.$$
$$\left.-2\lambda_mA_{16}\widetilde{F}_{mn}^{(1)} - 2\lambda_mA_{66}\widetilde{G}_{mn}^{(1)} - \frac{A_{16}\lambda_m^2}{8} - \frac{A_{26}\mu_n^2}{8}\right\}$$

$$N_{37} = \mu_n \left\{ \frac{1}{3}\left(2\mu_n A_{22}\widetilde{G}^{(6)}_{mn} + 2\lambda_m A_{12}\left(\widetilde{F}^{(1)}_{mn} + \widetilde{F}^{(5)}_{mn}\right) + 2\lambda_m A_{26}\widetilde{G}^{(1)}_{mn} - \mu_n^2 A_{22}\right)\right.$$

$$\left. -2\mu_n A_{22}\widetilde{G}^{(2)}_{mn} - 2\mu_n A_{26}\widetilde{F}^{(2)}_{mn} + \frac{A_{22}\mu_n^2}{8} + \frac{A_{12}\lambda_m^2}{8}\right\}$$

$$N_{36} = \lambda_m \left\{ \frac{1}{3}\left(2\lambda_m A_{11}\widetilde{F}^{(5)}_{mn} + 2\mu_n A_{12}\left(\widetilde{G}^{(2)}_{mn} + \widetilde{G}^{(6)}_{mn}\right) + 2\mu_n A_{16}\widetilde{F}^{(2)}_{mn} - \lambda_m^2 A_{11}\right)\right.$$

$$\left. -2\lambda_m A_{11}\widetilde{F}^{(1)}_{mn} - 2\lambda_m A_{16}\widetilde{G}^{(1)}_{mn} + \frac{A_{11}\lambda_m^2}{8} + \frac{A_{12}\mu_n^2}{8}\right\}$$

$$N_{35} = \frac{\mu_n}{3}\left\{2\lambda_m A_{16}\left(\widetilde{F}^{(1)}_{mn} + \widetilde{F}^{(5)}_{mn}\right) + 2\mu_n A_{26}\widetilde{G}^{(6)}_{mn} + 2\lambda_m A_{66}\widetilde{G}^{(1)}_{mn} - \mu_n^2 A_{26}\right\}$$

$$-\mu_n\left\{\frac{A_{16}\lambda_m^2}{8} + \frac{A_{26}\mu_n^2}{8} + 2\mu_n\left(A_{26}\widetilde{G}^{(2)}_{mn} + A_{66}\widetilde{F}^{(2)}_{mn}\right)\right\}$$

$$N_{34} = \mu_n\left(\frac{A_{12}}{R_1} + \frac{A_{22}}{R_2}\right), \quad N_{22} = N_{12} = \eta_1 = \eta_2 = M_{22} = v_3 = 0,$$

$$N_{32} = \mu_n\left(\frac{A_{16}}{R_1} + \frac{A_{26}}{R_2}\right)$$

$$N_{31} = \lambda_m\left(\frac{A_{11}}{R_1} + \frac{A_{12}}{R_2}\right), \quad N_{33} = \lambda_m\left(\frac{A_{16}}{R_1} + \frac{A_{26}}{R_2}\right) \tag{4.91a – t}$$

To determine the post-buckling behavior, Eq. (4.82) is put in a more suitable form which is conducive to generating results. Algebraically Eq. (4.82) can be put in the following form.

$$P^{(5)}_{mn} w^3_{mn} + \left(3P^{(5)}_{mn} w^\circ_{mn} + P^{(4)}_{mn} + P^{(3)}_{mn}\right) w^2_{mn} + \left(2P^{(5)}_{mn}\left(w^\circ_{mn}\right)^2 + P^{(4)}_{mn} w^\circ_{mn} + 2P^{(3)}_{mn} w^\circ_{mn} + \right.$$
$$\left. P^{(2)}_{mn} + P^{(1)}_{mn}\right) w_{mn} + P^{(2)}_{mn} w^\circ_{mn} + P^{(0)}_{mn} + q_{mn} = 0 \tag{4.92}$$

which can be further expressed as

$$L_1 w^3_{mn} + L_2 w^2_{mn} + L_3 w_{mn} + L_4 = 0 \tag{4.93}$$

where

$$L_1 = P^{(5)}_{mn} \tag{4.94a}$$

$$L_2 = 3P^{(5)}_{mn} w^{\circ}_{mn} + P^{(4)}_{mn} + P^{(3)}_{mn} \tag{4.94b}$$

$$L_3 = 2P^{(5)}_{mn} \left(w^{\circ}_{mn}\right)^2 + P^{(4)}_{mn} w^{\circ}_{mn} + 2P^{(3)}_{mn} w^{\circ}_{mn} + P^{(2)}_{mn} + P^{(1)}_{mn} \tag{4.94c}$$

$$L_4 = P^{(2)}_{mn} w^{\circ}_{mn} + P^{(0)}_{mn} + q_{mn} \tag{4.94d}$$

Equation (4.93) can be solved via Newton's method to solve for the various equilibrium configurations of an anisotropic laminated doubly curved sandwich panel with an incompressible (weak/soft) core. A more convenient form is to introduce the following Parameter δas

$$\delta = w_{mn} + w^{\circ}_{mn} \quad \Rightarrow w_{mn} = \delta - w^{\circ}_{mn} \tag{4.95}$$

Substituting Eq. (4.95) into Eq. (4.93) gives

$$L_1 \left(\delta - w^{\circ}_{mn}\right)^3 + L_2 \left(\delta - w^{\circ}_{mn}\right)^2 + L_3 \left(\delta - w^{\circ}_{mn}\right) + L_4 = 0 \tag{4.96}$$

Expanding, combining like terms, and simplifying gives

$$L_1 \delta^3_{mn} + \left(L_2 - 3L_1 w^{\circ}_{mn}\right)\delta^2_{mn} + \left(L_3 - 2L_2 w^{\circ}_{mn} - L_1 \left(w^{\circ}_{mn}\right)^2\right)\delta_{mn} + L_4 - L_1 \left(w^{\circ}_{mn}\right)^3 + L_2 \left(w^{\circ}_{mn}\right)^2 - L_3 w^{\circ}_{mn} = 0 \tag{4.97}$$

Solving Eq. (4.97) via Newton's method gives the compressive uniaxial load vs δ_{mn} (amplitude of deflection plus amplitude of imperfection).

4.5.3 Numerical Results and Discussion

With the theory in hand, several results are presented to better understand the effect of both the geometrical and material parameters on the post-buckling behavior of sandwich panels. For all of the numerical simulations, unless otherwise specified, Tables 4.5 and 4.6 contain the geometrical and material properties for the face sheets and the core for each of the figures. In addition, unless otherwise specified for the following numerical illustrations ($m = n = 1$).

Figure 4.11 depicts the effect of the geometric imperfections on the post-buckling behavior of a cylindrical sandwich panel. This figure reveals all of the stable and unstable directional loading paths that are theoretically possible for both the primary and secondary branches. The effect of the geometric imperfections shows the realistic direction the loading would take for compressive edge loading. This figure also shows that the larger imperfections whether negative or positive cause the

Table 4.5 The geometrical and material properties for the face sheets

h_f (mm)	L_1 (mm)	L_2 (mm)	E_1 (GPa)	E_2 (GPa)	G_{12} (GPa)	ν_{12}
0.02	609.6 (24in)	609.6 (24in)	180.987	10.342	7.239	0.28

Table 4.6 The geometrical material properties for the core

$\bar{h}$in (mm)	$\overline{G}_{13}$(MPa)	$\overline{G}_{23}$(MPa)
12.7 (0.5in)	1.437	0.651

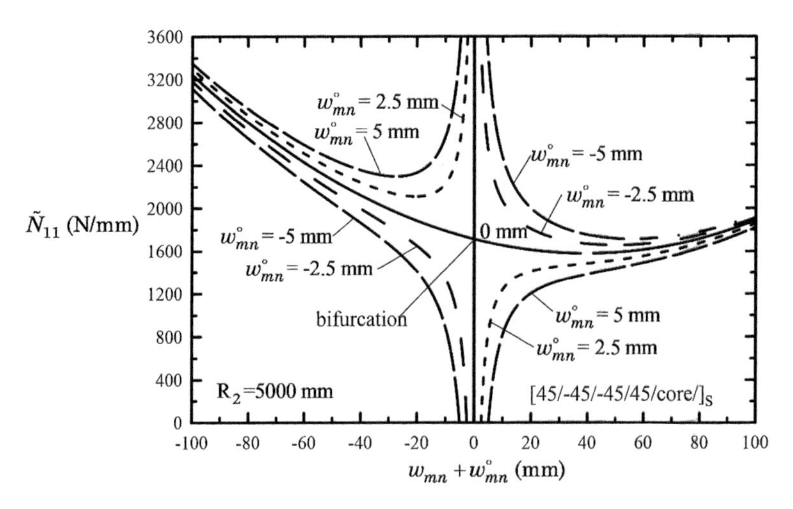

Fig. 4.11 The effect of geometric imperfections on the compressive uniaxial edge loading vs. amplitude of deflection for a four-layered symmetrical cylindrical sandwich panel

loading path proceed away from the ideal or perfect loading path shown as a solid line. The smaller the geometric imperfection, the closer the loading path is to the ideal or perfect one.

Figure 4.12 displays the counterpart of Fig. 4.11 in the compressive load-end shortening plane where the stable and unstable paths are displayed, while the presence of geometric imperfections shows the realistic behavior of the structure under a compressive edge loading. In Fig. 4.13, the effect of curvature on the post-buckling behavior of a cylindrical sandwich is shown. The typical response is seen where an increase in curvature results in a marginally higher buckling bifurcation point with an additional possibility of a snap-through-type behavior.

Figure 4.14 presents the effect of the curvature on the end-shortening of a cylindrical sandwich panel of the given stacking sequence in the facings. The typical one-to-one or linear behavior between the compressive edge loading and the end-shortening is seen up to the buckling bifurcation point beyond which varying amounts of snap-through behavior are seen depending on the amount of curvature

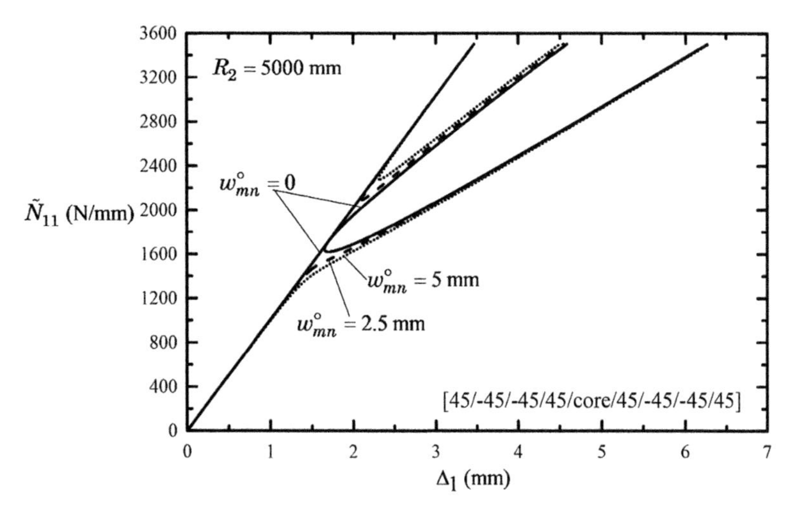

Fig. 4.12 The counterpart of Fig. 4.11 in the compressive edge load-end shortening plane for various geometric imperfections of a cylindrical sandwich panel

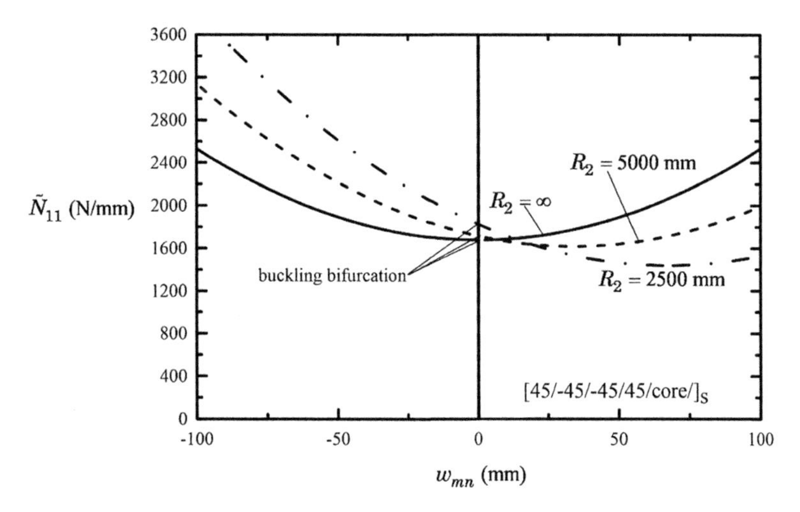

Fig. 4.13 The effect of the curvature on the compressive uniaxial edge loading vs. amplitude of deflection for the given layup of a cylindrical sandwich panel

present in the structure. Figure 4.15 highlights the effect of the ply angle for the given layup on the uniaxial compressive edge loading for a cylindrical sandwich panel. Similar behavior is seen as in the case of a flat panel. The buckling bifurcation rises with an increase of ply angle up to 45 degrees. Additionally, at a ply angle of 45 degrees there seems to be a diminishing capacity to sustain the compressive edge loading in contrast to the 0-degree layup where the load-carrying capacity of the

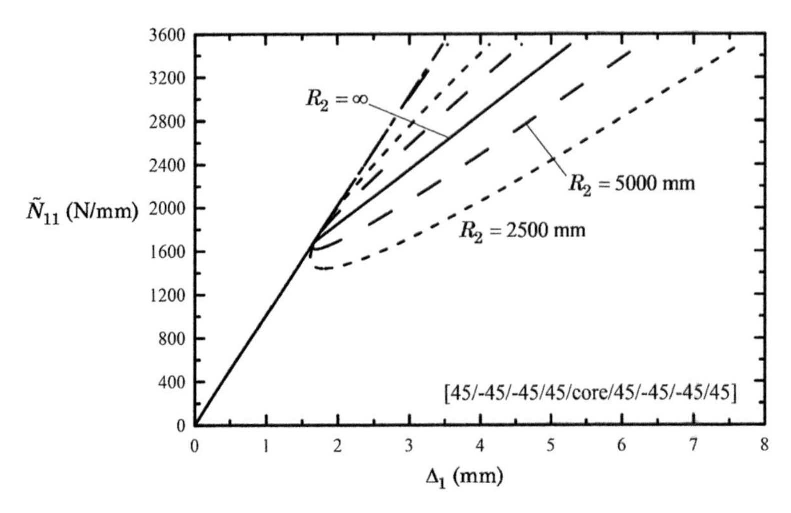

Fig. 4.14 The effect of imperfections on the end-shortening of a cylindrical sandwich panel for various curvatures under uniaxial compressive edge loading

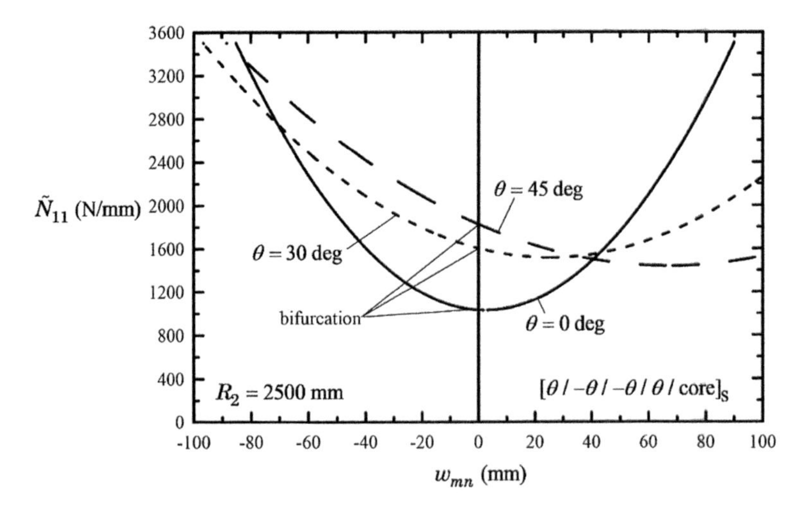

Fig. 4.15 The effect of the ply angle on the compressive uniaxial edge loading vs. amplitude of deflection for a cylindrical sandwich panel

structure seems to rise beyond the buckling bifurcation point. In Fig. 4.16, the effects of the geometric imperfections on a doubly curved sandwich panel are shown. The results reveal that for the case of a doubly curved sandwich panel, there is extraordinarily little effect on the ideal compressive edge load-displacement interaction of the structure. There is only a marginal effect with an increase in the amount of geometric imperfections. Figure 4.17 depicts the effects of the various magnitudes of curvature on the amplitude of deflection for a doubly curved sandwich panel for the

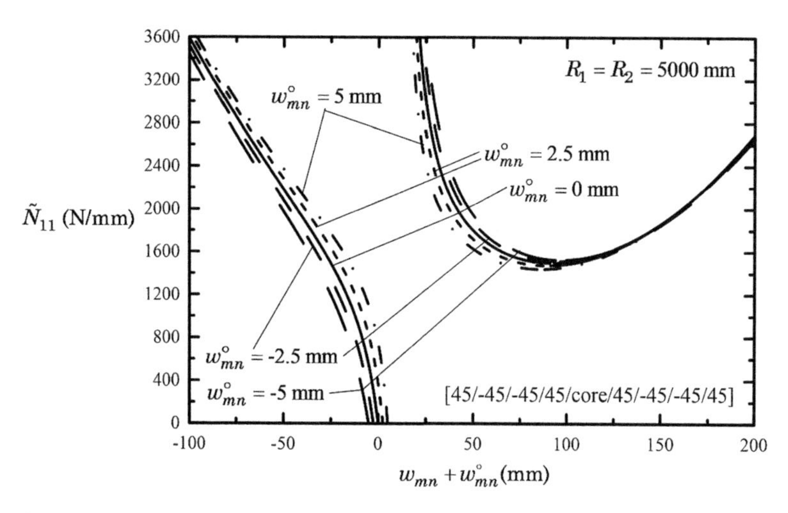

Fig. 4.16 The effect of geometric imperfections on the compressive uniaxial edge loading vs. amplitude of deflection for a doubly curved sandwich panel

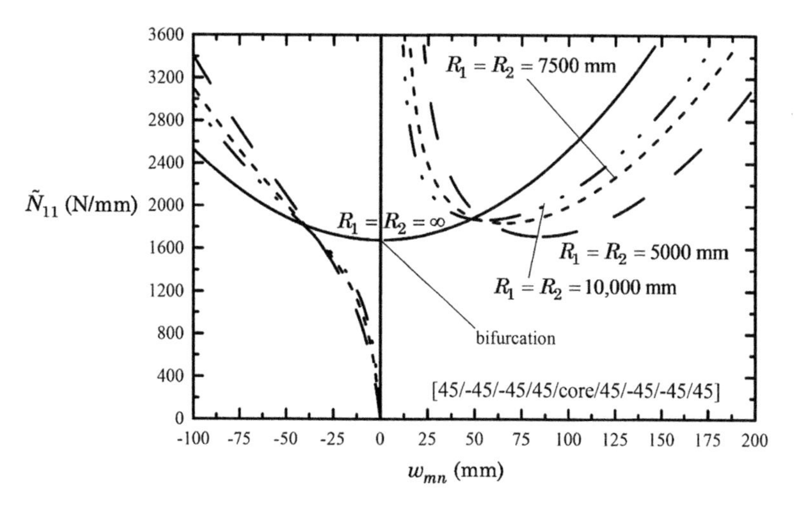

Fig. 4.17 The effect of the curvature on the compressive uniaxial edge loading vs. amplitude of deflection for a doubly curved sandwich panel

given layup. Considering the stable equilibrium paths, a marginal increase in the load-carrying capacity of the structure seems to follow an increase in curvature. Figure 4.18 illustrates the effect of the various ply angles for the given layup on the uniaxial compressive edge loading of a doubly curved sandwich panel. Along the stable loading direction, it is apparent that the ply angle configuration of 45 degrees provides the larger load-carrying capacity of the panel. In Fig. 4.19, the effect of the transverse pressure on the amplitude of deflection with an applied fixed-edge loading

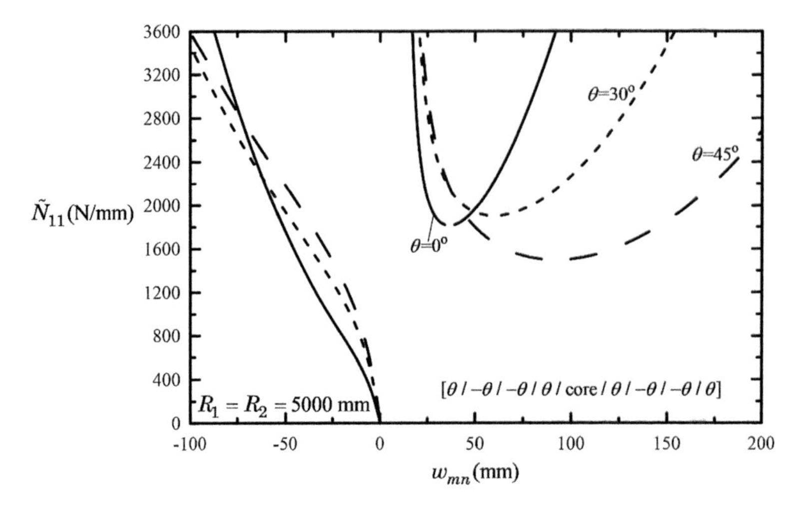

Fig. 4.18 The effect of the ply angle on the compressive uniaxial edge loading vs. amplitude of deflection for a doubly curved sandwich panel

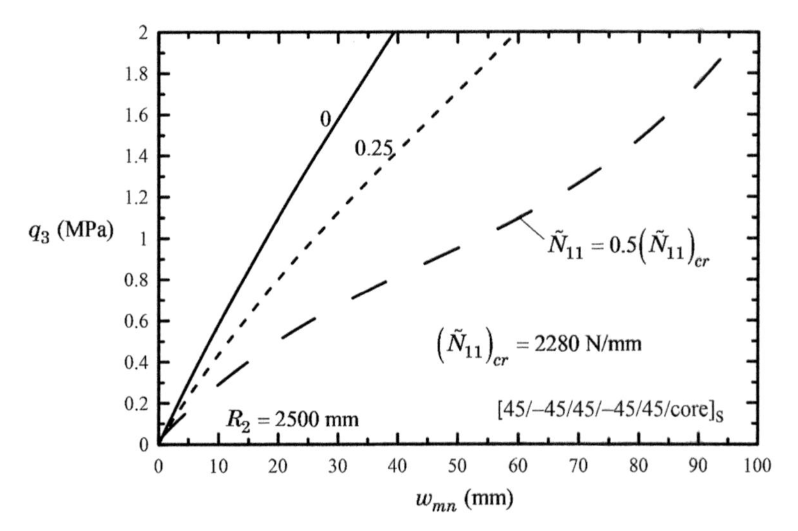

Fig. 4.19 Transverse pressure vs amplitude of deflection for various amounts of compressive uniaxial edge loading for cylindrical sandwich panel

as a percentage of the critical load for a cylindrical sandwich panel is displayed. With the transverse pressure loading present, as the compressive edge loading reaches within the vicinity of the critical load, the implications for the load-carrying capacity of the structure become more catastrophic. In Fig. 4.20, given a fixed compressive edge load, the implications of the transverse pressure on the amplitude of deflection are displayed. It appears that the construct with the larger curvature, from a structural standpoint, benefits from a larger load-carrying capacity.

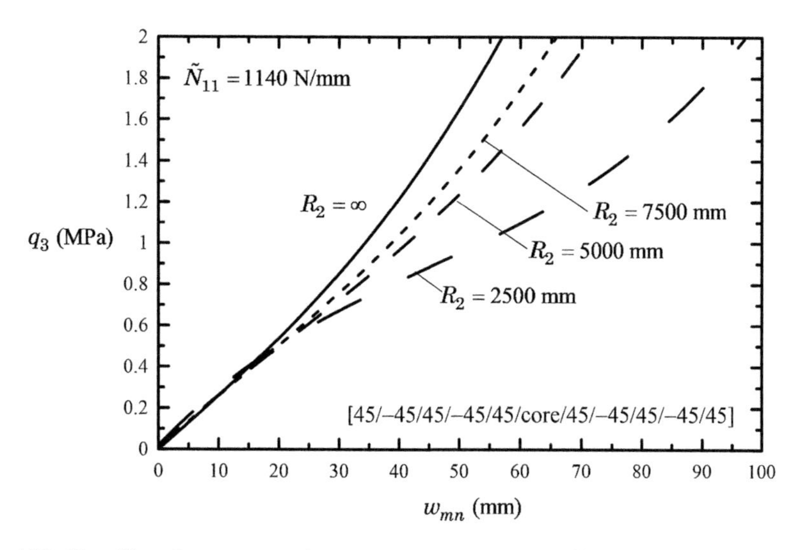

Fig. 4.20 The effect of curvature on the transverse pressure vs. amplitude of deflection for a fixed amount of compressive edge loading of cylindrical sandwich panel

4.6 Immovability of the Edges

So far only simply supported boundary conditions with freely movable edges have been considered. Some situations arise where all four edges are not necessarily movable such as two edges freely movable, and two edges immovable while under compressive edge loading. In such situations, the immovable edges are prevented from moving. To prevent these edges from moving some type of fictious edge loading is present to prevent these edges from moving. For the case that two edges are immovable along the edges ($x_1 = 0,\ L_1$) fulfillment of the following condition can provide the fictitious edge load $\widetilde{N}_{11}$ necessary to prevent the immovable edges from moving.

$$\int_0^{L_2}\int_0^{L_1}(\partial\xi_1/\partial x_1)dx_1dx_2 = 0 \tag{4.98}$$

Substituting Eq. (4.69a) into Eq. (4.98) carrying out the indicated operations results in the fictitious edge load. Once this is determined it can be substituted into the post-buckling solution to determine the response for two immovable and two freely movable edges. In the case, where two edges are immovable along the edges ($x_2 = 0,\ L_2$) the following condition needs to be fulfilled.

$$\int_0^{L_2}\int_0^{L_1}(\partial\xi_2/\partial x_2)dx_1dx_2=0 \tag{4.99}$$

By which the fictitious edge load $\widetilde{N}_{22}$ can be determined and then substituted into the post-buckling solution

4.7 End-Shortening

As a result of uniaxial or biaxial compression, the edges in both the x_1 and x_2 direction have a tendency to displace inward or outward depending on the loaded edges. This amount of edge displacement can be determined from the following expressions.

For displacement in the x_1 direction, the end-shortening, Δ_1 is determined from

$$\Delta_1=-\frac{1}{L_1L_2}\int_0^{L_2}\int_0^{L_1}\left(\frac{\partial\xi_1}{\partial x_1}\right)dx_1dx_2 \tag{4.100}$$

Substituting in the expression for ξ_1 from Eq. (4.69a) and integrating gives a quantifiable expression for Δ_1 as

$$\begin{aligned}\Delta_1&=\frac{\lambda_m^2}{8}\left(w_{mn}^2+2w_{mn}\overset{\circ}{w}_{mn}\right)+\frac{\Delta_n^m\widetilde{F}_3}{\mu_nL_1L_2}w_{mn}\\&-\widetilde{N}_{11}\left(\frac{A_{16}^2-A_{11}A_{66}}{\Omega}+\frac{A_{12}A_{66}-A_{16}A_{26}}{\Omega}L_R\right)\end{aligned} \tag{4.101}$$

For displacement in the x_2 direction, the end-shortening is determined from

$$\Delta_2=-\frac{1}{L_1L_2}\int_0^{L_2}\int_0^{L_1}\left(\frac{\partial\xi_2}{\partial x_2}\right)dx_1dx_2 \tag{4.102}$$

Substituting in the expression for ξ_2 from Eq. (4.69b) and integrating gives

$$\Delta_2=\frac{\mu_n^2}{8}\left(w_{mn}^2+2w_{mn}\overset{\circ}{w}_{mn}\right)+\frac{\Delta_n^m\widetilde{G}_3}{\lambda_mL_1L_2}w_{mn}-\widetilde{N}_{11}\left(\frac{A_{12}A_{66}-A_{16}A_{26}}{\Omega}+\frac{A_{26}^2-A_{22}A_{66}}{\Omega}L_R\right) \tag{4.103}$$

4.8 Summary

A comprehensive theoretical base governing the post-buckling response of flat and doubly curved sandwich panels has been presented in sufficient detail considering the geometric imperfections and the anisotropy of the face sheets. The theory considered three parts. The first part was concerned with the special case of flat and doubly curved sandwich panels with cross-ply laminated facings, while the second and third part was concerned with flat and doubly curved sandwich panels with symmetrically laminated angle ply face sheets, respectively. In contrast to the latter parts, a solution technique referred to as the stress potential method was used to solve the governing equations in conjunction with the extended Galerkin method and Newton's method. Following the theoretical developments, several results were presented considering the effects of curvature, geometric imperfections, panel face thickness, material directional properties (fiber orientation angles), biaxial edge loading, the transverse pressure and various stacking sequences and or layups. The results showed that all of these geometrical and material considerations play an important role in the post-buckling response of flat and doubly curved sandwich panels.

References

Fulton, R. E. (1961). Non-linear equations for a shallow unsymmetric sandwich shell of double curvature. In *Developments in mechanics, proceedings of 7th midwestern mechanics conference* (pp. 365–380). New York: Plenum.

Librescu, L. (1965). Aeroelasticity stability of orthotropic heterogeneous thin panels in the vicinity of the flutter critical panel. *Journal de Mecanique, 4*(1), 51–76.

Librescu, L. (1975). *Elastostatics and kinetics of anisotropic and heterogeneous shell-type structures*. Leyden: Noordhoff International Publishing.

Librescu, L., & Chang, M. J. (1992). Post-buckling and imperfection sensitivity of shear deformable composite doubly curved panels. *International Journal of Solids and Structures, 29*(9), 1065–1083.

Seide, P. (1974). A reexamination of Koiter's theory of initial post-buckling behavior and imperfection sensitivity of structures. In C. Y. Fung & E. E. Sechler (Eds.), *Thin shell structures: Theory, experiment, and design* (pp. 59–80). Englewood Cliffs: Prentice-Hall.

Simitses, G. J. (1986). Buckling and post-buckling of imperfect cylindrical shells: A review. *Applied Mechanics Reviews, 39*(10), 1517–1524.

Chapter 5
Free Vibration

Abstract An advanced model of sandwich plates and shells considering anisotropic laminated composite face sheets with a weak orthotropic core, in the context of determining the eigenfrequencies, is presented. Within this advanced model, adoption of the shallow shell theory is adhered to. The influence of several geometrical and material characteristics of the sandwich panel are considered with regard to the eigenfrequencies. The influence of these characteristics such as the panel geometry, the ply angle and stacking sequence of the face sheets, the orthotropicity ratio of the core, the panel face thickness, the aspect ratio, etc. are all determined to have an effect on the eigenfrequencies of the sandwich panel. Due to the nature of the governing equations, a closed form solution is obtainable with reasonable results. Several validations are made with several prominent author's results found within the literature. With this is hand, appropriate conclusions are made.

5.1 Introduction

This chapter is concerned with the free vibration behavior of doubly curved sandwich panels. This study is carried out in the framework of an advanced sandwich model, utilizing the linear equations from Chap. 2, which will allow for a closed-form solution despite the intricacy of the governing equations. The anisotropy of the face sheets, the stacking sequence, the fiber orientation, orthotropic properties of the core as well as other important geometrical and material properties of both the facings and the core are analyzed with respect to the effects on the eigenfrequencies of the structure. Numerical solutions and validations are presented which show the effects of the above-mentioned geometrical and material effects on the eigenfrequencies of the structure in a detailed optimized fashion through the use of the structural tailoring technique. Finally, this analysis is carried out in two parts. The first part is concerned with flat sandwich panels, while the second part addresses the analysis of doubly curved sandwich shells.

T. J. Hause, *Sandwich Structures: Theory and Responses*,
https://doi.org/10.1007/978-3-030-71895-4_5

5.2 Preliminaries and Basic Assumptions

In consideration of the free or natural vibration of sandwich structures, the incompressible core case will be considered for both the plate and shell configuration. With this in mind, the following basic assumptions are adopted:

1. The face sheets are orthotropic layers not necessarily coincident with the geometrical axes.
2. The core features orthotropic properties in the transverse direction and is considered the weak-type and much larger in thickness than the facings.
3. Perfect bonding between the face sheets and the facings and the core are assumed.
4. The shallow shell theory is assumed.
5. The transverse shear effects in the facings are discarded.
6. The face sheets are symmetric with respect to their local and global mid-surfaces.
7. The tangential and rotatory inertia terms are neglected. (Librescu 1987)

5.3 Flat Sandwich Panels

5.3.1 Governing System

The governing equations which apply to the free vibration problem of flat sandwich panels were developed in Chap. 2. The equations of interest are based on the linear theory for plates with symmetric laminated facings, a weak orthotropic core, and with the transverse shear effects discarded in the facings. Under these conditions, Eqs. (2.174)–(2.176) apply along with the set of simply supported boundary conditions with freely movable edges, from Eqs. (2.177a–d). This set of governing equations are the linearized counterpart of Eqs. (2.156)–(2.160). Also, as mentioned in Chap. 2, for flat panels, setting the curvatures to zero decouples the first two equations of motion from the last three. The last three equations of motion govern the bending problem separate from the stretching problem. This implies that, in the case of flat sandwich plates, the eigenfrequencies are only dependent on the bending problem. The governing system of equations for flat sandwich panels are given as

- *Equations of Motion*

$$A_{11}\frac{\partial^2\eta_1}{\partial x_1^2}+A_{66}\left(\frac{\partial^2\eta_1}{\partial x_2^2}+\frac{\partial^2\eta_2}{\partial x_1\partial x_2}\right)+A_{12}\frac{\partial^2\eta_2}{\partial x_1\partial x_2}+A_{16}\left(\frac{\partial^2\eta_2}{\partial x_1^2}+2\frac{\partial^2\eta_1}{\partial x_1\partial x_2}\right)$$
$$+A_{26}\frac{\partial^2\eta_2}{\partial x_2^2}-d_1\left(\eta_1+a\frac{\partial v_3}{\partial x_1}\right)=0 \tag{5.1a}$$

$$A_{22}\frac{\partial^2\eta_2}{\partial x_2^2}+A_{66}\left(\frac{\partial^2\eta_2}{\partial x_2^2}+\frac{\partial^2\eta_1}{\partial x_1\partial x_2}\right)+A_{12}\frac{\partial^2\eta_1}{\partial x_1\partial x_2}+A_{26}\left(\frac{\partial^2\eta_1}{\partial x_2^2}+2\frac{\partial^2\eta_2}{\partial x_1\partial x_2}\right)$$
$$+A_{16}\frac{\partial^2\eta_1}{\partial x_1^2}-d_2\left(\eta_2+a\frac{\partial v_3}{\partial x_2}\right)=0 \tag{5.1b}$$

$$-d_1a\left(\frac{\partial\eta_1}{\partial x_1}+a\frac{\partial^2 v_3}{\partial x_1^2}\right)-d_2a\left(\frac{\partial\eta_2}{\partial x_2}+a\frac{\partial^2 v_3}{\partial x_2^2}\right)+F_{11}\frac{\partial^4 v_3}{\partial x_1^4}+F_{22}\frac{\partial^4 v_3}{\partial x_2^4}$$
$$+2(F_{12}+2F_{66})\frac{\partial^4 v_3}{\partial x_1^2\partial x_2^2}+4F_{16}\frac{\partial^4 v_3}{\partial x_1^3\partial x_2}+4F_{26}\frac{\partial^4 v_3}{\partial x_1\partial x_2^3}+N_{11}^0\frac{\partial^2 v_3}{\partial x_1^2} \tag{5.1c}$$
$$+2N_{12}^0\frac{\partial^2 v_3}{\partial x_1\partial x_2}+N_{22}^0\frac{\partial^2 v_3}{\partial x_2^2}-m_0\frac{\partial^2 v_3}{\partial t^2}=0$$

- *Boundary Conditions*

The boundary conditions are reduced from six to only four required along each edge as

Along the edges $x_n = 0,\ L_n$

$$\begin{aligned}
&L_{nn}=\widetilde{L}_{nn} \quad \text{or} \quad \eta_n=\widetilde{\eta}_n\\
&L_{nt}=\widetilde{L}_{nt} \quad \text{or} \quad \eta_t=\widetilde{\eta}_t\\
&M_{nn}=\widetilde{M}_{nn} \quad \text{or} \quad \frac{\partial v_3}{\partial x_n}=\frac{\partial \widetilde{v}_3}{\partial x_n}\\
&N_{nt}\frac{\partial v_3}{\partial x_t}+N_{nn}\frac{\partial v_3}{\partial x_n}+\frac{\partial M_{nn}}{\partial x_n}+2\frac{\partial M_{nt}}{\partial x_t} \quad \text{or} \quad v_3=\widetilde{v}_3\\
&+(a/\overline{h})\overline{N}_{n3}=\frac{\partial \widetilde{M}_{nt}}{\partial x_t}+\widetilde{\overline{N}}_{n3}
\end{aligned} \tag{5.2a – d}$$

Because there are four boundary conditions prescribed along each edge, the governing system of equations should be of the eighth order. For simply supported boundary conditions freely movable on all edges

at $x_1 = 0,\ L_1$

$$\eta_1=\eta_2=M_{11}=v_3=0 \tag{5.3a – d}$$

at $x_2 = 0,\ L_2$

$$\eta_1=\eta_2=M_{22}=v_3=0 \tag{5.4a – d}$$

5.3.2 Solution Methodology

The methodology applied in Librescu et al. (1997) and Hause et al. (1998, 2000) will be utilized here. Due to the complexity of the governing equations, an approximate solution methodology such as the extended Galerkin method (EGM) will be adopted. The goal is to satisfy both the equations of motion and the boundary conditions in the Galerkin sense. To satisfy the kinematic boundary conditions, Eqs. (5.3d) and (5.4d), $v_3(x_1, x_2, t)$ can be expressed as

$$v_3(x_1, x_2, t) = (w_{mn} \sin \lambda_m x_1 \sin \mu_n x_2) \exp(i\omega_{mn} t) \tag{5.5}$$

where ω_{mn} are the undamped natural frequencies; $i = \sqrt{-1}$; $\lambda_m = m\pi/L_1$, $\mu_n = n\pi/L_2$; L_1 and L_2 are the panel length and width, respectively. With the first two equations of motion, Eqs. (5.1a and 5.1b), η_1 and η_2 are assumed in the following form

$$\eta_1(x_1, x_2, t) = \left(H_{mn}^{(1)} \cos \lambda_m x_1 \sin \mu_n x_2 + H_{mn}^{(2)} \sin \lambda_m x_1 \cos \mu_n x_2\right) \exp(i\omega_{mn} t) \tag{5.6a}$$

$$\eta_2(x_1, x_2, t) = \left(I_{mn}^{(1)} \cos \lambda_m x_1 \sin \mu_n x_2 + I_{mn}^{(2)} \sin \lambda_m x_1 \cos \mu_n x_2\right) \exp(i\omega_{mn} t) \tag{5.6b}$$

where $H_{mn}^{(1)}$, $H_{mn}^{(2)}$, $I_{mn}^{(1)}$, $I_{mn}^{(2)}$ are undetermined coefficients and. Substituting η_1 and η_2 into Eqs. (5.1a and 5.1b) and comparing the coefficients of the same trigonometric functions results in a system of equations in matrix form as

$$\begin{bmatrix} U_{11}^{(m,n)} & U_{12}^{(m,n)} & U_{13}^{(m,n)} & U_{14}^{(m,n)} \\ & U_{11}^{(m,n)} & U_{14}^{(m,n)} & U_{13}^{(m,n)} \\ & & U_{33}^{(m,n)} & U_{34}^{(m,n)} \\ \text{Symm.} & & & U_{33}^{(m,n)} \end{bmatrix} \begin{Bmatrix} H_1^{(m,n)} \\ H_2^{(m,n)} \\ I_1^{(m,n)} \\ I_2^{(m,n)} \end{Bmatrix} = \begin{Bmatrix} V_1^{(m,n)} \\ 0 \\ 0 \\ V_2^{(m,n)} \end{Bmatrix} \tag{5.7}$$

where the elements $U_{ij}^{(m,n)}$ of the matrix are expressions in terms of the geometrical and mechanical properties of the structure which are given as

$$U_{11}^{(m,n)} = A_{11}m^2 + A_{66}n^2\phi^2 + \frac{d_1 L_1^2}{\pi^2}, \quad U_{12}^{(m,n)} = 2A_{16}mn\phi, \quad U_{13}^{(m,n)} = A_{16}m^2 + A_{26}n^2\phi^2$$
$$U_{14}^{(m,n)} = (A_{12} + A_{66})mn\phi, \quad U_{33}^{(m,n)} = A_{22}n^2\phi^2 + A_{66}m^2 + \frac{d_2 L_1^2}{\pi^2}, \quad U_{34}^{(m,n)} = 2A_{26}mn\phi \tag{5.8a – f}$$

while

$$V_1^{(m,n)} = \frac{(-d_1 amL_1)}{\pi} w_{mn}, \qquad V_2^{(m,n)} = \frac{(-d_2 an\phi L_1)}{\pi} w_{mn} \tag{5.9a, b}$$

From Eq. (5.7), using Cramer's Rule, the expressions for $H_{mn}^{(\alpha)}$, $I_{mn}^{(\alpha)}$ $(\alpha = 1, 2)$ can be expressed as

$$H_{mn}^{(\alpha)} = \tilde{H}_{mn}^{(\alpha)} w_{mn}, \qquad I_{mn}^{(\alpha)} = \tilde{I}_{mn}^{(\alpha)} w_{mn}, \qquad /\sum_{m,n} \tag{5.10a, b}$$

where

$$\tilde{\mathrm{H}}_{\mathrm{mn}}^{(1)} = \frac{\det(U_1)}{\det(U)}, \qquad \tilde{\mathrm{H}}_{\mathrm{mn}}^{(2)} = \frac{\det(U_2)}{\det(U)}, \qquad \tilde{\mathrm{I}}_{\mathrm{mn}}^{(1)} = \frac{\det(U_3)}{\det(U)}, \qquad \tilde{\mathrm{I}}_{\mathrm{mn}}^{(2)} = \frac{\det(U_4)}{\det(U)} \tag{5.11a – d}$$

while

$$U_1 = \begin{pmatrix} V_1^{(m,n)} & U_{12}^{(m,n)} & U_{13}^{(m,n)} & U_{14}^{(m,n)} \\ 0 & U_{11}^{(m,n)} & U_{14}^{(m,n)} & U_{13}^{(m,n)} \\ 0 & U_{14}^{(m,n)} & U_{33}^{(m,n)} & U_{34}^{(m,n)} \\ V_2^{(m,n)} & U_{13}^{(m,n)} & U_{34}^{(m,n)} & U_{33}^{(m,n)} \end{pmatrix}, \quad U_2 = \begin{pmatrix} U_{11}^{(m,n)} & V_1^{(m,n)} & U_{13}^{(m,n)} & U_{14}^{(m,n)} \\ U_{12}^{(m,n)} & 0 & U_{14}^{(m,n)} & U_{13}^{(m,n)} \\ U_{13}^{(m,n)} & 0 & U_{33}^{(m,n)} & U_{34}^{(m,n)} \\ U_{14}^{(m,n)} & V_2^{(m,n)} & U_{34}^{(m,n)} & U_{33}^{(m,n)} \end{pmatrix} \tag{5.12a, b}$$

$$U_3 = \begin{pmatrix} U_{11}^{(m,n)} & U_{12}^{(m,n)} & V_1^{(m,n)} & U_{14}^{(m,n)} \\ U_{12}^{(m,n)} & U_{11}^{(m,n)} & 0 & U_{13}^{(m,n)} \\ U_{13}^{(m,n)} & U_{14}^{(m,n)} & 0 & U_{34}^{(m,n)} \\ U_{14}^{(m,n)} & U_{13}^{(m,n)} & V_2^{(m,n)} & U_{33}^{(m,n)} \end{pmatrix}, \quad U_4 = \begin{pmatrix} U_{11}^{(m,n)} & U_{12}^{(m,n)} & U_{13}^{(m,n)} & V_1^{(m,n)} \\ U_{12}^{(m,n)} & U_{11}^{(m,n)} & U_{14}^{(m,n)} & 0 \\ U_{13}^{(m,n)} & U_{14}^{(m,n)} & U_{33}^{(m,n)} & 0 \\ U_{14}^{(m,n)} & U_{13}^{(m,n)} & U_{34}^{(m,n)} & V_2^{(m,n)} \end{pmatrix} \tag{5.12c, d}$$

$$U = \begin{pmatrix} U_{11}^{(m,n)} & U_{12}^{(m,n)} & U_{13}^{(m,n)} & U_{14}^{(m,n)} \\ & U_{11}^{(m,n)} & U_{14}^{(m,n)} & U_{13}^{(m,n)} \\ & & U_{33}^{(m,n)} & U_{34}^{(m,n)} \\ \text{Symm.} & & & U_{33}^{(m,n)} \end{pmatrix} \tag{5.12e}$$

At this point, Eqs. (5.1a and 5.1b) are identically fulfilled. There remains the third equation of motion currently unfulfilled. This will be fulfilled in an average sense through the use of the already familiar extended Galerkin method. In addition, the boundary conditions, Eqs. (5.3d) and (5.4d) are fulfilled with the remaining ones unfulfilled. Retaining the unfulfilled third equation of motion and the remaining unfulfilled boundary conditions from Eqs. (5.3a–c) and (5.4a–c) in the energy functional (Hamilton's equation) results in

$$\begin{aligned} &\int_{t_0}^{t_1} \left\langle \int_0^{l_2} \int_0^{l_1} \left\{ -d_1 a \left(\frac{\partial \eta_1}{\partial x_1} + a \frac{\partial^2 v_3}{\partial x_1^2} \right) - d_2 a \left(\frac{\partial \eta_2}{\partial x_2} + a \frac{\partial^2 v_3}{\partial x_2^2} \right) + F_{11} \frac{\partial^4 v_3}{\partial x_1^4} + F_{22} \frac{\partial^4 v_3}{\partial x_2^4} + \right. \right. \\ &2(F_{12} + 2F_{66}) \frac{\partial^4 v_3}{\partial x_1^2 \partial x_2^2} + 4F_{16} \frac{\partial^4 v_3}{\partial x_1^3 \partial x_2} + 4F_{26} \frac{\partial^4 v_3}{\partial x_1 \partial x_2^3} + N_{11}^0 \frac{\partial^2 v_3}{\partial x_1^2} + N_{22}^0 \frac{\partial^2 v_3}{\partial x_2^2} - \\ &\left. \left. m_0 \frac{\partial^2 v_3}{\partial t^2} \right\} \delta v_3 dx_1 dx_2 \right\rangle dt + \int_{t_0}^{t_1} \left\langle \int_0^{l_2} \left[L_{11} \delta \eta_1 + L_{12} \delta \eta_2 + M_{11} \delta \left(\frac{\partial v_3}{\partial x_1} \right) \right] \Bigg|_0^{l_1} dx \right\rangle dt + \\ &\int_{t_0}^{t_1} \left\langle \int_0^{l_1} \left[L_{12} \delta \eta_1 + L_{22} \delta \eta_2 + M_{22} \delta \left(\frac{\partial v_3}{\partial x_2} \right) \right] \Bigg|_0^{l_2} dx \right\rangle dt = 0 \end{aligned} \tag{5.13}$$

Substituting the expressions for η_1, η_2, and v_3 into Eq. (5.13) and carrying out the necessary integrations and simplifying realizing the independent character of the variations results in an eigenvalue problem which gives the expression for the eigenfrequencies as

$$\begin{aligned} \omega_{mn}^2 = \frac{1}{m_0} \Big(& d_1 a \lambda_m H_{mn}^{(1)} + d_2 a \mu_n I_{mn}^{(2)} + d_1 a^2 \lambda_m^2 + d_2 a^2 \mu_n^2 + F_{11} \lambda_m^2 + F_{22} \mu_n^2 + 4F_{66} \lambda_m^2 \mu_n^2 + \\ & 2F_{12} \lambda_m^2 \mu_n^2 - N_{11}^0 \lambda_m^2 - N_{22}^0 \mu_n^2 \Big) \end{aligned} \tag{5.14}$$

Where the expression for the undamped natural frequency can be arranged in dimensionless form as

$$\begin{aligned} \Omega_{mn}^2 = m^4 + \frac{n^4 \phi^4 F_{22}}{F_{11}} + \frac{2m^2 n^2 \phi^2 (F_{12} + 2F_{66})}{F_{11}} + \frac{a^2 L_1^2}{\pi^2 F_{11}} \left(m^2 d_1 + n^2 \phi^2 d_2 \right) + \frac{a L_1^3}{\pi^3 F_{11}} \Big[m d_1 \widetilde{H}_{mn}^{(1)} \\ + n \phi d_2 \widetilde{I}_{mn}^{(2)} \Big] + K_x \left(m^2 + L_R n^2 \phi^2 \right) \end{aligned} \tag{5.15}$$

(Note: $\widetilde{H}_{mn}^{(1)}$ and $\widetilde{I}_{mn}^{(2)}$ are nondimensionalized also which are not provided here). The nondimensional parameters are defined as

$$\Omega_{mn}^2 = \frac{m_0 L_1^4 \omega_{mn}^2}{\pi^4 F_{11}}, \quad K_x = \frac{L_1^2 N_{11}^0}{\pi^4 F_{11}}, \quad L_R = \frac{N_{22}^0}{N_{11}^0}, \quad \phi = \frac{L_1}{L_2} \tag{5.15a – d}$$

5.3.3 Validation of the Theoretical Structural Model

Prior to the presentation of the numerical results validations are made to assess the accuracy of the theory. The case considered for the validations considers a three-layered flat sandwich panel whose material characteristics are listed in Table 5.1. The facings are aluminum while the core consists of an aluminum honeycomb–type construction. The length of the edges of the panel are $L_1 = 1.828$ m and $L_2 = 1.219$ m. In Table 5.2, the eigenfrequencies from three different sources are compared with the present theoretical results. Excellent agreement is seen. Some additional sources are listed in Table 5.3 for the same geometrical and material characteristics as in Table 5.1. Again, remarkable agreement is seen. Hohe et al. (2006) considered the transverse compressibility of the core. With this additional consideration, exceptional agreement is seen revealing that the compressibility of the core, for a flat sandwich panel, has extraordinarily little effect on the eigenfrequencies.

5.3.4 Results and Discussion

In consideration of the following numerical results, the material properties for the face sheets and the core are listed in Tables 5.4 and 5.5. Unless otherwise specified, the face sheets are of the F1 type and the core is of the C1 type while the following geometrical properties are given as $L_1 = 0.6096$ m, $\overline{h} = 0.0127$ m.

Figure 5.1 shows the effects of the compressive edge loading on the eigenfrequencies $\Omega_{mn}^{1/2}$ for various aspect ratios. It is known that when the

Table 5.1 Material and geometrical properties

	Thickness (mm)	Elastic modulus (GPa)	Poisson's ratio	Mass density $N \cdot s^2 \cdot m^{-4}$	Shear modulus (GPa)		
					G_{12}	$\overline{G}_{13}$	$\overline{G}_{23}$
Upper face	0.4064	68.95	0.33	2768.93	25.92	–	–
Bottom face	0.4064	68.95	0.33	2768.93	25.92	–	–
Core	6.35	0	0	121.83	–	0.0517	0.134

Table 5.2 Eigenfrequency comparisons (validation no. 1)

m	n					
	1	2	3	4	5	
1	23.5	71.0	146.5	245.3	362.5	Present study
	23.4	70.7	145.8	244.4	361.7	Hohe et al. (2006)
	–	69.0	152.0	246.0	381.0	Raville and Ueng (1967) (exp)
	–	71.0	146.0	244.0	360.0	Raville and Ueng (1967) (numerical)
2	45.1	92.1	166.7	264.5	–	Present study
	44.8	91.5	165.9	263.5	–	Hohe et al. (2006)
	45.0	92.0	169.0	262.0	–	Raville and Ueng (1967) (exp)
	45.0	91.0	165.0	263.0	–	Raville and Ueng (1967) (numerical)
3	80.7	126.8	200.1	–	–	Present study
	80.3	126.1	199.1	–	–	Hohe et al. (2006)
	78.0	129.0	199.0	–	–	Raville and Ueng (1967) (exp)
	80.0	126.0	195.0	–	–	Raville and Ueng (1967) (numerical)
4	130.0	174.9	–	–	–	Present study
	129.3	173.9	–	–	–	Hohe et al. (2006)
	133.0	177.0	–	–	–	Raville and Ueng (1967) (exp)
	129.0	174.0	–	–	–	Raville and Ueng (1967) (numerical)

Table 5.3 Eigenfrequency comparisons (validation no. 2)

ω_{ij}(Hz)	Experiment (R. and U.)	Exact (R. and U.)	FEM (M. and S.)	SFPM (Z. and L.)	Present
$\omega_1(=\omega_{11})$	-	23	23	23.3	23.40
$\omega_2(=\omega_{21})$	45	44	44	44.48	44.64
$\omega_3(=\omega_{12})$	69	71	70	71.36	71.51
$\omega_4(=\omega_{31})$	78	80	80	78.81	79.27
$\omega_5(=\omega_{22})$	92	91	90	91.90	92.2
$\omega_6(=\omega_{32})$	125	126	125	125.16	125.97

R. and U. → Raville and Ueng (1967); M. and S. → Monforton and Schmidt (1968); Z. and L. → Zhou and Li (1996)

Table 5.4 Material properties for the face sheets

Type	Material	E_1 (GPa)	E_2(GPa)	G_{12}(GPa)	ν_{12}
F1	HS Graph. Ep.	180.99	10.34	7.24	0.28
F2	IM7/977-2	79.98	75.15	9.65	0.06

Table 5.5 Core material properties

Type	Core type	$\overline{G}_{13}$ (GPa)	$\overline{G}_{23}$ (GPa)
C1	Titanium honeycomb	1.44	0.651

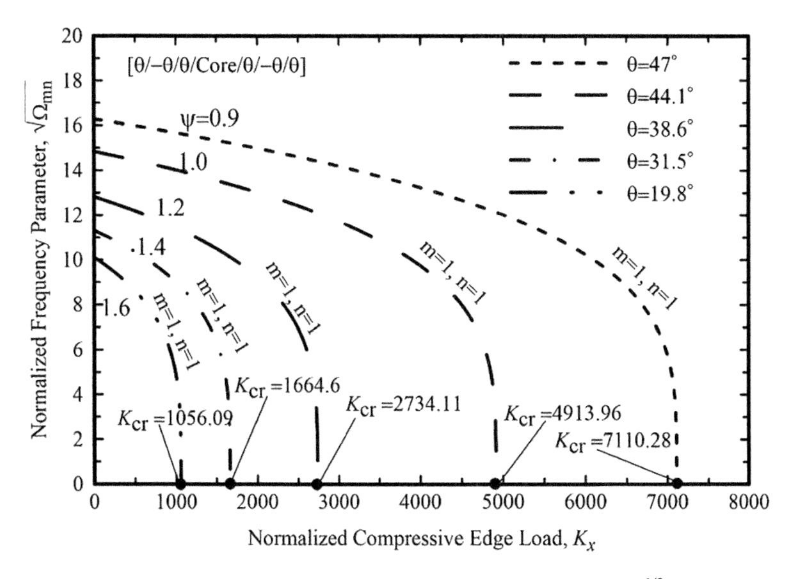

Fig. 5.1 The effect of the compressive edge load on the eigenfrequencies $\Omega^{1/2}$ of a flat sandwich panel for various aspect ratios

Table 5.6 Comparison of the critical buckling load

Layup $[\theta/-\theta/\theta/\text{core}/\theta/-\theta/\theta]$					
θ (deg)	$\phi(L_1/L_2)$	L_1(mm)	K_{cr}	$\left(N_{11}^{\circ}\right)_{cr}$ (N/m) (Chap. 3)	$\left(N_{11}^{\circ}\right)_{cr}$ (N/mm) (Present)
47	1.11	609.6	7110.28	2.08×10^6	2.089×10^6
44.1	1	609.6	4913.96	1.67×10^6	1.682×10^6
38.6	0.833	609.6	2734.11	1.22×10^6	1.222×10^6
31.5	0.714	609.6	1664.60	0.99×10^6	0.994×10^6
19.8	0.625	609.6	1056.09	0.88×10^6	0.889×10^6

eigenfrequency vanishes, the critical load has been reached. At this point the buckling load is identified. It can be seen that at smaller aspect ratios the buckling loads are higher. In Table 5.6, comparisons are made between the direct buckling approach as was presented in Chap. 3 and the present results where the buckling load is determined from the point at which the eigenfrequencies vanish. Excellent agreement is made between the two approaches.

In Fig. 5.2, the effect of the panel face thickness on the eigenfrequency $\Omega^{1/2}$ for various distances a is shown. It is revealed that the eigenfrequency decays as the panel face thickness increases. Also, as the distance a increases for a fixed face thickness, the eigenfrequencies become larger. In Fig. 5.3, the effect of the fiber orientation of the face sheets on the eigenfrequency $\Omega^{1/2}$ for various aspect ratios is depicted. At a larger aspect ratio, the range between the eigenfrequencies $\Omega^{1/2}$ between 0- and 90-degree ply angles is larger than at a small aspect ratio. The trend seems to increase for the larger aspect ratio where it appears flat for the small

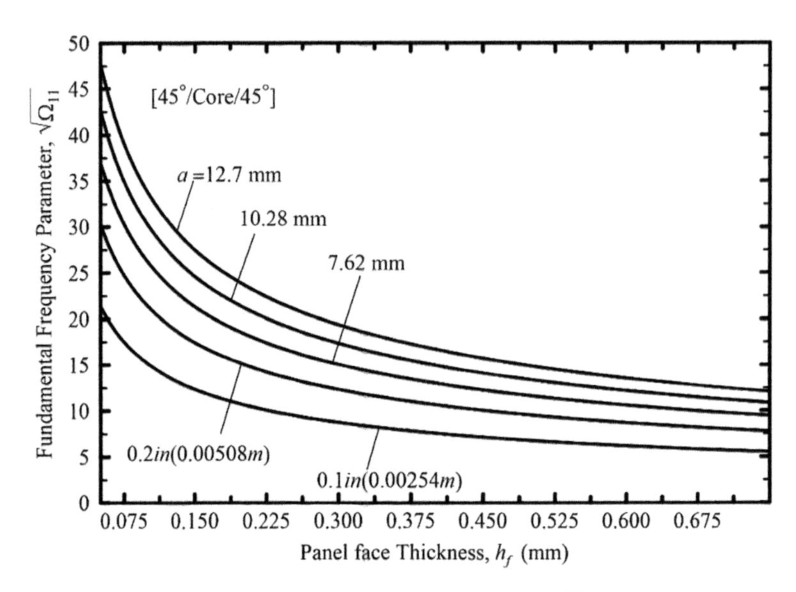

Fig. 5.2 The effect of the face thickness on the eigenfrequencies $\Omega^{1/2}$ of a flat sandwich panel for various distances from the mid-surface of the core to the mid-surface of the facings

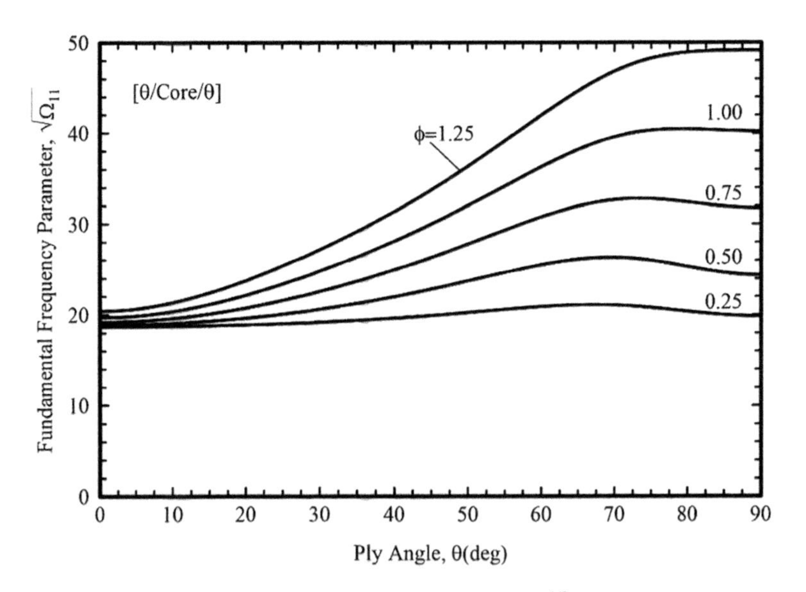

Fig. 5.3 The effect of the ply angle on the eigenfrequencies $\Omega^{1/2}$ of a flat sandwich panel for various aspect ratios

aspect ratio of around 0.25 or smaller. In Fig. 5.4, the effect of the ply angle on the eigenfrequency $\Omega^{1/2}$ at various mode numbers. It is evident that at the higher modes, the eigenfrequencies $\Omega^{1/2}$ are increased. In addition, for each mode the highest frequency appears to be around a ply angle of 45 degrees.

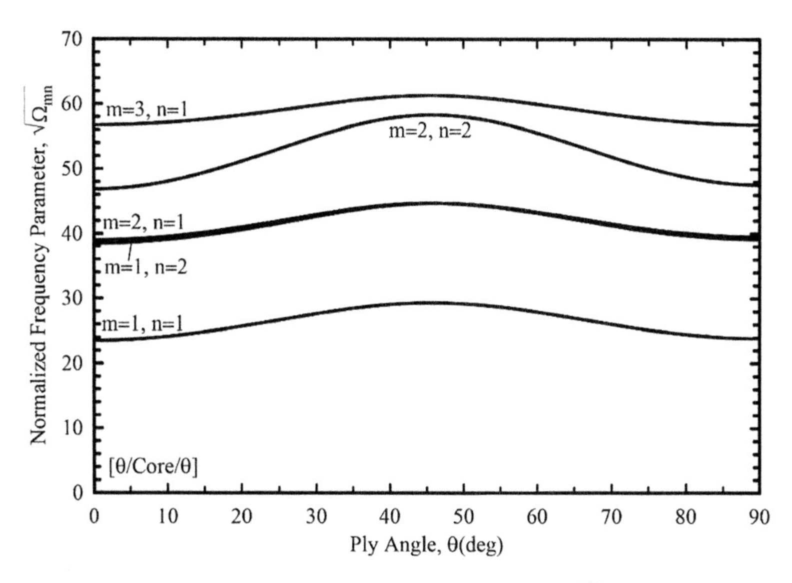

Fig. 5.4 The effect of ply angle on various mode eigenfrequencies $\Omega^{1/2}$ of a flat sandwich panel

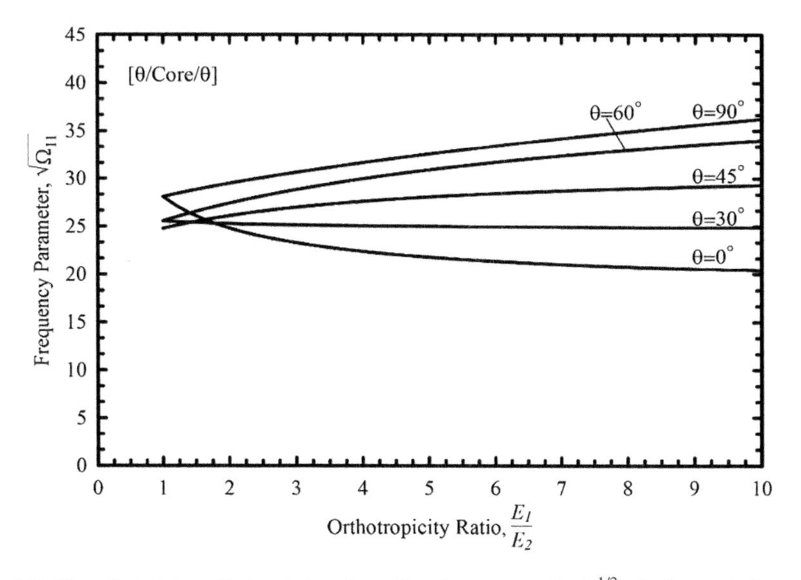

Fig. 5.5 The effect of the orthotropicity ratio on the eigenfrequencies $\Omega^{1/2}$ of a flat sandwich panel for various ply angles

In Fig. 5.5, the effect of the orthotropicity ratio on the eigenfrequencies $\Omega^{1/2}$ is displayed for various ply angles. Above 45 degrees the orthotropicity ratio seems to have a trend which slightly increases resulting in higher eigenfrequencies. In a similar sense, the effect of the core orthotropy ratio is to increase the eigenfrequency $\Omega^{1/2}$ with an increase in the ply angle as shown in Fig. 5.6. Larger ratios provide

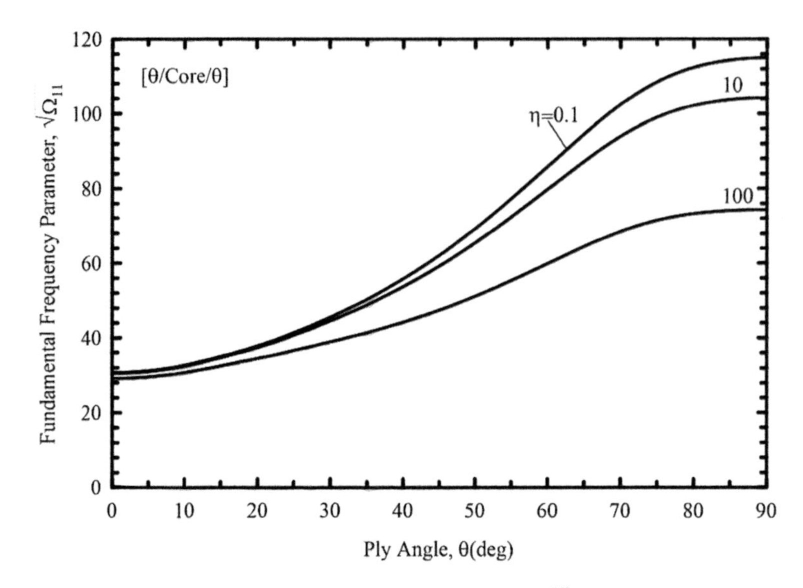

Fig. 5.6 The effect of the ply angle on the eigenfrequencies $\Omega^{1/2}$ of a flat sandwich panel for various orthotropy ratios $\eta\left(\equiv \overline{G}_{13}/\overline{G}_{23}\right)$ of the core

smaller eigenfrequencies, but the trends appear to be the same across the spectrum with an increase in ply angle.

5.4 Doubly Curved Sandwich Panels

5.4.1 Governing System

The governing equations which apply to the free vibration problem of doubly curved sandwich panels are an extension of the governing equations for the flat panels. In this case the governing equations are not decoupled, and six boundary conditions are required along each edge requiring the system of equations to be of the twelfth order. The equations of interest are based on the linear theory for sandwich shells assuming symmetric anisotropic laminated facings, a weak orthotropic core with the transverse shear effects discarded in the facings. Under these conditions, Eqs. (2.168)–(2.172) apply along with the corresponding set of simply supported boundary conditions Eqs. (2.173a–f). This set of governing equations are general in the sense that they can apply to a whole host of static and dynamic problems of doubly curved shallow sandwich shells. The governing equations for double-curved sandwich shells are

- *Equations of Motion*

$$A_{11}\left(\frac{\partial^2\xi_1}{\partial x_1^2}-\frac{1}{R_1}\frac{\partial v_3}{\partial x_1}\right)+A_{66}\left(\frac{\partial^2\xi_1}{\partial x_2^2}+\frac{\partial^2\xi_2}{\partial x_1\partial x_2}\right)+A_{12}\left(\frac{\partial^2\xi_2}{\partial x_1\partial x_2}-\frac{1}{R_2}\frac{\partial v_3}{\partial x_2}\right)$$
$$-A_{16}\left(2\frac{\partial^2\xi_1}{\partial x_1\partial x_2}+\frac{\partial^2\xi_2}{\partial x_1^2}-\frac{1}{R_1}\frac{\partial v_3}{\partial x_1}\right)-A_{26}\left(\frac{\partial^2\xi_2}{\partial x_2^2}-\frac{1}{R_2}\frac{\partial v_3}{\partial x_2}\right)=0 \tag{5.16a}$$

$$A_{22}\left(\frac{\partial^2\xi_2}{\partial x_2^2}-\frac{1}{R_2}\frac{\partial v_3}{\partial x_2}\right)+A_{66}\left(\frac{\partial^2\xi_2}{\partial x_1^2}+\frac{\partial^2\xi_1}{\partial x_1\partial x_2}\right)+A_{12}\left(\frac{\partial^2\xi_1}{\partial x_1\partial x_2}-\frac{1}{R_1}\frac{\partial v_3}{\partial x_1}\right)$$
$$-A_{26}\left(2\frac{\partial^2\xi_2}{\partial x_1\partial x_2}+\frac{\partial^2\xi_1}{\partial x_2^2}-\frac{1}{R_2}\frac{\partial v_3}{\partial x_2}\right)-A_{16}\left(\frac{\partial^2\xi_1}{\partial x_1^2}-\frac{1}{R_1}\frac{\partial v_3}{\partial x_1}\right)=0 \tag{5.16b}$$

$$A_{11}\frac{\partial^2\eta_1}{\partial x_1^2}+A_{66}\left(\frac{\partial^2\eta_1}{\partial x_2^2}+\frac{\partial^2\eta_2}{\partial x_1\partial x_2}\right)+A_{12}\frac{\partial^2\eta_2}{\partial x_1\partial x_2}+A_{16}\left(\frac{\partial^2\eta_2}{\partial x_1^2}+2\frac{\partial^2\eta_1}{\partial x_1\partial x_2}\right)$$
$$+A_{26}\frac{\partial^2\eta_2}{\partial x_2^2}-d_1\left(\eta_1+a\frac{\partial v_3}{\partial x_1}\right)=0 \tag{5.16c}$$

$$A_{22}\frac{\partial^2\eta_2}{\partial x_2^2}+A_{66}\left(\frac{\partial^2\eta_2}{\partial x_2^2}+\frac{\partial^2\eta_1}{\partial x_1\partial x_2}\right)+A_{12}\frac{\partial^2\eta_1}{\partial x_1\partial x_2}$$
$$+A_{26}\left(\frac{\partial^2\eta_1}{\partial x_2^2}+2\frac{\partial^2\eta_2}{\partial x_1\partial x_2}\right)+A_{16}\frac{\partial^2\eta_1}{\partial x_1^2}-d_2\left(\eta_2+a\frac{\partial v_3}{\partial x_2}\right)=0 \tag{5.16d}$$

$$-\left(\frac{A_{11}}{R_1}+\frac{A_{12}}{R_2}\right)\frac{\partial\xi_1}{\partial x_1}-\left(\frac{A_{16}}{R_1}+\frac{A_{26}}{R_2}\right)\frac{\partial\xi_1}{\partial x_2}-\left(\frac{A_{22}}{R_2}+\frac{A_{12}}{R_1}\right)\frac{\partial\xi_2}{\partial x_2}-\left(\frac{A_{26}}{R_2}+\frac{A_{16}}{R_1}\right)\frac{\partial\xi_2}{\partial x_1}$$
$$-\left(\frac{A_{11}}{R_1^2}+\frac{A_{22}}{R_2^2}+\frac{2A_{12}}{R_1R_2}\right)v_3-d_1a\left(\frac{\partial\eta_1}{\partial x_1}+a\frac{\partial^2v_3}{\partial x_1^2}\right)-d_2a\left(\frac{\partial\eta_2}{\partial x_2}+a\frac{\partial^2v_3}{\partial x_2^2}\right)$$
$$+F_{11}\frac{\partial^4v_3}{\partial x_1^4}+F_{22}\frac{\partial^4v_3}{\partial x_2^4}+2(F_{12}+2F_{66})\frac{\partial^4v_3}{\partial x_1^2\partial x_2^2}+4F_{16}\frac{\partial^4v_3}{\partial x_1^3\partial x_2}+4F_{26}\frac{\partial^4v_3}{\partial x_1\partial x_2^3}$$
$$+N_{11}^0\frac{\partial^2v_3}{\partial x_1^2}+2N_{12}^0\frac{\partial^2v_3}{\partial x_1\partial x_2}+N_{22}^0\frac{\partial^2v_3}{\partial x_2^2}-c\frac{\partial v_3}{\partial t}-m_0\frac{\partial^2v_3}{\partial t^2}=0 \tag{5.16e}$$

- *Boundary Conditions*

It should be noted that for the free vibration problem, the thermal, the moisture, and the transversal loading terms have been discarded. The boundary conditions are

Along the edges $x_n = 0,\ L_n$

$$
\begin{aligned}
& N_{nn} = \widetilde{N}_{nn} \quad \text{or} \quad \xi_n = \widetilde{\xi}_n \\
& N_{nt} = \widetilde{N}_{nt} \quad \text{or} \quad \xi_t = \widetilde{\xi}_t \\
& L_{nn} = \widetilde{L}_{nn} \quad \text{or} \quad \eta_n = \widetilde{\eta}_n \\
& L_{nt} = \widetilde{L}_{nt} \quad \text{or} \quad \eta_t = \widetilde{\eta}_t \\
& M_{nn} = \widetilde{M}_{nn} \quad \text{or} \quad \frac{\partial v_3}{\partial x_n} = \frac{\partial \widetilde{v}_3}{\partial x_n} \\
& N_{nt}\frac{\partial v_3}{\partial x_t} + N_{nn}\frac{\partial v_3}{\partial x_n} + \frac{\partial M_{nn}}{\partial x_n} + 2\frac{\partial M_{nt}}{\partial x_t} \quad \text{or} \quad v_3 = \widetilde{v}_3 \\
& +(a/\overline{h})\overline{N}_{n3} = \frac{\partial \widetilde{M}_{nt}}{\partial x_t} + \widetilde{\overline{N}}_{n3}
\end{aligned}
\tag{5.17a – f}
$$

For simply supported boundary conditions freely movable on all edges

at $x_1 = 0,\ L_1$

$$
N_{11} = N_{12} = \eta_1 = \eta_2 = M_{11} = v_3 = 0 \tag{5.18a – f}
$$

at $x_2 = 0,\ L_2$

$$
N_{22} = N_{12} = \eta_1 = \eta_2 = M_{22} = v_3 = 0 \tag{5.19a – f}
$$

It will be helpful to determine, further in the developments, if the boundary conditions are satisfied by expressing the boundary conditions in terms of displacements. In terms of displacements, the first, second, and fifth boundary conditions can be written as

$$
\begin{aligned}
N_{11} &= A_{11}\frac{\partial \xi_1}{\partial x_1} + A_{12}\frac{\partial \xi_2}{\partial x_2} + A_{16}\left(\frac{\partial \xi_2}{\partial x_1} + \frac{\partial \xi_1}{\partial x_2}\right) - \left(\frac{A_{11}}{R_1} + \frac{A_{12}}{R_2}\right)v_3 \\
&= 0, \qquad (1 \rightleftarrows 2)
\end{aligned}
\tag{5.20}
$$

$$
N_{12} = A_{66}\left(\frac{\partial \xi_2}{\partial x_1} + \frac{\partial \xi_1}{\partial x_2}\right) + A_{26}\frac{\partial \xi_2}{\partial x_2} + A_{16}\frac{\partial \xi_1}{\partial x_1} - \left(\frac{A_{16}}{R_1} + \frac{A_{26}}{R_2}\right)v_3 = 0 \tag{5.21}
$$

$$M_{11} = F_{11} \frac{\partial^2 v_3}{\partial x_1^2} + F_{12} \frac{\partial^2 v_3}{\partial x_2^2} + 2F_{16} \frac{\partial^2 v_3}{\partial x_1 \partial x_2} = 0 \tag{5.22}$$

$$M_{22} = F_{22} \frac{\partial^2 v_3}{\partial x_2^2} + F_{12} \frac{\partial^2 v_3}{\partial x_1^2} + 2F_{26} \frac{\partial^2 v_3}{\partial x_1 \partial x_2} = 0 \tag{5.23}$$

where the global stiffness quantities A_{ij}, $F_{ij}(i, j = 1, 2, 6)$ were defined in Chap. 2. The sign $(1 \rightleftarrows 2)$ in Eq. (5.20) indicates that N_{22} which is not explicitly supplied can be obtained by replacing the subscript 1 with 2 and vice versa.

5.4.2 Solution Methodology

The methodology applied in Librescu et al. (1997) and Hause et al. (1998, 2000) will be utilized here. To begin the process, attention is given to the first two equations of motion, Eqs. (5.16a and 5.16b). here ξ_1, ξ_2 are assumed in the following form

$$\xi_1(x_1, x_2, t) = \left(F_{mn}^{(1)} \cos \lambda_m x_1 \sin \mu_n x_2 + F_{mn}^{(2)} \sin \lambda_m x_1 \cos \mu_n x_2\right) \exp\left[(i\omega_{mn} - \alpha_{mn})t\right] \tag{5.24}$$

$$\xi_2(x_1, x_2, t) = \left(G_{mn}^{(1)} \cos \lambda_m x_1 \sin \mu_n x_2 + G_{mn}^{(2)} \sin \lambda_m x_1 \cos \mu_n x_2\right) \exp\left[(i\omega_{mn} - \alpha_{mn})t\right] \tag{5.25}$$

where $F_{mn}^{(1)}$, $F_{mn}^{(2)}$, $G_{mn}^{(1)}$, $G_{mn}^{(2)}$ are arbitrary constants to be determined; while $\lambda_m = m\pi/L_1$ and $\mu_n = n\pi/L_2$. $v_3(x_1, x_2, t)$ can be expressed as

$$v_3(x_1, x_2, t) = (w_{mn} \sin \lambda_m x_1 \sin \mu_n x_2) \exp\left[(i\omega_{mn} - \alpha_{mn})t\right] \tag{5.26}$$

ω_{mn} are the undamped natural frequencies and α_{mn} are constants which provide a measure of damping. Substituting Eqs. (5.24 and 5.25) into Eqs. (5.16a and 5.16b) and identifying the coefficients of the like trigonometric functions provides the coefficients $F_{mn}^{(1)}$, $F_{mn}^{(2)}$, $G_{mn}^{(1)}$, $G_{mn}^{(2)}$ as

$$\left(F_{mn}^{(1)},\ F_{mn}^{(2)},\ G_{mn}^{(1)},\ G_{mn}^{(2)}\right) = \left(\widetilde{F}_{mn}^{(1)},\ \widetilde{F}_{mn}^{(2)},\ \widetilde{G}_{mn}^{(1)},\ \widetilde{G}_{mn}^{(2)}\right) w_{mn} \tag{5.27}$$

The coefficients $\widetilde{F}_{mn}^{(1)}$, $\widetilde{F}_{mn}^{(2)}$, $\widetilde{G}_{mn}^{(1)}$, $\widetilde{G}_{mn}^{(2)}$ are solutions to the following matrix equation

$$\begin{bmatrix} Y_{11}^{(m,n)} & Y_{12}^{(m,n)} & Y_{13}^{(m,n)} & Y_{14}^{(m,n)} \\ & Y_{11}^{(m,n)} & Y_{14}^{(m,n)} & Y_{13}^{(m,n)} \\ & & Y_{33}^{(m,n)} & Y_{34}^{(m,n)} \\ \text{Symm.} & & & Y_{33}^{(m,n)} \end{bmatrix} \begin{Bmatrix} \widetilde{F}_{mn}^{(1)} \\ \widetilde{F}_{mn}^{(2)} \\ \widetilde{G}_{mn}^{(1)} \\ \widetilde{G}_{mn}^{(2)} \end{Bmatrix} = \begin{Bmatrix} Z_{mn}^{(1)} \\ Z_{mn}^{(2)} \\ Z_{mn}^{(3)} \\ Z_{mn}^{(4)} \end{Bmatrix} \tag{5.28}$$

where

$$\begin{aligned} &Y_{11}^{(m,n)} = A_{11}m^2 + A_{66}n^2\phi^2, \quad Y_{12}^{(m,n)} = 2A_{16}mn\phi, \quad Y_{13}^{(m,n)} = A_{16}m^2 + A_{26}n^2\phi^2 \\ &Y_{14}^{(m,n)} = (A_{12} + A_{66})mn\phi, \quad Y_{33}^{(m,n)} = A_{22}n^2\phi^2 + A_{66}m^2, \quad Y_{34}^{(m,n)} = 2A_{26}mn\phi \\ &Z_{mn}^{(1)} = -\frac{m}{n}(\psi_1 A_{16} + \psi_2\phi A_{12}), \quad Z_{mn}^{(2)} = -\frac{n\phi}{\pi}(\psi_1 A_{16} + \psi_2\phi A_{26}) \\ &Z_{mn}^{(3)} = -\frac{m}{\pi}(\psi_1 A_{16} + \psi_2\phi A_{26}), \quad Z_{mn}^{(4)} = -\frac{n\phi}{\pi}(\psi_1 A_{12} + \psi_2\phi A_{22}) \end{aligned} \tag{5.29a – j}$$

In the above expressions, $\phi = L_1/L_2$, $\psi_1 = L_1/R_1$, $\psi_2 = L_2/R_2$ where ϕ is the aspect ratio. From the above matrix equation, $\widetilde{F}_{mn}^{(1)}$, $\widetilde{F}_{mn}^{(2)}$, $\widetilde{G}_{mn}^{(1)}$, $\widetilde{G}_{mn}^{(2)}$ can be evaluated, using Cramer's rule such that

$$F_{mn}^{(1)} \frac{\det(Y_1)}{\det(Y)}, \quad F_{mn}^{(2)} \frac{\det(Y_2)}{\det(Y)}, \quad G_{mn}^{(1)} \frac{\det(Y_3)}{\det(Y)}, \quad G_{mn}^{(2)} \frac{\det(Y_4)}{\det(Y)} \tag{5.30a – d}$$

where

$$\begin{aligned} Y_1 &= \begin{pmatrix} Z_{mn}^{(1)} & Y_{12}^{(m,n)} & Y_{13}^{(m,n)} & Y_{14}^{(m,n)} \\ Z_{mn}^{(2)} & Y_{11}^{(m,n)} & Y_{14}^{(m,n)} & Y_{13}^{(m,n)} \\ Z_{mn}^{(3)} & Y_{14}^{(m,n)} & Y_{33}^{(m,n)} & Y_{34}^{(m,n)} \\ Z_{mn}^{(4)} & Y_{13}^{(m,n)} & Y_{34}^{(m,n)} & Y_{33}^{(m,n)} \end{pmatrix}, \\ Y_2 &= \begin{pmatrix} Y_{11}^{(m,n)} & Z_{mn}^{(1)} & Y_{13}^{(m,n)} & Y_{14}^{(m,n)} \\ Y_{12}^{(m,n)} & Z_{mn}^{(2)} & Y_{14}^{(m,n)} & Y_{13}^{(m,n)} \\ Y_{13}^{(m,n)} & Z_{mn}^{(3)} & Y_{33}^{(m,n)} & Y_{34}^{(m,n)} \\ Y_{14}^{(m,n)} & Z_{mn}^{(4)} & Y_{34}^{(m,n)} & Y_{33}^{(m,n)} \end{pmatrix} \end{aligned} \tag{5.31a, b}$$

$$Y_3 = \begin{pmatrix} Y_{11}^{(m,n)} & Y_{12}^{(m,n)} & Z_{mn}^{(1)} & Y_{14}^{(m,n)} \\ Y_{12}^{(m,n)} & Y_{11}^{(m,n)} & Z_{mn}^{(2)} & Y_{13}^{(m,n)} \\ Y_{13}^{(m,n)} & Y_{14}^{(m,n)} & Z_{mn}^{(3)} & Y_{34}^{(m,n)} \\ Y_{14}^{(m,n)} & Y_{13}^{(m,n)} & Z_{mn}^{(4)} & Y_{33}^{(m,n)} \end{pmatrix}, \quad Y_4 = \begin{pmatrix} Y_{11}^{(m,n)} & Y_{12}^{(m,n)} & Y_{13}^{(m,n)} & Z_{mn}^{(1)} \\ Y_{12}^{(m,n)} & Y_{11}^{(m,n)} & Y_{14}^{(m,n)} & Z_{mn}^{(2)} \\ Y_{13}^{(m,n)} & Y_{14}^{(m,n)} & Y_{33}^{(m,n)} & Z_{mn}^{(3)} \\ Y_{14}^{(m,n)} & Y_{13}^{(m,n)} & Y_{34}^{(m,n)} & Z_{mn}^{(4)} \end{pmatrix} \tag{5.31c, d}$$

$$Y = \begin{pmatrix} Y_{11}^{(m,n)} & Y_{12}^{(m,n)} & Y_{13}^{(m,n)} & Y_{14}^{(m,n)} \\ & Y_{11}^{(m,n)} & Y_{14}^{(m,n)} & Y_{13}^{(m,n)} \\ & & Y_{33}^{(m,n)} & Y_{34}^{(m,n)} \\ \text{Symm.} & & & Y_{33}^{(m,n)} \end{pmatrix} \tag{5.31e}$$

The expressions for ξ_1, ξ_2, and v_3 identically fulfill the first two equations of motion. A similar procedure to identically fulfill the third and fourth equation of motion is followed. With this in mind η_1 and η_2 are assumed in the following form

$$\eta_1(x_1, x_2, t) = \left(H_{mn}^{(1)} \cos \lambda_m x_1 \sin \mu_n x_2 + H_{mn}^{(2)} \sin \lambda_m x_1 \cos \mu_n x_2\right) \exp\left[(i\omega_{mn} - \alpha_{mn})t\right] \tag{5.32a}$$

$$\eta_2(x_1, x_2, t) = \left(I_{mn}^{(1)} \cos \lambda_m x_1 \sin \mu_n x_2 + I_{mn}^{(2)} \sin \lambda_m x_1 \cos \mu_n x_2\right) \exp\left[(i\omega_{mn} - \alpha_{mn})t\right] \tag{5.32b}$$

where $H_{mn}^{(1)}$, $H_{mn}^{(2)}$, $I_{mn}^{(1)}$, $I_{mn}^{(2)}$ are undetermined coefficients. Substituting η_1 and η_2 into Eqs. (5.16c and 5.16d) and comparing the coefficients of the same trigonometric functions yields the unknown coefficients. For doubly curved sandwich shells these coefficients are given in Eqs. (5.10a–d)–(5.12a–e). This leaves Eqs. (5.16c and 5.16d) identically fulfilled. There remains the fifth equilibrium equation currently unfulfilled. As previously, this will be fulfilled in an average sense through the use of the EGM. In addition, the boundary conditions, Eqs. (5.18f) and (5.19f) are fulfilled with the remaining ones unfulfilled. Retaining the unfulfilled fifth equation of motion and the remaining unfulfilled boundary conditions from Eqs. (5.18a–e) and (5.19a–e) in the energy functional (Hamilton's equation) results in

$$\int_{t_0}^{t_1}\left\langle\int_0^{l_2}\int_0^{l_1}\left\{-\left(\frac{A_{11}}{R_1}+\frac{A_{12}}{R_2}\right)\frac{\partial\xi_1}{\partial x_1}-\left(\frac{A_{16}}{R_1}+\frac{A_{26}}{R_2}\right)\frac{\partial\xi_1}{\partial x_2}-\left(\frac{A_{22}}{R_2}+\frac{A_{12}}{R_1}\right)\frac{\partial\xi_2}{\partial x_2}-\right.\right.$$
$$-\left(\frac{A_{26}}{R_2}+\frac{A_{16}}{R_1}\right)\frac{\partial\xi_2}{\partial x_1}\left(\frac{A_{11}}{R_1^2}+\frac{A_{22}}{R_2^2}+\frac{2A_{12}}{R_1R_2}\right)v_3-d_1a\left(\frac{\partial\eta_1}{\partial x_1}+a\frac{\partial^2 v_3}{\partial x_1^2}\right)-d_2a\left(\frac{\partial\eta_2}{\partial x_2}+a\frac{\partial^2 v_3}{\partial x_2^2}\right)$$
$$+F_{11}\frac{\partial^4 v_3}{\partial x_1^4}+F_{22}\frac{\partial^4 v_3}{\partial x_2^4}+2(F_{12}+2F_{66})\frac{\partial^4 v_3}{\partial x_1^2\partial x_2^2}+4F_{16}\frac{\partial^4 v_3}{\partial x_1^3\partial x_2}+4F_{26}\frac{\partial^4 v_3}{\partial x_1\partial x_2^3}$$
$$\left.\left.+N_{11}^0\frac{\partial^2 v_3}{\partial x_1^2}+N_{22}^0\frac{\partial^2 v_3}{\partial x_2^2}+C\frac{\partial v_3}{\partial t}+m_0\frac{\partial^2 v_3}{\partial t^2}\right\}\delta v_3 dx_1 dx_2\right\rangle dt+$$
$$\int_{t_0}^{t_1}\left\langle\int_0^{l_2}\left\{N_{11}\delta\xi_1+N_{12}\delta\xi_2+L_{11}\delta\eta_1+L_{12}\delta\eta_2+M_{11}\delta\left(\frac{\partial v_3}{\partial x_1}\right)\right\}\Bigg|_0^{l_2}dx_2\right\rangle dt+$$
$$\int_{t_0}^{t_1}\left\langle\int_0^{l_1}\left\{N_{22}\delta\xi_2+N_{12}\delta\xi_1+L_{22}\delta\eta_2+L_{12}\delta\eta_1+M_{22}\delta\left(\frac{\partial v_3}{\partial x_2}\right)\right\}\Bigg|_0^{l_1}dx_1\right\rangle dt=0 \tag{5.33}$$

Substituting the expressions for ξ_1, ξ_2, η_1, η_2, and v_3 into Eq. (5.33) and carrying out the necessary integrations and simplifying realizing the independent character of the variations gives the following characteristic equation

$$S_{mn}^2+2\Delta_{mn}\omega_{mn}S_{mn}+\omega_{mn}^2=0 \tag{5.34a}$$

where

$$\omega_{mn}^2=K_{mn}/m_0 \quad \text{and} \quad \Delta_{mn}=C/2m_0\omega_{mn} \tag{5.34b}$$

Denote the undamped natural frequencies squared and the modal damping, respectively. The expression for the undamped natural frequency is given by

$$\omega_{mn}^2=\frac{1}{m_0}\left(\lambda_m^4F_{11}+2\lambda_m^2\mu_n^2(F_{12}+2F_{66})+\mu_n^4F_{22}+\lambda_m ad_1\left(H_{mn}^{(1)}+a\lambda_m\right)+\mu_n ad_2\left(I_{mn}^{(2)}+a\mu_n\right)\right.$$
$$+\lambda_m\left(\frac{A_{11}}{R_1}+\frac{A_{12}}{R_2}\right)F_{mn}^{(1)}+\mu_n\left(\frac{A_{12}}{R_1}+\frac{A_{22}}{R_2}\right)G_{mn}^{(2)}+\mu_n\left(\frac{A_{16}}{R_1}+\frac{A_{26}}{R_2}\right)F_{mn}^{(2)}+\lambda_m\left(\frac{A_{16}}{R_1}+\right.$$
$$\left.\left.\frac{A_{26}}{R_2}\right)G_{mn}^{(1)}+\left(\frac{A_{11}}{R_1^2}+2\frac{A_{12}}{R_1R_2}+\frac{A_{22}}{R_2^2}\right)-\lambda_m^2N_{11}^0-\mu_n^2N_{22}^0\right) \tag{5.35}$$

The expression for undamped natural frequencies can be arranged in dimensionless form as

$$\begin{aligned}\Omega_{mn}^2 &= m^4 + \frac{n^4\phi^4 F_{22}}{F_{11}} + \frac{2m^2n^2\phi^2(F_{12}+2F_{66})}{F_{11}} + \frac{a^2L_1^2}{\pi^2F_{11}}\left(m^2d_1 + n^2\phi^2 d_2\right)\\ &+\frac{aL_1^3}{\pi^3F_{11}}\left[md_1\widetilde{H}_{mn}^{(1)} + n\phi d_2\widetilde{I}_{mn}^{(2)}\right] + \frac{L_1^2}{\pi^3F_{11}}\left[m(\psi_1A_{11}+\phi\psi_2A_{12})\widetilde{F}_{mn}^{(1)}\right.\\ &\left.+n\phi(\psi_1A_{12}+\phi\psi_2A_{22})\widetilde{G}_{mn}^{(2)} + \left(n\phi\widetilde{F}_{mn}^{(2)} + m\widetilde{G}_{mn}^{(1)}\right)(\psi_1A_{16}+\phi\psi_2A_{26})\right]\\ &+\frac{L_1^2}{\pi^4F_{11}}\left(\psi_1^2A_{11} + \psi_2^2\phi^2A_{22} + 2A_{12}\psi_1\psi_2\phi\right) - K_x\left(m^2\psi_1^2 + L_Rn^2\phi^2\psi_2^2\right)\end{aligned} \tag{5.36}$$

where the nondimensional parameters Ω_{mn}^2, K_x, L_R are defined by Eq. (5.15a–d).

5.4.3 Validation of the Theoretical Model

For the case of cylindrical shells, validations are made with Rahmani et al. (2010) who considered the free vibration of cylindrical sandwich shells with a flexible core using a higher order theoretical model. Although the current theoretical model does not incorporate the theoretical aspects of the flexible core, particularly good agreement is found between the current or present model and that of Rahmani et al. (2010). This shows, as far as the natural frequency is concerned, at least for the fundamental mode, that the compressibility of the core has only a minor effect on the fundamental natural frequency. For the following validations, Table 5.7 specifies the material properties for this particular case while Tables 5.8 and Table 5.9 contain the comparisons between the present model and that of Rahmani et al. (2010) of the fundamental frequencies of a cylindrical sandwich shell. Table 5.8 compares the

Table 5.7 Material and geometrical properties

	Elastic modulus (GPa)				Shear modulus (GPa)		
	E_1	E_2	Poisson's ratio	Mass density (kg/m^3)	G_{12}	$\overline{G}_{13}$	$\overline{G}_{23}$
Facings	24.51	7.77	0.078	1800	3.34	–	–
Core	–		0	130	–	0.05	0.05

Table 5.8 Natural frequencies of circular cylindrical sandwich shell: validation no. 1 $\left(H/L_2 = 1\ \text{m},\ 2\bar{h}/H = 0.88,\ L_1 = 1\ \text{m}\right)$

	Frequency (Hz)	$R_2 = 1$ m, $L_2 = 1$ m	$R_2 = 2$ m, $L_2 = 1$ m
R. et al.	ω_1	234.77	211.92
Present	ω_{11}	252.034	217.21
% Error		6.8	2.4

R. et al. → Rahmani et al. (2010)

Table 5.9 Natural frequencies of circular cylindrical sandwich shell: validation no. 2 $\left(H/L_2 = 1\ \text{m},\ 2\bar{h}/H = 0.88,\ L_1 = 1\ \text{m},\ R_2 = 1\ \text{m}\right)$

	Frequency (Hz)	$L_1 = 1$ m	$L_1 = 2$ m	$L_1 = 3$ m	$L_1 = 5$ m	$L_1 = 10$ m
R. et al.	ω_1	211.92	141.35	128.18	122.33	120.27
Present	ω_1	217.21	146.88	133.72	127.77	125.64
% Error		2.4	3.8	4.1	4.3	4.3

R. et al. → Rahmani et al. (2010)

Table 5.10 Material characteristics of the face sheets

Type	Material	E_1(GPa)	E_2(GPa)	G_{12}(GPa)	ν_{12}
F1	HS Graph. Ep.	180.99	10.34	7.24	0.28
F2	IM7/977-2	79.98	75.15	9.65	0.06

Table 5.11 Material characteristics of the core

Type	Core type	$\overline{G}_{13}$ (GPa)	$\overline{G}_{23}$ (GPa)
C1	Titanium honeycomb	1.44	0.651

Table 5.12 Comparisons of eigenfrequencies and buckling loads

		Plate (see Fig. 5.1)			Shell (see Fig. 5.7)		
θ	ψ	K_{cr}	N°_{11} (N/m)	$\sqrt{\Omega_{11}}$	K_{cr}	N°_{11} (N/m)	$\sqrt{\Omega_{11}}$
47	0.9	7110.28	2.08×10^6	16.28	8946.94	2.61×10^6	18.7631
44.1	1.0	4913.96	1.67×10^6	14.84	6945.93	2.37×10^6	17.5176
38.6	1.2	2734.11	1.22×10^6	12.82	4674.68	2.08×10^6	14.6559
31.5	1.4	1664.60	0.99×10^6	11.32	2289.93	1.36×10^6	12.2611
19.8	1.6	1056.09	0.88×10^6	10.10	1256.84	1.05×10^6	10.5534

fundamental frequencies ($m = n = 1$) for different radius of curvatures while Table 5.9 compares the fundamental frequencies for various lengths (L_1) of the side/edge of the sandwich panel. In both cases the facings consist of a [0/90/0/core/0/90/0] layup. Remarkably close agreement is seen between the present results and that of Rahmani et al. (2010) with discrepancies under 7%.

5.4.4 Present Results and Discussion

The material properties of the core and face sheets for the numerical results that follow are listed in Tables 5.10 and 5.11. Table 5.12 displays the comparison between the critical buckling loads for a flat and doubly curved sandwich panel where the eigenfrequency vanishes. Also, in comparison are the eigenfrequencies at

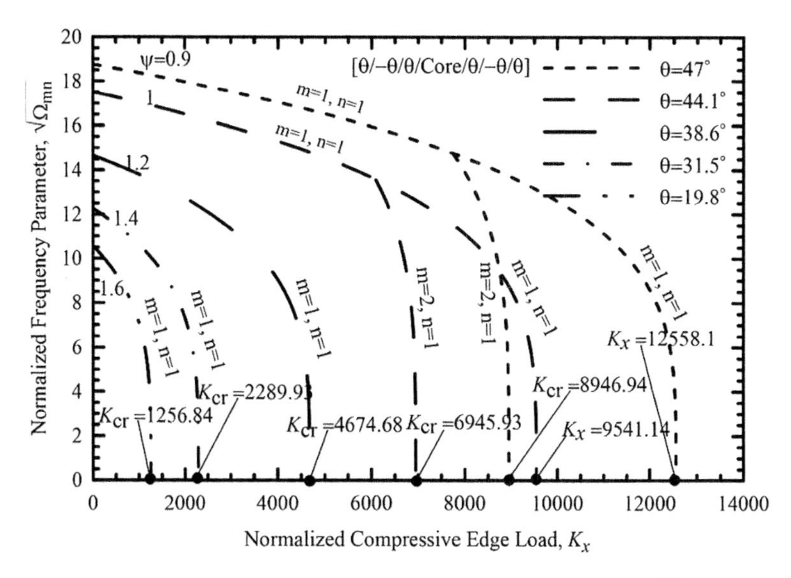

Fig. 5.7 The effect of the compressive edge load on the eigenfrequencies $\Omega^{1/2}$ of a cylindrical sandwich panel for various aspect ratios

zero compressive edge load. The comparison indicates that at the exclusion of the compressive edge load, higher eigenfrequencies are inherent for curved sandwich panel construction over flat sandwich panels. Figure 5.7 illustrates the vanishing of the eigenfrequencies at the critical buckling load for a cylindrical sandwich panel ($L_1/R_1 = L_2/R_2 = 0.4$) for various characteristic ply angles. At the larger ply angles, the structure benefits from its capacity to carry a larger compressive edge load prior to buckling. In Fig. 5.8, higher eigenfrequencies appear to be the norm at larger length-to-curvature ratios. There also appears to be a continual increase in the eigenfrequencies over the span from the 0- to 90-degree ply angles per length-to-curvature ratio. There also appears to be only a marginal difference between the eigenfrequencies, for the various curvature ratios, at the two extremes of 0- and 90-degrees. As illustrated in Fig. 5.8, similar behavior is seen in

Figure 5.9 with the increase of the length-to-curvature ratio. In addition, as the core thickness increases, providing a higher thickness ratio, the eigenfrequencies increase. An increase in the core thickness-to-face thickness ratio seems to act as a catalyst to boosting the eigenfrequencies.

Figure 5.10 highlights the effect of the orthotropicity ratio on the normalized eigenfrequencies for characteristic ply angles of a single-layered cylindrical sandwich panel. It appears that for ply angles less than 45 degrees there seems to be a marginal continuous increase in the eigenfrequencies over the orthotropicity ratio span of between 0 and 1. At the 45-degree level, the effect of the orthotropicity ratio seems to be flat or null. Above 45 degrees, as the orthotropicity ratio increases, the eigenfrequencies trend marginally downward. Figure 5.11 conveys the effect of the material directional properties on the frequency ratio for characteristic curvature

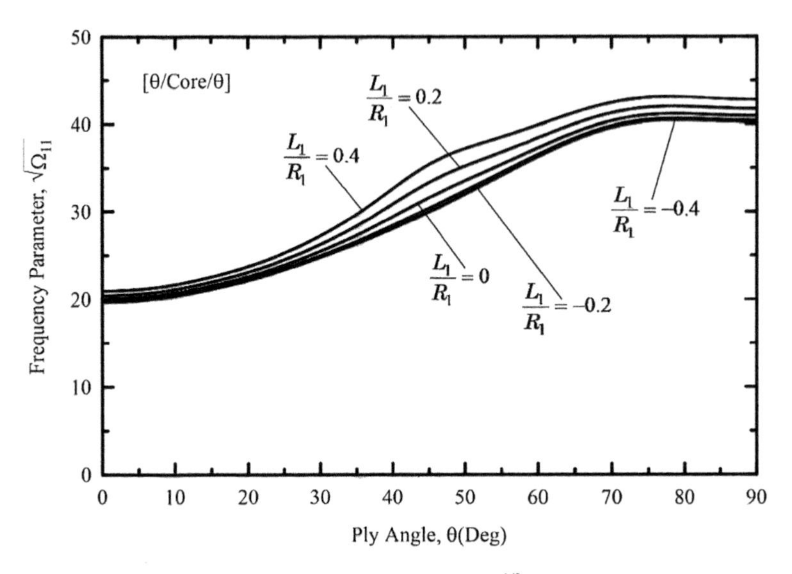

Fig. 5.8 The effect of the ply angle on the eigenfrequencies $\Omega^{1/2}$ of a cylindrical sandwich panel for various length-to-curvature ratios

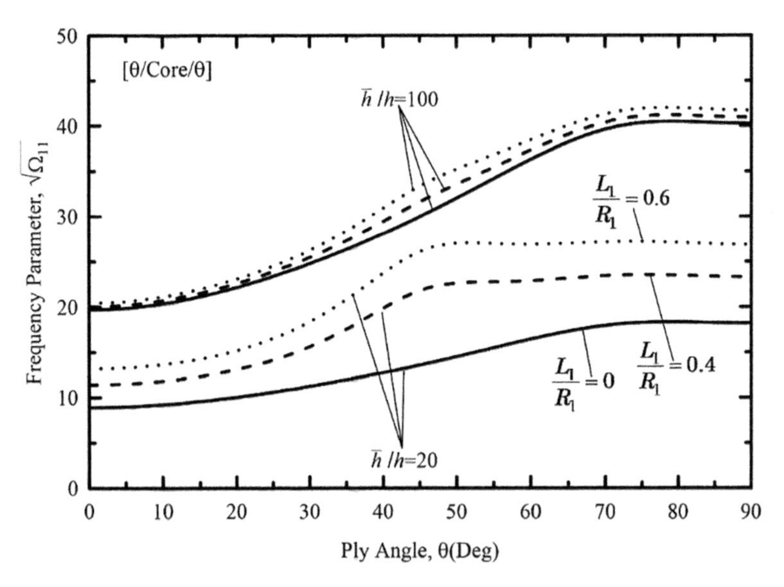

Fig. 5.9 The effect of the ply angle on the eigenfrequencies $\Omega^{1/2}$ of a cylindrical sandwich panel for various normalized curvatures and core-to-face thickness ratios

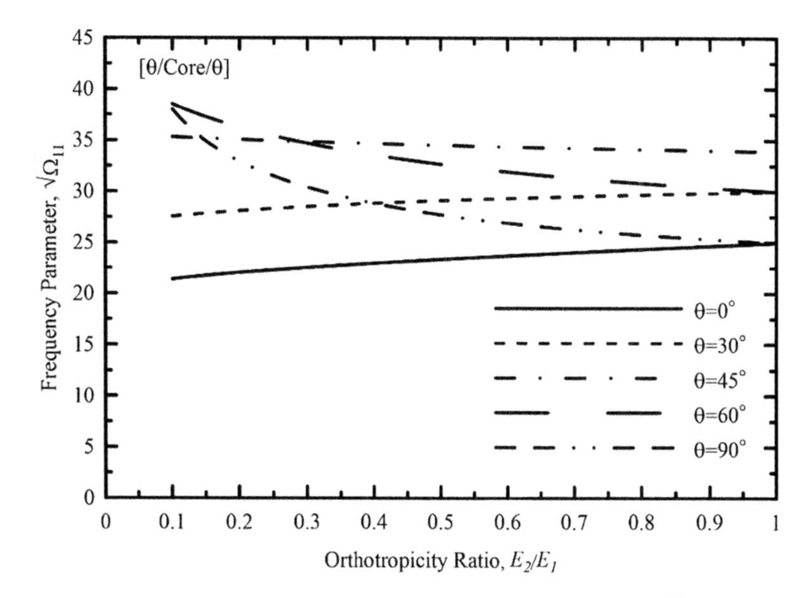

Fig. 5.10 The effect of the orthotropicity ratio on the eigenfrequencies $\Omega^{1/2}$ of a cylindrical sandwich panel

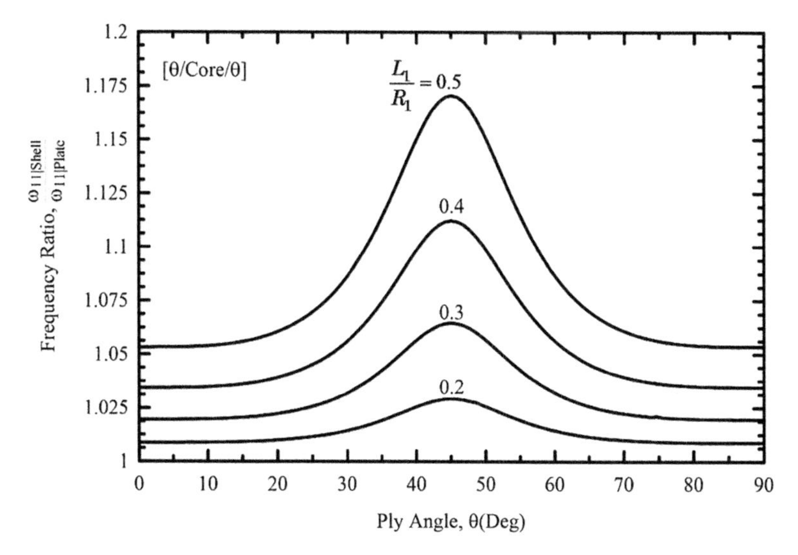

Fig. 5.11 The effect of the ply angle on the ratio of the shell-to-plate eigenfrequencies $\Omega^{1/2}$ of a cylindrical sandwich panel for various normalized curvatures

ratios. The larger curvature ratio exhibits higher frequency ratios which is consistently larger than one. This implies that the eigenfrequencies for a curved sandwich panel are higher than for flat sandwich panel. Additionally, evident, for a fixed

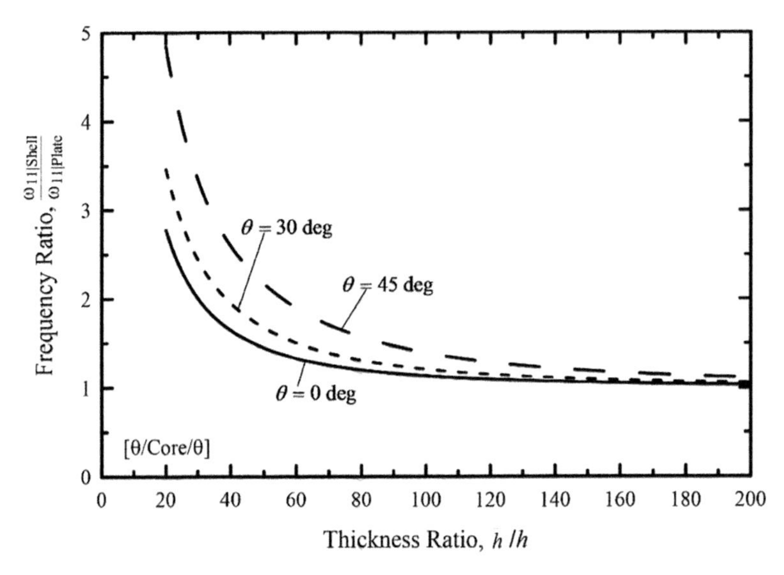

Fig. 5.12 The effect of the thickness ratio on the ratio of the shell-to-plate eigenfrequencies $\Omega^{1/2}$ of a cylindrical sandwich panel for various ply angles

curvature ratio, is the increase in the frequency ratio up to 45 degrees followed by a decrease until a ply angle of 90 degrees.

In Fig. 5.12 the effect of the core-to-face sheet thickness ratio on the frequency ratio for fixed ply angles concerning a single-layered curved sandwich panel is highlighted. At small core thickness-to- face thickness ratios there appears to be a much larger spread between the values of the frequency ratio for the various fixed ply angles as compared to the larger ones. At the extreme of the larger thickness ratios there is only a negligible difference in the values of the frequency ratio between the fixed ply angle constructs. It is in the vicinity of the larger thickness ratios that the flat and curved sandwich panels exhibit almost identical eigenfrequency values. It can also be seen that as the thickness ratio increases the frequency ratio decays exponentially. Figure 5.13 displays the effect of the normalized curvature on the frequency ratio of a curved sandwich panel for characteristic ply angles. It is seen that an increase in the curvature ratio results in higher frequency ratios for all ply angles. The larger the curvature ratio, the larger the margin between the frequency of the flat sandwich panel and a curved sandwich panel. Figure 5.14 exhibits the effect of the aspect ratio on the frequency ratio for fixed values of ply angles. There appears to be a continual increase in the frequency ratio for all cases considered up to a peak value followed by a decrease which asymptotically approaches a frequency ratio value of 1. Additionally, the larger peak values pertaining to the frequency ratio seem to rest with the fiber orientations of 30, 45, and 60 degrees.

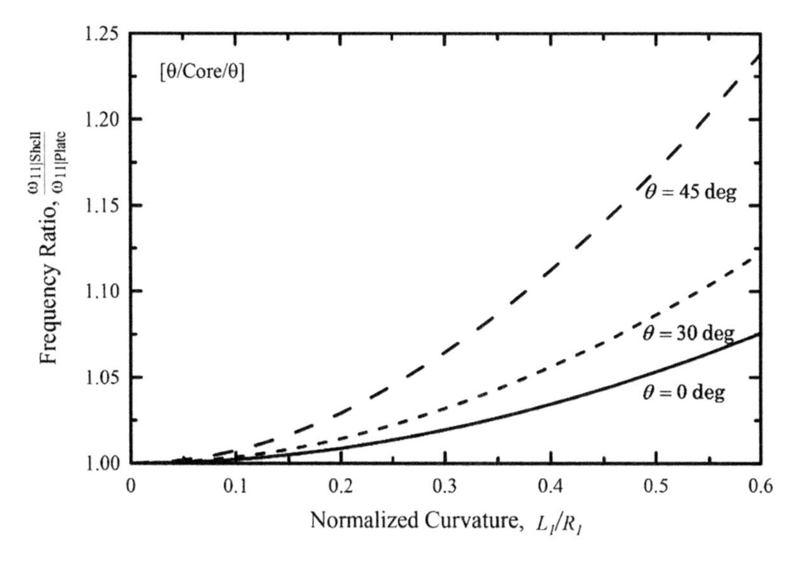

Fig. 5.13 The effect of the normalized curvature on the ratio of the shell-to-plate eigenfrequencies $\Omega^{1/2}$ of a cylindrical sandwich panel for various ply angles

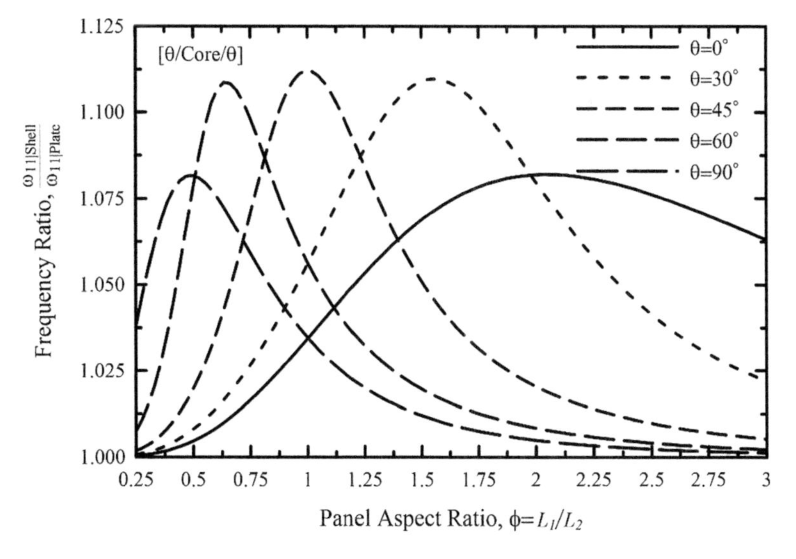

Fig. 5.14 The effect of the panel aspect ratio on the ratio of the shell-to-plate eigenfrequencies $\Omega^{1/2}$ of a cylindrical sandwich panel for various curvatures

5.5 Summary

A theoretical model applied to the free vibration of flat and curved sandwich panels with anisotropic laminated face sheets and a weak/soft core has been presented. Flat panels were considered in the first part which then followed with a discussion of the free vibration of curved sandwich panels in the second part. In both cases, very intricate analytical solution methodologies were applied. The theory was then validated against other cases found in the literature. Remarkable close agreement was seen. Finally numerical results were presented which included the effects of the orthotropicity ratios of both the core and the face sheets, fiber orientation angles within the face sheets, the aspect ratio of the panel, the curvature ratio, the core thickness-to face thickness ratio, and other additional geometrical parameters. Also, the concept of the eigenfrequencies vanishing at the critical buckling load was discussed. Additionally, all of these effects on the frequency ratio were addressed. It was found that the eigen-frequencies of a sandwich shell are higher than for a sandwich plate.

References

Hause, T., Librescu, L., & Johnson, T. F. (1998). Thermomechanical load-carrying capacity of sandwich flat panels. *Journal of Thermal Stresses, 21*(6), 627–653.

Hause, T., Johnson, T. F., & Librescu, L. (2000). Effect of face-sheet anisotropy on buckling and post buckling of flat sandwich panels. *Journal of Spacecraft and Rockets, 37*(3), 331–341.

Hohe, J., Librescu, L., & Oh, S. Y. (2006). Dynamic buckling of flat and curved sandwich panels with transversely compressible core. *Composite Structures, 74*(1), 10–24.

Librescu, L. (1987). Refined geometrically non-linear theories of anisotropic laminated shells. *Quarterly of Applied Mathematics, 45*(1), 1–22.

Librescu, L., Hause, T., & Camarda, C. J. (1997). Geometrically nonlinear theory of initially imperfect sandwich plates and shells incorporating non-classical effects. *AIAA Journal, 35*(8), 1393–1403.

Monforton, G. R., & Schmidt, L. A., Jr. (1968). Finite element analysis of sandwich plates and cylindrical shells with laminate faces. *Proceedings of the conference on matrix methods on structural mechanics*, TR-68-150, Air Force Flight Dynamics Lab., Wright Patterson Air Force Base, OH, pp. 573–616.

Rahmani, O., Khalili, S. M. R., & Malekzadeh, K. (2010). Free vibration response of composite sandwich cylindrical shell with flexible core. *Composite Structures, 92*, 1269–1281.

Raville, M. E., & Ueng, C. E. S. (1967). Determination of natural frequencies of vibration of a sandwich plate. *Experimental Mechanics, 7*(4), 490–493.

Zhou, H. B., & Li, G. Y. (1996). Free vibration analysis of sandwich plates with laminated faces using spline finite point method. *Computers & Structures, 59*(2), 257–263.

Chapter 6
Dynamic Response to Time-Dependent External Excitations

Abstract An advanced theoretical model pertaining to sandwich structures characterized by anisotropic laminated facings and a weak orthotropic core is addressed, in regard to the dynamic response problem. In this regard, several time-dependent external pressure pulses are considered which lend themself to a closed form solution, for both the eigenfrequencies and the dynamic response problem, utilizing the Laplace transform and the extended Galerkin method. Finally, a detailed analysis considering the influence of several parameters is conducted, with numerical results covering a broad spectrum of responses to various material or geometrical configurations presented with a thorough discussion of the results.

6.1 Introduction

This chapter is concerned with the dynamic response behavior of doubly curved sandwich panels experiencing forced vibration. The first part of the chapter is concerned with the dynamic response of sandwich plates followed by doubly curved sandwich shells. Solution methodologies are applied which allows for a closed form solution for both the eigenfrequencies and the dynamic response. Such closed form solutions involve the Laplace Transform and the extended Galerkin method. The dynamic response considers various time-dependent external loadings for both plates and shells with anisotropic laminated face sheets. The implications of the panel curvature, anisotropy and stacking sequence of the face sheets, the orthotropy of the core, and the structural damping are considered as to their effects on the structural response to time-dependent stimuli, of the structure. Various types of time-dependent loads are considered such as the sonic boom, triangular pulse, Heaviside step function, rectangular pulse, the sine pulse, a tangential traveling air blast, and the Friedlander in-Air explosive pulse, numerical results are presented which covers a broad spectrum of responses to various material or geometrical configurations.

T. J. Hause, *Sandwich Structures: Theory and Responses*,
https://doi.org/10.1007/978-3-030-71895-4_6

6.2 Preliminaries and Basic Assumptions

In consideration of the dynamic response of sandwich structures, the incompressible core case will be considered for both the plate and shell configuration. With this in mind the following basic assumptions are adopted:

1. The face sheets are orthotropic layers not necessarily coincident with the geometrical axes.
2. The core features orthotropic properties in the transverse direction and is considered the weak-type and much larger in thickness than the facings.
3. Perfect bonding between the face sheets and the facings and the core are assumed.
4. The shallow shell theory is assumed.
5. The transverse shear effects in the facings are discarded.
6. The face sheets are symmetric with respect to their local and global mid-surfaces.
7. The linearized counterpart of the equations of motion, Eqs. (2.156)–(2.160) are adopted.
8. The tangential and rotatory inertia terms are neglected (Librescu 1987).

6.3 Flat Sandwich Panels

6.3.1 *Governing System*

In this section, the governing equations which are considered for the dynamic response problem of flat sandwich panels are based on the linear theory for flat sandwich panels assuming symmetric laminated facings, a weak orthotropic core, and with the transverse shear effects discarded in the facings. With these considerations in mind, Eqs. (2.174)–(2.176) are applicable along with the set of simply supported boundary conditions expressed in Eqs. (2.177a–d). As mentioned in Chap. 2, for flat panels, setting the curvatures to zero decouples the first two equations of motion from the last three. The last three equations of motion govern the bending problem separate from the stretching problem. This implies that, in the case of flat sandwich plates, the dynamic response problem is only dependent on the bending problem. This chapter will be concerned with both flat and curved sandwich panels. The dynamic response problem is an extension of the free vibration problem in that the transverse pressure loading term $p_3 \neq 0$. The same governing equations apply except with the transverse pressure loading term p_3 included. This allows for the ability to include several time-dependent loading scenarios into the equations. Neglecting the thermal terms, the governing equations of motion are

- *Equations of Motion*

$$A_{11}\frac{\partial^2\eta_1}{\partial x_1^2}+A_{66}\left(\frac{\partial^2\eta_1}{\partial x_2^2}+\frac{\partial^2\eta_2}{\partial x_1\partial x_2}\right)+A_{12}\frac{\partial^2\eta_2}{\partial x_1\partial x_2}+A_{16}\left(\frac{\partial^2\eta_2}{\partial x_1^2}+2\frac{\partial^2\eta_1}{\partial x_1\partial x_2}\right)+A_{26}\frac{\partial^2\eta_2}{\partial x_2^2}$$
$$-d_1\left(\eta_1+a\frac{\partial v_3}{\partial x_1}\right)=0 \tag{6.1a}$$

$$A_{22}\frac{\partial^2\eta_2}{\partial x_2^2}+A_{66}\left(\frac{\partial^2\eta_2}{\partial x_2^2}+\frac{\partial^2\eta_1}{\partial x_1\partial x_2}\right)+A_{12}\frac{\partial^2\eta_1}{\partial x_1\partial x_2}+A_{26}\left(\frac{\partial^2\eta_1}{\partial x_2^2}+2\frac{\partial^2\eta_2}{\partial x_1\partial x_2}\right)+A_{16}\frac{\partial^2\eta_1}{\partial x_1^2}$$
$$-d_2\left(\eta_2+a\frac{\partial v_3}{\partial x_2}\right)=0 \tag{6.1b}$$

$$-d_1a\left(\frac{\partial\eta_1}{\partial x_1}+a\frac{\partial^2 v_3}{\partial x_1^2}\right)-d_2a\left(\frac{\partial\eta_2}{\partial x_2}+a\frac{\partial^2 v_3}{\partial x_2^2}\right)+F_{11}\frac{\partial^4 v_3}{\partial x_1^4}+F_{22}\frac{\partial^4 v_3}{\partial x_2^4}+2(F_{12}+$$
$$2F_{66})\frac{\partial^4 v_3}{\partial x_1^2\partial x_2^2}+4F_{16}\frac{\partial^4 v_3}{\partial x_1^3\partial x_2}+4F_{26}\frac{\partial^4 v_3}{\partial x_1\partial x_2^3}+N_{11}^0\frac{\partial^2 v_3}{\partial x_1^2}+2N_{12}^0\frac{\partial^2 v_3}{\partial x_1\partial x_2}+$$
$$N_{22}^0\frac{\partial^2 v_3}{\partial x_2^2}-c\frac{\partial v_3}{\partial t}-m_0\frac{\partial^2 v_3}{\partial t^2}=-P_3(t) \tag{6.1c}$$

This leaves three equations in terms of three unknowns, η_1, η_2, v_3. These unknown variables need to be determined in such a manner as to satisfy the corresponding equations of motion, while at the same time satisfying the boundary conditions. The governing equations of motion and the boundary conditions constitute the governing system for the dynamic response of plates. For simply supported boundary the following conditions hold.

at $x_1 = 0,\ L_1$

$$\eta_1=\eta_2=M_{11}=v_3=0 \tag{6.2a – d}$$

at $x_2 = 0,\ L_2$

$$\eta_1=\eta_2=M_{22}=v_3=0 \tag{6.3a – d}$$

6.3.2 Solution Methodology

To identically satisfy the fourth and fifth boundary conditions, Eq. (6.2-2f) and Eq. (6.2-3f), the transverse displacement $v_3(t)$ is assumed in the following form

$$v_3(x_1, x_2, t) = w_{mn}(t) \sin \lambda_m x_1 \sin \mu_n x_2 \tag{6.4}$$

where $\lambda_m = m\pi/L_1$, $\mu_n = n\pi/L_2$ and $w_{mn}(t)$ is the amplitude of deflection as a function of time. Regarding η_1, η_2, they are assumed in the following form.

$$\eta_1(x_1, x_2, t) = w_{mn}(t)\left(H_{mn}^{(1)} \cos \lambda_m x_1 \sin \mu_n x_2 + H_{mn}^{(2)} \sin \lambda_m x_1 \cos \mu_n x_2\right) \tag{6.5}$$

$$\eta_2(x_1, x_2, t) = w_{mn}(t)\left(I_{mn}^{(1)} \cos \lambda_m x_1 \sin \mu_n x_2 + I_{mn}^{(2)} \sin \lambda_m x_1 \cos \mu_n x_2\right) \tag{6.6}$$

where, $H_{mn}^{(1)}, H_{mn}^{(2)}, I_{mn}^{(1)}$ and $I_{mn}^{(2)}$ are undetermined coefficients. These coefficients are easily determined by the use of Eqs. (6.5)–(6.6) in Eqs. (6.1a, b) and comparing coefficients of like trigonometric functions. These coefficients are identical to the ones previously determined in earlier sections. The first two equations of motion, Eqs. (6.1a, b) and the boundary conditions, Eqs. (6.2d) and (6.3d) are identically fulfilled. There remains the third equation of motion and the remaining unfulfilled boundary conditions. These will be retained in the energy functional and thus by performing the necessary operations will result in fulfilling the last equation of motion and the remaining boundary conditions in an average sense. Inserting these unfilled expressions back into Hamilton's equation from gives

$$\begin{aligned}
&\int_{t_0}^{t_1}\left\langle\int_0^{l_2}\int_0^{l_1}\left\{-d_1 a\left(\frac{\partial \eta_1}{\partial x_1}+a\frac{\partial^2 v_3}{\partial x_1^2}\right)-d_2 a\left(\frac{\partial \eta_2}{\partial x_2}+a\frac{\partial^2 v_3}{\partial x_2^2}\right)+F_{11}\frac{\partial^4 v_3}{\partial x_1^4}+F_{22}\frac{\partial^4 v_3}{\partial x_2^4}+2(F_{12}+\right.\right.\\
&2F_{66})\frac{\partial^4 v_3}{\partial x_1^2\partial x_2^2}+4F_{16}\frac{\partial^4 v_3}{\partial x_1^3\partial x_2}+4F_{26}\frac{\partial^4 v_3}{\partial x_1\partial x_2^3}+N_{11}^0\frac{\partial^2 v_3}{\partial x_1^2}+2N_{12}^0\frac{\partial^2 v_3}{\partial x_1\partial x_2}+N_{22}^0\frac{\partial^2 v_3}{\partial x_2^2}+\\
&\qquad\left.\left.C\frac{\partial v_3}{\partial t}+m_0\frac{\partial^2 v_3}{\partial t^2}-P_3(t)\right\}\delta v_3 dx_1 dx_2\right\rangle dt+\\
&\qquad\int_{t_0}^{t_1}\left\langle\int_0^{l_2}\left\{L_{11}\delta\eta_1+L_{12}\delta\eta_2+M_{11}\delta\left(\frac{\partial v_3}{\partial x_1}\right)\right\}\Bigg|_0^{l_1}dx\right\rangle dt+\\
&\qquad\int_{t_0}^{t_1}\left\langle\int_0^{l_1}\left\{L_{12}\delta\eta_1+L_{22}\delta\eta_2+M_{22}\delta\left(\frac{\partial v_3}{\partial x_2}\right)\right\}\Bigg|_0^{l_2}dx\right\rangle dt=0
\end{aligned} \tag{6.7}$$

Substituting in the expressions for L_{11}, L_{22}, L_{12} along with $\delta\eta_1$, $\delta\eta_2$, δv_3 into Eq. (6.7) and carrying out the indicated operations, results in a second-order differential equation which governs the dynamic response of sandwich plate structures. This governing differential equation is presented as

$$\ddot{w}_{mn} + 2\Delta_{mn}\omega_{mn}\dot{w}_{mn} + \omega_{mn}^2 w_{mn} = F_{mn}(t) \tag{6.8}$$

where the expression for the natural undamped frequencies, ω_{mn}^2 is given by

$$\omega_{mn}^2 = \frac{1}{m_0}\left(d_1 a\lambda_m H_{mn}^{(1)} + d_2 a\mu_n I_{mn}^{(2)} + d_1 a^2\lambda_m^2 + d_2 a^2\mu_n^2 + F_{11}\lambda_m^2 + F_{22}\mu_n^2 + 4F_{66}\lambda_m^2\mu_n^2 + 2F_{12}\lambda_m^2\mu_n^2 - N_{11}^0\lambda_m^2 - N_{22}^0\mu_n^2\right) \tag{6.9}$$

Or in dimensionless form as

$$\Omega_{mn}^2 = m^4 + \frac{F_{22}n^4\phi^4}{F_{11}} + \frac{2(F_{12}+2F_{66})m^2n^2\phi^2}{F_{11}} + \frac{a^2L_1^2}{F_{11}\pi^2}\left(d_1m^2 + d_2n^2\phi^2\right) + \frac{aL_1^3}{F_{11}\pi^3}\left(d_1 m H_{mn}^{(1)} + d_2 n\phi I_{mn}^{(2)}\right) + K_x\left(m^2 + L_R n^2\phi^2\right) \tag{6.10}$$

The nondimensional parameters $\phi,\ K_x,\ L_R,\ \Omega_{mn}^2$ were defined previously in Chap. 5. Typically, the undamped natural frequency is expressed as

$$\omega_{mn} = \sqrt{\frac{K_{mn}}{m_0}} \tag{6.11}$$

Leaving K_{mn} determined as

$$K_{mn} = d_1 a\lambda_m H_{mn}^{(1)} + d_2 a\mu_n I_{mn}^{(2)} + d_1 a^2\lambda_m^2 + d_2 a^2\mu_n^2 + F_{11}\lambda_m^2 + F_{22}\mu_n^2 + 4F_{66}\lambda_m^2\mu_n^2 + 2F_{12}\lambda_m^2\mu_n^2 - N_{11}^0\lambda_m^2 - N_{22}^0\mu_n^2 \tag{6.12}$$

In addition, $\Delta_{mn} = C/2m_0\omega_{mn}$ which denotes the modal viscous damping ratio, while the expression for the generalized load, $F_{mn}(t)$ is given by

$$F_{mn}(t) = \frac{16\delta_{m,2s-1}\delta_{n,2q-1}}{m_0(2s-1)(2q-1)\pi^2}p_3(t) \tag{6.13}$$

where,

$$\delta_{m,2s-1} = \begin{cases} 1 & m \text{ odd} \quad (s = 0,1,2,\ldots..) \\ 0 & m \text{ even} \end{cases} \tag{6.14}$$

The same holds true for $\delta_{n,\,2q\,-\,1}$. From Eq. (6.13), $p_3(t)$ can represent any one of a number of different types of external time-dependent loadings. This will be discussed in sect. 6.5.

6.4 Doubly Curved Sandwich Panels

6.4.1 Governing System

For doubly curved sandwich shells there is no decoupling present among the governing equations of motion. For this case coupling exists between stretching and bending. With this in mind, there are five coupled equations of motion which constitute the governing system of equations, in addition to the boundary conditions which needs to be fulfilled. For the dynamic response problems simply supported boundary conditions which are freely movable in both the tangential and normal directions will be assumed. There are five unknown displacement functions which are ξ_1, ξ_2, η_1, η_2, v_3 that need to be determined which satisfy both the chosen boundary conditions and the equations of motion. With this in mind the governing system of equations are

- *Equations of Motion.*

$$A_{11}\left(\frac{\partial^2 \xi_1}{\partial x_1^2}-\frac{1}{R_1}\frac{\partial v_3}{\partial x_1}\right)+A_{66}\left(\frac{\partial^2 \xi_1}{\partial x_2^2}+\frac{\partial^2 \xi_2}{\partial x_1 \partial x_2}\right)+A_{12}\left(\frac{\partial^2 \xi_2}{\partial x_1 \partial x_2}-\frac{1}{R_2}\frac{\partial v_3}{\partial x_2}\right)$$
$$-A_{16}\left(2\frac{\partial^2 \xi_1}{\partial x_1 \partial x_2}+\frac{\partial^2 \xi_2}{\partial x_1^2}-\frac{1}{R_1}\frac{\partial v_3}{\partial x_1}\right)-A_{26}\left(\frac{\partial^2 \xi_2}{\partial x_2^2}-\frac{1}{R_2}\frac{\partial v_3}{\partial x_2}\right)=0 \tag{6.15a}$$

$$A_{22}\left(\frac{\partial^2 \xi_2}{\partial x_2^2}-\frac{1}{R_2}\frac{\partial v_3}{\partial x_2}\right)+A_{66}\left(\frac{\partial^2 \xi_2}{\partial x_1^2}+\frac{\partial^2 \xi_1}{\partial x_1 \partial x_2}\right)+A_{12}\left(\frac{\partial^2 \xi_1}{\partial x_1 \partial x_2}-\frac{1}{R_1}\frac{\partial v_3}{\partial x_1}\right)$$
$$-A_{26}\left(2\frac{\partial^2 \xi_2}{\partial x_1 \partial x_2}+\frac{\partial^2 \xi_1}{\partial x_2^2}-\frac{1}{R_2}\frac{\partial v_3}{\partial x_2}\right)-A_{16}\left(\frac{\partial^2 \xi_1}{\partial x_1^2}-\frac{1}{R_1}\frac{\partial v_3}{\partial x_1}\right)=0 \tag{6.15b}$$

$$A_{11}\frac{\partial^2 \eta_1}{\partial x_1^2}+A_{66}\left(\frac{\partial^2 \eta_1}{\partial x_2^2}+\frac{\partial^2 \eta_2}{\partial x_1 \partial x_2}\right)+A_{12}\frac{\partial^2 \eta_2}{\partial x_1 \partial x_2}+A_{16}\left(\frac{\partial^2 \eta_2}{\partial x_1^2}+2\frac{\partial^2 \eta_1}{\partial x_1 \partial x_2}\right)$$
$$+A_{26}\frac{\partial^2 \eta_2}{\partial x_2^2}-d_1\left(\eta_1+a\frac{\partial v_3}{\partial x_1}\right)=0 \tag{6.15c}$$

$$A_{22}\frac{\partial^2 \eta_2}{\partial x_2^2}+A_{66}\left(\frac{\partial^2 \eta_2}{\partial x_2^2}+\frac{\partial^2 \eta_1}{\partial x_1 \partial x_2}\right)+A_{12}\frac{\partial^2 \eta_1}{\partial x_1 \partial x_2}+A_{26}\left(\frac{\partial^2 \eta_1}{\partial x_2^2}+2\frac{\partial^2 \eta_2}{\partial x_1 \partial x_2}\right)$$
$$+A_{16}\frac{\partial^2 \eta_1}{\partial x_1^2}-d_2\left(\eta_2+a\frac{\partial v_3}{\partial x_2}\right)=0 \tag{6.15d}$$

$$-\left(\frac{A_{11}}{R_1}+\frac{A_{12}}{R_2}\right)\frac{\partial \xi_1}{\partial x_1}-\left(\frac{A_{16}}{R_1}+\frac{A_{26}}{R_2}\right)\frac{\partial \xi_1}{\partial x_2}-\left(\frac{A_{22}}{R_2}+\frac{A_{12}}{R_1}\right)\frac{\partial \xi_2}{\partial x_2}-\left(\frac{A_{26}}{R_2}+\frac{A_{16}}{R_1}\right)\frac{\partial \xi_2}{\partial x_1}$$
$$-\left(\frac{A_{11}}{R_1^2}+\frac{A_{22}}{R_2^2}+\frac{2A_{12}}{R_1R_2}\right)v_3-d_1a\left(\frac{\partial \eta_1}{\partial x_1}+a\frac{\partial^2 v_3}{\partial x_1^2}\right)-d_2a\left(\frac{\partial \eta_2}{\partial x_2}+a\frac{\partial^2 v_3}{\partial x_2^2}\right)+F_{11}\frac{\partial^4 v_3}{\partial x_1^4}$$
$$+F_{22}\frac{\partial^4 v_3}{\partial x_2^4}+2(F_{12}+2F_{66})\frac{\partial^4 v_3}{\partial x_1^2\partial x_2^2}+4F_{16}\frac{\partial^4 v_3}{\partial x_1^3\partial x_2}+4F_{26}\frac{\partial^4 v_3}{\partial x_1\partial x_2^3}+N_{11}^0\frac{\partial^2 v_3}{\partial x_1^2}+$$
$$+2N_{12}^0\frac{\partial^2 v_3}{\partial x_1\partial x_2}+N_{22}^0\frac{\partial^2 v_3}{\partial x_2^2}-c\frac{\partial v_3}{\partial t}-m_0\frac{\partial^2 v_3}{\partial t^2}=-P_3(t)$$
(6.15e)

- *Boundary Conditions.*

  Along the edges $x_n = 0,\ L_n$

$$N_{nn}=\widetilde{N}_{nn} \quad \text{or} \quad \xi_n=\widetilde{\xi}_n$$
$$N_{nt}=\widetilde{N}_{nt} \quad \text{or} \quad \xi_t=\widetilde{\xi}_t$$
$$L_{nn}=\widetilde{L}_{nn} \quad \text{or} \quad \eta_n=\widetilde{\eta}_n$$
$$L_{nt}=\widetilde{L}_{nt} \quad \text{or} \quad \eta_t=\widetilde{\eta}_t$$
$$M_{nn}=\widetilde{M}_{nn} \quad \text{or} \quad \frac{\partial v_3}{\partial x_n}=\frac{\partial \widetilde{v}_3}{\partial x_n}$$
$$N_{nt}\left(\frac{\partial v_3}{\partial x_t}+\frac{\partial \overset{\circ}{v}_3}{\partial x_t}\right)+N_{nn}\left(\frac{\partial v_3}{\partial x_n}+\frac{\partial \overset{\circ}{v}_3}{\partial x_n}\right)+\frac{\partial M_{nn}}{\partial x_n}+2\frac{\partial M_{nt}}{\partial x_t} \quad \text{or} \quad v_3=\widetilde{v}_3$$
$$+(a/\overline{h})\overline{N}_{n3}=\frac{\partial \widetilde{M}_{nt}}{\partial x_t}+\widetilde{\overline{N}}_{n3}$$
(6.16a – f)

For simply supported boundary conditions freely movable on all edges

at $x_1 = 0,\ L_1$

$$N_{11}=N_{12}=\eta_1=\eta_2=M_{11}=v_3=0 \qquad (6.17a-f)$$

at $x_2 = 0,\ L_2$

$$N_{22}=N_{12}=\eta_1=\eta_2=M_{22}=v_3=0 \qquad (6.18a-f)$$

In terms of displacements, the first, second, and fifth boundary conditions can be written as

$$N_{11} = A_{11}\frac{\partial \xi_1}{\partial x_1} + A_{12}\frac{\partial \xi_2}{\partial x_2} + A_{16}\left(\frac{\partial \xi_2}{\partial x_1} + \frac{\partial \xi_1}{\partial x_2}\right) - \left(\frac{A_{11}}{R_1} + \frac{A_{12}}{R_2}\right)v_3$$
$$= 0, \qquad (1 \rightleftarrows 2) \tag{6.19}$$

$$N_{12} = A_{66}\left(\frac{\partial \xi_2}{\partial x_1} + \frac{\partial \xi_1}{\partial x_2}\right) + A_{26}\frac{\partial \xi_2}{\partial x_2} + A_{16}\frac{\partial \xi_1}{\partial x_1} - \left(\frac{A_{16}}{R_1} + \frac{A_{26}}{R_2}\right)v_3 = 0 \tag{6.20}$$

$$M_{11} = F_{11}\frac{\partial^2 v_3}{\partial x_1^2} + F_{12}\frac{\partial^2 v_3}{\partial x_2^2} + 2F_{16}\frac{\partial^2 v_3}{\partial x_1 \partial x_2} = 0 \tag{6.21}$$

$$M_{22} = F_{22}\frac{\partial^2 v_3}{\partial x_2^2} + F_{12}\frac{\partial^2 v_3}{\partial x_1^2} + 2F_{26}\frac{\partial^2 v_3}{\partial x_1 \partial x_2} = 0 \tag{6.22}$$

6.4.2 Solution Methodology

The transverse displacement can be assumed in the following form

$$v_3(x_1, x_2, t) = w_{mn}(t)\sin\lambda_m x_1 \sin\mu_n x_2 \tag{6.23}$$

$v_3(x_1, x_2, t)$ identically fulfills the sixth boundary conditions provided in Eqs. (6.17f) and (6.18f). To fulfill the first two equations of motion, ξ_1 and ξ_2 can be assumed in the following form

$$\xi_1(x_1, x_2, t) = w_{mn}(t)\left(F_{mn}^{(1)}\cos\lambda_m x_1 \sin\mu_n x_2 + F_{mn}^{(2)}\sin\lambda_m x_1 \cos\mu_n x_2\right) \tag{6.24a}$$

$$\xi_2(x_1, x_2, t) = w_{mn}(t)\left(G_{mn}^{(1)}\cos\lambda_m x_1 \sin\mu_n x_2 + G_{mn}^{(2)}\sin\lambda_m x_1 \cos\mu_n x_2\right) \tag{6.24b}$$

Where $F_{mn}^{(1)},\ F_{mn}^{(2)},\ G_{mn}^{(1)},\ G_{mn}^{(2)}$ are coefficients that have been previously determined and provided in Eqs. (5.27) – (5.31a–e). Following in a similar manner, η_1 and η_2 can be assumed in the following form.

$$\eta_1(x_1, x_2, t) = w_{mn}(t)\left(H_{mn}^{(1)}\cos\lambda_m x_1 \sin\mu_n x_2 + H_{mn}^{(2)}\sin\lambda_m x_1 \cos\mu_n x_2\right) \tag{6.25a}$$

$$\eta_2(x_1, x_2, t) = w_{mn}(t)\left(I_{mn}^{(1)}\cos\lambda_m x_1 \sin\mu_n x_2 + I_{mn}^{(2)}\sin\lambda_m x_1 \cos\mu_n x_2\right) \tag{6.25b}$$

where, the coefficients $H_{mn}^{(1)}, H_{mn}^{(2)}, I_{mn}^{(1)}$, and $I_{mn}^{(2)}$ have previously been determined in earlier sections. The first four equations of motion, Eqs. (6.15a–d) and the boundary conditions, Eqs. (6.18f) and Eqs. (6.18f) are identically fulfilled. There remains the fifth equation of motion and the remaining unfulfilled boundary conditions. The identical procedure as was carried out for flat plates will be duplicated here for the case of doubly curved sandwich shells. The unfulfilled quantities will be retained in the energy functional and thus by performing the necessary operations will result in fulfilling the last equation of motion and the remaining boundary conditions in an average sense. Inserting these unfilled expressions back into Hamilton's equation gives

$$\begin{aligned}
&\int_{t_0}^{t_1}\left\langle\int_0^{l_2}\int_0^{l_1}\left\{-\left(\frac{A_{11}}{R_1}+\frac{A_{12}}{R_2}\right)\frac{\partial\xi_1}{\partial x_1}-\left(\frac{A_{16}}{R_1}+\frac{A_{26}}{R_2}\right)\frac{\partial\xi_1}{\partial x_2}+\left(\frac{A_{22}}{R_2}+\frac{A_{12}}{R_1}\right)\frac{\partial\xi_2}{\partial x_2}-\left(\frac{A_{26}}{R_2}+\frac{A_{16}}{R_1}\right)\frac{\partial\xi_2}{\partial x_1}\right.\right.\\
&-\left(\frac{A_{11}}{R_1^2}+\frac{A_{22}}{R_2^2}+\frac{2A_{12}}{R_1R_2}\right)v_3-d_1a\left(\frac{\partial\eta_1}{\partial x_1}+a\frac{\partial^2v_3}{\partial x_1^2}\right)-d_2a\left(\frac{\partial\eta_2}{\partial x_2}+a\frac{\partial^2v_3}{\partial x_2^2}\right)+F_{11}\frac{\partial^4v_3}{\partial x_1^4}+F_{22}\frac{\partial^4v_3}{\partial x_2^4}\\
&+2(F_{12}+2F_{66})\frac{\partial^4v_3}{\partial x_1^2\partial x_2^2}+4F_{16}\frac{\partial^4v_3}{\partial x_1^3\partial x_2^1}+4F_{26}\frac{\partial^4v_3}{\partial x_1^1\partial x_2^3}+N_{11}\frac{\partial^2v_3}{\partial x_1^2}++2N_{12}^0\frac{\partial^2v_3}{\partial x_1\partial x_2}+N_{22}\frac{\partial^2v_3}{\partial x_2^2}\\
&\left.\left.+C\frac{\partial v_3}{\partial t}+m_0\frac{\partial^2v_3}{\partial t^2}-P_3(t)\right\}\delta v_3dx_1dx_2\right\rangle dt+\\
&\int_{t_0}^{t_1}\left\langle\int_0^{l_2}\left\{N_{11}\delta\xi_1+N_{12}\delta\xi_2+L_{11}\delta\eta_1+L_{12}\delta\eta_2+M_{11}\delta\left(\frac{\partial v_3}{\partial x_1}\right)\right\}\Bigg|_0^{l_1}dx_2\right\rangle dt+\\
&\int_{t_0}^{t_1}\left\langle\int_0^{l_1}\left\{N_{22}\delta\xi_2+N_{12}\delta\xi_1+L_{22}\delta\eta_2+L_{12}\delta\eta_1+M_{22}\delta\left(\frac{\partial v_3}{\partial x_2}\right)\right\}\Bigg|_0^{l_2}dx_1\right\rangle dt=0
\end{aligned} \tag{6.26}$$

Substituting in the expressions for $\xi_1,\ \xi_2,\ \eta_1,\ \eta_2,\ v_3$ into Eq. (6.26) and carrying out the indicated operations results in a second-order differential equation which governs the dynamic response of shallow sandwich shell structures. This governing differential equation is expressed as

$$\ddot{w}_{mn}+2\Delta_{mn}\omega_{mn}\dot{w}_{mn}+\omega_{mn}^2w_{mn}=F_{mn}(t) \tag{6.27}$$

It can be seen that this is the same result as was shown in Eq. (6.8) where the undamped natural frequency squared ω_{mn}^2 is expressed by

$$\begin{aligned}
\omega_{mn}^2=\frac{1}{m_0}&\left(\lambda_m^4F_{11}+2\lambda_m^2\mu_n^2(F_{12}+2F_{66})+\mu_n^4F_{22}+\lambda_mad_1\left(H_{mn}^{(1)}+a\lambda_m\right)+\mu_nad_2\left(I_{mn}^{(2)}+a\mu_n\right)\right.\\
&+\lambda_m\left(\frac{A_{11}}{R_1}+\frac{A_{12}}{R_2}\right)F_{mn}^{(1)}+\mu_n\left(\frac{A_{12}}{R_1}+\frac{A_{22}}{R_2}\right)G_{mn}^{(2)}+\mu_n\left(\frac{A_{16}}{R_1}+\frac{A_{26}}{R_2}\right)F_{mn}^{(2)}+\lambda_m\left(\frac{A_{16}}{R_1}+\right.\\
&\left.\left.\frac{A_{26}}{R_2}\right)G_{mn}^{(1)}+\left(\frac{A_{11}}{R_1^2}+2\frac{A_{12}}{R_1R_2}+\frac{A_{22}}{R_2^2}\right)-\lambda_m^2N_{11}^0-\mu_n^2N_{22}^0\right)
\end{aligned} \tag{6.28}$$

Or in dimensionless form as

$$\Omega_{mn}^2 = m^4 + \frac{F_{22}n^4\phi^4}{F_{11}} + \frac{2(F_{12}+2F_{66})m^2n^2\phi^2}{F_{11}} + \frac{a^2L_1^2}{F_{11}\pi^2}\left(d_1m^2 + d_2n^2\phi^2\right) +$$
$$\frac{aL_1^3}{F_{11}\pi^3}\left(d_1mH_{mn}^{(1)} + d_2n\phi I_{mn}^{(2)}\right) + \frac{L_1^2}{F_{11}\pi^3}\Big[m(\psi_1A_{11} + \psi_2\phi A_{12})F_{mn}^{(1)} +$$
$$n\phi(\psi_1A_{12} + \psi_2\phi A_{22})G_{mn}^2 + \left(n\phi F_{mn}^{(2)} + mG_{mn}^{(1)}\right)(\psi_1A_{16} + \psi_2\phi A_{26})\Big] +$$
$$\frac{L_1^2}{F_{11}\pi^4}\left(\psi_1^2A_{11} + \psi_2^2\phi^2A_{22} + 2A_{12}\psi_1\psi_2\phi\right) - K_x\left(m^2\psi_1^2 + L_Rn^2\psi_2^2\phi^2\right) \tag{6.29}$$

Where the nondimensional parameters ψ_1, ψ_2, ϕ, K_x, L_R, Ω_{mn}^2 were defined previously in Chap. 5. Equation (6.27) can be solved for various pressure pulses via the Laplace transform method. First a discussion and understanding of the various pressure pulses needs to be ascertained. In the next section various types of pressure pulses are discussed.

6.5 Explosive Pressure Pulses and Numerical Results

The dynamic response of structures is an especially important and essential area for understanding the effects of rapid-type, time-dependent loadings on the structural response and how to design the structure to withstand such large stresses generated within the structure under such loadings. Some examples would include aircraft exposed to shockwaves or space vehicles exposed to blast pulses. These types of external stimuli can be very damaging to the structure. It is therefore imperative to gain an understanding of the structural response within these kinds of environments to design against catastrophic failure. Several numerical simulations are presented for each of the various types of pressure pulses which are discussed next. Unless otherwise specified, Tables 6.1 and 6.2 contain the material and geometrical characteristics for the results which follow.

The first type of external pulse is the Sonic Boom. This is expressed as follows.

Table 6.1 Material properties for the face sheets and the core

Face sheets				
E_1(GPa)	E_2(GPa)	G_{12}(GPa)	ρ (kg/m^3)	ν_{12}
207	5.17	2.55	1588.22	0.25
Core				
$\overline{G}_{13}$ (GPa)	$\overline{G}_{23}$ (GPa)	–	$\overline{\rho}$ (kg/m^3)	
0.1027	0.0621	–	16	

Table 6.2 Geometrical properties

Ply Thickness, t_k(m)	h_f (m)	L_1 (m)	L_2 (m)	$2\overline{h}$ (m)	Δ_{mn}
0.0005	0.002	0.420	0.420	0.0250	0.05

6.5.1 Sonic Boom

$$p_3(t) = \begin{cases} p_0(1 - t/t_p) & 0 < t < rt_p \\ 0 & t < 0, \ t > rt_p \end{cases} \tag{6.30}$$

p_0 denotes the peak reflected pressure in excess of the ambient, t_p denotes the positive phase duration of the pulse measured from the time of impact of the structure, and r denotes the shock pulse length factor. If $r = 1$, the sonic boom becomes a triangular pulse, if $r = 2$ a symmetric sonic boom is generated, for $r \neq 2$ a nonsymmetric N-Shaped pulse results. If $r = 1$, $t_p \to \infty$ the N-shaped pulse degenerates into a step pulse. $p_3(t)$ can be expressed (see Marzocca et al. 2001; Librescu and Nosier 1990) in terms of the Heaviside function as

$$p_3(t) = p_0\left(1 - \frac{t}{t_p}\right)\left[H(t) - H\left(t - rt_p\right)\right] \tag{6.31}$$

Substituting this expression into Eq. (6.27) gives

$$\ddot{w}_{mn} + 2\Delta_{mn}\omega_{mn}\dot{w}_{mn} + \omega_{mn}^2 w_{mn} = \frac{\widetilde{F}}{m_0}\left(1 - \frac{t}{t_p}\right)\left[H(t) - H\left(t - rt_p\right)\right] \tag{6.32}$$

where,

$$\widetilde{F} = \frac{16 p_0 \delta_{m,2s-1}\delta_{n,2q-1}}{(2s-1)(2q-1)\pi^2} \tag{6.33}$$

The Laplace Transform Method in time is used to solve Eq. (6.32), assuming zero initial conditions. The Laplace transform operator is defined as

$$\mathcal{L}\{\cdot\} = \int_0^\infty \{\cdot\} e^{-st} dt \tag{6.34}$$

And is applied to Eq. (6.32) where s is the transform variable. For the case of the sonic boom type pulse, applying the Laplace transform to Eq. (6.32) results in

$$W_{mn}(s) = \frac{\widetilde{F}}{m_0\left[s^2 + 2\Delta_{mn}\omega_{mn}s + \omega_{mn}^2\right]}\left\{\frac{1}{s} - \frac{1}{t_p s^2} - \frac{(1-r)e^{-rt_p s}}{s} + \frac{e^{-rt_p s}}{t_p s^2}\right\} \quad (6.35)$$

where the following zero initial conditions were assumed.

$$w_{mn}(0) = \dot{w}_{mn}(t) = 0 \quad (6.36)$$

Taking the inverse Laplace Transform of Eq. (6.35), to arrive back in the time domain, one obtains

$$\begin{aligned} w_{mn}(t) = & \frac{\widetilde{F}}{m_0\omega_{mn}^2}\left\{1 + \frac{2\Delta_{mn}}{t_p\omega_{mn}} - \frac{t}{t_p} - \left(1 + \frac{2\Delta_{mn}}{t_p\omega_{mn}}\right)e^{-\Delta_{mn}\omega_{mn}t}\cos\Omega_{mn}t - \left(\frac{2\Delta_{mn}^2 + \Delta_{mn}\omega_{mn}t_p - 1}{\Omega_{mn}t_p}\right)\cdot\right. \\ & \cdot e^{-\Delta_{mn}\omega_{mn}t}\sin\Omega_{mn}t - \left[1 + \frac{2\Delta_{mn}}{t_p\omega_{mn}} - \frac{t}{t_p} - \left((1-r) + \frac{2\Delta_{mn}}{t_p\omega_{mn}}\right)e^{-\Delta_{mn}\omega_{mn}(t-rt_p)}\cos\Omega_{mn}(t-rt_p) - \right. \\ & \left.\left. - \left(\frac{2\Delta_{mn}^2 + (1-r)\Delta_{mn}\omega_{mn}t_p - 1}{\Omega_{mn}t_p}\right)e^{-\Delta_{mn}\omega_{mn}(t-rt_p)}\sin\Omega_{mn}(t-rt_p)\right]H(t-rt_p)\right\} \end{aligned} \quad (6.37)$$

In Eq. (6.37),

$$\Omega_{mn} = \omega_{mn}\sqrt{1 - \Delta_{mn}^2} \quad (6.38)$$

denotes the damped natural frequency. Figure 6.1 illustrates the effect of curvature on the deflection–time response of a doubly curved sandwich panel due to a sonic

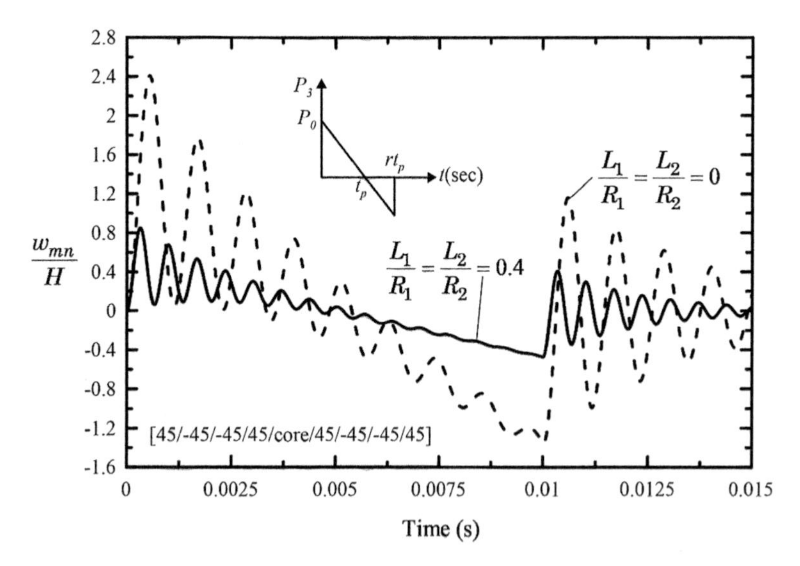

Fig. 6.1 The effects of panel curvature on the dynamic response of a sandwich panel due to a sonic boom ($r = 2,\ t_p = 0.005,\ P_0 = 5\ MPa$)

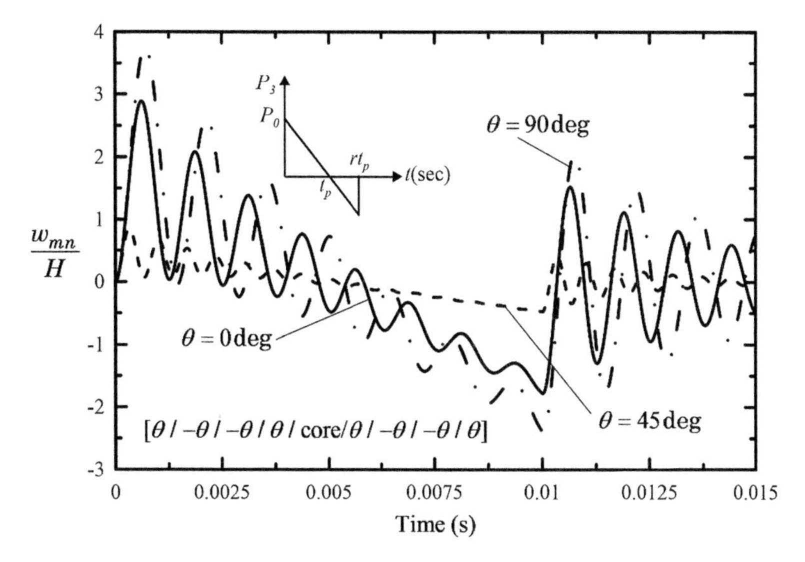

Fig. 6.2 The effects of the ply angle on the dynamic response due to a sonic boom ($r = 2$, $t_p = 0.005$, $P_0 = 5\ MPa$, $L_1/R_1 = L_2/R_2 = 0.4$)

boom for a given fixed layup. It appears that the larger curvature ratios are less detrimental providing smaller amplitudes of oscillations and a more rapid decaying response. Figure 6.2 shows the response due to a sonic boom for a doubly curved sandwich panel for various ply angles under a specific stacking sequence in the facings. It clearly shows that the ply angle plays an important role in the amplitude and decay of the response in both the free and forced regimes. The results show that a ply angle of $\theta = 45$ deg appears to be the most beneficial from an amplitude of oscillation standpoint.

In Fig. 6.3, the response to a sonic boom for a doubly curved sandwich panel for various amounts of damping is depicted. It is apparent that an increase in the amount of damping leads to a more rapid decay in the response.

6.5.2 Rectangular Pulse

$$p_3(t) = \begin{cases} p_0 & 0 < t < t_p \\ 0 & t > t_p \end{cases} \tag{6.39}$$

In terms of the Heaviside step function, $p_3(t)$ is expressed as

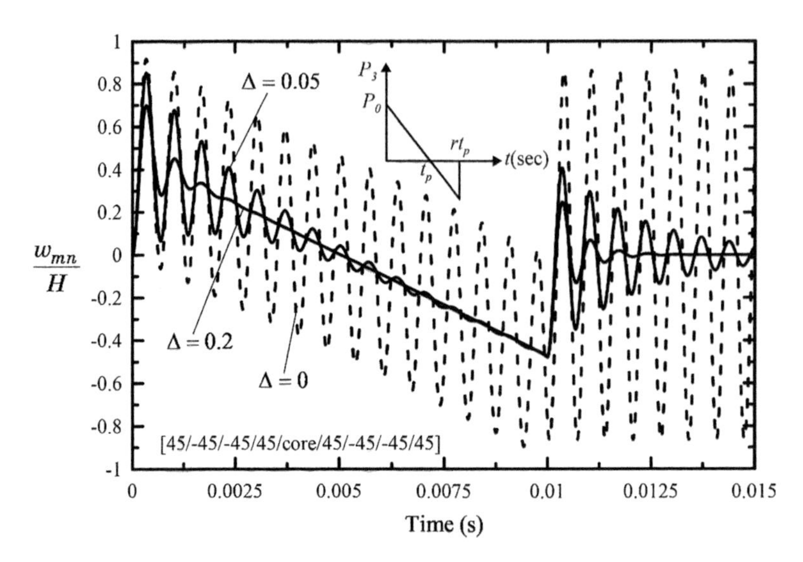

Fig. 6.3 The implications of damping on the dynamic response due to a sonic boom ($r = 2$, $t_p = 0.005$, $P_0 = 5\ MPa$, $L_1/R_1 = L_2/R_2 = 0.4$)

$$p_3(t) = p_0\{H(t) - H(t - t_p)\} \tag{6.40}$$

Substituting Eq. (6.40) into Eq. (6.27) results in

$$\ddot{w}_{mn} + 2\Delta_{mn}\omega_{mn}\dot{w}_{mn} + \omega_{mn}^2 w_{mn} = \frac{\widetilde{F}}{m_0}\left[H(t) - H(t - t_p)\right] \tag{6.41}$$

Taking the Laplace Transform and assuming zero initial conditions, the response in the Laplace domain is

$$W_{mn}(s) = \frac{\widetilde{F}_{mn}}{m_0}\frac{1}{s\left(s^2 + 2\Delta_{mn}\omega_{mn}s + \omega_{mn}^2\right)}\left(1 - e^{-t_p s}\right) \tag{6.42}$$

Arriving back in the time domain by taking the inverse Laplace transform gives

$$w_{mn}(t) = \frac{\widetilde{F}_{mn}}{m_0\omega_{mn}^2}\left\{1 - e^{-\Delta_{mn}\omega_{mn}t}\cos\Omega_{mn}t - \frac{\Delta_{mn}}{\sqrt{1-\Delta_{mn}^2}}e^{-\Delta_{mn}\omega_{mn}t}\sin\Omega_{mn}t - \left[1 - \right.\right.$$

$$\left.\left. e^{-\Delta_{mn}\omega_{mn}(t-t_p)}\cos\Omega_{mn}(t - t_p) - \frac{\Delta_{mn}}{\sqrt{1-\Delta_{mn}^2}}e^{-\Delta_{mn}\omega_{mn}(t-t_p)}\sin\Omega_{mn}(t - t_p)\right]\right\}H(t - t_p) \tag{6.43}$$

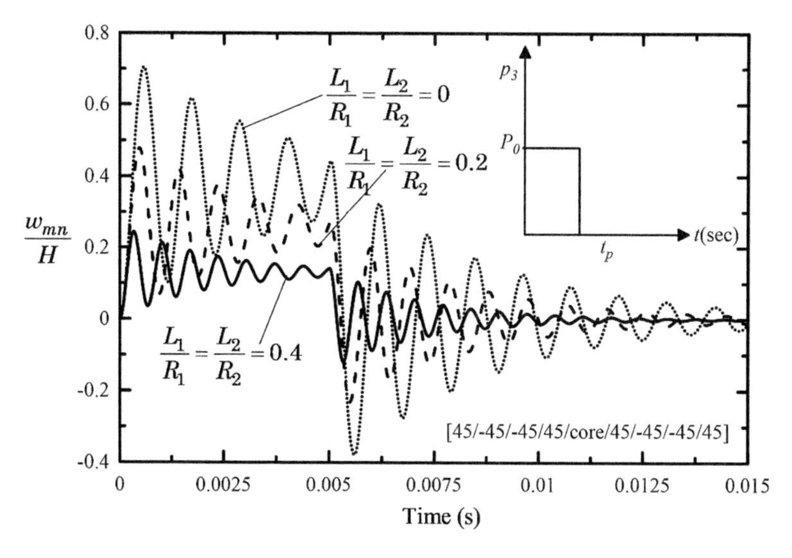

Fig. 6.4 The implications of curvature on the dynamic response of a doubly curved sandwich panel under a rectangular pulse ($t_p = 0.005,\ P_0 = 1.38$ MPa)

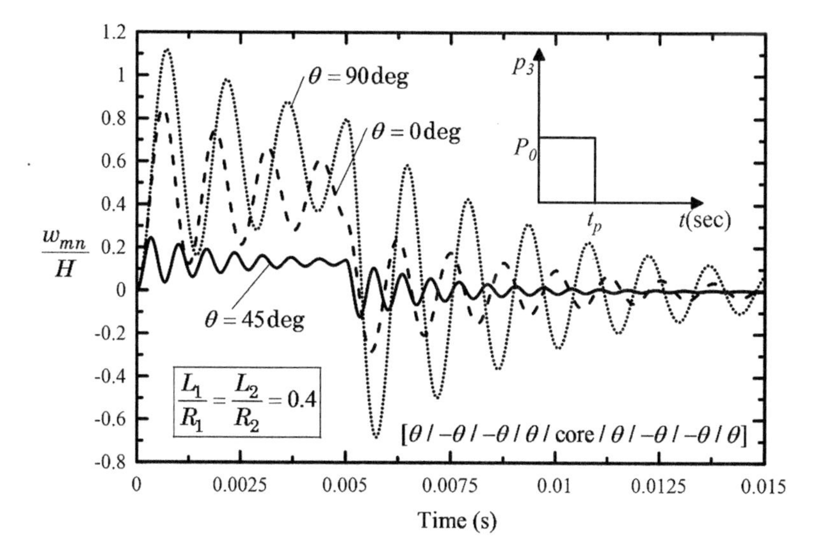

Fig. 6.5 The implications of the fiber orientation in the face sheets for a given layup on the dynamic response of a doubly curved sandwich panel under a rectangular pulse ($t_p = 0.005$, $P_0 = 1.38$ MPa)

The results show in Fig. 6.4 that as the curvature ratio increases, the amplitude of oscillation diminishes and decays more rapidly over time as compared to the smaller curved panel counterparts. Such was the case for the sonic boom. Figure 6.5 which highlights the effect of the fiber orientation in the face sheets on the deflection–time response of a doubly curved sandwich panel due to a rectangular pulse. The results reveal that the response seems to benefit largely from a ply angle of 45 degrees.

6.5.3 Heaviside Pulse

$$p_3(t) = p_0, \qquad t > 0 \tag{6.44}$$

Substituting $p_3(t)$ into Eq. (6.27) gives

$$\ddot{w}_{mn} + 2\Delta_{mn}\omega_{mn}\dot{w}_{mn} + \omega_{mn}^2 w_{mn} = \frac{\widetilde{F}}{m_0} \tag{6.45}$$

Taking the Laplace transform gives

$$W_{mn}(s) = \frac{\widetilde{F}_{mn}}{m_0} \frac{1}{s\left(s^2 + 2\Delta_{mn}\omega_{mn}s + \omega_{mn}^2\right)} \tag{6.46}$$

Performing an inverse Laplace transform results in

$$w_{mn}(t) = \frac{\widetilde{F}_{mn}}{m_0\omega_{mn}^2} \times \left(1 - e^{-\Delta_{mn}\omega_{mn}t}\cos\Omega_{mn}t - \frac{\Delta_{mn}}{\sqrt{1-\Delta_{mn}^2}} e^{-\Delta_{mn}\omega_{mn}t}\sin\Omega_{mn}t\right) \tag{6.47}$$

Figures 6.6 and 6.7 illustrate the effects of the fiber orientation in the face sheets and the effect of the curvature ratio on the deflection–time response of a doubly curved sandwich panel exposed to a Heaviside pulse, respectively. The responses in this case are identical to the behavior seen in Figs. 6.4 and 6.5, where the effects of the curvature and the fiber orientation under a rectangular pulse are highlighted. The results are said to be identical in nature.

6.5.4 Sine Pulse

The sine pulse is represented mathematically as

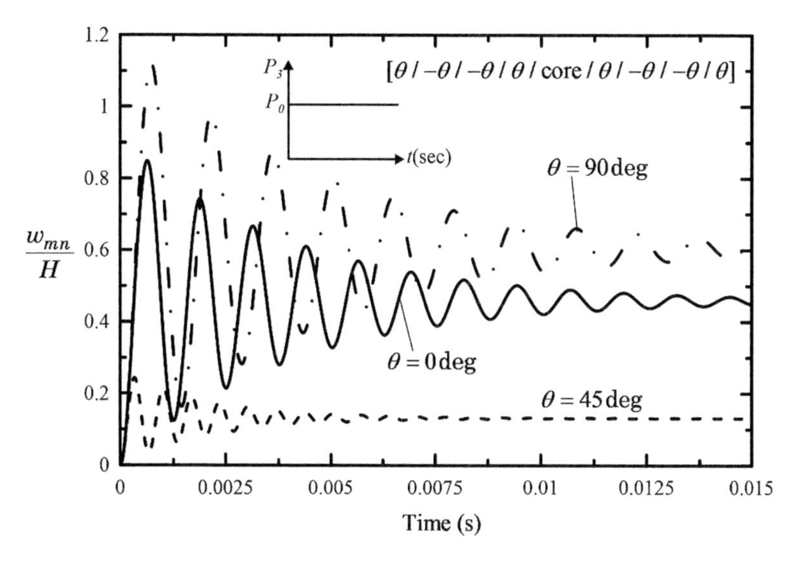

Fig. 6.6 The implications of the fiber orientation in the face sheets for a given layup on the dynamic response of a doubly curved sandwich panel under a Heaviside pulse ($P_0 = 1.38$ MPa)

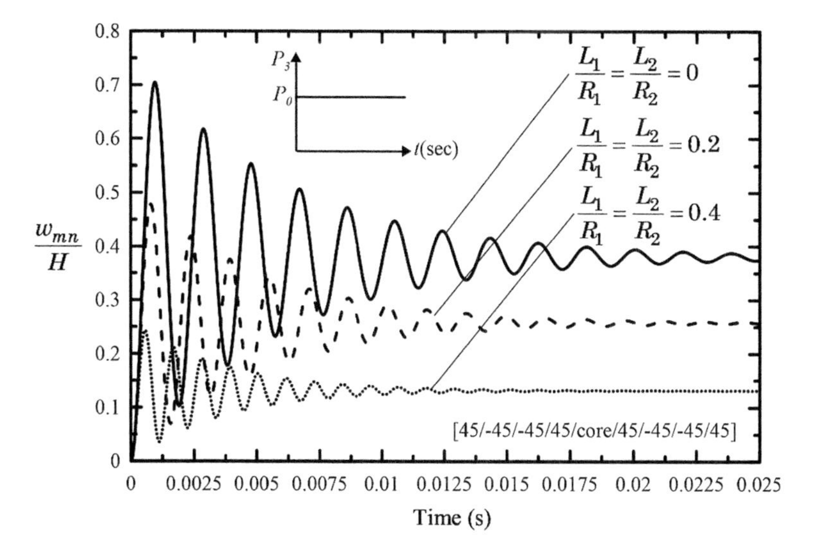

Fig. 6.7 The implications of curvature on the dynamic response of a doubly curved sandwich panel under a Heaviside pulse ($P_0 = 1.38$ MPa)

$$p_3(t) = \begin{cases} p_0 \sin\left(\dfrac{\pi t}{t_p}\right) & 0 < t < t_p \\ 0 & t > t_p \end{cases} \tag{6.48}$$

In terms of the Heaviside function $p_3(t)$ is expressed as

$$p_3(t) = p_0 \sin\left(\frac{\pi t}{t_p}\right)\{H(t) - H(t - t_p)\} \tag{6.49}$$

Substituting into Eq. (6.27) results in

$$\ddot{w}_{mn} + 2\Delta_{mn}\omega_{mn}\dot{w}_{mn} + \omega_{mn}^2 w_{mn} = \frac{\widetilde{F}}{m_0}\sin\left(\frac{\pi t}{t_p}\right)\left[H(t) - H(t - t_p)\right] \tag{6.50}$$

Applying the Laplace transform with zero initial conditions to Eq. (6.50) results in the Laplace domain as

$$W_{mn}(s) = \frac{\widetilde{F}_{mn}}{m_0}\frac{\pi/t_p(1 + e^{-t_p s})}{\left[s^2 + (\pi/t_p)^2\right]\left[s^2 + 2\Delta_{mn}\omega_{mn}s + \omega_{mn}^2\right]} \tag{6.51}$$

Taking the inverse Laplace transform gives the response in the time domain as

$$\begin{aligned} w_{mn}(t) = \frac{\widetilde{F}_{mn}\pi}{m_0 t_p}\Bigg\{ & a_1\cos(\pi t/t_p) + b_1\sin(\pi t/t_p) + a_2 e^{-\Delta_{mn}\omega_{mn}t}\cos\Omega_{mn}t + \\ & \left(\frac{b_2 - a_2\Delta_{mn}\omega_{mn}}{\Omega_{mn}}\right)e^{-\Delta_{mn}\omega_{mn}t}\sin\Omega_{mn}t + \Bigg[a_1\cos\pi(t - t_p)/t_p + \\ & \frac{b_1 t_p}{\pi}\sin\pi(t - t_p)/t_p + a_2 e^{-\Delta_{mn}\omega_{mn}t}\cos\Omega_{mn}(t - t_p) + \\ & \left(\frac{b_2 - a_2\Delta_{mn}\omega_{mn}}{\Omega_{mn}}\right)e^{-\Delta_{mn}\omega_{mn}(t - t_p)}\sin\Omega_{mn}(t - t_p)\Bigg]H(t - t_p)\Bigg\} \end{aligned} \tag{6.52}$$

where

$$a_1 = -a_2 = \frac{2t_p^2\Delta_{mn}\omega_{mn}^2}{\pi^2}\left\{\left(\omega_{mn}^2 - \pi^2/t_p^2\right)\left(1 - \frac{\omega_{mn}^2 t_p^2}{\pi^2}\right) - 4\Delta_{mn}^2\omega_{mn}^2\right\}^{-1} \tag{6.53a}$$

$$b_1 = \frac{1 - \frac{\omega_{mn}^2 t_p^2}{\pi^2}}{\left(\omega_{mn}^2 - \pi^2/t_p^2\right)\left(1 - \frac{\omega_{mn}^2 t_p^2}{\pi^2}\right) - 4\Delta_{mn}^2\omega_{mn}^2} \tag{6.53b}$$

$$b_2 = -\left[\left(1 - \frac{\omega_{mn}^2 t_p^2}{\pi^2}\right) - 4\Delta_{mn}^2\omega_{mn}^2 t_p^2/\pi^2\right]\left\{\left(\omega_{mn}^2 - \pi^2/t_p^2\right)\left(1 - \frac{\omega_{mn}^2 t_p^2}{\pi^2}\right) - 4\Delta_{mn}^2\omega_{mn}^2\right\}^{-1} \tag{6.53c}$$

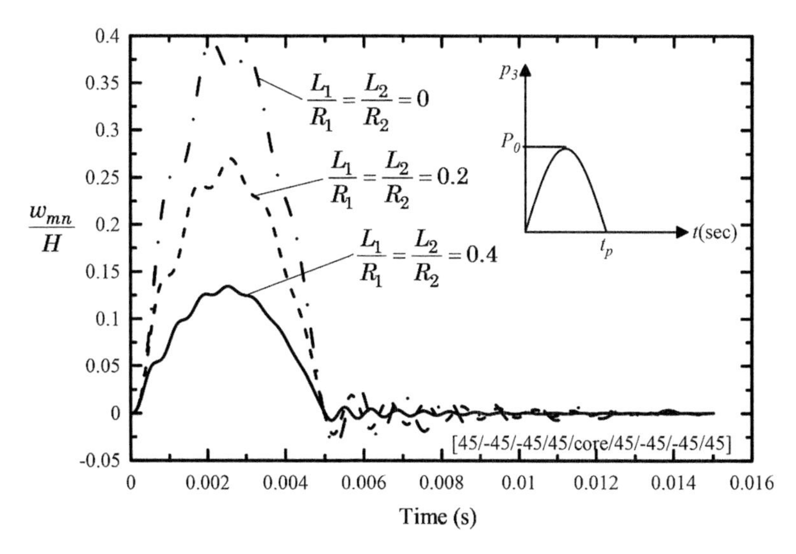

Fig. 6.8 The implications of curvature on the dynamic response of a doubly curved sandwich panel under a sine pulse ($t_p = 0.005$, $P_0 = 1.38$ MPa)

Figure 6.8 displays the effect of the curvature on a doubly curved sandwich panel subjected to a sine pulse. The results clearly show that the construct with the larger curvature ratio exhibits a lower amplitude of deflection during the positive phase duration of the applied pulse. Shortly thereafter, the decay process occurs rapidly. A similar trend is seen in Fig. 6.9 where the ply angle for a fixed layup portrays the same role that the curvature ratio played in Fig. 6.8. The 90-degree ply angle construct reveals the least beneficial response where the 45-degree ply angle favors the most beneficial response.

6.5.5 *Tangential Blast Pulse*

This case concerns a tangential air-blast traveling in the tangential direction to the panel span. This type of rapid time-dependent loading is represented by (see Kim and Han 2006)

$$p_3(t) = p_0 e^{-\eta(ct - x_1)} H(ct - x_1) \tag{6.54}$$

where c is the wave speed in the medium surrounding the structure, while η is an exponent which determines the measure of the blast decay. Substituting into Eq. (6.27) results in

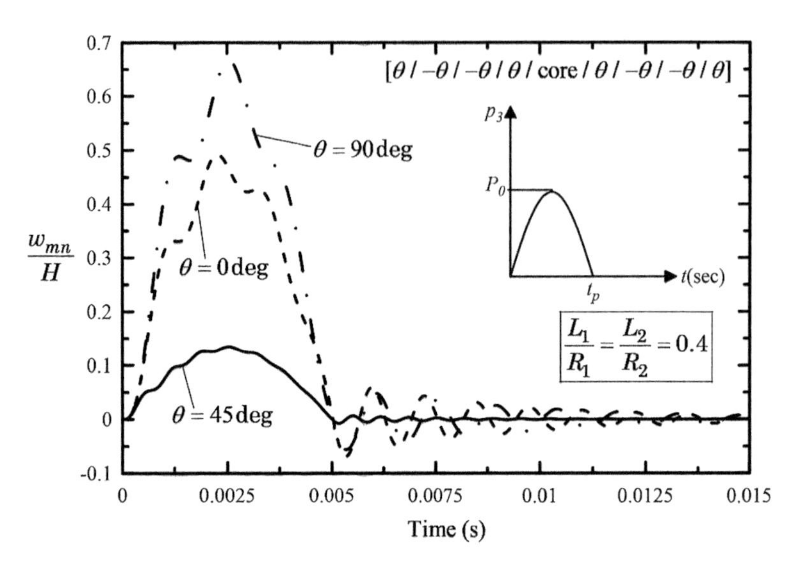

Fig. 6.9 The implications of ply angle in the face sheets on the dynamic response of a doubly curved sandwich panel under a sine pulse ($t_p = 0.005$, $P_0 = 1.38$ MPa)

$$\ddot{w}_{mn} + 2\Delta_{mn}\omega_{mn}\dot{w}_{mn} + \omega_{mn}^2 w_{mn} = \frac{\widetilde{F}}{m_0} e^{-\eta(ct - x_1)} H(ct - x_1) \tag{6.55}$$

Taking the Laplace transform provides

$$W_{mn}(s) = \frac{\widetilde{F}}{m_0} \frac{e^{-sx_1/c}}{(s + \eta c)(s^2 + 2\Delta_{mn}\omega_{mn}s + \omega_{mn}^2)} \tag{6.56}$$

By taking the inverse Laplace transform, the response becomes

$$w_{mn}(t) = \frac{\widetilde{F}}{m_0} \left\{ Ae^{-\eta c(t - x_1/c)} + B\left[e^{-\Delta_{mn}\omega_{mn}(t - x_1/c)} \cos \Omega_{mn} t - \frac{e^{-\Delta_{mn}\omega_{mn}(t - x_1/c)}}{\Omega_{mn}} \Delta_{mn}\omega_{mn} \sin \Omega_{mn}(t - x_1/c) \right] + \frac{C}{\Omega_{mn}} e^{-\Delta_{mn}\omega_{mn}(t - x_1/c)} \sin \Omega_{mn}(t - x_1/c) \right\} \cdot H(t - x_1/c) \tag{6.57}$$

where

$$A = \frac{1}{\omega_{mn}^2 + \eta c(\eta c - 2\Delta_{mn}\omega_{mn})} \tag{6.58a}$$

$$B = \frac{-1}{\omega_{mn}^2 + \eta c(\eta c - 2\Delta_{mn}\omega_{mn})} \tag{6.58b}$$

$$C = \frac{\eta c - 2\Delta_{mn}\omega_{mn}}{\omega_{mn}^2 + \eta c(\eta c - 2\Delta_{mn}\omega_{mn})} \tag{6.58c}$$

Figure 6.10 illustrates the effect of curvature on a doubly curved sandwich panel subjected to an air-blast traveling tangentially to the panel span, also known as a tangential blast. Typically, this type of pressure pulse models a gun blast pressure. This figure highlights the effect of the curvature as previously seen in prior results favoring the larger curvature ratios by diminishing the amplitude of deflection while decaying much more rapidly. Figure 6.11 shows the effect of the wave speed of the traveling blast within the surrounding medium of a doubly curved panel. The air-blast with the larger wave speed seems to decay much more rapidly.

It is clearly seen in Fig. 6.12 that the 45-degree ply construction for the fixed layup provides the most beneficial time–deflection response, while the 0-degree configuration provides the least beneficial response. Figure 6.13 illustrates the counterpart of Fig. 6.10 with the addition of a fixed compressive edge applied along the edges. When comparing the results from both figures, clearly it can be seen that with a compressive edge load present the response is more detrimental to the dynamic response of the panel.

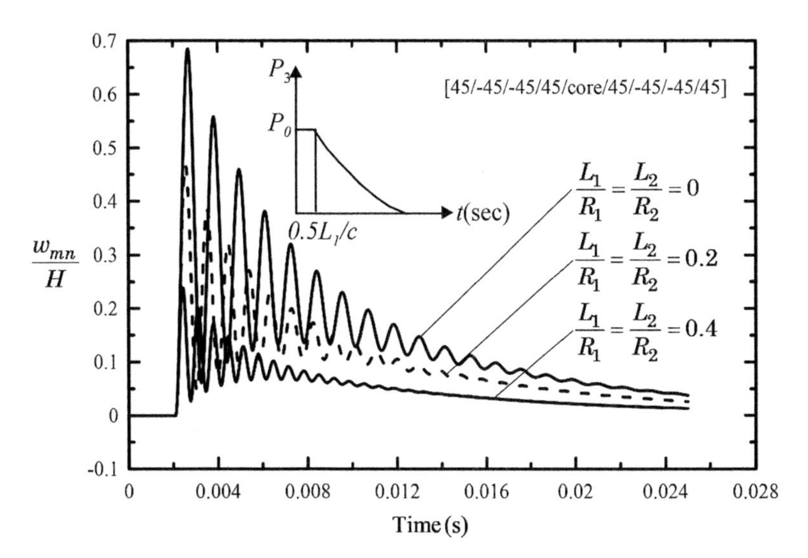

Fig. 6.10 The implications of the curvature on the time deflection response of a doubly curved sandwich panel exposed to a tangential blast pulse ($c = 100$ m/s, $P_0 = 1.38$ MPa)

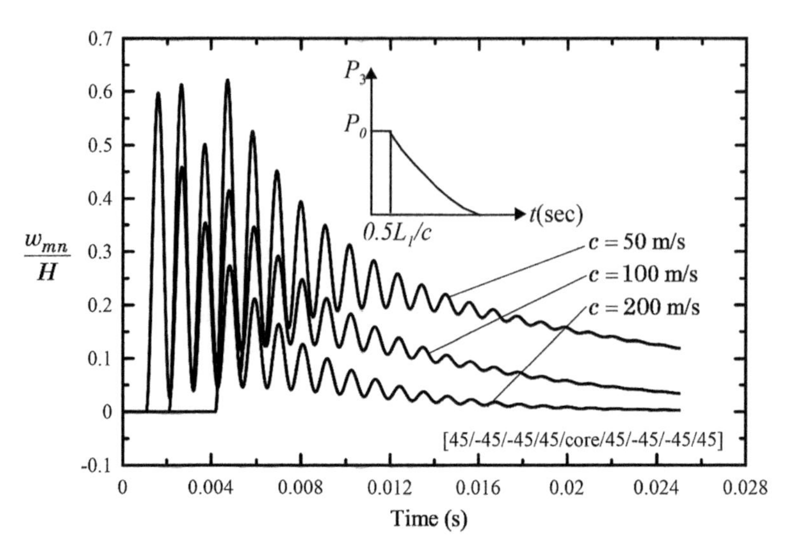

Fig. 6.11 The implications of the wave speed on the time deflection response of a doubly curved sandwich panel exposed to a tangential blast pulse ($P_0 = 1.38$ MPa)

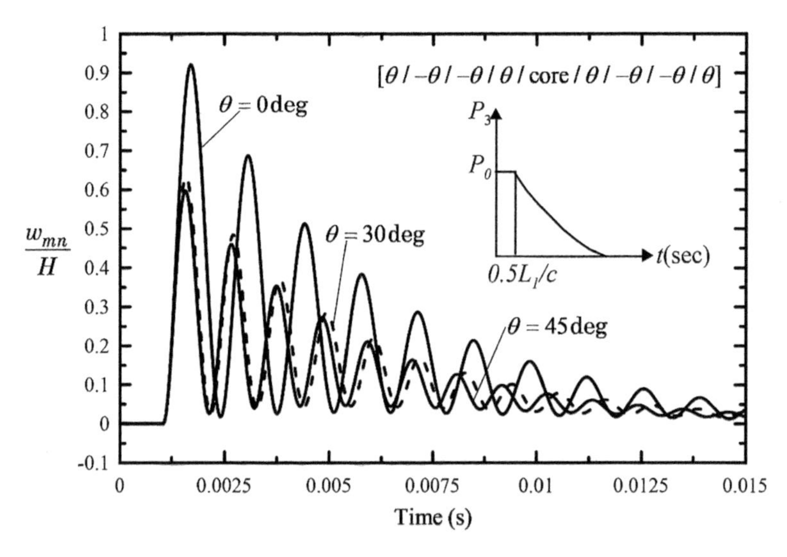

Fig. 6.12 The effect of the ply angle on the time deflection response of a doubly curved sandwich panel exposed to a tangential blast pulse ($c = 100$ m/s, $P_0 = 1.38$ MPa)

In contrast to the results seen in Figs. 6.13, and 6.14 highlights the comparison of compressive edge loading and tensile edge loading and its effect on the response. It can clearly be seen that tensile loading improves the dynamic response of a doubly curved sandwich structure.

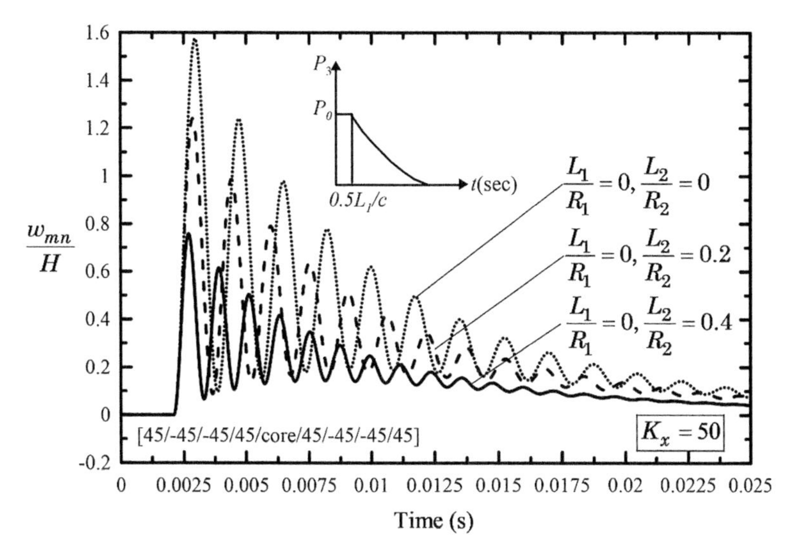

Fig. 6.13 The effect of the curvature on the time deflection response of a doubly curved sandwich panel exposed to a tangential blast pulse with a fixed compressive edge load ($c = 100$ m/s, $P_0 = 1.38$ MPa)

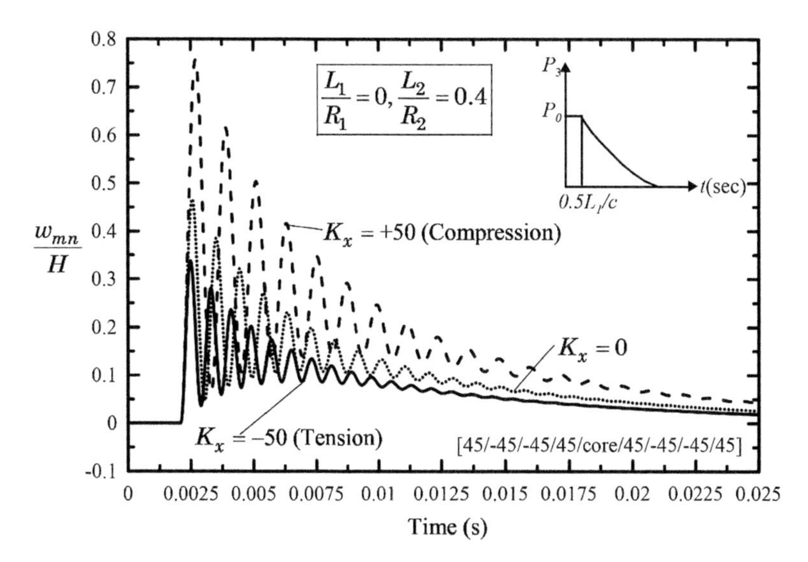

Fig. 6.14 The effect of the types of edge loading on the time deflection response of a doubly curved sandwich panel exposed to a tangential blast pulse ($c = 100$ m/s, $P_0 = 1.38$ MPa)

6.5.6 Friedlander Explosive Blast Pulse

This type of explosive blast is a free in-air spherical air blast. This type of explosion creates a spherical shockwave which travels radially outward in all directions that diminishes with time. The mathematical expression for this type of pressure pulse is given by

$$p_3(t) = P_0\left(1 - \frac{t}{t_p}\right)e^{-a^{/} t/t_p} \tag{6.59}$$

$a^{/}$ represents a decay parameter which is adjusted to an actual pressure curve from a blast test. t is measured from the time of arrival and t_p is the positive phase duration of the blast pulse. Substituting into Eq. (6.27) gives

$$\ddot{w}_{mn} + 2\Delta_{mn}\omega_{mn}\dot{w}_{mn} + \omega_{mn}^2 w_{mn} = \widetilde{f}e^{-\alpha t}(t_p - t)\{H(t) - H(t - t_p)\} \tag{6.60}$$

where

$$\alpha = \frac{a^{/}}{t_p} \text{ and } \widetilde{f} = \frac{4\Delta_n^m P_0}{mn\pi^2 m_0 t_p} \tag{6.61a, b}$$

Taking the Laplace Transform results in

$$W_{mn}(s) = \widetilde{f}\left\{\frac{t_p}{(s+\alpha)(s^2 + 2\Delta\omega_{mn}s + \omega_{mn}^2)} + \frac{e^{-(s+\alpha)t_p}}{(s+\alpha)(s^2 + 2\Delta\omega_{mn}s + \omega_{mn}^2)} - \frac{1}{(s+\alpha)(s^2 + 2\Delta\omega_{mn}s + \omega_{mn}^2)}\right\} \tag{6.62}$$

The inverse Laplace Transform provides the response in the time domain as

$$w_{mn}(t) = \widetilde{f}\left\langle (A_1 - D_1 - E_1 t)e^{-\alpha t} + e^{-\Delta\omega_{mn}t}\left\{(B_1 - F_1)\cos\Omega_{mn}t + \frac{1}{\Omega_{mn}}[-G_1 + C_1 - \Delta\omega_{mn}(-F_1 + B_1)]\sin\Omega_{mn}t\right\} + e^{-\alpha t_p}\left\{\left[D_1 + E_1(t - t_p)\right]e^{-\alpha(t-t_p)} + e^{-\Delta\omega_{mn}(t-t_p)}\left[F_1\cos\Omega_{mn}(t-t_p) + \frac{1}{\Omega_{mn}}(G_1 - \Delta\omega_{mn}F_1)\sin\Omega_{mn}(t-t_p)\right]\right\}H(t-t_p)\right\rangle \tag{6.63}$$

where

$$A_1 = \frac{t_p}{\alpha(\alpha - 2\Delta\omega_{mn}) + \omega_{mn}^2}, \quad \infty B_1 = \frac{-t_p}{\alpha(\alpha - 2\Delta\omega_{mn}) + \omega_{mn}^2}, \quad \infty C_1 = \frac{t_p(\alpha - 2\Delta\omega_{mn})}{\alpha(\alpha - 2\Delta\omega_{mn}) + \omega_{mn}^2} \tag{6.64a – c}$$

$$D_1 = \frac{\det\begin{pmatrix} 1 & \omega_{mn}^2 & 0 & \alpha^2 \\ 0 & 2\Delta\omega_{mn} & \alpha^2 & 2\alpha \\ 0 & 1 & 2\alpha & 1 \\ 0 & 0 & 1 & 0 \end{pmatrix}}{\det\begin{pmatrix} \alpha\omega_{mn}^2 & \omega_{mn}^2 & 0 & \alpha^2 \\ \omega_{mn}^2 + 2\alpha\Delta\omega_{mn} & 2\Delta\omega_{mn} & \alpha^2 & 2\alpha \\ \alpha + 2\Delta\omega_{mn} & 1 & 2\alpha & 1 \\ 1 & 0 & 1 & 0 \end{pmatrix}}, E_1 = \frac{\det\begin{pmatrix} \alpha\omega_{mn}^2 & 1 & 0 & \alpha^2 \\ \omega_{mn}^2 + 2\alpha\Delta\omega_{mn} & 0 & \alpha^2 & 2\alpha \\ \alpha + 2\Delta\omega_{mn} & 0 & 2\alpha & 1 \\ 1 & 0 & 1 & 0 \end{pmatrix}}{\det\begin{pmatrix} \alpha\omega_{mn}^2 & \omega_{mn}^2 & 0 & \alpha^2 \\ \omega_{mn}^2 + 2\alpha\Delta\omega_{mn} & 2\Delta\omega_{mn} & \alpha^2 & 2\alpha \\ \alpha + 2\Delta\omega_{mn} & 1 & 2\alpha & 1 \\ 1 & 0 & 1 & 0 \end{pmatrix}} \tag{6.65a, b}$$

$$F_1 = \frac{\det\begin{pmatrix} \alpha\omega_{mn}^2 & \omega_{mn}^2 & 1 & \alpha^2 \\ \omega_{mn}^2 + 2\alpha\Delta\omega_{mn} & 2\Delta\omega_{mn} & 0 & 2\alpha \\ \alpha + 2\Delta\omega_{mn} & 1 & 0 & 1 \\ 1 & 0 & 0 & 0 \end{pmatrix}}{\det\begin{pmatrix} \alpha\omega_{mn}^2 & \omega_{mn}^2 & 0 & \alpha^2 \\ \omega_{mn}^2 + 2\alpha\Delta\omega_{mn} & 2\Delta\omega_{mn} & \alpha^2 & 2\alpha \\ \alpha + 2\Delta\omega_{mn} & 1 & 2\alpha & 1 \\ 1 & 0 & 1 & 0 \end{pmatrix}}, G_1 = \frac{\det\begin{pmatrix} \alpha\omega_{mn}^2 & \omega_{mn}^2 & 0 & 1 \\ \omega_{mn}^2 + 2\alpha\Delta\omega_{mn} & 2\Delta\omega_{mn} & \alpha^2 & 0 \\ \alpha + 2\Delta\omega_{mn} & 1 & 2\alpha & 0 \\ 1 & 0 & 1 & 0 \end{pmatrix}}{\det\begin{pmatrix} \alpha\omega_{mn}^2 & \omega_{mn}^2 & 0 & \alpha^2 \\ \omega_{mn}^2 + 2\alpha\Delta\omega_{mn} & 2\Delta\omega_{mn} & \alpha^2 & 2\alpha \\ \alpha + 2\Delta\omega_{mn} & 1 & 2\alpha & 1 \\ 1 & 0 & 1 & 0 \end{pmatrix}} \tag{6.65c, d}$$

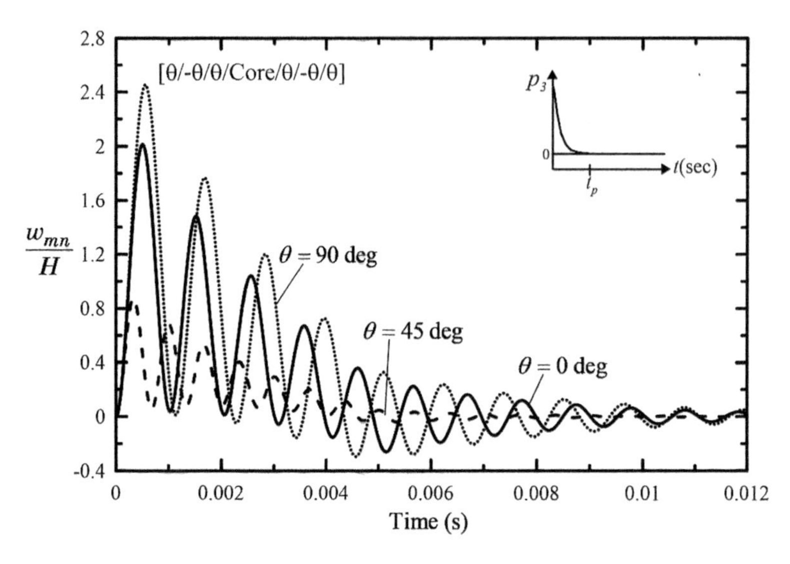

Fig. 6.15 The effect of the ply angle on the time deflection response of a doubly curved sandwich panel exposed to a Friedlander-type explosive blast pulse

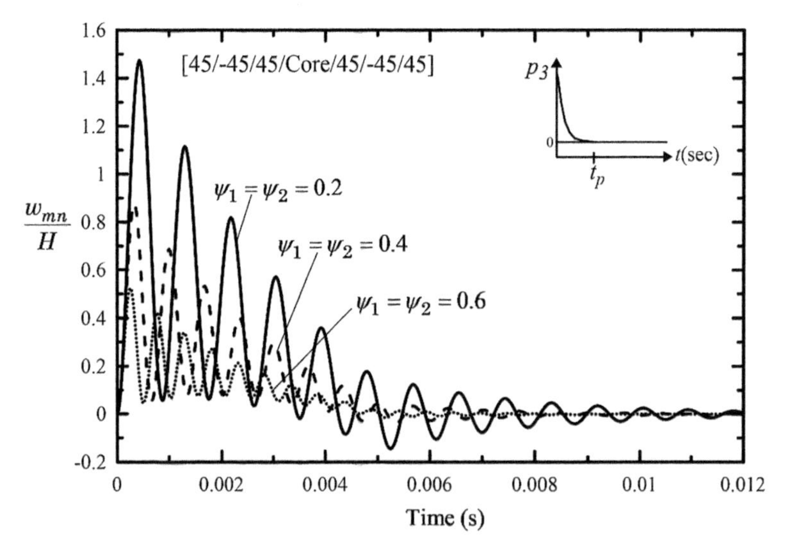

Fig. 6.16 The effect of the curvature ratio on the time–deflection response of a doubly curved sandwich panel exposed to a Friedlander-type explosive blast pulse

Figure 6.15 displays the effect of various ply angles for a given stacking sequence on the dynamic response of a doubly curved sandwich panel subjected to a Friedlander type of explosive pulse. It is clearly shown that the 45-degree ply angle provides the best response while a ply angle of 90 degrees is the least beneficial. Figure 6.16 shows the effect of the curvature ratio on the dynamic response of a doubly curved sandwich panel exposed to a Friedlander-type explosive blast. Clearly as seen in prior results the same sort of trend is seen where the larger curvature ratio is the least detrimental to the response.

6.5.7 Underwater Shock Pulse

While the previous pulses were considered to be in-air, a brief discussion on underwater explosive-type shock loading is presented. It is known that the total pressure due to an explosive underwater shock loading consists of two components, the incident shock wave pulse and the reflected shock wave pulse. This is expressed mathematically as

$$P(t) = P_i(t) + P_r(t) \tag{6.66}$$

where $P_i(t)$ is the incident pressure and $P_r(t)$ is the reflected pressure. The expression for the incident pressure is given by

$$P_i(t) = q_m e^{-(t-t_1)/\alpha}, \qquad t \geq t_1 \tag{6.67}$$

where $t - t_1$ (ms) is the time elapsed after the arrival of the shock pulse, where $t_1 = (R + z)/c$ is the time it takes to reach a point where the pressure is measured. c is the speed of sound in seawater; q_m represents the peak pressure of the shock front and α is the exponential decay constant in (ms). It should be noted that both and are dependent on the charge weight (kg) and the standoff distance (m) from the charge to the target.

The reflected pressure is given by the following expression

$$P_r(t) = \frac{q_m}{1 - m_r}\left[2e^{-(t+t_1)/m_r\alpha} - (1 + m_r)e^{-(t+t_1)/\alpha}\right] \tag{6.68}$$

where is known as the mass ratio. Herein, is the panel mass per unit area and is the density in seawater given as 1000 . Due to experimental data, are given by the approximate expressions (Shin and Geers 1994; Jian and Olsen 1994)

$$q_m = 56.6\left(\frac{Q^{1/3}}{R}\right)^{1.15} \text{ (MPa)} \tag{6.69a}$$

$$\alpha = 0.08 + Q^{1/3}\left(\frac{Q^{1/3}}{R}\right)^{-0.23} \text{ (ms)} \tag{6.69b}$$

It should be noted that the incident pressure is not valid if the sandwich panel is near to the explosive charge and that the pressure wave penetrating the core was not accounted for. Also, the effects of cavitation have not been included. With this model and following the same procedure with the Laplace Transform, the dynamic response to an underwater shock pulse can be determined. The results are not within the scope of this text.

6.6 Summary

In summary a comprehensive theoretical model has been presented which governs the dynamic response of both flat and curved sandwich panels. The governing equations were solved via the extended Galerkin method which resulted in a governing differential equation that was general in the sense that it was applicable to any type of pressure pulse. Several mathematical models of various types of time-dependent blast pulses were introduced where the Laplace transform method was applied to produce the dynamic response for each loading scenario. The theoretical response to several types of blast pulses were presented considering such material and geometrical characteristics as the curvature ratio, the material directional properties, the orthotropy ratio, edge loading (compression and tension), and the effects of damping. As similar trends were found among the various pulses, it was found that the larger curvature ratio provided the most beneficial effect in regard to the dynamic response. It was also determined that the 45-degree ply angle was the least detrimental to the dynamic response. Additionally, it was found that tensile edge loading improved the dynamic response of doubly curved sandwich panels in contrast to the compressive edge load which negatively impacted the dynamic response.

References

Jian, J., & Olsen, M. D. (1994). *Modeling of underwater shock-induced response of thin plate structures* (Report No. 39). Vancouver: Department of Civil Engineering, University of British Columbia.

Kim, D. H., & Han, J. H. (2006). Establishment of gun blast wave model and structural analysis for blast load. *Journal of Aircraft, 43*(4), 1159–1168.

Librescu, L. (1987). Refined geometrically non-linear theories of anisotropic laminated shells. *Quarterly of Applied Mathematics, 45*(1), 1–22.

Librescu, L., & Nosier, A. (1990). Response of shear deformable elastic laminated composite panels to sonic boom and explosive blast loadings. *AIAA Journal, 28*(2), 345–352.

Marzocca, P., Librescu, L., & Chiocchia, G. (2001). Aeroelastic response of 2D lifting surfaces to gust and arbitrary explosive loading signature. *International Journal of Impact Engineering, 25* (1), 67–85.

Shin, Y. S., & Geers, T. L. (1994). *Response of marine structures to underwater explosions.* Monterey: International Short Course, Shock and Vibration Research.

Chapter 7
Theory of Sandwich Plates and Shells with an Transversely Compressible Core – Theory One

Abstract A comprehensive detailed nonlinear theoretical model of asymmetric anisotropic doubly curved laminated composite sandwich plates and shells considering the case for the transversely compressible core, thereby capturing the wrinkling phenomenon of the sandwich panel, is presented. The theory assumes a transversal displacement of the core in terms of a first-order power series expansion, thus theory one. The governing equations are developed by way of an energy approach known as Hamilton's principle. Finally, an application of the governing equations is provided to demonstrate the solution approach to these equations through the use of the extended Galerkin method.

7.1 Introduction

In this chapter, a detailed comprehensive nonlinear theoretical treatment of the governing equations of asymmetric anisotropic doubly curved laminated sandwich shells considering a transversely compressible weak orthotropic core are presented. The tangential core displacements are represented by a second-order power series expansion, while the transverse core displacement is represented by a power series expansion to the first power. The theory includes the geometrical imperfections, transverse inertia, and external loadings such as the prescribed edge loadings and the transverse loading. The governing equations are developed via an energy approach through the application of Hamilton's principle. The face sheets are considered to be thin and initially asymmetric adopting the Love–Kirchhoff assumptions which are later reduced for the case of anisotropic symmetric laminated facings. The core is considered to be compressible (extensible in the transverse (normal) direction) thereby capturing wrinkling and any global instabilities. Finally, a chosen application of these theoretical developments is presented concerning the dynamic response where the solution methodology is highlighted and briefly discussed.

T. J. Hause, *Sandwich Structures: Theory and Responses*,
https://doi.org/10.1007/978-3-030-71895-4_7

7.2 Preliminaries and Basic Assumptions

As in Chap. 2, the middle surface of the core becomes the global mid-surface of the structure which is referred to an orthogonal curvilinear coordinate system x_i ($i = 1,2,3$). The coordinate x_3 is considered positive when measured in the downward normal direction from the mid-surface of the core. For convenience sake and simplicity, the thickness of the core is t_c which is uniform throughout. The thicknesses of the bottom and top facings are t^b_f and t^t_f, respectively. This implies that $H\left(\equiv t^b_f + t_c + t^t_f\right)$ is the total thickness of the structure.

For the compressible core case, some of the notations and terminology have been revised for simplicity. This change in notation provides for an easier understanding of the developments for the compressible core case as shown in Fig. 7.1. The notation changes are as follows:

1. Single and double prime notation (/, //) has been changed to (b, t), where b represents the bottom face and t represents the top face.
2. The thickness of the core has been changed from $2\overline{h}$ to t_c.
3. The thicknesses of the top and bottom facings have been changed to $h^{//} \Rightarrow t^t_f$, $h^{/} \Rightarrow t^b_f$, respectively.
4. The average and half difference quantities have been changed to $\xi_\alpha \Rightarrow u^a_\alpha$, $\eta_\alpha \Rightarrow u^d_\alpha$.
5. σ_{ij}will be replaced by τ_{ij}.
6. e_{ij}will be replaced by γ_{ij}.

Also, with regard to the half-difference quantities, the subtraction is reversed from the incompressible core case. The geometrically nonlinear theory of doubly curved sandwich panels with a compressible core is based on a series of assumptions which are as follows:

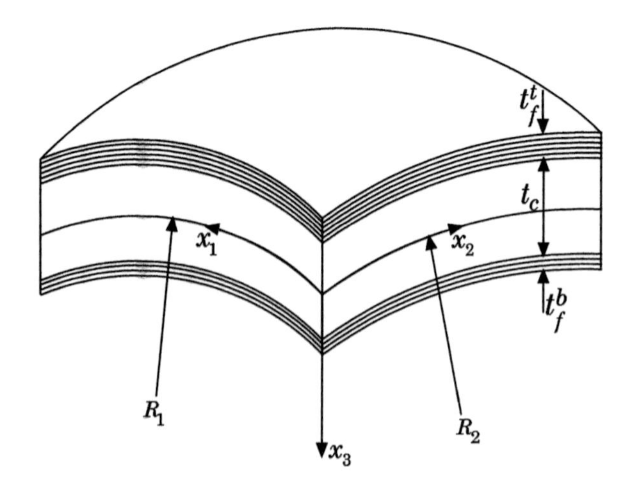

Fig. 7.1 The geometry of a doubly curved sandwich panel the compressible core case

1. The Love–Kirchhoff assumptions are assumed for the facings which are thin compared with the core layer.
2. The face sheets are constructed of a number of orthotropic material layers, the axes of orthotropy of the individual plies being not necessarily coincident with the geometrical axes x_α ($\alpha = 1{,}2$) of the structure.
3. The thickness of the core is much larger than those of the face sheets, i.e., $t_c \gg, t_f^b, t_f^t$.
4. The core is considered to be a weak orthotropic transversely compressible core carrying only the transverse strains and normal strains.
5. A perfect bonding between the face sheets and between the faces and the core exists.
6. The geometrical nonlinearities in the von Kármán sense with the geometrical imperfections are included in the sandwich structure model.
7. The principles of shallow shell theory apply.

Note: Unless otherwise stated, the Greek indices have the range 1, 2, while the Latin indices have the range 1, 2, 3, and Einstein's summation convention over repeated indices is assumed.

7.3 Basic Equations

7.3.1 Displacement Field

Consistent with standard plate and shell theory the displacement field for the top and bottom facings are given as

Bottom Face Sheets $\left(\frac{t_c}{2} \leq x_3 \leq \frac{t_c}{2} + t_f^b\right)$

$$v_1^b = u_1^b + \left(x_3 - \frac{t_c + t_f^b}{2}\right)\psi_1^b \tag{7.1a}$$

$$v_2^b = u_2^b + \left(x_3 - \frac{t_c + t_f^b}{2}\right)\psi_2^b \tag{7.1b}$$

$$v_3^b = u_3^b \tag{7.1c}$$

Top Face Sheets $\left(-\frac{t_c}{2} - t_f^t \leq x_3 \leq -\frac{t_c}{2}\right)$

$$v_1^t = u_1^t + \left(x_3 + \frac{t_c + t_f^t}{2}\right)\psi_1^t \tag{7.2a}$$

$$v_2^t = u_2^t + \left(x_3 + \frac{t_c + t_f^t}{2}\right)\psi_2^t \tag{7.2b}$$

$$v_3^t = u_3^t \tag{7.2c}$$

where, in the above equations, ψ_1^b, ψ_2^b, ψ_1^t, ψ_2^t represent the shear angles, while t and b represent the association with the top and bottom facings, respectively. A second-order power series is assumed for the core tangential displacements, while a first-order polynomial is assumed for the core transverse displacements, as shown below see Hohe and Librescu (2006).

Core $\left(-\frac{t_c}{2} \leq x_3 \leq \frac{t_c}{2}\right)$

$$v_1^c = u_1^c + x_3\psi_1^c + x_3^2\gamma_1^c \tag{7.3a}$$

$$v_2^c = u_2^c + x_3\psi_2^c + x_3^2\gamma_2^c \tag{7.3b}$$

$$v_3^c = u_3^c + x_3\psi_3^c \tag{7.3c}$$

Next, the interfacial continuity conditions at the interfaces need to be satisfied. These are stating that at the interfaces between the core and the facings, the displacements must be equal for both the core and the facings. As for the incompressible core case, in Chap. 2, they are expressed mathematically as

Top face sheet–core interface ($x_3 = -t_c/2$)

$$v_\alpha^c = v_\alpha^t, \qquad v_3^c = v_3^t \tag{7.4a, b}$$

Bottom face sheet–core interface ($x_3 = t_c/2$)

$$v_\alpha^c = v_\alpha^b, \qquad v_3^c = v_3^b \tag{7.4c, d}$$

Also, in addition to the continuity conditions, the Love–Kirchhoff assumptions for thin facings are adopted where the shear strains are assumed to be negligible or zero. By setting the mathematical expressions for the shear strains to zero and solving for the shear angles ψ_1^b, ψ_2^b, ψ_1^t, ψ_2^t results in

$$\psi_1^t = -\frac{\partial u_3^t}{\partial x_1}, \qquad \psi_2^t = -\frac{\partial u_3^t}{\partial x_2}, \qquad \psi_1^b = -\frac{\partial u_3^b}{\partial x_1}, \qquad \psi_2^b = -\frac{\partial u_3^b}{\partial x_2} \tag{7.5a, b}$$

With these assumptions, the displacement field for the facings and the core become

Top face sheets $\left(-\frac{t_c}{2} - t_f^t \leq x_3 \leq -\frac{t_c}{2}\right)$

$$v_1^t = u_1^a + u_1^d - \left(x_3 + \frac{t_c + t_f^t}{2}\right)\frac{\partial u_3^a}{\partial x_1} - \left(x_3 + \frac{t_c + t_f^t}{2}\right)\frac{\partial u_3^d}{\partial x_1} \tag{7.6a}$$

$$v_2^t = u_2^a + u_2^d - \left(x_3 + \frac{t_c + t_f^t}{2}\right)\frac{\partial u_3^a}{\partial x_2} - \left(x_3 + \frac{t_c + t_f^t}{2}\right)\frac{\partial u_3^d}{\partial x_2} \tag{7.6b}$$

$$v_3^t = u_3^a + u_3^d \tag{7.6c}$$

Bottom face sheets $\left(\frac{t_c}{2} \leq x_3 \leq \frac{t_c}{2} + t_f^b\right)$

$$v_1^b = u_1^a - u_1^d - \left(x_3 - \frac{t_c + t_f^b}{2}\right)\frac{\partial u_3^a}{\partial x_1} + \left(x_3 - \frac{t_c + t_f^b}{2}\right)\frac{\partial u_3^d}{\partial x_1} \tag{7.7a}$$

$$v_2^b = u_2^a - u_2^d - \left(x_3 - \frac{t_c + t_f^b}{2}\right)\frac{\partial u_3^a}{\partial x_2} + \left(x_3 - \frac{t_c + t_f^b}{2}\right)\frac{\partial u_3^d}{\partial x_2} \tag{7.7b}$$

$$v_3^b = u_3^a - u_3^d \tag{7.7c}$$

Core $\left(-\frac{t_c}{2} \leq x_3 \leq \frac{t_c}{2}\right)$

$$v_1^c = u_1^a - \left(\frac{t_f^t - t_f^b}{4}\right)\frac{\partial u_3^a}{\partial x_1} - \left(\frac{t_f^t + t_f^b}{4}\right)\frac{\partial u_3^d}{\partial x_1} - \frac{2x_3}{t_c}u_1^d + \frac{1}{t_c}\left(\frac{t_f^t + t_f^b}{2}\right)x_3\frac{\partial u_3^a}{\partial x_1} + \frac{1}{t_c}\left(\frac{t_f^t - t_f^b}{2}\right)x_3\frac{\partial u_3^d}{\partial x_1} + \frac{4}{t_c^2}\left(x_3^2 - \frac{t_c^2}{4}\right)\Phi_1^c \tag{7.8a}$$

$$v_2^c = u_2^a - \left(\frac{t_f^t - t_f^b}{4}\right)\frac{\partial u_3^a}{\partial x_2} - \left(\frac{t_f^t + t_f^b}{4}\right)\frac{\partial u_3^d}{\partial x_2} - \frac{2x_3}{t_c}u_2^d + \frac{1}{t_c}\left(\frac{t_f^t + t_f^b}{2}\right)x_3\frac{\partial u_3^a}{\partial x_2} + \frac{1}{t_c}\left(\frac{t_f^t - t_f^b}{2}\right)x_3\frac{\partial u_3^d}{\partial x_2} + \frac{4}{t_c^2}\left(x_3^2 - \frac{t_c^2}{4}\right)\Phi_2^c \tag{7.8b}$$

$$v_3^c = u_3^a - \frac{2x_3}{t_c}u_3^d \tag{7.8c}$$

If the facings are symmetric with respect to the global mid-surface, then $t^t_f = t^b_f = t^f$. In this case the displacement equations reduce to the same displacement equations found in Hohe and Librescu (2006). In the above displacement equations, the average displacement functions are defined as

$$u_i^a = \frac{1}{2}\left(u_i^t + u_i^b\right), \qquad u_i^d = \frac{1}{2}\left(u_i^t - u_i^b\right) \tag{7.9a, b}$$

which represent the average and half difference of the mid-surface displacements. ψ^a_α, ψ^d_α $(\alpha, \beta = 1, 2)$ represent the rotation angles, while the core quantities Φ^c_1, Φ^c_2, Ω^c_1, and Ω^c_2 represent the warping functions. In the following section, a stress-free initial geometric imperfection is introduced which remains constant during deformation. These expressions for the geometric imperfections were introduced by Hohe and Librescu (2003) and Hohe and Librescu (2006). These geometric imperfections are introduced into the Green–Lagrange strain tensor as

$$\overset{\circ}{v}{}^t_3 = \overset{\circ}{u}{}^a_3 + \overset{\circ}{u}{}^d_3 \tag{7.10}$$

$$\overset{\circ}{v}{}^b_3 = \overset{\circ}{u}{}^a_3 - \overset{\circ}{u}{}^d_3 \tag{7.11}$$

$$\overset{\circ}{v}{}^c_3 = \overset{\circ}{u}{}^a_3 - \frac{2x_3}{t_c}\overset{\circ}{u}{}^d_3 \tag{7.12}$$

7.3.2 Nonlinear Strain–Displacement Equations

The Green–Lagrange strain tensor with the geometric imperfections, from Eqs. (2.18)–(2.23), in conjunction with the von Kármán assumptions is (see Amabili 2004)

$$\gamma_{11} = \frac{\partial v_1}{\partial x_1} - \frac{v_3}{R_1} + \frac{1}{2}\left(\frac{\partial v_3}{\partial x_1}\right)^2 + \frac{\partial v_3}{\partial x_1}\frac{\partial v_3^\circ}{\partial x_1} \tag{7.13a}$$

$$\gamma_{22} = \frac{\partial v_2}{\partial x_2} - \frac{v_3}{R_2} + \frac{1}{2}\left(\frac{\partial v_3}{\partial x_2}\right)^2 + \frac{\partial v_3}{\partial x_2}\frac{\partial v_3^\circ}{\partial x_2} \tag{7.13b}$$

$$\gamma_{12} = \frac{1}{2}\left(\frac{\partial v_2}{\partial x_1} + \frac{\partial v_1}{\partial x_2}\right) + \frac{1}{2}\frac{\partial v_3}{\partial x_1}\frac{\partial v_3}{\partial x_2} + \frac{1}{2}\frac{\partial v_3}{\partial x_1}\frac{\partial v_3^\circ}{\partial x_2} + \frac{1}{2}\frac{\partial v_3}{\partial x_2}\frac{\partial v_3^\circ}{\partial x_1} \tag{7.13c}$$

$$\gamma_{13} = \frac{1}{2}\left(\frac{\partial v_1}{\partial x_3} + \frac{\partial v_3}{\partial x_1}\right) + \frac{1}{2}\frac{\partial v_3}{\partial x_1}\frac{\partial v_3}{\partial x_3} + \frac{1}{2}\frac{\partial v_3}{\partial x_1}\frac{\partial v_3^\circ}{\partial x_3} + \frac{1}{2}\frac{\partial v_3}{\partial x_3}\frac{\partial v_3^\circ}{\partial x_1} \tag{7.13d}$$

$$\gamma_{23} = \frac{1}{2}\left(\frac{\partial v_2}{\partial x_3} + \frac{\partial v_3}{\partial x_2}\right) + \frac{1}{2}\frac{\partial v_3}{\partial x_2}\frac{\partial v_3}{\partial x_3} + \frac{1}{2}\frac{\partial v_3}{\partial x_2}\frac{\partial v_3^\circ}{\partial x_3} + \frac{1}{2}\frac{\partial v_3}{\partial x_3}\frac{\partial v_3^\circ}{\partial x_2} \tag{7.13e}$$

$$\gamma_{33} = \frac{\partial v_3}{\partial x_3} + \frac{1}{2}\left(\frac{\partial v_3}{\partial x_3}\right)^2 + \frac{\partial v_3}{\partial x_3}\frac{\partial v_3^\circ}{\partial x_3} \tag{7.13f}$$

Substituting the displacement equations, Eqs. (7.6a–c), (7.7a–c), and (7.8a–c) into the nonlinear strain–displacement relationships Eqs. (7.13a–f) gives for each layer

Top face sheets $\left(-\frac{t_c}{2} - t_f^t \le x_3 \le -\frac{t_c}{2}\right)$

$$\begin{aligned}
\gamma_{11}^t &= \overline{\gamma}_{11}^a + \overline{\gamma}_{11}^d + \left(x_3 + \frac{t_c + t_f^t}{2}\right)\kappa_{11}^a + \left(x_3 + \frac{t_c + t_f^t}{2}\right)\kappa_{11}^d \\
\gamma_{22}^t &= \overline{\gamma}_{22}^a + \overline{\gamma}_{22}^d + \left(x_3 + \frac{t_c + t_f^t}{2}\right)\kappa_{22}^a + \left(x_3 + \frac{t_c + t_f^t}{2}\right)\kappa_{22}^d \\
\gamma_{12}^t &= \overline{\gamma}_{12}^a + \overline{\gamma}_{12}^d + \left(x_3 + \frac{t_c + t_f^t}{2}\right)\kappa_{12}^a + \left(x_3 + \frac{t_c + t_f^t}{2}\right)\kappa_{12}^d
\end{aligned} \tag{7.14a – c}$$

Core $\left(-\frac{t_c}{2} \le x_3 \le \frac{t_c}{2}\right)$

$$\begin{aligned}
\gamma_{13}^c &= \overline{\gamma}^c{}_{13} + x_3\kappa_{13}^c \\
\gamma_{23}^c &= \overline{\gamma}^c{}_{23} + x_3\kappa_{23}^c \\
\gamma_{33}^c &= \overline{\gamma}_{33}^c + x_3\kappa_{33}^c
\end{aligned} \tag{7.15a – c}$$

Bottom face sheets $\left(\frac{t_c}{2} \le x_3 \le \frac{t_c}{2} + t_f^b\right)$

$$\begin{aligned}
\gamma_{11}^b &= \overline{\gamma}_{11}^a - \overline{\gamma}_{11}^d + \left(x_3 - \frac{t_c + t_f^b}{2}\right)\kappa_{11}^a - \left(x_3 - \frac{t_c + t_f^b}{2}\right)\kappa_{11}^d \\
\gamma_{22}^b &= \overline{\gamma}_{22}^a - \overline{\gamma}_{22}^d + \left(x_3 - \frac{t_c + t_f^b}{2}\right)\kappa_{22}^a - \left(x_3 - \frac{t_c + t_f^b}{2}\right)\kappa_{22}^d \\
\gamma_{12}^b &= \overline{\gamma}_{12}^a - \overline{\gamma}_{12}^d + \left(x_3 - \frac{t_c + t_f^b}{2}\right)\kappa_{12}^a - \left(x_3 - \frac{t_c + t_f^b}{2}\right)\kappa_{12}^d
\end{aligned} \tag{7.16a – c}$$

where

$$\overline{\gamma}^a_{\alpha\beta} = \frac{1}{2}\left(\overline{\gamma}^t_{\alpha\beta} + \overline{\gamma}^b_{\alpha\beta}\right), \quad \overline{\gamma}^d_{\alpha\beta} = \frac{1}{2}\left(\overline{\gamma}^t_{\alpha\beta} - \overline{\gamma}^b_{\alpha\beta}\right) \tag{7.17a}$$

$$\kappa^a_{\alpha\beta} = \frac{1}{2}\left(\kappa^t_{\alpha\beta} + \kappa^b_{\alpha\beta}\right), \quad \kappa^d_{\alpha\beta} = \frac{1}{2}\left(\kappa^t_{\alpha\beta} - \kappa^b_{\alpha\beta}\right) \tag{7.17b}$$

In the above strain displacement equations, $\overline{\gamma}^a_{ij}$ and $\overline{\gamma}^d_{ij}$ are the in-plane average and the half difference of the tangential strains of the top and bottom facings, while $\kappa^a_{\alpha\beta}$ and $\kappa^d_{\alpha\beta}$ are the average and half-difference bending strains of the top and bottom facings. For the core, $\overline{\gamma}^c_{i3}$ and κ^c_{i3} are the tangential and bending strains, respectively. The expressions for these strains are given as

$$\overline{\gamma}^a_{11} = \frac{\partial u^a_1}{\partial x_1} - \frac{u^a_3}{R_1} + \frac{1}{2}\left(\frac{\partial u^a_3}{\partial x_1}\right)^2 + \frac{\partial u^a_3}{\partial x_1}\frac{\partial \mathring{u}^a_3}{\partial x_1} + \frac{1}{2}\left(\frac{\partial u^d_3}{\partial x_1}\right)^2 + \frac{\partial u^d_3}{\partial x_1}\frac{\partial \mathring{u}^d_3}{\partial x_1} \tag{7.18}$$

$$\overline{\gamma}^a_{22} = \frac{\partial u^a_2}{\partial x_2} - \frac{u^a_3}{R_2} + \frac{1}{2}\left(\frac{\partial u^a_3}{\partial x_2}\right)^2 + \frac{\partial u^a_3}{\partial x_2}\frac{\partial \mathring{u}^a_3}{\partial x_2} + \frac{1}{2}\left(\frac{\partial u^d_3}{\partial x_2}\right)^2 + \frac{\partial u^d_3}{\partial x_2}\frac{\partial \mathring{u}^d_3}{\partial x_2} \tag{7.19}$$

$$\begin{aligned}\overline{\gamma}^a_{12} = {} & \frac{1}{2}\frac{\partial u^a_1}{\partial x_2} + \frac{1}{2}\frac{\partial u^a_2}{\partial x_1} + \frac{1}{2}\frac{\partial u^a_3}{\partial x_1}\frac{\partial u^a_3}{\partial x_2} + \frac{1}{2}\frac{\partial u^d_3}{\partial x_1}\frac{\partial u^d_3}{\partial x_2} + \frac{1}{2}\frac{\partial u^a_3}{\partial x_1}\frac{\partial \mathring{u}^a_3}{\partial x_2} + \frac{1}{2}\frac{\partial \mathring{u}^a_3}{\partial x_1}\frac{\partial u^a_3}{\partial x_2} + \frac{1}{2}\frac{\partial u^d_3}{\partial x_1}\frac{\partial \mathring{u}^d_3}{\partial x_2} \\ & + \frac{1}{2}\frac{\partial \mathring{u}^d_3}{\partial x_1}\frac{\partial u^d_3}{\partial x_2}\end{aligned} \tag{7.20}$$

$$\overline{\gamma}^d_{11} = \frac{\partial u^d_1}{\partial x_1} - \frac{u^d_3}{R_1} + \frac{\partial u^a_3}{\partial x_1}\frac{\partial u^d_3}{\partial x_1} + \frac{\partial u^a_3}{\partial x_1}\frac{\partial \mathring{u}^d_3}{\partial x_1} + \frac{\partial \mathring{u}^a_3}{\partial x_1}\frac{\partial u^d_3}{\partial x_1} \tag{7.21}$$

$$\overline{\gamma}^d_{22} = \frac{\partial u^d_2}{\partial x_2} - \frac{u^d_3}{R_2} + \frac{\partial u^a_3}{\partial x_2}\frac{\partial u^d_3}{\partial x_2} + \frac{\partial u^a_3}{\partial x_2}\frac{\partial \mathring{u}^d_3}{\partial x_2} + \frac{\partial \mathring{u}^a_3}{\partial x_2}\frac{\partial u^d_3}{\partial x_2} \tag{7.22}$$

$$\begin{aligned}\overline{\gamma}^d_{12} = {} & \frac{1}{2}\frac{\partial u^d_1}{\partial x_2} + \frac{1}{2}\frac{\partial u^d_2}{\partial x_1} + \frac{1}{2}\frac{\partial u^a_3}{\partial x_1}\frac{\partial u^d_3}{\partial x_2} + \frac{1}{2}\frac{\partial u^a_3}{\partial x_2}\frac{\partial u^d_3}{\partial x_1} + \frac{1}{2}\frac{\partial u^a_3}{\partial x_1}\frac{\partial \mathring{u}^d_3}{\partial x_2} + \frac{1}{2}\frac{\partial \mathring{u}^a_3}{\partial x_1}\frac{\partial u^d_3}{\partial x_2} + \frac{1}{2}\frac{\partial u^a_3}{\partial x_2}\frac{\partial \mathring{u}^d_3}{\partial x_1} \\ & + \frac{1}{2}\frac{\partial \mathring{u}^a_3}{\partial x_2}\frac{\partial u^d_3}{\partial x_1}\end{aligned} \tag{7.23}$$

$$\kappa_{11}^{a} = -\frac{\partial^2 u_3^a}{\partial x_1^2} \tag{7.24}$$

$$\kappa_{22}^{a} = -\frac{\partial^2 u_3^a}{\partial x_2^2} \tag{7.25}$$

$$\kappa_{12}^{a} = -\frac{\partial^2 u_3^a}{\partial x_1 \partial x_2} \tag{7.26}$$

$$\kappa_{11}^{d} = -\frac{\partial^2 u_3^d}{\partial x_1^2} \tag{7.27}$$

$$\kappa_{22}^{d} = -\frac{\partial^2 u_3^d}{\partial x_2^2} \tag{7.28}$$

$$\kappa_{12}^{d} = -\frac{\partial^2 u_3^d}{\partial x_1 \partial x_2} \tag{7.29}$$

$$\overline{\gamma}_{33}^{c} = -\frac{2}{t_c} u_3^d + \frac{2}{t_c^2}\left(u_3^d\right)^2 + \frac{4}{t_c^2} u_3^d \left(u_3^d\right)^{\circ} \tag{7.30}$$

$$\overline{\gamma}_{13}^{c} = -\frac{u_1^d}{t_c} + \left(\frac{1}{2} + \frac{1}{t_c}\left(\frac{t_f^t + t_f^b}{4}\right)\right)\frac{\partial u_3^a}{\partial x_1} + \frac{1}{t_c}\left(\frac{t_f^t - t_f^b}{4}\right)\frac{\partial u_3^d}{\partial x_1} - \frac{u_3^d}{t_c}\frac{\partial u_3^a}{\partial x_1}$$
$$-\frac{\mathring{u}_3^a}{t_c}\frac{\partial u_3^a}{\partial x_1} - \frac{u_3^d}{t_c}\frac{\partial \mathring{u}_3^a}{\partial x_1} \tag{7.31}$$

$$\overline{\gamma}_{23}^{c} = -\frac{u_2^d}{t_c} + \left(\frac{1}{2} + \frac{1}{t_c}\left(\frac{t_f^t + t_f^b}{4}\right)\right)\frac{\partial u_3^a}{\partial x_2} + \frac{1}{t_c}\left(\frac{t_f^t - t_f^b}{4}\right)\frac{\partial u_3^d}{\partial x_2} - \frac{u_3^d}{t_c}\frac{\partial u_3^a}{\partial x_2}$$
$$-\frac{\mathring{u}_3^a}{t_c}\frac{\partial u_3^a}{\partial x_2} - \frac{u_3^d}{t_c}\frac{\partial \mathring{u}_3^a}{\partial x_2} \tag{7.32}$$

$$\kappa_{33}^{c} = 0 \tag{7.33}$$

$$\kappa_{13}^{c} = \frac{4}{t_c^2}\Phi_1^c - \frac{1}{t_c}\frac{\partial u_3^d}{\partial x_1} + \frac{2}{t_c^2} u_3^d \frac{\partial u_3^d}{\partial x_1} + \frac{2}{t_c^2}\left(u_3^{\circ}\right)^d \frac{\partial u_3^d}{\partial x_1} + \frac{2}{t_c^2} u_3^d \frac{\partial \mathring{u}_3^d}{\partial x_1} \tag{7.34}$$

$$\kappa_{23}^{c} = \frac{4}{t_c^2}\Phi_2^c - \frac{1}{t_c}\frac{\partial u_3^d}{\partial x_2} + \frac{2}{t_c^2} u_3^d \frac{\partial u_3^d}{\partial x_2} + \frac{2}{t_c^2}\left(u_3^{\circ}\right)^d \frac{\partial u_3^d}{\partial x_2} + \frac{2}{t_c^2} u_3^d \frac{\partial \mathring{u}_3^d}{\partial x_2} \tag{7.35}$$

7.3.3 Constitutive Equations

The top and bottom facings are considered to be constructed from unidirectional fiber-reinforced anisotropic laminated composites. The stress–strain relationships per lamina of the facings is repeated here as

Top face sheets $\left(-\frac{t_c}{2} - t_f^t \le x_3 \le -\frac{t_c}{2}\right)$

$$\begin{pmatrix} \tau_{11}^t \\ \tau_{22}^t \\ \tau_{12}^t \end{pmatrix}_k = \begin{bmatrix} \widehat{Q}_{11}^t & \widehat{Q}_{11}^t & \widehat{Q}_{11}^t \\ & \widehat{Q}_{22}^t & \widehat{Q}_{26}^t \\ \text{Sym} & & \widehat{Q}_{66}^t \end{bmatrix}_k \begin{pmatrix} \gamma_{11}^t \\ \gamma_{22}^t \\ \gamma_{12}^t \end{pmatrix}_k - \begin{Bmatrix} \widehat{\lambda}_{11}^t \\ \widehat{\lambda}_{22}^t \\ \widehat{\lambda}_{12}^t \end{Bmatrix}_k \Delta T - \begin{Bmatrix} \widehat{\mu}_{11}^t \\ \widehat{\mu}_{22}^t \\ \widehat{\mu}_{12}^t \end{Bmatrix}_k \Delta M \quad (7.36)$$

where k represents the kth lamina in the facing and $\widehat{Q}_{ij}$ for $i, j = (1, 2, 6)$ are the transformed plane-stress reduced stiffness measures (Reddy (2004) and Jones (1999)). These were given in Chap. 2.

Bottom face sheets $\left(\frac{t_c}{2} \le x_3 \le \frac{t_c}{2} + t_f^b\right)$

$$\begin{pmatrix} \tau_{11}^b \\ \tau_{22}^b \\ \tau_{12}^b \end{pmatrix}_k = \begin{bmatrix} \widehat{Q}_{11}^b & \widehat{Q}_{11}^b & \widehat{Q}_{11}^b \\ & \widehat{Q}_{22}^b & \widehat{Q}_{26}^b \\ \text{Sym} & & \widehat{Q}_{66}^b \end{bmatrix}_k \begin{pmatrix} \gamma_{11}^b \\ \gamma_{22}^b \\ \gamma_{12}^b \end{pmatrix}_k - \begin{Bmatrix} \widehat{\lambda}_{11}^b \\ \widehat{\lambda}_{22}^b \\ \widehat{\lambda}_{12}^b \end{Bmatrix}_k \Delta T - \begin{Bmatrix} \widehat{\mu}_{11}^b \\ \widehat{\mu}_{22}^b \\ \widehat{\mu}_{12}^b \end{Bmatrix}_k \Delta M \quad (7.37)$$

The stress–strain relationships for the orthotropic weak (soft) core with the geometrical and material axes coincident are expressed as

Core $\left(-\frac{t_c}{2} \le x_3 \le \frac{t_c}{2}\right)$

$$\begin{pmatrix} \tau_{33}^c \\ \tau_{23}^c \\ \tau_{13}^c \end{pmatrix} = \begin{bmatrix} Q_{33}^c & 0 & 0 \\ 0 & Q_{44}^c & 0 \\ 0 & 0 & Q_{55}^c \end{bmatrix} \begin{pmatrix} \gamma_{33}^c \\ \gamma_{23}^c \\ \gamma_{13}^c \end{pmatrix} \quad (7.38)$$

The core transverse and normal moduli are given as

$$Q_{33}^c = E^c, \quad Q_{44}^c = G_{23}^c, \quad Q_{55}^c = G_{13}^c \quad (7.39\text{a} - \text{c})$$

where E^c is the Young's modulus and G_{13}^c, G_{23}^c are the shear moduli for the core which allows the core stresses to be expressed as

$$\tau_{33}^c = E^c \overline{\gamma}_{33}^c, \qquad \tau_{13}^c = G_{13}\overline{\gamma}_{13}^c, \qquad \tau_{23}^c = G_{23}\overline{\gamma}_{23}^c \tag{7.40a – c}$$

7.4 Hamilton's Principle

An energy approach using Hamilton's Principle is used to derive the equations of motion, as was shown in Chap. 2. Letting U represent the strain energy, W represent the work done by external forces, and T represent the kinetic energy, Hamilton's Principle (see Soedel (2004)) is expressed as

$$\delta J = \int_{t_0}^{t_1} (\delta U - \delta W - \delta T)dt = 0 \tag{7.41}$$

7.4.1 *Strain Energy*

Utilizing Eq. (2.98) from Chap. 2 and assuming a weak compressible core where the core carries only the transverse shear stresses, the variation in the strain energy is given by

$$\delta U = \int_A \left(\int_{\frac{-t_c}{2}-t^t_f}^{-\frac{t_c}{2}} \tau_{\alpha\beta}^t \delta\gamma_{\alpha\beta}^t dx_3 + \int_{\frac{-t_c}{2}}^{+\frac{t_c}{2}} \tau_{i3}^c \delta\gamma_{i3}^c dx_3 + \int_{\frac{t_c}{2}}^{\frac{t_c}{2}+t^b_f} \tau_{\alpha\beta}^b \delta\gamma_{\alpha\beta}^b dx_3 \right) dA \tag{7.42}$$

where τ_{ij} are the tensorial components of the second Piola–Kirchhoff stress tensor, while A is attributed to the planar area of the sandwich shell. The strain energy in the tensorial form is expanded out and shown as

$$\begin{aligned} \delta U = \int_A \Bigg\{ & \int_{\frac{t_c}{2}}^{\frac{t_c}{2}+t^b_f} \left(\tau_{11}^b\delta\gamma_{11}^b + \tau_{22}^b\delta\gamma_{22}^b + 2\tau_{12}^b\delta\gamma_{12}^b\right)dx_3 + \\ & + \int_{-\frac{t_c}{2}}^{\frac{t_c}{2}} \left(2\tau_{13}^c\delta\gamma_{13}^c + 2\tau_{23}^c\delta\gamma_{23}^c + \tau_{33}^c\delta\gamma_{33}^c\right)dx_3 + \\ & + \int_{-\frac{t_c}{2}-t^t_f}^{-\frac{t_c}{2}} \left(\tau_{11}^t\delta\gamma_{11}^t + \tau_{22}^t\delta\gamma_{22}^t + 2\tau_{12}^t\delta\gamma_{12}^t\right)dx_3 \Bigg\} dA \end{aligned} \tag{7.43}$$

Substituting in the expressions for the strain relationships, Eqs. (7.14a–c)–(7.16a–c) results in

$$\delta U = \int_A \Bigg\{ \int_{\frac{t_c}{2}}^{\frac{t_c}{2}+t_f^b} \Bigg[\tau_{11}^b \left(\delta\overline{\gamma}_{11}^a - \delta\overline{\gamma}_{11}^d + \left(x_3 - \frac{t_c + t_f^b}{2} \right) \delta\kappa_{11}^a - \left(x_3 - \frac{t_c + t_f^b}{2} \right) \delta\kappa_{11}^d \right)$$
$$+ \tau_{22}^b \left(\delta\overline{\gamma}_{22}^a - \delta\overline{\gamma}_{22}^d + \left(x_3 - \frac{t_c + t_f^b}{2} \right) \delta\kappa_{22}^a - \left(x_3 - \frac{t_c + t_f^b}{2} \right) \delta\kappa_{22}^d \right)$$
$$+ 2\tau_{12}^b \left(\delta\overline{\gamma}_{12}^a - \delta\overline{\gamma}_{12}^d + \left(x_3 - \frac{t_c + t_f^b}{2} \right) \delta\kappa_{12}^a - \left(x_3 - \frac{t_c + t_f^b}{2} \right) \delta\kappa_{12}^d \right) \Bigg] dx_3$$
$$+ \int_{-\frac{t_c}{2}}^{\frac{t_c}{2}} \left[2\tau_{13}^c \left(\delta\overline{\gamma}^c{}_{13} + x_3 \delta\kappa_{13}^c \right) + 2\tau_{23}^c \left(\delta\overline{\gamma}^c{}_{23} + x_3 \delta\kappa_{23}^c \right) + \tau_{33}^c \left(\delta\overline{\gamma}_{33}^c + x_3 \delta\kappa_{33}^c \right) \right] dx_3$$
$$+ \int_{-\frac{t_c}{2}-t^t{}_f}^{-\frac{t_c}{2}} \Bigg[\tau_{11}^t \left(\delta\overline{\gamma}_{11}^a + \delta\overline{\gamma}_{11}^d + \left(x_3 + \frac{t_c + t_f^t}{2} \right) \delta\kappa_{11}^a + \left(x_3 + \frac{t_c + t_f^t}{2} \right) \delta\kappa_{11}^d \right)$$
$$+ \tau_{22}^t \left(\delta\overline{\gamma}_{22}^a + \delta\overline{\gamma}_{22}^d + \left(x_3 + \frac{t_c + t_f^t}{2} \right) \delta\kappa_{22}^a + \left(x_3 + \frac{t_c + t_f^t}{2} \right) \delta\kappa_{22}^d \right) + 2\tau_{12}^t \Bigg(\delta\overline{\gamma}_{12}^a$$
$$+ \delta\overline{\gamma}_{12}^d + \left(x_3 + \frac{t_c + t_f^t}{2} \right) \delta\kappa_{12}^a + \left(x_3 + \left(x_3 + \frac{t_c + t_f^t}{2} \right) \delta\kappa_{12}^a \right) \delta\kappa_{12}^d \Bigg) \Bigg] dx_3 \Bigg\} dA \tag{7.44}$$

which can be expressed in the following form as

$$\delta U = \int_A \left[N_{11}^b \delta\overline{\gamma}_{11}^a - N_{11}^b \delta\overline{\gamma}_{11}^d + M_{11}^b \delta\kappa_{11}^a - M_{11}^b \delta\kappa_{11}^d + N_{22}^b \delta\overline{\gamma}_{22}^a - N_{22}^b \delta\overline{\gamma}_{22}^d \right.$$
$$+ M_{22}^b \delta\kappa_{22}^a - M_{22}^b \delta\kappa_{22}^d + 2N_{12}^b \delta\overline{\gamma}_{12}^a - 2N_{12}^b \delta\overline{\gamma}_{12}^d + 2M_{12}^b \delta\kappa_{12}^a - 2M_{12}^b \delta\kappa_{12}^d$$
$$+ 2N_{13}^c \delta\overline{\gamma}_{13}^c + 2M_{13}^c \delta\kappa_{13}^c + 2N_{23}^c \delta\overline{\gamma}_{23}^c + 2M_{23}^c \delta\kappa_{23}^c + N_{33}^c \delta\overline{\gamma}_{33}^c + M_{33}^c \delta\kappa_{33}^c$$
$$+ N_{11}^t \delta\overline{\gamma}_{11}^a + N_{11}^t \delta\overline{\gamma}_{11}^d + M_{11}^t \delta\kappa_{11}^a + M_{11}^t \delta\kappa_{11}^d + N_{22}^t \delta\overline{\gamma}_{22}^a + N_{22}^t \delta\overline{\gamma}_{22}^d + M_{22}^t \delta\kappa_{22}^a$$
$$\left. + M_{22}^t \delta\kappa_{22}^d + 2N_{12}^t \delta\overline{\gamma}_{12}^a + 2N_{12}^t \delta\overline{\gamma}_{12}^d + 2M_{12}^t \delta\kappa_{12}^a + 2M_{12}^t \delta\kappa_{12}^d \right] dA \tag{7.45}$$

where the local stress resultants $N_{\alpha\beta}^t$, $N_{\alpha\beta}^b$, N_{i3}^c and the stress couples $M_{\alpha\beta}^t$, $M_{\alpha\beta}^b$, M_{i3}^c are defined as

$$\left\{ N^t{}_{\alpha\beta} . M^t{}_{\alpha\beta} \right\} = \int_{-\frac{t_c}{2}-t^t{}_f}^{-\frac{t_c}{2}} \tau^t{}_{\alpha\beta} \left\{ 1, \left(x_3 + \frac{t_c + t^t{}_f}{2} \right) \right\} dx_3 \tag{7.46a}$$

$$\left\{N^b_{\alpha\beta}, M^b_{\alpha\beta}\right\} = \int_{\frac{t_c}{2}}^{\frac{t_c}{2}+t^b_f} \tau^b_{\alpha\beta}\left\{1, \left(x_3 - \frac{t_c + t^b_f}{2}\right)\right\} dx_3 \tag{7.46b}$$

$$\left\{N^c_{i3}, M^c_{i3}\right\} = \int_{-\frac{t_c}{2}}^{\frac{t_c}{2}} \tau^c_{i3}(1, x_3) dx_3 \tag{7.46c}$$

Combining like terms gives

$$\begin{aligned}
\delta U = \int_A \{ & \left(N^t_{11} + N^b_{11}\right)\delta\overline{\gamma}^a_{11} + \left(N^t_{11} - N^b_{11}\right)\delta\overline{\gamma}^d_{11} + \left(M^t_{11} + M^b_{11}\right)\delta\kappa^a_{11} + \left(M^t_{11} - M^b_{11}\right)\delta\kappa^d_{11} \\
& + \left(N^t_{22} + N^b_{22}\right)\delta\overline{\gamma}^a_{22} + \left(N^t_{22} - N^b_{22}\right)\delta\overline{\gamma}^d_{22} + \left(M^t_{22} + M^b_{22}\right)\delta\kappa^a_{22} + \left(M^t_{22} - M^b_{22}\right)\delta\kappa^d_{22} \\
& + 2\left(N^t_{12} + N^b_{12}\right)\delta\overline{\gamma}^a_{12} + 2\left(N^t_{12} - N^b_{12}\right)\delta\overline{\gamma}^d_{12} + 2\left(M^t_{12} + M^b_{12}\right)\delta\kappa^a_{12} + 2\left(M^t_{12} - M^b_{12}\right)\delta\kappa^d_{12} \\
& + N^c_{33}\delta\overline{\gamma}^c_{33} + M^c_{33}\delta\kappa^c_{33} + 2N^c_{13}\delta\overline{\gamma}^c_{13} + 2M^c_{13}\delta\kappa^c_{13} + 2N^c_{23}\delta\overline{\gamma}^c_{23} + 2M^c_{23}\delta\kappa^c_{23} \} dA
\end{aligned} \tag{7.47}$$

Utilizing the definition of global stress resultants and global stress couples, as defined below, in Eqs. (7.49a and 7.49b), allows the variation in the strain energy to be written as

$$\begin{aligned}
\delta U = \int_A & \left[2N^a_{11}\delta\overline{\gamma}^a_{11} + 2N^d_{11}\delta\overline{\gamma}^d_{11} + 2M^a_{11}\delta\kappa^a_{11} + 2M^d_{11}\delta\kappa^d_{11} + 2N^a_{22}\delta\overline{\gamma}^a_{22} + 2N^d_{22}\delta\overline{\gamma}^d_{22} + \right. \\
& 2M^a_{22}\delta\kappa^a_{22} + 2M^d_{22}\delta\kappa^d_{22} + 4N^a_{12}\delta\overline{\gamma}^a_{12} + 4N^d_{12}\delta\overline{\gamma}^d_{12} + 4M^a_{12}\delta\kappa^a_{12} + 4M^d_{12}\delta\kappa^d_{12} + \\
& \left. N^c_{33}\delta\overline{\gamma}^c_{33} + M^c_{33}\delta\kappa^c_{33} + 2N^c_{13}\delta\overline{\gamma}^c_{13} + 2M^c_{13}\delta\kappa^c_{13} + 2N^c_{23}\delta\overline{\gamma}^c_{23} + 2M^c_{23}\delta\kappa^c_{23}\right] dA
\end{aligned} \tag{7.48}$$

where the global stress resultants and global stress couples are defined as

$$\left(N^a_{\alpha\beta}, M^a_{\alpha\beta}\right) = \frac{1}{2}\left\{\left(N^t_{\alpha\beta} + N^b_{\alpha\beta}\right), \left(M^t_{\alpha\beta} + M^b_{\alpha\beta}\right)\right\} \tag{7.49a}$$

$$\left(N^d_{\alpha\beta}, M^d_{\alpha\beta}\right) = \frac{1}{2}\left\{\left(N^t_{\alpha\beta} - N^b_{\alpha\beta}\right), \left(M^t_{\alpha\beta} - M^b_{\alpha\beta}\right)\right\} \tag{7.49b}$$

Substituting in the strain–displacement equations, Eqs. (7.18)–(7.35) into Eq. (7.48) results in

$$\delta U=\int_A\Bigg[2N_{11}^a\left\{\frac{\partial\left(\delta u_1^a\right)}{\partial x_1}-\frac{\delta u_3^a}{R_1}+\frac{\partial u_3^a}{\partial x_1}\frac{\partial\left(\delta u_3^a\right)}{\partial x_1}+\frac{\partial\left(\delta u_3^a\right)}{\partial x_1}\frac{\partial \mathring{u}_3^a}{\partial x_1}+\frac{\partial u_3^d}{\partial x_1}\frac{\partial\left(\delta u_3^d\right)}{\partial x_1}\right.$$

$$\left.+\frac{\partial\left(\delta u_3^d\right)}{\partial x_1}\frac{\partial \mathring{u}_3^d}{\partial x_1}\right\}+2N_{11}^d\left\{\frac{\partial\left(\delta u_1^d\right)}{\partial x_1}-\frac{\delta u_3^d}{R_1}+\frac{\partial u_3^d}{\partial x_1}\frac{\delta\left(u_3^a\right)}{\partial x_1}+\frac{\partial u_3^a}{\partial x_1}\frac{\partial\left(\delta u_3^d\right)}{\partial x_1}\right.$$

$$\left.+\frac{\partial\left(\delta u_3^a\right)}{\partial x_1}\frac{\partial \mathring{u}_3^d}{\partial x_1}+\frac{\partial\left(\delta u_3^d\right)}{\partial x_1}\frac{\partial \mathring{u}_3^a}{\partial x_1}\right\}--2M_{11}^a\frac{\partial^2\left(\delta u_3^a\right)}{\partial x_1^2}-2M_{11}^d\frac{\partial^2\left(\delta u_3^d\right)}{\partial x_1^2}$$

$$+2N_{22}^a\left\{\frac{\partial\left(\delta u_2^a\right)}{\partial x_2}-\frac{\delta u_3^a}{R_2}+\frac{\partial u_3^a}{\partial x_2}\frac{\partial\left(\delta u_3^a\right)}{\partial x_2}+\frac{\partial\left(\delta u_3^a\right)}{\partial x_2}\frac{\partial \mathring{u}_3^a}{\partial x_2}+\frac{\partial u_3^d}{\partial x_2}\frac{\partial\left(\delta u_3^d\right)}{\partial x_2}\right.$$

$$\left.+\frac{\partial\left(\delta u_3^a\right)}{\partial x_2}\frac{\partial \mathring{u}_3^d}{\partial x_2}\right\}+2N_{22}^d\left\{\frac{\partial\left(\delta u_2^d\right)}{\partial x_2}-\frac{\delta u_3^d}{R_2}+\frac{\partial u_3^d}{\partial x_2}\frac{\partial\left(\delta u_3^a\right)}{\partial x_2}+\frac{\partial u_3^a}{\partial x_2}\frac{\partial\left(\delta u_3^d\right)}{\partial x_2}+\right.$$

$$\left.+\frac{\partial\left(\delta u_3^a\right)}{\partial x_2}\frac{\partial \mathring{u}_3^d}{\partial x_2}+\frac{\partial\left(\delta u_3^d\right)}{\partial x_2}\frac{\partial \mathring{u}_3^a}{\partial x_2}\right\}-2M_{22}^a\frac{\partial^2\left(\delta u_3^a\right)}{\partial x_2^2}-2M_{22}^d\frac{\partial^2\left(\delta u_3^d\right)}{\partial x_2^2}$$

$$+4N_{12}^a\left\{\frac{\partial\left(\delta u_1^a\right)}{\partial x_2}++\frac{\partial\left(\delta u_2^a\right)}{\partial x_1}+\frac{\partial\left(\delta u_3^a\right)}{\partial x_1}\frac{\partial u_3^a}{\partial x_2}+\frac{\partial u_3^a}{\partial x_1}\frac{\partial\left(\delta u_3^a\right)}{\partial x_2}+\frac{\partial\left(\delta u_3^d\right)}{\partial x_1}\frac{\partial u_3^d}{\partial x_2}\right.$$

$$\left.+\frac{\partial u_3^d}{\partial x_1}\frac{\partial\left(\delta u_3^d\right)}{\partial x_2}+\frac{\partial\left(\delta u_3^a\right)}{\partial x_1}\frac{\partial \mathring{u}_3^a}{\partial x_2}+\frac{\partial \mathring{u}_3^a}{\partial x_1}\frac{\partial\left(\delta u_3^a\right)}{\partial x_2}+\frac{\partial\left(\delta u_3^d\right)}{\partial x_1}\frac{\partial \mathring{u}_3^a}{\partial x_2}+\frac{\partial \mathring{u}_3^d}{\partial x_1}\frac{\partial\left(\delta u_3^d\right)}{\partial x_2}\right\}$$

$$+4N_{12}^d\left\{\frac{\partial\left(\delta u_1^d\right)}{\partial x_2}+\frac{\partial\left(\delta u_2^d\right)}{\partial x_1}+\frac{\partial\left(\delta u_3^a\right)}{\partial x_1}\frac{\partial u_3^d}{\partial x_2}+\frac{\partial u_3^a}{\partial x_1}\frac{\partial\left(\delta u_3^d\right)}{\partial x_2}+\frac{\partial\left(\delta u_3^a\right)}{\partial x_2}\frac{\partial u_3^d}{\partial x_1}\right.$$

$$\left.+\frac{\partial u_3^a}{\partial x_2}\frac{\partial\left(\delta u_3^d\right)}{\partial x_1}+\frac{\partial\left(\delta u_3^a\right)}{\partial x_1}\frac{\partial \mathring{u}_3^d}{\partial x_2}+\frac{\partial \mathring{u}_3^a}{\partial x_1}\frac{\partial\left(\delta u_3^d\right)}{\partial x_2}+\frac{\partial\left(\delta u_3^a\right)}{\partial x_2}\frac{\partial \mathring{u}_3^d}{\partial x_1}\right\}-4M_{12}^a\frac{\partial^2\left(\delta u_3^a\right)}{\partial x_1\partial x_2}$$

$$-4M_{12}^d\frac{\partial^2\left(\delta u_3^d\right)}{\partial x_1\partial x_2}+N_{33}^c\left\{-\frac{2}{t_c}\delta u_3^d+\frac{4}{t_c^2}u_3^d\delta u_3^d+\frac{4}{t_c^2}\mathring{u}_3^d\delta u_3^d\right\}$$

$$+2N_{13}^c\left\{-\frac{1}{t_c}\delta u_1^d+\left(\frac{1}{2}+\frac{1}{t_c}\left(\frac{t_f^t+t_f^b}{4}\right)\right)\frac{\partial\left(\delta u_3^a\right)}{\partial x_1}+\frac{1}{t_c}\left(\frac{t_f^t-t_f^b}{4}\right)\frac{\partial\left(\delta u_3^d\right)}{\partial x_1}-\right.$$

$$\left.-\frac{1}{t_c}\frac{\partial u_3^a}{\partial x_1}\delta u_3^d-\frac{1}{t_c}u_3^d\frac{\partial\left(\delta u_3^a\right)}{\partial x_1}-\frac{1}{t_c}\mathring{u}_3^a\frac{\partial\left(\delta u_3^a\right)}{\partial x_1}-\frac{1}{t_c}\frac{\partial \mathring{u}_3^a}{\partial x_1}\delta u_3^d\right\}$$

$$+2M_{13}^c\left\{\frac{4}{t_c^2}\delta\Phi_1^c-\frac{1}{t_c}\frac{\partial\left(\delta u_3^d\right)}{\partial x_1}+\frac{2}{t_c^2}\frac{\partial u_3^d}{\partial x_1}\delta u_3^d+\frac{2}{t_c^2}u_3^d\frac{\partial\left(\delta u_3^d\right)}{\partial x_1}+\frac{2}{t_c^2}\mathring{u}_3^d\frac{\partial\left(\delta u_3^d\right)}{\partial x_1}\right.$$

$$+\frac{2}{t_c^2}\frac{\partial \mathring{u}_3^d}{\partial x_1}\delta u_3^d\Bigg\} + 2N_{23}^c\Bigg\{-\frac{1}{t_c}\delta u_2^d + \left(\frac{1}{2}+\frac{1}{t_c}\left(\frac{t_f^t + t_f^b}{4}\right)\right)\frac{\partial\left(\delta u_3^a\right)}{\partial x_2}$$

$$+\frac{1}{t_c}\left(\frac{t_f^t - t_f^b}{4}\right)\frac{\partial\left(\delta u_3^d\right)}{\partial x_2} - \frac{1}{t_c}\frac{\partial u_3^a}{\partial x_2}\delta u_3^d - \frac{1}{t_c}u_3^d\frac{\partial\left(\delta u_3^a\right)}{\partial x_2} - \frac{1}{t_c}\mathring{u}_3^a\frac{\partial\left(\delta u_3^a\right)}{\partial x_2}$$

$$-\frac{1}{t_c}\frac{\partial \mathring{u}_3^a}{\partial x_2}\delta u_3^d\Bigg\} + 2M_{23}^c\Bigg\{\frac{4}{t_c^2}\delta\Phi_2^c - \frac{1}{t_c}\frac{\partial\left(\delta u_3^d\right)}{\partial x_2} + \frac{2}{t_c^2}\frac{\partial u_3^d}{\partial x_2}\delta u_3^d + \frac{2}{t_c^2}u_3^d\frac{\partial\left(\delta u_3^d\right)}{\partial x_2}$$

$$+\frac{2}{t_c^2}\mathring{u}_3^d\frac{\partial\left(\delta u_3^d\right)}{\partial x_2} + \frac{2}{t_c^2}\frac{\partial \mathring{u}_3^d}{\partial x_2}\delta u_3^d\Bigg\}\Bigg]dA \qquad (7.50)$$

Integrating each appropriate term by parts, combining coefficients of like variational displacements and simplifying gives the variation in the strain energy as

$$\delta U = \int_0^{l_2}\int_0^{l_1}\Bigg\langle -2\left(\frac{\partial N_{11}^a}{\partial x_1}+\frac{\partial N_{12}^a}{\partial x_2}\right)\delta u_1^a - 2\left(\frac{\partial N_{22}^a}{\partial x_2}+\frac{\partial N_{12}^a}{\partial x_1}\right)\delta u_2^a$$

$$-2\left(\frac{\partial N_{11}^a}{\partial x_1}+\frac{\partial N_{12}^a}{\partial x_2}+\frac{N_{13}^c}{t_c}\right)\delta u_1^d - 2\left(\frac{\partial N_{22}^a}{\partial x_2}+\frac{\partial N_{12}^a}{\partial x_1}+\frac{N_{23}^c}{t_c}\right)\delta u_2^d$$

$$+\left(\frac{8}{t_c^2}M_{13}^c\right)\delta\Phi_1^c + \left(\frac{8}{t_c^2}M_{23}^c\right)\delta\Phi_2^c + \Bigg\{-2N_{11}^a\left(\frac{\partial^2 u_3^a}{\partial x_1^2}+\frac{\partial^2 \mathring{u}_3^a}{\partial x_1^2}+\frac{1}{R_1}\right)$$

$$-4N_{12}^a\left(\frac{\partial^2 u_3^a}{\partial x_1\partial x_2}+\frac{\partial^2 \mathring{u}_3^a}{\partial x_1\partial x_2}\right) - 2N_{22}^a\left(\frac{\partial^2 u_3^a}{\partial x_2^2}+\frac{\partial^2 \mathring{u}_3^a}{\partial x_2^2}+\frac{1}{R_2}\right) - 2\frac{\partial^2 M_{11}^a}{\partial x_1^2}$$

$$-4\frac{\partial^2 M_{12}^a}{\partial x_1\partial x_2} - 2\frac{\partial^2 M_{22}^a}{\partial x_2^2} - 2N_{11}^d\left(\frac{\partial^2 u_3^a}{\partial x_1^2}+\frac{\partial^2 \mathring{u}_3^a}{\partial x_1^2}\right) - 4N_{12}^d\left(\frac{\partial^2 u_3^d}{\partial x_1\partial x_2}+\frac{\partial^2 \mathring{u}_3^d}{\partial x_1\partial x_2}\right)$$

$$-2N_{22}^d\left(\frac{\partial^2 u_3^a}{\partial x_2^2}+\frac{\partial^2 \mathring{u}_3^a}{\partial x_2^2}\right) - \frac{2}{t_c}\left(\frac{t_c}{2}+\frac{t_f^t + t_f^b}{4} - u_3^d - \circ u_3^d\right)\left(\frac{\partial N_{13}^c}{\partial x_1}+\frac{\partial N_{23}^c}{\partial x_2}\right)$$

$$+\frac{4}{t_c}\left(\frac{\partial u_3^d}{\partial x_2}+\frac{\circ\partial u_3^d}{\partial x_2}\right)N_{23}^c + \frac{4}{t_c}\left(\frac{\partial u_3^d}{\partial x_1}+\frac{\circ\partial u_3^d}{\partial x_1}\right)N_{13}^c - 2\left(\frac{\partial u_3^a}{\partial x_1}+\frac{\circ\partial u_3^a}{\partial x_1}\right)$$

$$\times\left(\frac{\partial N_{11}^a}{\partial x_1}+\frac{\partial N_{12}^a}{\partial x_2}\right) - 2\left(\frac{\partial u_3^a}{\partial x_2}+\frac{\circ\partial u_3^a}{\partial x_2}\right)\left(\frac{\partial N_{22}^a}{\partial x_2}+\frac{\partial N_{12}^a}{\partial x_1}\right) - 2\left(\frac{\partial u_3^d}{\partial x_1}+\frac{\circ\partial u_3^d}{\partial x_1}\right)$$

$$\times\left(\frac{\partial N_{11}^d}{\partial x_1}+\frac{\partial N_{12}^d}{\partial x_2}\right) - 2\left(\frac{\partial u_3^d}{\partial x_2}+\frac{\circ\partial u_3^d}{\partial x_2}\right)\left(\frac{\partial N_{22}^d}{\partial x_2}+\frac{\partial N_{12}^d}{\partial x_1}\right)\Bigg\}\delta u_3^a$$

$$+\left\{-2N_{11}^{d}\left(\frac{\partial^{2}u_{3}^{a}}{\partial x_{1}^{2}}+\frac{\partial^{2}\mathring{u}_{3}^{a}}{\partial x_{1}^{2}}+\frac{1}{R_{1}}\right)-4N_{12}^{d}\left(\frac{\partial^{2}u_{3}^{a}}{\partial x_{1}\partial x_{2}}+\frac{\partial^{2}\mathring{u}_{3}^{a}}{\partial x_{1}\partial x_{2}}\right)\right.$$

$$-2N_{22}^{d}\left(\frac{\partial^{2}u_{3}^{a}}{\partial x_{2}^{2}}+\frac{\partial^{2}\mathring{u}_{3}^{a}}{\partial x_{2}^{2}}+\frac{1}{R_{2}}\right)-2\frac{\partial^{2}M_{11}^{d}}{\partial x_{1}^{2}}-4\frac{\partial^{2}M_{12}^{d}}{\partial x_{1}\partial x_{2}}-2\frac{\partial^{2}M_{22}^{d}}{\partial x_{2}^{2}}$$

$$-2N_{11}^{a}\left(\frac{\partial^{2}u_{3}^{d}}{\partial x_{1}^{2}}+\frac{\partial^{2}\mathring{u}_{3}^{d}}{\partial x_{1}^{2}}\right)-4N_{12}^{a}\left(\frac{\partial^{2}u_{3}^{d}}{\partial x_{1}\partial x_{2}}+\frac{\partial^{2}\mathring{u}_{3}^{d}}{\partial x_{1}\partial x_{2}}\right)-2N_{22}^{a}\left(\frac{\partial^{2}u_{3}^{d}}{\partial x_{2}^{2}}+\frac{\partial^{2}\mathring{u}_{3}^{d}}{\partial x_{2}^{2}}\right)$$

$$-\frac{4}{t_{c}}\left(\frac{t_{c}}{2}-u_{3}^{d}-\circ u_{3}^{d}\right)N_{33}^{c}+\frac{4}{t_{c}}\left(\frac{t_{f}^{t}-t_{f}^{b}}{16}\right)\left(\frac{\partial N_{13}^{c}}{\partial x_{1}}+\frac{\partial N_{23}^{c}}{\partial x_{2}}\right)+$$

$$-2\left(\frac{\partial u_{3}^{d}}{\partial x_{1}}+\frac{\partial\mathring{u}_{3}^{d}}{\partial x_{1}}\right)\left(\frac{\partial N_{11}^{a}}{\partial x_{1}}+\frac{\partial N_{12}^{a}}{\partial x_{2}}\right)-2\left(\frac{\partial u_{3}^{d}}{\partial x_{2}}+\frac{\partial\mathring{u}_{3}^{d}}{\partial x_{2}}\right)\left(\frac{\partial N_{22}^{a}}{\partial x_{2}}+\frac{\partial N_{12}^{a}}{\partial x_{1}}\right)$$

$$-2\left(\frac{\partial u_{3}^{a}}{\partial x_{1}}+\frac{\partial\mathring{u}_{3}^{a}}{\partial x_{1}}\right)\left(\frac{\partial N_{11}^{d}}{\partial x_{1}}+\frac{\partial N_{12}^{d}}{\partial x_{2}}+\frac{N_{13}^{c}}{t_{c}}\right)-2\left(\frac{\partial u_{3}^{a}}{\partial x_{2}}+\frac{\partial\mathring{u}_{3}^{a}}{\partial x_{2}}\right)$$

$$\left.\left.\times\left(\frac{\partial N_{22}^{d}}{\partial x_{2}}+\frac{\partial N_{12}^{d}}{\partial x_{1}}+\frac{N_{23}^{c}}{t_{c}}\right)+\frac{4}{t_{c}^{2}}\left(\frac{t_{c}}{2}-u_{3}^{d}-\circ u_{3}^{d}\right)\left(\frac{\partial M_{13}^{c}}{\partial x_{1}}+\frac{\partial M_{23}^{c}}{\partial x_{2}}\right)\right\}\delta u_{3}^{d}\right\rangle dx_{1}dx_{2}$$

$$+\int_{0}^{l_{2}}\left\langle 2N_{11}^{a}\delta u_{1}^{a}+2N_{12}^{a}\delta u_{2}^{a}+2N_{11}^{d}\delta u_{2}^{d}+2N_{12}^{d}\delta u_{2}^{d}+\left\{2N_{11}^{a}\left(\frac{\partial u_{3}^{a}}{\partial x_{1}}+\frac{\partial\mathring{u}_{3}^{a}}{\partial x_{1}}\right)\right.\right.$$

$$+2N_{12}^{a}\left(\frac{\partial u_{3}^{a}}{\partial x_{2}}+\frac{\partial\mathring{u}_{3}^{a}}{\partial x_{2}}\right)+2N_{11}^{d}\left(\frac{\partial u_{3}^{d}}{\partial x_{1}}+\frac{\partial\mathring{u}_{3}^{d}}{\partial x_{1}}\right)+2N_{12}^{d}\left(\frac{\partial u_{3}^{d}}{\partial x_{2}}+\frac{\partial\mathring{u}_{3}^{d}}{\partial x_{2}}\right)$$

$$\left.+2\frac{\partial M_{11}^{a}}{\partial x_{1}}+4\frac{\partial M_{12}^{a}}{\partial x_{2}}+\frac{2}{t_{c}}\left[\frac{t_{c}}{2}+\frac{t_{f}^{t}+t_{f}^{b}}{4}-u_{3}^{d}-\circ u_{3}^{d}\right]N_{13}^{c}\right\}\delta u_{3}^{a}$$

$$+\left\{2N_{11}^{a}\left(\frac{\partial u_{3}^{d}}{\partial x_{1}}+\frac{\partial\mathring{u}_{3}^{d}}{\partial x_{1}}\right)+2N_{11}^{d}\left(\frac{\partial u_{3}^{a}}{\partial x_{1}}+\frac{\partial\mathring{u}_{3}^{a}}{\partial x_{1}}\right)+2N_{12}^{a}\left(\frac{\partial u_{3}^{d}}{\partial x_{2}}+\frac{\partial\mathring{u}_{3}^{d}}{\partial x_{2}}\right)\right.$$

$$+2N_{12}^{d}\left(\frac{\partial u_{3}^{a}}{\partial x_{2}}+\frac{\partial\mathring{u}_{3}^{a}}{\partial x_{2}}\right)+2\frac{\partial M_{11}^{d}}{\partial x_{1}}+4\frac{\partial M_{12}^{d}}{\partial x_{2}}-\frac{1}{t_{c}}\left(\frac{t_{f}^{t}-t_{f}^{b}}{4}\right)N_{13}^{c}$$

$$\left.\left.-\frac{2}{t_{c}^{2}}\left(\frac{t_{c}}{2}-u_{3}^{d}-\circ u_{3}^{d}\right)M_{13}^{c}\right\}\delta u_{3}^{d}+2M_{11}^{a}\delta\left(\frac{\partial u_{3}^{a}}{\partial x_{1}}\right)+2M_{11}^{d}\delta\left(\frac{\partial u_{3}^{d}}{\partial x_{1}}\right)\right\rangle\Bigg|_{0}^{l_{1}}dx_{2}$$

$$+\int_{0}^{l_{1}}\left\langle 2N_{22}^{a}\delta u_{2}^{a}+2N_{21}^{a}\delta u_{1}^{a}+2N_{22}^{d}\delta u_{1}^{d}+2N_{21}^{d}\delta u_{1}^{d}+\left\{2N_{22}^{a}\left(\frac{\partial u_{3}^{a}}{\partial x_{2}}+\frac{\partial\mathring{u}_{3}^{a}}{\partial x_{2}}\right)\right.\right.$$

$$+2N_{21}^{a}\left(\frac{\partial u_3^a}{\partial x_1}\frac{\partial \mathring{u}_3^a}{\partial x_1}\right)+2N_{22}^{d}\left(\frac{\partial u_3^d}{\partial x_2}+\frac{\partial \mathring{u}_3^d}{\partial x_2}\right)+2N_{21}^{d}\left(\frac{\partial u_3^d}{\partial x_1}+\frac{\partial \mathring{u}_3^d}{\partial x_1}\right)$$

$$+2\frac{\partial M_{22}^a}{\partial x_2}+4\frac{\partial M_{21}^a}{\partial x_1}+\frac{2}{t_c}\left[\frac{t_c}{2}+\frac{t_f^t+t_f^b}{4}--u_3^d-\circ u_3^d\right]N_{23}^c\Bigg\}\delta u_3^a$$

$$+\Bigg\{2N_{22}^{a}\left(\frac{\partial u_3^d}{\partial x_2}+\frac{\partial \mathring{u}_3^d}{\partial x_2}\right)+2N_{22}^{d}\left(\frac{\partial u_3^a}{\partial x_2}+\frac{\partial \mathring{u}_3^a}{\partial x_2}\right)+2N_{21}^{a}\left(\frac{\partial u_3^d}{\partial x_1}+\frac{\partial \mathring{u}_3^d}{\partial x_1}\right)$$

$$+2N_{21}^{d}\left(\frac{\partial u_3^a}{\partial x_1}+\frac{\partial \mathring{u}_3^a}{\partial x_1}\right)+2\frac{\partial M_{22}^d}{\partial x_2}+4\frac{\partial M_{21}^d}{\partial x_1}-\frac{1}{t_c}\left(\frac{t_f^t-t_f^b}{4}\right)N_{23}^c$$

$$-\frac{2}{t_c^2}\left(\frac{t_c}{2}-u_3^d-\circ u_3^d\right)M_{23}^c\Bigg\}\delta u_3^d+2M_{22}^a\delta\left(\frac{\partial u_3^a}{\partial x_2}\right)+2M_{22}^d\delta\left(\frac{\partial u_3^d}{\partial x_2}\right)\Bigg\rangle\Bigg|_0^{l_2}dx_1 \tag{7.51}$$

7.4.2 *Work Done by External Loads*

The total work considered in this text that is performed on an elastic body is the summation of the work due to body forces, edge loads, surface tractions, and damping. This is expressed mathematically as

$$W_{\text{total}}=W_{\text{body forces}}+W_{\text{edge loads}}+W_{\text{surface tractions}}+W_{\text{Damping}} \tag{7.52}$$

- *Work due to body forces*

$$\delta W_b=\int_\sigma\left\{\int_{\bar{h}}^{\bar{h}+h^{/}}\rho^{/}H_i^{/}\delta V_i^{/}dx_3+\int_{-\bar{h}}^{\bar{h}}\bar{\rho}\bar{H}_i\delta\bar{V}_i dx_3+\int_{-\bar{h}-h^{//}}^{-\bar{h}}\rho^{//}H_i^{//}\delta V_i^{//}dx_3\right\}d\sigma \tag{7.53}$$

where ρ is the mass density and H_i is the body force vector. It should be noted that the body forces will not be included in the following developments. Typical body forces are gravity, electrical and magnetic forces which will be neglected.

- *Work due to edge loads*

$$W_{\text{edge loads}}=W_{\text{edge loads}}^{t}+W_{\text{edge loads}}^{c}+W_{\text{edge loads}}^{b} \tag{7.54}$$

By definition, the work due to edge loads starts with the basic expression below. Considering the contribution from each layer, the total work due to edge loads is expressed as

$$\begin{aligned}\delta W_{el} = &\int_{x_1}\left(\int_{\frac{-t_c}{2}-t^t_f}^{\frac{-t_c}{2}}\left(\widehat{\tau}^t_{22}\delta v^t_2+\widehat{\tau}^t_{21}\delta v^t_1\right)dx_3+\int_{\frac{-t_c}{2}}^{\frac{t_c}{2}}\widehat{\tau}^c_{23}\delta v^c_3\right.\\ &\left.+\int_{\frac{t_c}{2}}^{\frac{t_c}{2}+t^b_f}\left(\widehat{\tau}^b_{22}\delta v^b_2+\widehat{\tau}^b_{21}\delta v^b_1\right)dx_3\right)dx_1\\ &+\int_{x_2}\left(\int_{\frac{-t_c}{2}-t^t_f}^{\frac{-t_c}{2}}\left(\widehat{\tau}^t_{11}\delta v^t_1+\widehat{\tau}^t_{12}\delta v^t_2\right)dx_3\right.\\ &\left.+\int_{\frac{-t_c}{2}}^{\frac{t_c}{2}}\widehat{\tau}^c_{13}\delta v^c_3+\int_{\frac{t_c}{2}}^{\frac{t_c}{2}+t^b_f}\left(\widehat{\tau}^b_{11}\delta v^b_1+\widehat{\tau}^b_{12}\delta v^b_2\right)dx_3\right)dx_2\end{aligned} \tag{7.55}$$

Substituting in the expressions for the displacements, Eqs. (7.6a–c)–(7.8a–c) while considering the expressions for the local stress resultants and stress couples, defined earlier in Eqs. (7.46a–c), give

$$\begin{aligned}\delta W_{el} = &\int_{x_1}\left(\widehat{N}^t_{22}\delta u^a_2+\widehat{N}^t_{22}\delta u^d_2-\widehat{M}^t_{22}\delta\left(\frac{\partial u^a_3}{\partial x_2}\right)-\widehat{M}^t_{22}\delta\left(\frac{\partial u^d_3}{\partial x_2}\right)+\widehat{N}^t_{21}\delta u^a_1+\widehat{N}^t_{21}\delta u^d_1+\right.\\ &-\widehat{M}^t_{21}\delta\left(\frac{\partial u^a_3}{\partial x_1}\right)-\widehat{M}^t_{21}\delta\left(\frac{\partial u^d_3}{\partial x_1}\right)+\widehat{N}^c_{23}\delta u^a_3-\frac{2}{t_c}\widehat{M}^c_{23}\delta u^d_3+\widehat{N}^b_{22}\delta u^a_2-\widehat{N}^b_{22}\delta u^d_2\\ &-\widehat{M}^b_{22}\delta\left(\frac{\partial u^a_3}{\partial x_2}\right)+\widehat{M}^b_{22}\delta\left(\frac{\partial u^d_3}{\partial x_2}\right)+\widehat{N}^b_{21}\delta u^a_1-\widehat{N}^b_{21}\delta u^d_1-\widehat{M}^b_{21}\delta\left(\frac{\partial u^a_3}{\partial x_1}\right)+\\ &\left.+\widehat{M}^b_{21}\delta\left(\frac{\partial u^d_3}{\partial x_1}\right)\right)dx_1\\ &\int_{x_2}\left(\widehat{N}^t_{11}\delta u^a_1+\widehat{N}^t_{11}\delta u^d_1-\widehat{M}^t_{11}\delta\left(\frac{\partial u^a_3}{\partial x_1}\right)-\widehat{M}^t_{11}\delta\left(\frac{\partial u^d_3}{\partial x_1}\right)+\widehat{N}^t_{12}\delta u^a_2+\widehat{N}^t_{12}\delta u^d_2+\right.\\ &-\widehat{M}^t_{12}\delta\left(\frac{\partial u^a_3}{\partial x_2}\right)-\widehat{M}^t_{12}\delta\left(\frac{\partial u^d_3}{\partial x_2}\right)+\widehat{N}^c_{13}\delta u^a_3-\frac{2}{t_c}\widehat{M}^c_{13}\delta u^d_3+\widehat{N}^b_{11}\delta u^a_1-\widehat{N}^b_{11}\delta u^d_1\\ &-\widehat{M}^b_{11}\delta\left(\frac{\partial u^a_3}{\partial x_1}\right)+\widehat{M}^b_{11}\delta\left(\frac{\partial u^d_3}{\partial x_1}\right)+\widehat{N}^b_{12}\delta u^a_2-\widehat{N}^b_{12}\delta u^d_2-\widehat{M}^b_{12}\delta\left(\frac{\partial u^a_3}{\partial x_2}\right)+\\ &\left.+\widehat{M}^b_{12}\delta\left(\frac{\partial u^d_3}{\partial x_2}\right)\right)dx_2\end{aligned} \tag{7.56}$$

In the above equation, Eq. (7.56), gathering coefficients of identical variational displacements utilizing the definition of global stress and couple resultants from Eqs. (7.49a and 7.49b) while simplifying gives

$$W_{el} = \int_{x_1} \left\{ 2\widehat{N}^a_{22}\delta u^a_2 + 2\widehat{N}^d_{22}\delta u^d_2 + 2\widehat{N}^a_{21}\delta u^a_1 + 2\widehat{N}^d_{21}\delta u^d_1 + \left(2\frac{\partial \widehat{M}^a_{21}}{\partial x_1} + \widehat{N}^c_{23}\right)\delta u^a_3 \right.$$
$$\left. + \left(2\frac{\partial \widehat{M}^d_{21}}{\partial x_1} - \frac{\widehat{M}^c_{23}}{t_c}\right)\delta u^d_3 - 2\widehat{M}^a_{22}\delta\left(\frac{\partial u^a_3}{\partial x_2}\right) - 2\widehat{M}^d_{22}\delta\left(\frac{\partial u^d_3}{\partial x_2}\right)\right\} dx_1$$
$$+ \int_{x_2} \left\{ 2\widehat{N}^a_{11}\delta u^a_1 + 2\widehat{N}^d_{11}\delta u^d_1 + 2\widehat{N}^a_{12}\delta u^a_2 + 2\widehat{N}^d_{12}\delta u^d_2 + \left(2\frac{\partial \widehat{M}^a_{12}}{\partial x_2} + \widehat{N}^c_{13}\right)\delta u^a_3 \right.$$
$$\left. + \left(2\frac{\partial \widehat{M}^d_{12}}{\partial x_2} - \frac{\widehat{M}^c_{13}}{t_c}\right)\delta u^d_3 - 2\widehat{M}^a_{11}\delta\left(\frac{\partial u^a_3}{\partial x_1}\right) - 2\widehat{M}^d_{11}\delta\left(\frac{\partial u^d_3}{\partial x_1}\right)\right\} dx_2 \tag{7.57}$$

where

$$\left(\widehat{N}^a_{\alpha\beta}, \widehat{M}^a_{\alpha\beta}\right) = \frac{1}{2}\left\{\left(\widehat{N}^t_{\alpha\beta} + \widehat{N}^b_{\alpha\beta}\right), \left(\widehat{M}^t_{\alpha\beta} + \widehat{M}^b_{\alpha\beta}\right)\right\} \tag{7.58}$$

$$\left(\widehat{N}^d_{\alpha\beta}, \widehat{M}^d_{\alpha\beta}\right) = \frac{1}{2}\left\{\left(\widehat{N}^t_{\alpha\beta} - \widehat{N}^b_{\alpha\beta}\right), \left(\widehat{M}^t_{\alpha\beta} - \widehat{M}^b_{\alpha\beta}\right)\right\} \tag{7.59}$$

- *Work due to surface tractions*

The expression for the work done by lateral vertical loading such as an external pressure is by definition given as

$$W_{st} = \int_A \left(\widehat{q}^t_3 \delta v^t_3 + \widehat{q}^b_3 \delta v^b_3\right) dA \tag{7.60}$$

Substituting Eqs. (7.6c) and (7.7c) into Eq. (7.60) gives

$$W_{st} = \int_A \left(\widehat{q}^t_3\left(\delta u^a_3 + \delta u^d_3\right) + \widehat{q}^b_3\left(\delta u^a_3 - \delta u^d_3\right)\right) dA \tag{7.61}$$

Simplifying results in

$$W_{st} = \int_A \left\{\left(\widehat{q}^t_3 + \widehat{q}^b_3\right)\delta u^a_3 + \left(\widehat{q}^t_3 - \widehat{q}^b_3\right)\delta u^d_3\right\} dA = 2\int_A \left(\widehat{q}^a_3 \delta u^a_3 + \widehat{q}^d_3 \delta u^d_3\right) dA \tag{7.62}$$

where

$$\widehat{q}^a_3 = \frac{\widehat{q}^t_3 + \widehat{q}^b_3}{2}, \qquad \widehat{q}^d_3 = \frac{\widehat{q}^t_3 - \widehat{q}^b_3}{2} \tag{7.63a, b}$$

- *Work due to damping*

By definition, the work due to damping for the transverse direction is given by

$$W_d = -\int_A \left(C^t \dot{v}_3^t \delta v_3^t + C^c \dot{v}_3^c \delta v_3^c + C^b \dot{v}_3^b \delta v_3^b\right) dA \tag{7.64}$$

where C^t, C^c and C^b are the structural damping coefficients per unit area of the facings and the core. Substituting in v_3^t, v_3^c and v_3^t from Eqs. (7.6c), (7.7c), and (7.8c) gives

$$W_d = -\int_A \left(C^t\left(\dot{u}_3^a + \dot{u}_3^d\right)\left(\delta u_3^a + \delta u_3^d\right) + C^c \dot{u}_3^a \delta u_3^a + C^b\left(\dot{u}_3^a - \dot{u}_3^d\right)\left(\delta u_3^a - \delta u_3^d\right)\right) dA \tag{7.65}$$

It should be mentioned that damping is assumed to be constant throughout the thickness of the core. As a result, simplifying and gathering like variational terms gives

$$W_d = -\int_A \left[\left(C^t + C^b + C^c\right)\dot{u}_3^a \delta u_3^a + \left(C^t - C^b\right)\dot{u}_3^a \delta u_3^d + \left(C^t - C^b\right)\dot{u}_3^d \delta u_3^a + \left(C^t + C^b\right)\dot{u}_3^d \delta u_3^d\right] dA \tag{7.66}$$

which can be written as

$$W_d = 2\int_A \left[\left(\left(C^a + \frac{C^c}{2}\right)\dot{u}_3^a + C^d \dot{u}_3^d\right)\delta u_3^a + \left(C^d \dot{u}_3^a + C^a \dot{u}_3^d\right)\delta u_3^d\right] dA \tag{7.67}$$

where

$$C^a = \frac{C^t + C^b}{2}$$
$$C^d = \frac{C^t - C^b}{2} \tag{7.68a, b}$$

C^a and C^d are the average and half difference of the damping coefficients with regard to the top and bottom facings.

7.4.3 Kinetic Energy

By definition, the kinetic energy is given by

$$T = \frac{1}{2}\int_{x_3}\int_{x_2}\int_{x_1}\rho\left[\left(\frac{\partial V_1}{\partial t}\right)^2 + \left(\frac{\partial V_2}{\partial t}\right)^2\left(\frac{\partial V_3}{\partial t}\right)^2\right]dx_1dx_2dx_3 \tag{7.69}$$

Taking a variation in the expression for the kinetic energy and integrating with respect to time gives

$$\begin{aligned}\int_{t_0}^{t_1}\delta T dt = &-\int_{t_0}^{t_1}\int_{\sigma}\left[\int_{\frac{t_c}{2}}^{\frac{t_c}{2}+t_f^b}\rho^b\left(\ddot{V}_1^b\delta V_1^b + \ddot{V}_2^b\delta V_2^b + \ddot{V}_3^b\delta V_3^b\right)dx_3 + \int_{-\frac{t_c}{2}}^{\frac{t_c}{2}}\rho^c\left(\ddot{V}_1^c\delta V_1^c + \ddot{V}_2^c\delta V_2^c + \ddot{V}_3^c\delta V_3^c\right)dx_3\right.\\ &\left.+\int_{-\frac{t_c}{2}-t_f^t}^{-\frac{t_c}{2}}\rho^t\left(\ddot{V}_1^t\delta V_1^t + \ddot{V}_2^t\delta V_2^t + \ddot{V}_3^t\delta V_3^t\right)dx_3\right]d\sigma dt\end{aligned} \tag{7.70}$$

If the in-plane inertias are neglected and only the transverse inertias are considered, the kinetic energy can be expressed as

$$\int_{t_0}^{t_1}\delta T dt = -\int_{t_0}^{t_1}\int_A\left[\int_{\frac{t_c}{2}}^{\frac{t_c}{2}+t_f^b}\rho^b\ddot{V}_3^b\delta V_3^b dx_3 + \int_{-\frac{t_c}{2}}^{\frac{t_c}{2}}\rho^c\ddot{V}_3^c\delta V_3^c dx_3 + \int_{-\frac{t_c}{2}-t_f^t}^{-\frac{t_c}{2}}\rho^t\ddot{V}_3^t\delta V_3^t dx_3\right]dA dt \tag{7.71}$$

Using the expressions for the displacements for each respective layer, from Eqs. (7.6c), (7.7c), and (7.8c) give

$$\begin{aligned}\int_{t_0}^{t_1}\delta T dt = \int_{t_0}^{t_1} &- \int_A\left\{\int_{-\frac{t_c}{2}-t_f^t}^{-\frac{t_c}{2}}\rho^t\left(\ddot{u}_3^a + \ddot{u}_3^d\right)\left(\delta u_3^a + \delta u_3^d\right)dx_3 + \int_{-\frac{t_c}{2}}^{\frac{t_c}{2}}\rho^c\left(\ddot{u}_3^a - \frac{2}{t_c}x_3\ddot{u}_3^d\right)\right.\\ &\left.\cdot\left(\delta u_3^a - \frac{2}{t_c}x_3\delta u_3^d\right)dx_3 + \int_{\frac{t_c}{2}}^{\frac{t_c}{2}+t_f^b}\rho^b\left(\ddot{u}_3^a - \ddot{u}_3^d\right)\left(\delta u_3^a - \delta u_3^d\right)dx_3\right)dA dt\end{aligned} \tag{7.72}$$

Integrating and simplifying give

$$\begin{aligned}\int_{t_0}^{t_1}\delta T dt = -2\int_{t_0}^{t_1}\int_A&\left[\left(m^a + \frac{m^c}{2}\right)\ddot{u}_3^a\delta u_3^a + m^d\ddot{u}_3^a\delta u_3^d + m^d\ddot{u}_3^d\delta u_3^a\right.\\ &\left.+\left(m^a + \frac{m^c}{3}\right)\ddot{u}_3^d\delta u_3^d\right]dA dt\end{aligned} \tag{7.73}$$

where

$$m^a = \frac{m_f^t + m_f^b}{2}, \qquad m^d = \frac{m_f^t - m_f^b}{2} \tag{7.74a, b}$$

and

$$m^c = \rho^c t_c, \qquad m_f^t = \int_{-\frac{t_c}{2}-t_f^t}^{-\frac{t_c}{2}} \rho_{(k)}^t dx_3, \qquad m_f^b = \int_{\frac{t_c}{2}}^{\frac{t_c}{2}+t_f^b} \rho_{(k)}^b dx_3 \tag{7.75a – c}$$

7.5 Governing Equations

7.5.1 Equations of Motion

Substituting δU, δW, δT back into Hamilton's equation gives

$$\int_{t_0}^{t_1} \Bigg\langle \int_0^{l_2} \int_0^{l_1} \Bigg[-2\left(\frac{\partial N_{11}^a}{\partial x_1} + \frac{\partial N_{12}^a}{\partial x_2}\right)\delta u_1^a - 2\left(\frac{\partial N_{22}^a}{\partial x_2} + \frac{\partial N_{12}^a}{\partial x_1}\right)\delta u_2^a$$

$$-2\left(\frac{\partial N_{11}^a}{\partial x_1} + \frac{\partial N_{12}^a}{\partial x_2} + \frac{N_{13}^c}{t_c}\right)\delta u_1^d - 2\left(\frac{\partial N_{22}^a}{\partial x_2} + \frac{\partial N_{12}^a}{\partial x_1} + \frac{N_{23}^c}{t_c}\right)\delta u_2^d + \left(\frac{8}{t_c^2} M_{13}^c\right)\delta\Phi_1^c$$

$$+\left(\frac{8}{t_c^2} M_{23}^c\right)\delta\Phi_2^c + \Bigg\{ -2N_{11}^a\left(\frac{\partial^2 u_3^a}{\partial x_1^2} + \frac{\partial^2 \mathring{u}_3^a}{\partial x_1^2} + \frac{1}{R_1}\right) - 4N_{12}^a\left(\frac{\partial^2 u_3^a}{\partial x_1 \partial x_2} + \frac{\partial^2 \mathring{u}_3^a}{\partial x_1 \partial x_2}\right)$$

$$-2N_{22}^a\left(\frac{\partial^2 u_3^a}{\partial x_2^2} + \frac{\partial^2 \mathring{u}_3^a}{\partial x_2^2} + \frac{1}{R_2}\right) - 2\frac{\partial^2 M_{11}^a}{\partial x_1^2} - 4\frac{\partial^2 M_{12}^a}{\partial x_1 \partial x_2} - 2\frac{\partial^2 M_{22}^a}{\partial x_2^2}$$

$$-2N_{11}^d\left(\frac{\partial^2 u_3^a}{\partial x_1^2} + \frac{\partial^2 \mathring{u}_3^a}{\partial x_1^2}\right) - 4N_{12}^d\left(\frac{\partial^2 u_3^d}{\partial x_1 \partial x_2} + \frac{\partial^2 \mathring{u}_3^d}{\partial x_1 \partial x_2}\right) - 2N_{22}^d\left(\frac{\partial^2 u_3^a}{\partial x_2^2} + \frac{\partial^2 \mathring{u}_3^a}{\partial x_2^2}\right)$$

$$-\frac{2}{t_c}\left(\frac{t_c}{2} + \frac{t_f^t + t_f^b}{4} - u_3^d - \mathring{u}_3^d\right)\left(\frac{\partial N_{13}^c}{\partial x_1} + \frac{\partial N_{23}^c}{\partial x_2}\right) + \frac{4}{t_c}\left(\frac{\partial u_3^d}{\partial x_2} + \frac{\partial \mathring{u}_3^d}{\partial x_2}\right)N_{23}^c$$

$$+\frac{4}{t_c}\left(\frac{\partial u_3^d}{\partial x_1} + \frac{\partial \mathring{u}_3^d}{\partial x_1}\right)N_{13}^c - 2\left(\frac{\partial u_3^a}{\partial x_1} + \frac{\partial \mathring{u}_3^a}{\partial x_1}\right)\left(\frac{\partial N_{11}^a}{\partial x_1} + \frac{\partial N_{12}^a}{\partial x_2}\right)$$

$$-2\left(\frac{\partial u_3^a}{\partial x_2} + \frac{\partial \mathring{u}_3^a}{\partial x_2}\right)\left(\frac{\partial N_{22}^a}{\partial x_2} + \frac{\partial N_{12}^a}{\partial x_1}\right) - 2\left(\frac{\partial u_3^d}{\partial x_1} + \frac{\partial \mathring{u}_3^d}{\partial x_1}\right)\left(\frac{\partial N_{11}^d}{\partial x_1} + \frac{\partial N_{12}^d}{\partial x_2}\right)$$

$$
-2\left(\frac{\partial u_3^d}{\partial x_2}+\frac{\partial \mathring{u}_3^d}{\partial x_2}\right)\left(\frac{\partial N_{22}^d}{\partial x_2}+\frac{\partial N_{12}^d}{\partial x_1}\right)-2\widehat{q}_3^a-2\left(C^a+\frac{C^c}{2}\right)\dot{u}_3^a-2C^d\dot{u}_3^d
$$

$$
-2\left(m^a+\frac{m_c}{2}\right)\ddot{u}_3^a-2m^d\ddot{u}_3^d\Bigg\}\delta u_3^a+\Bigg\{-2N_{11}^d\left(\frac{\partial^2 u_3^a}{\partial x_1^2}+\frac{\partial^2 \mathring{u}_3^a}{\partial x_1^2}+\frac{1}{R_1}\right)
$$

$$
-4N_{12}^d\left(\frac{\partial^2 u_3^a}{\partial x_1\partial x_2}+\frac{\partial^2 \mathring{u}_3^a}{\partial x_1\partial x_2}\right)-2N_{22}^d\left(\frac{\partial^2 u_3^a}{\partial x_2^2}+\frac{\partial^2 \mathring{u}_3^a}{\partial x_2^2}+\frac{1}{R_2}\right)-2\frac{\partial^2 M_{11}^d}{\partial x_1^2}-4\frac{\partial^2 M_{12}^d}{\partial x_1\partial x_2}
$$

$$
-2\frac{\partial^2 M_{22}^d}{\partial x_2^2}-2N_{11}^a\left(\frac{\partial^2 u_3^d}{\partial x_1^2}+\frac{\partial^2 \mathring{u}_3^d}{\partial x_1^2}\right)-4N_{12}^a\left(\frac{\partial^2 u_3^d}{\partial x_1\partial x_2}+\frac{\partial^2 \mathring{u}_3^d}{\partial x_1\partial x_2}\right)
$$

$$
-2N_{22}^a\left(\frac{\partial^2 u_3^d}{\partial x_2^2}+\frac{\partial^2 \mathring{u}_3^d}{\partial x_2^2}\right)-\frac{4}{t_c}\left(\frac{t_c}{2}-u_3^d-\mathring{u}_3^d\right)N_{33}^c+\frac{4}{t_c}\left(\frac{t_f^t-t_f^b}{16}\right)\left(\frac{\partial N_{13}^c}{\partial x_1}+\frac{\partial N_{23}^c}{\partial x_2}\right)
$$

$$
-2\left(\frac{\partial u_3^d}{\partial x_1}+\frac{\partial^2 \mathring{u}_3^d}{\partial x_1}\right)\left(\frac{\partial N_{11}^a}{\partial x_1}+\frac{\partial N_{12}^a}{\partial x_2}\right)-2\left(\frac{\partial u_3^d}{\partial x_2}+\frac{\partial \mathring{u}_3^d}{\partial x_2}\right)\left(\frac{\partial N_{22}^a}{\partial x_2}+\frac{\partial N_{12}^a}{\partial x_1}\right)
$$

$$
-2\left(\frac{\partial u_3^a}{\partial x_1}+\frac{\partial \mathring{u}_3^a}{\partial x_1}\right)\left(\frac{\partial N_{11}^d}{\partial x_1}+\frac{\partial N_{12}^d}{\partial x_2}+\frac{N_{13}^c}{t_c}\right)-2\left(\frac{\partial u_3^a}{\partial x_2}+\frac{\partial \mathring{u}_3^a}{\partial x_2}\right)\left(\frac{\partial N_{22}^d}{\partial x_2}+\frac{\partial N_{12}^d}{\partial x_1}+\frac{N_{23}^c}{t_c}\right)
$$

$$
+\frac{4}{t_c^2}\left(\frac{t_c}{2}-u_3^d-\circ u_3^d\right)\left(\frac{\partial M_{13}^c}{\partial x_1}+\frac{\partial M_{23}^c}{\partial x_2}\right)-2\widehat{q}_3^d-2C^d\dot{u}_3^a-2C^a\dot{u}_3^d
$$

$$
-2\left(m^a+\frac{m_c}{6}\right)\ddot{u}_3^d-2m^d\ddot{u}_3^a\Bigg\}\delta u_3^d\Bigg]dx_1dx_2\Bigg\rangle dt
$$

$$
+\int_{t_0}^{t_1}\int_0^{l_2}\Bigg\langle 2N_{11}^a\delta u_1^a+2N_{12}^a\delta u_2^a+2N_{11}^d\delta u_2^d+2N_{12}^d\delta u_2^d+\Bigg\{2N_{11}^a\left(\frac{\partial u_3^a}{\partial x_1}+\frac{\partial \mathring{u}_3^a}{\partial x_1}\right)
$$

$$
+2N_{12}^a\left(\frac{\partial u_3^a}{\partial x_2}+\frac{\partial \mathring{u}_3^a}{\partial x_2}\right)+2N_{11}^d\left(\frac{\partial u_3^d}{\partial x_1}+\frac{\partial \mathring{u}_3^d}{\partial x_1}\right)+2N_{12}^d\left(\frac{\partial u_3^d}{\partial x_2}+\frac{\partial \mathring{u}_3^d}{\partial x_2}\right)+2\frac{\partial M_{11}^a}{\partial x_1}
$$

$$
+4\frac{\partial M_{12}^a}{\partial x_2}+\frac{2}{t_c}\left[\frac{t_c}{2}+\frac{t_f^t+t_f^b}{4}--u_3^d-\mathring{u}_3^d\right]N_{13}^c\Bigg\}\delta u_3^a
$$

$$
+\Bigg\{2N_{11}^a\left(\frac{\partial u_3^d}{\partial x_1}+\frac{\partial \mathring{u}_3^d}{\partial x_1}\right)+2N_{11}^d\left(\frac{\partial u_3^a}{\partial x_1}+\frac{\partial \mathring{u}_3^a}{\partial x_1}\right)+2N_{12}^a\left(\frac{\partial u_3^d}{\partial x_2}+\frac{\partial \mathring{u}_3^d}{\partial x_2}\right)
$$

$$
+2N_{12}^d\left(\frac{\partial u_3^a}{\partial x_2}+\frac{\partial \mathring{u}_3^a}{\partial x_2}\right)+2\frac{\partial M_{11}^d}{\partial x_1}+4\frac{\partial M_{12}^d}{\partial x_2}-\frac{1}{t_c}\left(\frac{t_f^t-t_f^b}{4}\right)N_{13}^c+
$$

$$
-\frac{2}{t_c^2}\left(\frac{t_c}{2}-u_3^d-\circ u_3^d\right)M_{13}^c\Bigg\}\delta u_3^d 2M_{11}^a\delta\left(\frac{\partial u_3^a}{\partial x_1}\right)+2M_{11}^d\delta\left(\frac{\partial u_3^d}{\partial x_1}\right)\Bigg\rangle\Bigg|_0^{l_1}dx_2dt
$$

$$
+\int_{t_0}^{t_1}\int_{0}^{l_1}\left\langle 2N_{22}^{a}\delta u_2^a + 2N_{21}^{a}\delta u_1^a + 2N_{22}^{d}\delta u_1^d + 2N_{21}^{d}\delta u_1^d + \left\{ 2N_{22}^{a}\left(\frac{\partial u_3^a}{\partial x_2} + \frac{\partial \overset{\circ}{u}_3^a}{\partial x_2}\right)\right.\right.
$$

$$
+2N_{21}^{a}\left(\frac{\partial u_3^a}{\partial x_1} + \frac{\partial \overset{\circ}{u}_3^a}{\partial x_1}\right) + 2N_{22}^{d}\left(\frac{\partial u_3^d}{\partial x_2} + \frac{\partial \overset{\circ}{u}_3^a}{\partial x_2}\right) + 2N_{21}^{d}\left(\frac{\partial u_3^d}{\partial x_1} + \frac{\partial \overset{\circ}{u}_3^d}{\partial x_1}\right) + 2\frac{\partial M_{22}^a}{\partial x_2}
$$

$$
\left. +4\frac{\partial M_{21}^a}{\partial x_1} + \frac{2}{t_c}\left[\frac{t_c}{2} + \frac{t_f^t + t_f^b}{4} - -u_3^d - \circ u_3^d\right] N_{23}^c \right\}\delta u_3^a
$$

$$
+\left\{ 2N_{22}^{a}\left(\frac{\partial u_3^d}{\partial x_2} + \frac{\partial \overset{\circ}{u}_3^d}{\partial x_2}\right) + 2N_{22}^{d}\left(\frac{\partial u_3^a}{\partial x_2} + \frac{\partial \overset{\circ}{u}_3^a}{\partial x_2}\right) + 2N_{21}^{a}\left(\frac{\partial u_3^d}{\partial x_1} + \frac{\partial \overset{\circ}{u}_3^d}{\partial x_1}\right) + 2N_{21}^{d}\right.
$$

$$
\left.\left(\frac{\partial u_3^a}{\partial x_1} + \frac{\partial \overset{\circ}{u}_3^a}{\partial x_1}\right) + 2\frac{\partial M_{22}^d}{\partial x_2} + 4\frac{\partial M_{21}^d}{\partial x_1} - \frac{1}{t_c}\left(\frac{t_f^t - t_f^b}{4}\right)N_{23}^c - \frac{2}{t_c^2}\left(\frac{t_c}{2} - u_3^d - \mathring{u}_3^d\right)M_{23}^c\right\}\delta u_3^d
$$

$$
\left. +2M_{22}^a\delta\left(\frac{\partial u_3^a}{\partial x_2}\right) + 2M_{22}^d\delta\left(\frac{\partial u_3^d}{\partial x_2}\right)\right\rangle\Bigg|_{0}^{l_2} dx_1 dt = 0 \tag{7.76}
$$

This equation can only be satisfied by setting the coefficients of the variational displacements equal to zero since the variational displacements are arbitrary. Setting the coefficients of $\delta u_1^a, \delta u_2^a, \delta u_1^d$, δu_2^d, $\delta\Phi_1^c$, $\delta\Phi_2^c$, δu_3^a, and δu_3^d to zero provides the equations of motion along with the corresponding boundary conditions. These are provided as

$$
\delta u_1^a: \quad \frac{\partial N_{11}^a}{\partial x_1} + \frac{\partial N_{12}^a}{\partial x_2} = 0 \tag{7.77}
$$

$$
\delta u_2^a: \quad \frac{\partial N_{12}^a}{\partial x_1} + \frac{\partial N_{22}^a}{\partial x_2} = 0 \tag{7.78}
$$

$$
\delta u_1^d: \quad \frac{\partial N_{11}^d}{\partial x_1} + \frac{\partial N_{12}^d}{\partial x_2} + \frac{N_{13}^c}{t_c} = 0 \tag{7.79}
$$

$$
\delta u_2^d: \quad \frac{\partial N_{12}^d}{\partial x_1} + \frac{\partial N_{22}^d}{\partial x_2} + \frac{N_{23}^c}{t_c} = 0 \tag{7.80}
$$

$$
\delta\Phi_1^c: \quad M_{13}^c = 0 \tag{7.81}
$$

$$
\delta\Phi_2^c: \quad M_{23}^c = 0 \tag{7.82}
$$

δu_3^a :

$$
\begin{aligned}
& N_{11}^a\left(\frac{\partial^2 u_3^a}{\partial x_1^2}+\frac{\partial^2 \mathring{u}_3^a}{\partial x_1^2}+\frac{1}{R_1}\right)+2N_{12}^a\left(\frac{\partial^2 u_3^a}{\partial x_1 \partial x_2}+\frac{\partial^2 \mathring{u}_3^a}{\partial x_1 \partial x_2}\right)+N_{22}^a\left(\frac{\partial^2 u_3^a}{\partial x_2^2}+\frac{\partial^2 \mathring{u}_3^a}{\partial x_2^2}+\frac{1}{R_2}\right) \\
& +\frac{\partial^2 M_{11}^a}{\partial x_1^2}+2\frac{\partial^2 M_{12}^a}{\partial x_1 \partial x_2}+\frac{\partial^2 M_{22}^a}{\partial x_2^2}+N_{11}^d\left(\frac{\partial^2 u_3^d}{\partial x_1^2}+\frac{\partial^2 \mathring{u}_3^d}{\partial x_1^2}\right)+2N_{12}^d\left(\frac{\partial^2 u_3^d}{\partial x_1 \partial x_2}+\frac{\partial^2 \mathring{u}_3^d}{\partial x_1 \partial x_2}\right) \\
& +N_{22}^d\left(\frac{\partial^2 u_3^d}{\partial x_2^2}+\frac{\partial^2 \mathring{u}_3^d}{\partial x_2^2}\right)+\frac{1}{t_c}\left(\frac{t_c}{2}+\frac{t_f^t+t_f^b}{4}-u_3^d-\mathring{u}_3^d\right)\left(\frac{\partial N_{13}^c}{\partial x_1}+\frac{\partial N_{23}^c}{\partial x_2}\right) \\
& -\frac{2}{t_c}\left(\frac{\partial u_3^d}{\partial x_2}+\frac{\partial \mathring{u}_3^d}{\partial x_2}\right)N_{23}^c-\frac{2}{t_c}\left(\frac{\partial u_3^d}{\partial x_1}+\frac{\partial \mathring{u}_3^d}{\partial x_1}\right)N_{13}^c+\widehat{q}_3^a+\left(C^a+\frac{C^c}{2}\right)\dot{u}_3^a+C^d\dot{u}_3^d \\
& +\left(m^a+\frac{m_c}{2}\right)\ddot{u}_3^a+m^d\ddot{u}_3^d=0
\end{aligned}
\tag{7.83}
$$

δu_3^d :

$$
\begin{aligned}
& N_{11}^d\left(\frac{\partial^2 u_3^a}{\partial x_1^2}+\frac{\partial^2 \mathring{u}_3^a}{\partial x_1^2}+\frac{1}{R_1}\right)+2N_{12}^d\left(\frac{\partial^2 u_3^a}{\partial x_1 \partial x_2}+\frac{\partial^2 \mathring{u}_3^a}{\partial x_1 \partial x_2}\right)+N_{22}^d\left(\frac{\partial^2 u_3^a}{\partial x_2^2}+\frac{\partial^2 \mathring{u}_3^a}{\partial x_2^2}+\frac{1}{R_2}\right) \\
& +\frac{\partial^2 M_{11}^d}{\partial x_1^2}+2\frac{\partial^2 M_{12}^d}{\partial x_1 \partial x_2}+\frac{\partial^2 M_{22}^d}{\partial x_2^2}+N_{11}^a\left(\frac{\partial^2 u_3^d}{\partial x_1^2}+\frac{\partial^2 \mathring{u}_3^d}{\partial x_1^2}\right)+2N_{12}^a\left(\frac{\partial^2 u_3^d}{\partial x_1 \partial x_2}+\frac{\partial^2 \mathring{u}_3^d}{\partial x_1 \partial x_2}\right) \\
& +N_{22}^a\left(\frac{\partial^2 u_3^d}{\partial x_2^2}+\frac{\partial^2 \mathring{u}_3^d}{\partial x_2^2}\right)+\frac{2}{t_c}\left(\frac{t_c}{2}-u_3^d-\mathring{u}_3^d\right)N_{33}^c\underline{-\frac{1}{3}\left(\frac{t_f^t-t_f^b}{8t_c}\right)\left(\frac{\partial N_{13}^c}{\partial x_1}+\frac{\partial N_{23}^c}{\partial x_2}\right)} \\
& +\widehat{q}_3^d+C^d\dot{u}_3^a+C^a\dot{u}_3^d+\left(m^a+\frac{m_c}{6}\right)\ddot{u}_3^d+m^d\ddot{u}_3^a=0
\end{aligned}
\tag{7.84}
$$

The corresponding boundary conditions become

$$u_n^a=\widehat{u}_n^a \quad \text{or} \quad N_{nn}^a=\widehat{N}_{nn}^a \tag{7.85}$$

$$u_t^a=\widehat{u}_t^a \quad \text{or} \quad N_{nt}^a=\widehat{N}_{nt}^a \tag{7.86}$$

$$u_n^d=\widehat{u}_n^d \quad \text{or} \quad N_{nn}^d=\widehat{N}_{nn}^d \tag{7.87}$$

$$u_t^d = \widehat{u}_t^d \quad \text{or} \quad N_{nt}^d = \widehat{N}_{nt}^d \tag{7.88}$$

$$\begin{aligned} u_3^a = \widehat{u}_3^a \quad \text{or} \quad & \left(\frac{\partial u_3^a}{\partial x_n} + \frac{\partial \mathring{u}_3^a}{\partial x_n}\right) N_{nn}^a + \left(\frac{\partial u_3^a}{\partial x_t} + \frac{\partial \mathring{u}_3^a}{\partial x_t}\right) N_{nt}^a + \left(\frac{\partial u_3^d}{\partial x_n} + \frac{\partial \mathring{u}_3^d}{\partial x_n}\right) N_{nn}^d + \\ & \left(\frac{\partial u_3^d}{\partial x_t} + \frac{\partial \mathring{u}_3^d}{\partial x_t}\right) N_{nt}^d + \frac{\partial M_{nn}^a}{\partial x_n} + 2\frac{\partial M_{nt}^a}{\partial x_t} + \frac{1}{t_c}\left(\frac{t_c}{2} + \frac{t_f^t + t_f^b}{4} - u_3^d - \mathring{u}_3^d\right) N_{13}^c \\ & = \frac{\partial \widehat{M}_{nt}^a}{\partial x_t} + \frac{1}{2}\widehat{N}_{n3}^c \end{aligned} \tag{7.89}$$

$$\begin{aligned} u_3^d = \widehat{u}_3^d \quad \text{or} \quad & \left(\frac{\partial u_3^d}{\partial x_n} + \frac{\partial \mathring{u}_3^d}{\partial x_n}\right) N_{nn}^a + \left(\frac{\partial u_3^d}{\partial x_t} + \frac{\partial \mathring{u}_3^d}{\partial x_t}\right) N_{nt}^a + \left(\frac{\partial u_3^a}{\partial x_n} + \frac{\partial \mathring{u}_3^a}{\partial x_n}\right) N_{nn}^d + \\ & \left(\frac{\partial u_3^a}{\partial x_t} + \frac{\partial \mathring{u}_3^a}{\partial x_t}\right) N_{nt}^d + \frac{\partial M_{nn}^d}{\partial x_n} + 2\frac{\partial M_{nt}^d}{\partial x_t} - \frac{2}{t_c}\left(\frac{\Phi_3^c}{3} - \frac{t_f^t - t_f^b}{8}\right) N_{n3}^c = \\ & \frac{\partial \widehat{M}_{nt}^d}{\partial x_t} - \frac{\widehat{M}_{n3}^c}{t_c} \end{aligned} \tag{7.90}$$

$$\frac{\partial u_3^a}{\partial x_n} = \frac{\partial \widehat{u}_3^a}{\partial x_n} \quad \text{or} \quad M_{nn}^a = \widehat{M}_{nn}^a \tag{7.91}$$

$$\frac{\partial u_3^d}{\partial x_n} = \frac{\partial \widehat{u}_3^d}{\partial x_n} \quad \text{or} \quad M_{nn}^d = \widehat{M}_{nn}^d \tag{7.92}$$

7.5.2 Boundary Conditions

Where n and t are the normal and tangential directions to the boundary, when $n = 1$, $t = 2$ and when $n = 2,\ t = 1$. For the case of **simply supported** boundary conditions the following conditions must be satisfied.

Along the edges $x_n = 0,\ L_n$

$$N_{nn}^a = N_{nn}^d = N_{nt}^a = N_{nt}^d = M_{nn}^a = M_{nn}^d = u_3^a = u_3^d = 0 \tag{7.93}$$

For the case of **clamped** boundary conditions, the following conditions must be satisfied.

$$u_n^a = u_t^a = u_n^d = u_t^d = u_3^a = u_3^d = \frac{\partial u_3^a}{\partial x_n} = \frac{\partial u_3^d}{\partial x_n} = 0 \tag{7.94}$$

7.6 The Stress Resultants, Stress Couples, and Stiffnesses

The local stress resultants and stress couples can be expressed in terms of the strain measures by substituting Eqs. (7.36)–(7.38) into Eqs. (7.46a–c). For the top face, the matrix form is

$$\begin{pmatrix} N_{11}^t \\ N_{22}^t \\ N_{12}^t \\ M_{11}^t \\ M_{22}^t \\ M_{12}^t \end{pmatrix} = \begin{pmatrix} A_{11}^t & A_{12}^t & A_{16}^t & B_{11}^t & B_{12}^t & B_{16}^t \\ & A_{22}^t & A_{26}^t & B_{12}^t & B_{22}^t & B_{26}^t \\ & & A_{66}^t & B_{16}^t & B_{26}^t & B_{66}^t \\ & & & D_{11}^t & D_{12}^t & D_{16}^t \\ & & & & D_{22}^t & D_{26}^t \\ \text{Sym} & & & & & D_{66}^t \end{pmatrix} \begin{pmatrix} \overline{\gamma}_{11}^t \\ \overline{\gamma}_{22}^t \\ 2\overline{\gamma}_{12}^t \\ \kappa_{11}^t \\ \kappa_{22}^t \\ 2\kappa_{12}^t \end{pmatrix} - \begin{pmatrix} N_{11}^t \\ N_{22}^t \\ N_{12}^t \\ M_{11}^t \\ M_{22}^t \\ M_{12}^t \end{pmatrix}^T - \begin{pmatrix} N_{11}^t \\ N_{22}^t \\ N_{12}^t \\ M_{11}^t \\ M_{22}^t \\ M_{12}^t \end{pmatrix}^m \tag{7.95}$$

where the local stiffnesses $A_{ij}^t, B_{ij}^t, D_{ij}^t$ are defined as

$$\left(A_{ij}^t, B_{ij}^t, D_{ij}^t\right) = \int_{-\frac{t_c}{2}-t_f^t}^{-\frac{t_c}{2}} \widehat{Q}_{ij}^t \left\{1, \left(x_3 + \frac{t_c + t_f^t}{2}\right), \left(x_3 + \frac{t_c + t_f^t}{2}\right)^2\right\} dx_3 \tag{7.96}$$

For the bottom face, the local stress resultants and couples in matrix form become

$$\begin{pmatrix} N_{11}^b \\ N_{22}^b \\ N_{12}^b \\ M_{11}^b \\ M_{22}^b \\ M_{12}^b \end{pmatrix} = \begin{pmatrix} A_{11}^b & A_{12}^b & A_{16}^b & B_{11}^b & B_{12}^b & B_{16}^b \\ & A_{22}^b & A_{26}^b & B_{12}^b & B_{22}^b & B_{26}^b \\ & & A_{66}^b & B_{16}^b & B_{26}^b & B_{66}^b \\ & & & D_{11}^b & D_{12}^b & D_{16}^b \\ & & & & D_{22}^b & D_{26}^b \\ \text{Sym} & & & & & D_{66}^b \end{pmatrix} \begin{pmatrix} \overline{\gamma}_{11}^b \\ \overline{\gamma}_{22}^b \\ 2\overline{\gamma}_{12}^b \\ \kappa_{11}^b \\ \kappa_{22}^b \\ 2\kappa_{12}^b \end{pmatrix} - \begin{pmatrix} N_{11}^b \\ N_{22}^b \\ N_{12}^b \\ M_{11}^b \\ M_{22}^b \\ M_{12}^b \end{pmatrix}^T - \begin{pmatrix} N_{11}^b \\ N_{22}^b \\ N_{12}^b \\ M_{11}^b \\ M_{22}^b \\ M_{12}^b \end{pmatrix}^m \tag{7.97}$$

where the local stiffnesses $A_{ij}^b, B_{ij}^b, D_{ij}^b$ are defined as

$$\left(A_{ij}^b, B_{ij}^b, D_{ij}^b\right) = \int_{\frac{t_c}{2}}^{\frac{t_c}{2}+t_f^b} \widehat{Q}_{ij}^b \left\{1, \left(x_3 - \frac{t_c + t_f^b}{2}\right), \left(x_3 - \frac{t_c + t_f^b}{2}\right)^2\right\} dx_3 \tag{7.98}$$

If the structural tailoring of the faces is such that the stacking sequence is symmetric with respect to its local mid-surface, then $\left[B_{ij}^{(t,b)}\right] = 0$. The stress resultants and stress couples for the core are expressed as

$$\begin{pmatrix} N_{33}^c \\ N_{23}^c \\ N_{13}^c \end{pmatrix} = \begin{pmatrix} A_{33}^c & 0 & 0 \\ 0 & A_{44}^c & 0 \\ 0 & 0 & A_{55}^c \end{pmatrix} \begin{pmatrix} \overline{\gamma}_{33}^c \\ 2\overline{\gamma}_{23}^c \\ 2\overline{\gamma}_{13}^c \end{pmatrix} \tag{7.99a}$$

$$\begin{pmatrix} M_{33}^c \\ M_{23}^c \\ M_{13}^c \end{pmatrix} = \begin{pmatrix} D_{33}^c & 0 & 0 \\ 0 & D_{44}^c & 0 \\ 0 & 0 & D_{55}^c \end{pmatrix} \begin{pmatrix} \kappa_{33}^c \\ 2\kappa_{23}^c \\ 2\kappa_{13}^c \end{pmatrix} \tag{7.99b}$$

where the core stiffnesses are given by

$$\left(A_{ii}^c, D_{ii}^c\right) = \int_{-\frac{t_c}{2}}^{\frac{t_c}{2}} Q_{ii}^c\left(1, x_3^2\right) dx_3, \qquad i = (3, 4, 5) \tag{7.100}$$

Note: No summation with respect to the repeated indices is assumed. The core transverse and normal moduli are given in Eqs. (7.39a–c).

7.7 Comments

To-date only a few applications have been addressed utilizing these state-of-the-art theoretical developments. Such areas include buckling, dynamic buckling, post-buckling, and the transient dynamic response problem (Hohe and Librescu (2003, 2006) and Hause (2012)). As an example, on how to apply these equations, the dynamic response problem is presented in the next section. This example is not meant to be an exhaustive treatment of possible applications of this theory nor is it meant to be an exhaustive treatment of the dynamic response. Several results have been compiled by Hohe and Librescu (2003, 2006), where they considered the buckling, post-buckling, and the transient dynamic buckling problem. The reader is referred to these authors to see additional applications of this theory as well as the results.

7.8 Application – Dynamic Response of Flat Sandwich Panels

7.8.1 *Preliminaries*

With the theoretical foundation in hand some basic assumptions are in order toward the dynamic response of flat sandwich panels. The following assumptions are made. The tangential deformations are assumed negligible in contrast to the transverse direction where the deformations can be appreciable. The face sheets are assumed incompressible, while the core can exhibit extensibility in the transverse direction. Likewise, the tangential and rotary inertias are assumed to be negligible, whereas the transverse inertia is retained. With this in mind, the following additional basic assumptions are adopted:

1. The face sheets are orthotropic layers not necessarily coincident with the geometrical axes.
2. The core features orthotropic properties in the transverse direction and is considered the weak-type and much larger in thickness than the facings.
3. Perfect bonding between the face sheets and the facings and the core are assumed.
4. The transverse shear effects in the facings are discarded.
5. The face sheets are symmetric with respect to their local and global mid-surfaces.
6. All time-dependent external pressures are uniformly distributed over the panel face.

7.8.2 Governing Equations

The equations of motion, Eqs. (7.77)–(7.84), are provided below which will be referred to for the solution methodology. These equations have been modified for the case of symmetric facings with respect to the global and their respective local mid-surfaces. For this specialized case of symmetry, the following relationships hold.

$$\begin{aligned} t^d &= \frac{t_f^t - t_f^b}{2} = 0, \; t^a = \frac{t_f^t + t_f^b}{2} = \frac{2t_f^t \left(= 2t_f^b\right)}{2} = t_f, \; a^t = a^b = a \\ &= \frac{t_c + t_f}{2} \end{aligned} \tag{7.101}$$

In addition, the transverse pressure, $\widehat{q}_3^b$ has been discarded. As a result of these relationships, the nonlinear equations of motion become

$$\delta u_1^a : \frac{\partial N_{11}^a}{\partial x_1} + \frac{\partial N_{12}^a}{\partial x_2} = 0 \tag{7.102a}$$

$$\delta u_2^a : \frac{\partial N_{12}^a}{\partial x_1} + \frac{\partial N_{22}^a}{\partial x_2} = 0 \tag{7.102b}$$

$$\delta u_1^d : \frac{\partial N_{11}^d}{\partial x_1} + \frac{\partial N_{12}^d}{\partial x_2} + \frac{N_{13}^c}{t_c} = 0 \tag{7.102c}$$

$$\delta u_2^d : \frac{\partial N_{12}^d}{\partial x_1} + \frac{\partial N_{22}^d}{\partial x_2} + \frac{N_{23}^c}{t_c} = 0 \tag{7.102d}$$

$$\delta \Phi_1^c : M_{13}^c = 0 \tag{7.102e}$$

$$\delta \Phi_2^c : M_{23}^c = 0 \tag{7.102f}$$

$$\begin{aligned}
&\delta u_3^a : \\
&\frac{\partial^2 u_3^a}{\partial x_1^2} N_{11}^a + 2\frac{\partial^2 u_3}{\partial x_1 \partial x_2} N_{12}^a + \frac{\partial^2 u_3^a}{\partial x_2^2} N_{22}^a + \frac{\partial^2 M_{11}^a}{\partial x_1^2} + 2\frac{\partial^2 M_{12}^a}{\partial x_1 \partial x_2} + \frac{\partial^2 M_{22}^a}{\partial x_2^2} + \\
&\frac{\partial^2 u_3^a}{\partial x_1^2} N_{11}^d + 2\frac{\partial^2 u_3^a}{\partial x_1 \partial x_2} N_{12}^d + \frac{\partial^2 u_3^a}{\partial x_2^2} N_{22}^d + \frac{1}{t_c}\left(\frac{t_c + t_f}{2} - u_3^d\right)\left(\frac{\partial N_{13}^c}{\partial x_1} + \frac{\partial N_{23}^c}{\partial x_2}\right) \\
&-\frac{2}{t_c}\frac{\partial u_3^d}{\partial x_1} N_{13}^c - \frac{2}{t_c}\frac{\partial u_3^d}{\partial x_2} N_{23}^c - \left(m^f + \frac{m^c}{2}\right)\ddot{u}_3^a - \left(C^f + C^c\right)\dot{u}_3^a + \widehat{q}_3 = 0
\end{aligned} \tag{7.102g}$$

$$\begin{aligned}
&\delta u_3^d : \\
&\frac{\partial^2 u_3^a}{\partial x_1^2} N_{11}^d + 2\frac{\partial^2 u_3}{\partial x_1 \partial x_2} N_{12}^d + \frac{\partial^2 u_3^a}{\partial x_2^2} N_{22}^d + \frac{\partial^2 M_{11}^d}{\partial x_1^2} + 2\frac{\partial^2 M_{12}^d}{\partial x_1 \partial x_2} + \frac{\partial^2 M_{22}^d}{\partial x_2^2} + \frac{\partial^2 u_3^d}{\partial x_1^2} N_{11}^a \\
&+2\frac{\partial^2 u_3^d}{\partial x_1 \partial x_2} N_{12}^a + \frac{\partial^2 u_3^d}{\partial x_2^2} N_{22}^a + \frac{2}{t_c}\left(\frac{t_c}{2} - u_3^d\right) N_{33}^c - \left(m^f + \frac{m^c}{3}\right)\ddot{u}_3^d - C^f \dot{u}_3^d + \widehat{q}_3^t = 0
\end{aligned} \tag{7.102h}$$

Simply supported boundary conditions will be assumed. These are specified as Along the edges $x_1 = (0, L_1)$

$$N_{11}^a = N_{11}^d = N_{12}^a = N_{12}^d = M_{11}^a = M_{11}^d = u_3^a = u_3^d = 0 \tag{7.103a – h}$$

Along the edges $x_2 = (0, L_2)$

$$N_{22}^a = N_{22}^d = N_{21}^a = N_{21}^d = M_{22}^a = M_{22}^d = u_3^a = u_3^d = 0 \tag{7.104a – h}$$

Equations (7.8-2a–h)–(7.104a–h) constitute the governing system to determine the dynamic response for flat sandwich panels with symmetric thin facings and a thick orthotropic weak transversely compressible core.

7.8.3 Solution Methodology

As a beginning to the solution process for the dynamic response, u_3^a, u_3^d can be assumed in the following form as

$$u_3^a = w_{mn}^a(t) \sin \lambda_m x_1 \sin \mu_n x_2 \tag{7.105a}$$

$$u_3^d = w_{mn}^d(t) \sin \lambda_m x_1 \sin \mu_n x_2 \tag{7.105b}$$

where $\lambda_m = m\pi/L_1$, $\mu_n = n\pi/L_2$, m and n are the number of half sine waves in the respective directions, whereas $w_{mn}^a(t)$ and $w_{mn}^d(t)$ denote the modal amplitudes as a function of time of the transverse displacement functions. It should be noted that Hohe and Librescu considered the mode numbers and modal amplitudes as independent of one another. For the current case they are not considered to be independent. For the case where they are assumed independent, they would take the following form

$$u_3^a = w_{mn}^a \sin\left(\lambda_m^a x_1\right) \sin\left(\mu_n^a x_2\right)$$
$$u_3^d = w_{pq}^d \sin\left(\lambda_p^d x_1\right) \sin\left(\mu_q^d x_2\right)$$

where

$$\lambda_m^a = \frac{m\pi}{L_1}, \quad \mu_n^a = \frac{n\pi}{L_2}, \quad \lambda_p^d = \frac{p\pi}{L_1}, \quad \mu_q^d = \frac{q\pi}{L_2}$$

With the current representation, the transverse geometric boundary conditions are identically fulfilled. The transverse pressure is represented by

$$\widehat{q}_3^t(x_1, x_2, t) = q_{mn}(t) \sin \lambda_m x_1 \sin \mu_n x_2 \tag{7.106}$$

Integrating both sides of Eq. (7.106) over the plate area such that

$$q_{mn}(t) = \frac{4}{L_1 L_2} \int_0^{L_2} \int_0^{L_1} q_t(x_1, x_2, t) \sin \lambda_m x_1 \sin \mu_n x_2 dx_1 dx_2 \tag{7.107}$$

gives

$$q_{mn}(t) = \frac{16 q_t}{mn\pi^2} \tag{7.108}$$

Equations (7.8-2a, b) can be fulfilled by assuming a stress potential for $N_{11}^a, N_{22}^a, N_{12}^a$ in the following form.

$$N_{11}^a = \frac{\partial^2 \varphi}{\partial x_2^2}, \quad N_{22}^a = \frac{\partial^2 \varphi}{\partial x_1^2}, \quad N_{12}^a = -\frac{\partial^2 \varphi}{\partial x_1 \partial x_2} \tag{7.109a – c}$$

Another variable φ has been thrown into the governing system of equations. This variable will be determined from a compatibility equation. Eliminating the tangential strains from the in-plane displacements u_1^a and u_2^a results in

$$\frac{\partial^2 \overline{\gamma}_{11}^a}{\partial x_2^2} - 2\frac{\partial^2 \overline{\gamma}_{12}^a}{\partial x_1 \partial x_2} + \frac{\partial^2 \overline{\gamma}_{22}^a}{\partial x_1^2} = \left(\frac{\partial u_3^a}{\partial x_1 \partial x_2}\right)^2 + \left(\frac{\partial u_3^d}{\partial x_1 \partial x_2}\right)^2 - \frac{\partial^2 u_3^a}{\partial x_1^2}\frac{\partial^2 u_3^a}{\partial x_2^2} - \frac{\partial^2 u_3^d}{\partial x_1^2}\frac{\partial^2 u_3^d}{\partial x_2^2} \tag{7.110}$$

Since symmetry exists in the facings locally and globally the stress resultants from Eqs. (7.95) and (7.97) can be expressed in decoupled form as

$$\begin{pmatrix} N_{11}^a \\ N_{22}^a \\ N_{12}^a \end{pmatrix} = \begin{bmatrix} A_{11}^a & A_{12}^a & A_{16}^a \\ & A_{22}^a & A_{26}^a \\ \text{Sym} & & A_{66}^a \end{bmatrix} \begin{pmatrix} \overline{\gamma}_{11}^a \\ \overline{\gamma}_{22}^a \\ 2\overline{\gamma}_{12}^a \end{pmatrix} \tag{7.111}$$

By performing a matrix inversion of Eq. (7.111) utilizing Eqs. (7.109a–c) and substituting $\overline{\gamma}_{11}^a, \overline{\gamma}_{22}^a$, and $\overline{\gamma}_{12}^a$ into Eq. (7.110) results in the compatibility equation in terms of the variables φ, u_3^a, and u_3^d. This appears as

$$\begin{aligned} &A_{11}^* \frac{\partial^4 \varphi}{\partial x_2^4} + \left(2A_{12}^* + A_{66}^*\right)\frac{\partial^4 \varphi}{\partial x_1^2 \partial x_2^2} + A_{22}^* \frac{\partial^4 \varphi}{\partial x_1^4} - 2A_{16}^* \frac{\partial^4 \varphi}{\partial x_1 \partial x_2^3} - 2A_{26}^* \frac{\partial^4 \varphi}{\partial x_1^3 \partial x_2} \\ &= \left(\frac{\partial u_3^a}{\partial x_1 \partial x_2}\right)^2 + \left(\frac{\partial u_3^d}{\partial x_1 \partial x_2}\right)^2 - \frac{\partial^2 u_3^a}{\partial x_1^2}\frac{\partial^2 u_3^a}{\partial x_2^2} - \frac{\partial^2 u_3^d}{\partial x_1^2}\frac{\partial^2 u_3^d}{\partial x_2^2} \end{aligned} \tag{7.112}$$

where, $[A^*] = [A]^{-1}$. Since u_3^a and u_3^d are known, φ can be determined by assuming the following form in terms of unknown coefficients then substituting into Eq. (7.112) and solving for the unknown coefficients. This assumed form is expressed in terms of trigonometric functions as

$$\begin{aligned} \varphi(x_1, x_2, t) = &\left(w_{mn}^a\right)^2 C_1 \cos 2\lambda_m x_1 + \left(w_{mn}^a\right)^2 C_2 \cos 2\mu_n x_2 + \left(w_{mn}^d\right)^2 C_3 \cos 2\lambda_m x_1 + \\ &\left(w_{mn}^d\right)^2 C_4 \cos 2\mu_n x_2 \end{aligned} \tag{7.113}$$

After substituting Eq. (7.113) into Eq. (7.112) and comparing coefficients, $C_1 - C_4$ are determined to be

$$C_1 = C_3 = \frac{\mu_n^2 \left\{ A_{66}^f \left(A_{11}^f A_{22}^f - \left(A_{12}^f \right)^2 \right) + 2A_{12}^f A_{16}^f A_{26}^f - A_{11}^f \left(A_{26}^f \right)^2 - A_{22}^f \left(A_{16}^f \right)^2 \right\}}{32\lambda_m^2 \left[A_{11}^f A_{66}^f - \left(A_{16}^f \right)^2 \right]} \tag{7.114}$$

$$C_2 = C_4 = \frac{\lambda_m^2 \left\{ A_{66}^f \left(A_{11}^f A_{22}^f - \left(A_{12}^f \right)^2 \right) + 2A_{12}^f A_{16}^f A_{26}^f - A_{11}^f \left(A_{26}^f \right)^2 - A_{22}^f \left(A_{16}^f \right)^2 \right\}}{32\mu_n^2 \left[A_{22}^f A_{66}^f - \left(A_{26}^f \right)^2 \right]} \tag{7.115}$$

Thus far, φ, u_3^a, u_3^d are known. Next u_1^a and u_2^a will be determined from the stress resultants expressed in terms of displacements. From Eq. (7.111), N_{11}^a and N_{22}^a expressed in terms of displacements are

$$\begin{aligned}
N_{11}^a &= \frac{\partial^2 \varphi}{\partial x_2^2} = A_{11}^a \left\{ \frac{\partial u_1^a}{\partial x_1} + \frac{1}{2}\left(\frac{\partial u_3^a}{\partial x_1}\right)^2 + \frac{1}{2}\left(\frac{\partial u_3^d}{\partial x_1}\right)^2 \right\} \\
&+ A_{12}^a \left\{ \frac{\partial u_2^a}{\partial x_2} + \frac{1}{2}\left(\frac{\partial u_3^a}{\partial x_2}\right)^2 + \frac{1}{2}\left(\frac{\partial u_3^d}{\partial x_2}\right)^2 \right\} \\
&+ A_{16}^a \left\{ \frac{\partial u_1^a}{\partial x_2} + \frac{\partial u_2^a}{\partial x_1} + \frac{\partial u_3^a}{\partial x_1}\frac{\partial u_3^a}{\partial x_2} + \frac{\partial u_3^d}{\partial x_1}\frac{\partial u_3^d}{\partial x_2} \right\} \\
N_{22}^a &= \frac{\partial^2 \varphi}{\partial x_1^2} = A_{22}^a \left\{ \frac{\partial u_2^a}{\partial x_2} + \frac{1}{2}\left(\frac{\partial u_3^a}{\partial x_2}\right)^2 + \frac{1}{2}\left(\frac{\partial u_3^d}{\partial x_2}\right)^2 \right\} \\
&+ A_{12}^a \left\{ \frac{\partial u_1^a}{\partial x_1} + \frac{1}{2}\left(\frac{\partial u_3^a}{\partial x_1}\right)^2 + \frac{1}{2}\left(\frac{\partial u_3^d}{\partial x_1}\right)^2 \right\} \\
&+ A_{26}^a \left\{ \frac{\partial u_1^a}{\partial x_2} + \frac{\partial u_2^a}{\partial x_1} + \frac{\partial u_3^a}{\partial x_1}\frac{\partial u_3^a}{\partial x_2} + \frac{\partial u_3^d}{\partial x_1}\frac{\partial u_3^d}{\partial x_2} \right\}
\end{aligned} \tag{7.116a, b}$$

Equations (7.8-16a, 7.116a, b) are two coupled inhomogeneous partial differential equations in terms of two unknowns u_1^a and u_2^a which can be assumed in the following form

$$u_1^a(x_1,x_2,t) = \left(\left(w_{mn}^a\right)^2 + \left(w_{mn}^d\right)^2\right)D_1x_1 + \left(\left(w_{mn}^a\right)^2 + \left(w_{mn}^d\right)^2\right)D_2 \sin 2\lambda_m x_1 +$$
$$\left(\left(w_{mn}^a\right)^2 + \left(w_{mn}^d\right)^2\right)D_3 \sin 2\mu_n x_2 + \left(\left(w_{mn}^a\right)^2 + \left(w_{mn}^d\right)^2\right)D_4 \sin 2\lambda_m x_1 \cos 2\mu_n x_2$$
$$\left(\left(w_{mn}^a\right)^2 + \left(w_{mn}^d\right)^2\right)D_5 \cos 2\lambda_m x_1 \sin 2\mu_n x_2$$
$$u_2^a(x_1,x_2,t) = \left(\left(w_{mn}^a\right)^2 + \left(w_{mn}^d\right)^2\right)E_1x_2 + \left(\left(w_{mn}^a\right)^2 + \left(w_{mn}^d\right)^2\right)E_2 \sin 2\mu_n x_2 +$$
$$\left(\left(w_{mn}^a\right)^2 + \left(w_{mn}^d\right)^2\right)E_3 \sin 2\lambda_m x_1 + \left(\left(w_{mn}^a\right)^2 + \left(w_{mn}^d\right)^2\right)E_4 \cos 2\lambda_m x_1 \sin 2\mu_n x_2$$
$$\left(\left(w_{mn}^a\right)^2 + \left(w_{mn}^d\right)^2\right)E_5 \sin 2\lambda_m x_1 \cos 2\mu_n x_2$$
(7.117a, b)

Substituting the expressions for u_1^a and u_2^a into Eqs. (7.116a, b) and comparing coefficients of like trigonometric functions give $D_1 - D_5$ and $E_1 - E_5$. These coefficients are provided as the solution to the following matrix equations.

$$\begin{pmatrix} N_1 & N_2 \\ N_3 & N_4 \end{pmatrix}\begin{pmatrix} D_1 \\ E_1 \end{pmatrix} = \begin{pmatrix} S_1 \\ S_2 \end{pmatrix}, \quad \begin{pmatrix} N_4 & N_5 \\ N_6 & N_7 \end{pmatrix}\begin{pmatrix} D_2 \\ E_3 \end{pmatrix} = \begin{pmatrix} S_3 \\ S_4 \end{pmatrix}, \quad \begin{pmatrix} N_8 & N_9 \\ N_{10} & N_{11} \end{pmatrix}\begin{pmatrix} D_3 \\ E_2 \end{pmatrix} = \begin{pmatrix} S_5 \\ S_6 \end{pmatrix}$$
(7.118a – c)

$$\begin{pmatrix} N_{12} & N_{13} & N_{14} & N_{15} \\ -N_{13} & -N_{12} & -N_{15} & -N_{14} \\ N_{16} & N_{17} & N_{18} & N_{19} \\ -N_{17} & -N_{16} & -N_{19} & -N_{18} \end{pmatrix}\begin{pmatrix} D_4 \\ D_5 \\ E_4 \\ E_5 \end{pmatrix} = \begin{pmatrix} S_7 \\ S_8 \\ S_9 \\ S_{10} \end{pmatrix} \quad (7.119)$$

where

$$N_1 = A_{11}^f, \quad N_2 = A_{12}^f, \quad N_3 = A_{22}^f$$
$$N_4 = 2\lambda_m A_{11}^f, \quad N_5 = 2\lambda_m A_{16}^f, \quad N_6 = 2\lambda_m A_{12}^f, \quad N_7 = 2\lambda_m A_{26}^f$$
$$N_8 = 2\mu_n A_{16}^f, \quad N_9 = 2\mu_n A_{12}^f, \quad N_{10} = 2\mu_n A_{26}^f, \quad N_{11} = 2\mu_n A_{22}^f$$
$$N_{12} = 2\lambda_m A_{11}^f, \quad N_{13} = 2\mu_n A_{16}^f, \quad N_{14} = 2\mu_n A_{12}^f, \quad N_{15} = 2\lambda_m A_{16}^f,$$
$$N_{16} = 2\lambda_m A_{12}^f \quad N_{17} = 2\mu_n A_{26}^f, \quad N_{18} = 2\mu_n A_{22}^f, \quad N_{19} = 2\lambda_m A_{26}^f$$
(7.120a – s)

$$
\begin{aligned}
S_1 &= -\frac{\lambda_m^2 A_{11}^f + \mu_n^2 A_{12}^f}{8}, \qquad S_2 = -\frac{\lambda_m^2 A_{12}^f + \mu_n^2 A_{22}^f}{8} \\
S_3 &= \frac{\mu_n^2 A_{12}^f - \lambda_m^2 A_{11}^f}{8}, \qquad S_4 = \frac{\mu_n^2 A_{22}^f - \lambda_m^2 A_{12}^f}{8} - 4\lambda_m^2 C_1 \\
S_5 &= -\frac{\lambda_m^2 A_{11}^f - \mu_n^2 A_{12}^f}{8} - 4\mu_n^2 C_2, \qquad S_6 = -\frac{\lambda_m^2 A_{12}^f - \mu_n^2 A_{22}^f}{8} \\
S_7 &= \frac{\lambda_m^2 A_{11}^f + \mu_n^2 A_{12}^f}{8}, \qquad S_8 = -\frac{\lambda_m \mu_n A_{16}^f}{4}, \qquad S_9 = \frac{\lambda_m^2 A_{12}^f + \mu_n^2 A_{22}^f}{8} \\
S_{10} &= -\frac{\lambda_m \mu_n A_{26}^f}{4}
\end{aligned}
\tag{7.121a – j}
$$

The third and fourth equations of motion are now addressed neglecting the thermal terms. These equations in terms of the displacement quantities are expressed as

$$
\begin{aligned}
& A_{11}^a \left(\frac{\partial^2 u_1^d}{\partial x_1^2} + \frac{\partial u_3^d}{\partial x_1}\frac{\partial^2 u_3^a}{\partial x_1^2} + \frac{\partial u_3^a}{\partial x_1}\frac{\partial^2 u_3^d}{\partial x_1^2} \right) + A_{12}^a \left(\frac{\partial^2 u_2^d}{\partial x_1 \partial x_2} + \frac{\partial u_3^d}{\partial x_2}\frac{\partial^2 u_3^a}{\partial x_1 \partial x_2} + \frac{\partial u_3^a}{\partial x_2}\frac{\partial^2 u_3^d}{\partial x_1 \partial x_2} \right) \\
& + A_{16}^a \left(\frac{\partial^2 u_2^d}{\partial x_1^2} + 2\frac{\partial^2 u_1^d}{\partial x_1 \partial x_2} + 2\frac{\partial u_3^a}{\partial x_1}\frac{\partial^2 u_3^d}{\partial x_1 \partial x_2} + 2\frac{\partial u_3^d}{\partial x_1}\frac{\partial^2 u_3^a}{\partial x_1 \partial x_2} + \frac{\partial^2 u_3^d}{\partial x_1^2}\frac{\partial u_3^a}{\partial x_2} + \frac{\partial^2 u_3^a}{\partial x_1^2}\frac{\partial u_3^d}{\partial x_2} \right) \\
& + A_{26}^a \left(\frac{\partial^2 u_2^d}{\partial x_2^2} + \frac{\partial^2 u_3^a}{\partial x_2^2}\frac{\partial u_3^d}{\partial x_2} + \frac{\partial u_3^a}{\partial x_2}\frac{\partial^2 u_3^d}{\partial x_2^2} \right) + A_{66}^a \left(\frac{\partial^2 u_1^d}{\partial x_2^2} + \frac{\partial^2 u_2^d}{\partial x_1 \partial x_2} + \frac{\partial u_3^a}{\partial x_2}\frac{\partial^2 u_3^d}{\partial x_1 \partial x_2} \right. \\
& \left. + \frac{\partial u_3^d}{\partial x_2}\frac{\partial^2 u_3^a}{\partial x_1 \partial x_2} + \frac{\partial u_3^a}{\partial x_1}\frac{\partial^2 u_3^d}{\partial x_2^2} + \frac{\partial u_3^d}{\partial x_1}\frac{\partial^2 u_3^a}{\partial x_2^2} \right) + \frac{A_{55}^c}{t_c^2}\left\{ -2u_1^d + \left[\left(t_c + t_f\right) - 2u_3^d \right] \frac{\partial u_3^a}{\partial x_1} \right\} \\
& = 0 \\
& A_{22}^a \left(\frac{\partial^2 u_2^d}{\partial x_2^2} + \frac{\partial u_3^d}{\partial x_2}\frac{\partial^2 u_3^a}{\partial x_2^2} + \frac{\partial u_3^a}{\partial x_2}\frac{\partial^2 u_3^d}{\partial x_2^2} \right) + A_{12}^a \left(\frac{\partial^2 u_1^d}{\partial x_1 \partial x_2} + \frac{\partial u_3^d}{\partial x_1}\frac{\partial^2 u_3^a}{\partial x_1 \partial x_2} + \frac{\partial u_3^a}{\partial x_1}\frac{\partial^2 u_3^d}{\partial x_1 \partial x_2} \right) \\
& + A_{26}^a \left(\frac{\partial^2 u_1^d}{\partial x_2^2} + 2\frac{\partial^2 u_2^d}{\partial x_1 \partial x_2} + 2\frac{\partial u_3^a}{\partial x_2}\frac{\partial^2 u_3^d}{\partial x_1 \partial x_2} + 2\frac{\partial u_3^d}{\partial x_2}\frac{\partial^2 u_3^a}{\partial x_1 \partial x_2} + \frac{\partial^2 u_3^d}{\partial x_2^2}\frac{\partial u_3^a}{\partial x_1} + \frac{\partial^2 u_3^a}{\partial x_2^2}\frac{\partial u_3^d}{\partial x_1} \right) \\
& + A_{16}^a \left(\frac{\partial^2 u_1^d}{\partial x_1^2} + \frac{\partial^2 u_3^a}{\partial x_1^2}\frac{\partial u_3^d}{\partial x_1} + \frac{\partial u_3^a}{\partial x_1}\frac{\partial^2 u_3^d}{\partial x_1^2} \right) + A_{66}^a \left(\frac{\partial^2 u_2^d}{\partial x_1^2} + \frac{\partial^2 u_1^d}{\partial x_1 \partial x_2} + \frac{\partial u_3^a}{\partial x_1}\frac{\partial^2 u_3^d}{\partial x_1 \partial x_2} \right. \\
& \left. + \frac{\partial u_3^d}{\partial x_1}\frac{\partial^2 u_3^a}{\partial x_1 \partial x_2} + \frac{\partial u_3^a}{\partial x_2}\frac{\partial^2 u_3^d}{\partial x_1^2} + \frac{\partial u_3^d}{\partial x_2}\frac{\partial^2 u_3^a}{\partial x_1^2} \right) + \frac{A_{55}^c}{t_c^2}\left\{ -2u_2^d + \left[\left(t_c + t_f^a\right) - 2u_3^d \right] \frac{\partial u_3^a}{\partial x_2} \right\} \\
& = 0
\end{aligned}
\tag{7.122a, b}
$$

It can be seen that Eqs. (7.121a, b) are two couple partial differential equations in terms of two unknowns u_1^d and u_2^d. These displacement quantities can be assumed in the following form following the same procedure as before by inserting the unknown functions into the two coupled partial differential equations, Eqs. (7.122a, b), followed by comparing coefficients of like trigonometric functions. u_1^d and u_2^d are assumed as follows

$$
\begin{aligned}
u_1^d(x_1, x_2, t) &= w_{mn}^a w_{mn}^d A_1 \sin 2\lambda_m x_1 + w_{mn}^a w_{mn}^d A_2 \sin 2\mu_n x_2 \\
&\quad + w_{mn}^a w_{mn}^d A_3 \sin 2\lambda_m x_1 \cos 2\mu_n x_2 + w_{mn}^a w_{mn}^d A_4 \cos 2\lambda_m x_1 \sin 2\mu_n x_2 \\
&\quad + w_{mn}^a A_5 \cos \lambda_m x_1 \sin \mu_n x_2 + w_{mn}^a A_6 \sin \lambda_m x_1 \cos \mu_n x_2 \\
u_2^d(x_1, x_2, t) &= w_{mn}^a w_{mn}^d B_1 \sin 2\mu_n x_2 + w_{mn}^a w_{mn}^d B_2 \sin 2\lambda_m x_1 \\
&\quad + w_{mn}^a w_{mn}^d B_3 \cos 2\lambda_m x_1 \sin 2\mu_n x_2 + w_{mn}^a w_{mn}^d B_4 \sin 2\lambda_m x_1 \cos 2\mu_n x_2 \\
&\quad + w_{mn}^a B_5 \sin \lambda_m x_1 \cos \mu_n x_2 + w_{mn}^a B_6 \cos \lambda_m x_1 \sin \mu_n x_2
\end{aligned}
\tag{7.123a, b}
$$

Following the described procedure just mentioned provides the constants, $A_1 - A_6$ and $B_1 - B_6$ which are presented as the solution to the following matrix equations.

$$
\begin{pmatrix} M_1 & M_2 \\ M_2 & M_3 \end{pmatrix} \begin{pmatrix} A_1 \\ B_2 \end{pmatrix} = \begin{pmatrix} -R_1 \\ -R_2 \end{pmatrix} \tag{7.124}
$$

$$
\begin{pmatrix} M_7 & M_8 & M_9 & M_{10} \\ & M_{11} & M_{10} & M_{12} \\ & & M_{13} & M_8 \\ \text{Sym} & & & M_{11} \end{pmatrix} \begin{pmatrix} A_3 \\ B_3 \\ A_4 \\ B_4 \end{pmatrix}
= \begin{pmatrix} -R_5 \\ -R_6 \\ -R_7 \\ -R_8 \end{pmatrix}, \quad
\begin{pmatrix} M_{14} & M_{15} & M_{16} & M_{17} \\ & M_{18} & M_{17} & M_{19} \\ & & M_{14} & M_{15} \\ \text{Sym} & & & M_{18} \end{pmatrix} \begin{pmatrix} A_5 \\ B_5 \\ A_6 \\ B_6 \end{pmatrix} = \begin{pmatrix} -R_9 \\ -R_{10} \\ 0 \\ 0 \end{pmatrix} \tag{7.125}
$$

where

$$
\begin{aligned}
&M_1 = 4\lambda_m^2 A_{11}^f + \frac{2G_{13}^c}{t_c}, \quad M_2 = 4\lambda_m^2 A_{16}^f, \quad M_3 = 4\lambda_m^2 A_{66}^f + \frac{2G_{23}^c}{t_c} \\
&M_4 = 4\mu_n^2 A_{66}^f + \frac{2G_{13}^c}{t_c}, \quad M_5 = 4\mu_n^2 A_{26}^f, \quad M_6 = 4\mu_n^2 A_{22}^f + \frac{2G_{23}^c}{t_c} \\
&M_7 = 8\lambda_m \mu_n A_{16}^f, \quad M_8 = 4\left(\lambda_m^2 A_{16}^f + \mu_n^2 A_{26}^f\right), \quad M_9 = 4\left(\lambda_m^2 A_{11}^f + \mu_n^2 A_{66}^f + \frac{G_{13}^c}{2t_c}\right) \\
&M_{10} = 4\lambda_m \mu_n \left(A_{12}^f + A_{66}^f\right), \quad M_{11} = 8\lambda_m \mu_n A_{26}^f, \quad M_{12} = 4\left(\lambda_m^2 A_{66}^f + \mu_n^2 A_{22}^f + \frac{G_{23}^c}{2t_c}\right) \\
&M_{13} = 8\lambda_m \mu_n A_{16}^f, \quad M_{14} = \lambda_m^2 A_{11}^f + \mu_n^2 A_{66}^f + \frac{G_{13}^c}{t_c}, \quad M_{15} = \lambda_m \mu_n \left(A_{12}^f + A_{66}^f\right) \\
&M_{16} = 2\lambda_m \mu_n A_{16}^f, \quad M_{17} = \lambda_m^2 A_{16}^f + \mu_n^2 A_{26}^f, \quad M_{18} = \lambda_m^2 A_{66}^f + \mu_n^2 A_{22}^f + \frac{2G_{23}^c}{t_c} \\
&M_{19} = 2\lambda_m \mu_n A_{26}^f
\end{aligned}
\tag{7.126a – s}
$$

$$
\begin{aligned}
&R_1 = \frac{\lambda_m}{2}\left(\lambda_m^2 A_{11}^f - \mu_n^2 A_{12}^f + \frac{G_{13}^c}{t_c}\right), \quad R_2 = \frac{\lambda_m}{2}\left(\lambda_m^2 A_{16}^f - \mu_n^2 A_{26}^f\right), \\
&R_3 = \frac{\mu_n}{2}\left(\mu_n^2 A_{26}^f - \lambda_m^2 A_{16}^f\right) \quad R_4 = \frac{\mu_n}{2}\left(\mu_n^2 A_{22}^f - \lambda_m^2 A_{12}^f + \frac{G_{23}^c}{t_c}\right), \\
&R_5 = -\frac{\mu_n}{2}\left(3\lambda_m^2 A_{16}^f + \mu_n^2 A_{26}^f\right) \quad R_6 = -\frac{\lambda_m}{2}\left(\lambda_m^2 A_{16}^f + 3\mu_n^2 A_{26}^f\right), \\
&R_7 = -\frac{\lambda_m}{2}\left(\lambda_m^2 A_{11}^f + \mu_n^2\left(A_{12}^f + 2A_{66}^f\right) + \frac{G_{13}^c}{t_c}\right) \\
&R_8 = -\frac{\mu_n}{2}\left(\mu_n^2 A_{22}^f + \lambda_m^2\left(A_{12}^f + 2A_{66}^f\right) + \frac{G_{13}^c}{t_c}\right), \quad R_9 = -\frac{2a\lambda_m G_{13}^c}{t_c} \\
&R_{10} = -\frac{2a\mu_n G_{23}^c}{t_c}
\end{aligned}
\tag{7.127a – j}
$$

Addressing the fifth and sixth equations of motion, Eqs. (7.102e, f), by substituting the core deformation components κ_{i3}^c $(i = 1, 2, 3)$ into Eqs. (7.99b) then substituting into the respective equations of motion provides two simultaneous equations in terms of Φ_α^c $(\alpha = 1, 2)$. The result is given as

$$
\Phi_1^c = \frac{t_c}{4}\frac{\partial u_3^d}{\partial x_1} - \frac{1}{2}u_3^d\frac{\partial u_3^d}{\partial x_1}, \quad \Phi_2^c = \frac{t_c}{4}\frac{\partial u_3^d}{\partial x_2} - \frac{1}{2}u_3^d\frac{\partial u_3^d}{\partial x_2} \tag{7.128a, b}
$$

Up to this point the entire governing system for the dynamic response has been reduced from a nine-parameter system $\left(u_i^a, u_i^d, \Phi_\alpha^c, \varphi\right)$ to a seven-parameter system $\left(u_i^a, u_i^d, \varphi\right)$ by eliminating Φ_α^c from the governing system. At this juncture, the first six equations of motion are satisfied, and all of the displacement functions are known. Also in agreement with this fulfillment are the transverse boundary conditions and the natural boundary conditions M_{nn}^a and M_{nn}^d. The remaining natural boundary conditions with respect to the stress resultants $N_{nn}^a, N_{nt}^a, N_{nn}^d, N_{nt}^d$ will be satisfied through the use of the extended-Galerkin method in an average sense. The only unknowns at this point are the modal amplitudes $w_{mn}^a(t)$ and $w_{mn}^d(t)$ which will be determined through the use of the extended-Galerkin method. By expressing the last two equations of motion, Eqs. (7.102g, h), and the unfulfilled boundary conditions in terms of displacements and retaining these expressions in Hamilton's energy functional carrying out the indicated operations and collecting identical variational coefficients of the modal amplitudes $w_{mn}^a(t)$ and $w_{mn}^d(t)$ noting that the variations of δw_{mn}^a and δw_{mn}^d are arbitrary and independent from each other and that the corresponding coefficients must vanish results in two nonlinear coupled second-order ordinary differential equations in terms of the modal amplitudes. These are solved via fourth-order Runge–Kutta numerical procedure for a system of differential equations. This system of differential equations is presented as

$$m_1\frac{\partial^2 w_{mn}^a}{\partial t^2}+C\frac{\partial w_{mn}^a}{\partial t}+C_{10}^a w_{mn}^a+C_{11}^a w_{mn}^a w_{mn}^d+C_{12}^a w_{mn}^a\left(w_{mn}^d\right)^2+C_{30}^a\left(w_{mn}^a\right)^3=\frac{q_{mn}}{2}$$

$$m_2\frac{\partial^2 w_{mn}^d}{\partial t^2}+C\frac{\partial w_{mn}^d}{\partial t}+C_{01}^d w_{mn}^d+C_{02}^d\left(w_{mn}^d\right)^2+C_{03}^d\left(w_{mn}^d\right)^3+C_{20}^d\left(w_{mn}^a\right)^2$$

$$+C_{21}^d\left(w_{mn}^a\right)^2 w_{mn}^d=\frac{q_{mn}}{2}$$

(7.129a, b)

These coupled differential equations can be used in conjunction with any transient dynamical transverse pressure pulse q_{mn}. The coefficients $C_{10}-C_{12}$, C_{30}, $C_{01}-C_{03}$, C_{20}, C_{21} which depend on the material and geometric properties are given as

$$C_{10}^a=\lambda_m^4 D_{11}^f+2\lambda_m^2\mu_n^2\left(D_{12}^f+2D_{66}^f\right)+\mu_n^4 D_{22}^f+\frac{2a\lambda_m G_{xz}^c}{t_c}(a\lambda_m-A_3)+\frac{2a\mu_n G_{yz}^c}{t_c}(a\mu_n-B_3)$$

$$C_{11}^a=-\frac{32}{9\pi^2}\Big[2\lambda_m^3 A_{11}^f A_5+\lambda_m\mu_n\left(A_{66}^f+2A_{12}^f\right)(\lambda_m B_5+\mu_n A_5)+2\mu_n^3 A_{22}^f B_5+$$
$$\lambda_m^2 A_{16}^f(3\mu_n A_6+2\lambda_m B_6)+\mu_n^2 A_{26}^f(3\lambda_m B_6+2\mu_n A_6)-\frac{2a}{t_c}\left(\lambda_m^2 G_{xz}^c+\mu_n^2 G_{yz}^c\right)\Big]+$$
$$+\frac{64a}{9\pi^2 t_c}\left[\lambda_m G_{xz}^c(A_2-3A_1)+\mu_n G_{yz}^c(B_2-3B_1)\right]$$

$$C_{12}^a=2\lambda_m^2\mu_n^2(C_1+C_2)+\frac{\lambda_m^3 A_{11}^f}{16}[3\lambda_m+8(A_3-2A_1)]+\frac{\mu_n^3 A_{22}^f}{16}[3\mu_n+8(B_3-2B_1)]+$$
$$\frac{\lambda_m\mu_n A_{12}^f}{8}[4\lambda_m(B_3-2B_1)+4\mu_n(A_3-2A_1)+3\lambda_m\mu_n]+\frac{\lambda_m^2 A_{16}^f}{2}[\mu_n(3A_4-2A_2)+]$$
$$\lambda_m(B_4-2B_2)]+\frac{\mu_n^2 A_{26}^f}{2}[\mu_n(A_4-2A_2)+\lambda_m(3B_4-2B_2)]+\frac{\lambda_m\mu_n A_{66}^f}{4}[4(\mu_n A_3+$$
$$\lambda_m B_3)-\lambda_m\mu_n]$$

$$C_{30}^{a} = 2\lambda_m^2\mu_n^2(C_1 + C_2)$$

$$C_{01}^{d} = \lambda_m^4 D_{11}^f + 2\lambda_m^2\mu_n^2\left(D_{12}^f + 2D_{66}^f\right) + \mu_n^4 D_{22}^f + 2E^c$$

$$C_{02}^{d} = -\frac{128E^c}{3\pi^2 t_c}$$

$$C_{03}^{d} = 2\lambda_m^2\mu_n^2(C_1 + C_2) + \frac{9E^c}{4t_c^2}$$

$$C_{20}^{d} = -\frac{64}{9\pi^2}\left[\lambda_m^3 A_{11}^f A_5 + \lambda_m\mu_n A_{12}^f(\lambda_m B_5 + \mu_n A_5) + \mu_n^3 A_{22}^f B_5 + \frac{\lambda_m\mu_n A_{66}^f}{2}(\lambda_m B_5 + \mu_n A_5) + \right.$$
$$\left. \frac{\lambda_m^2 A_{16}^f}{2}(3\mu_n A_6 + 2\lambda_m B_6) + \frac{\mu_n^2 A_{26}^f}{2}(2\mu_n A_6 + 3\lambda_m B_6)\right]$$

$$C_{21}^{d} = \frac{1}{2}\left[\lambda_m^3 A_{11}^f\left(A_3 - 2A_1 + \frac{3\lambda_m}{8}\right) + \lambda_m\mu_n A_{12}^f\left[\mu_n(A_3 - 2A_1) + \lambda_m(B_3 - 2B_1) + \frac{3}{4}\lambda_m\mu_n\right] + \right.$$
$$\mu_n^3 A_{22}^f\left(B_3 - 2B_1 + \frac{3}{8}\mu_n\right) + 2\lambda_m\mu_n A_{66}^f\left(\mu_n A_3 + \lambda_m B_3 - \frac{1}{4}\lambda_m\mu_n\right)$$
$$\lambda_m^2 A_{16}^f[\lambda_m(B_4 - 2B_2) + \mu_n(3A_4 - 2A_2)] + \mu_n^2 A_{26}^f[\lambda_m(3B_4 - 2B_2) + \mu_n(A_4 - 2A_2)]$$
$$\left. + \ 4\lambda_m^2\mu_n^2(C_1 + C_2)\right] \qquad (7.130\text{a} - \text{i})$$

7.9 Comments

A very comprehensive theoretical treatment of doubly curved sandwich shells considering a transversely compressible core has been presented. In addition, an application was introduced to demonstrate how the equations can be applied toward solving the many types of typical structural response problems in engineering mechanics. The presented application is not meant to be an exhaustive treatment of the typical response problems such as buckling, post-buckling, free vibration, dynamic response, thermal buckling, etc. As a matter of fact, there are many response problems that yet need to be addressed regarding these theoretical equations. The researcher and/or scientist is encouraged to further advance the field with these equations. There are a series of results by Hohe and Librescu (2003, 2006) that the reader is referred to.

References

Amabili, M. (2004). *Nonlinear vibrations and stability of shells and plates*. Cambridge University Press.

Hohe, J., & Librescu, L. (2003). A nonlinear theory for doubly curved anisotropic sandwich shells with transversely compressible core. *International Journal of Solids and Structures, 40*, 1059–1088.

Hohe, J., & Librescu, L. (2006). Dynamic buckling of flat and curved sandwich panels with transversely compressible core. *Composite Structures, 74*, 10–24.

Jones, R. M. (1999). *Mechanics of composite materials* (2nd ed.). New York, London: Taylor and Francis.

Reddy, J. N. (2004). *Mechanics of laminated composite plates and shells-theory and analysis* (2nd ed.). Boca Raton: CRC Press.

Soedel, W. (2004). *Vibrations of shells and plates* (3rd ed.). New York: Marcel Dekker, Inc.

Hause, T. (2012). Elastic structural response of anisotropic sandwich plates with a first-order compressible core impacted by a Friedlander-type shock loading. *Composite Structures, 94*, 1634–1645.

Chapter 8
Theory of Sandwich Plates and Shells with an Transversely Compressible Core – Theory Two

Abstract A further comprehensive detailed nonlinear theoretical model of asymmetric anisotropic doubly curved laminated composite sandwich plates and shells considering the case for the transversely compressible core in sufficient detail, thereby capturing the wrinkling phenomenon of the sandwich panel, is presented. The theory assumes a transversal displacement of the core in terms of a second-order power series expansion, thus theory two. The governing equations are developed by way of an energy approach known as Hamilton's principle. Finally, an application of the governing equations with regard to buckling and postbuckling is provided to demonstrate the solution approach to these equations through the use of the extended Galerkin method.

8.1 Introduction

This chapter considers many of the same theoretical aspects of a sandwich panel as was presented in Chap. 7. The principal difference resides with the transverse displacement representation for the core. In Chap. 7, a linear or first-order power series was assumed for the transverse displacement function, whereas in the present case a second-order polynomial is assumed for the transverse displacement thereby capturing the local and global instability modes on a higher-order basis. Some of the theoretical considerations include the Kirchhoff assumptions in the facings, a weak core, geometric imperfections, anisotropic laminated face sheets, and large displacements in the transverse direction. The governing equations are derived via an energy approach using Hamilton's principle. The result is 11 equations of motion and 9 boundary conditions prescribed along each edge. This is in comparison with 8 equations of motion and 8 prescribed boundary conditions required along each edge from the first theory (Chap. 7). At the conclusion of the chapter, an application of the buckling and post-buckling response is presented demonstrating how to apply and solve these theoretical equations.

T. J. Hause, *Sandwich Structures: Theory and Responses*,
https://doi.org/10.1007/978-3-030-71895-4_8

8.2 Preliminaries and Basic Assumptions

The geometrically nonlinear theory of doubly curved sandwich panels with a transversely compressible core considering a higher-order representation for the core transverse displacement function contains the same theoretical assumptions as was considered in the previous chapter along with the same terminology. Without repeating the same preliminaries and basic assumptions the reader is referred back to the Chap. 7.

8.3 Basic Equations

8.3.1 Displacement Field

The displacement field for the top and bottom facings are given as

Bottom face sheets $\left(\frac{t_c}{2} \leq x_3 \leq \frac{t_c}{2} + t_f^b\right)$

$$v_1^b = u_1^b + \left(x_3 - \frac{t_c + t_f^b}{2}\right)\psi_1^b \tag{8.1a}$$

$$v_2^b = u_2^b + \left(x_3 - \frac{t_c + t_f^b}{2}\right)\psi_2^b \tag{8.1b}$$

$$v_3^b = u_3^b \tag{8.1c}$$

Top face sheets $\left(-\frac{t_c}{2} - t_f^t \leq x_3 \leq -\frac{t_c}{2}\right)$

$$v_1^t = u_1^t + \left(x_3 + \frac{t_c + t_f^t}{2}\right)\psi_1^t \tag{8.2a}$$

$$v_2^t = u_2^t + \left(x_3 + \frac{t_c + t_f^t}{2}\right)\psi_2^t \tag{8.2b}$$

$$v_3^t = u_3^t \tag{8.2c}$$

where, in the above equations, ψ_1^b, ψ_2^b, ψ_1^t, ψ_2^t represent the shear angles, while $t$and b represent the association with the top and bottom facings, respectively. In contrast to Chap. 3 a third-order power series is assumed for the core tangential displacements while a second-order polynomial is assumed for the core transverse displacements as shown below. This is referred to as the {3, 2}-order theory (Barut et al. (2001).

Core $\left(-\frac{t_c}{2} \leq x_3 \leq \frac{t_c}{2}\right)$

$$v_1^c = u_1^c + x_3\psi_1^c + x_3^2\gamma_1^c + x_3^3\theta_1 \tag{8.3a}$$

$$v_2^c = u_2^c + x_3\psi_2^c + x_3^2\gamma_2^c + x_3^3\theta_2 \tag{8.3b}$$

$$v_3^c = u_3^c + x_3\psi_3^c + \underline{x_3^2\gamma_3} \tag{8.3c}$$

The underlined term in Eq. (8.3c) is a second-order term added to capture a more enhanced behavior of the compressible core. This is in contrast to Chap. 3 where the expression for the transverse displacement was only carried to the first order (linear). Next the interfacial continuity conditions at the interfaces need to be satisfied. This was shown and presented in Chap. 3. Applying the interfacial continuity conditions and at the same time adopting the Love–Kirchhoff assumptions results in the following displacement field for the facings and the core.

Top face sheets $\left(-\frac{t_c}{2} - t_f^t \leq x_3 \leq -\frac{t_c}{2}\right)$

$$v_1^t = u_1^a + u_1^d - \left(x_3 + \frac{t_c + t_f^t}{2}\right)\frac{\partial u_3^a}{\partial x_1} - \left(x_3 + \frac{t_c + t_f^t}{2}\right)\frac{\partial u_3^d}{\partial x_1} \tag{8.6a}$$

$$v_2^t = u_2^a + u_2^d - \left(x_3 + \frac{t_c + t_f^t}{2}\right)\frac{\partial u_3^a}{\partial x_2} - \left(x_3 + \frac{t_c + t_f^t}{2}\right)\frac{\partial u_3^d}{\partial x_2} \tag{8.6b}$$

$$v_3^t = u_3^a + u_3^d \tag{8.6c}$$

Bottom face sheets $\left(\frac{t_c}{2} \leq x_3 \leq \frac{t_c}{2} + t_f^b\right)$

$$v_1^b = u_1^a - u_1^d - \left(x_3 - \frac{t_c + t_f^b}{2}\right)\frac{\partial u_3^a}{\partial x_1} + \left(x_3 - \frac{t_c + t_f^b}{2}\right)\frac{\partial u_3^d}{\partial x_1} \tag{8.7a}$$

$$v_2^b = u_2^a - u_2^d - \left(x_3 - \frac{t_c + t_f^b}{2}\right)\frac{\partial u_3^a}{\partial x_2} + \left(x_3 - \frac{t_c + t_f^b}{2}\right)\frac{\partial u_3^d}{\partial x_2} \tag{8.7b}$$

$$v_3^b = u_3^a - u_3^d \tag{8.7c}$$

Core $\left(-\frac{t_c}{2} \le x_3 \le \frac{t_c}{2}\right)$

$$v_1^c = u_1^a - \left(\frac{t_f^t - t_f^b}{4}\right)\frac{\partial u_3^a}{\partial x_1} - \left(\frac{t_f^t + t_f^b}{4}\right)\frac{\partial u_3^d}{\partial x_1} - \frac{2x_3}{t_c}u_1^d + \frac{1}{t_c}\left(\frac{t_f^t + t_f^b}{4}\right)x_3\frac{\partial u_3^a}{\partial x_1}$$
$$+\frac{1}{t_c}\left(\frac{t_f^t - t_f^b}{2}\right)x_3\frac{\partial u_3^d}{\partial x_1} + \left(\frac{4x_3^2}{t_c^2} - 1\right)\Phi_1^c + \left(\frac{4x_3^2}{t_c^2} - 1\right)x_3\Omega_1^c \tag{8.8a}$$

$$v_2^c = u_2^a - \left(\frac{t_f^t - t_f^b}{4}\right)\frac{\partial u_3^a}{\partial x_2} - \left(\frac{t_f^t + t_f^b}{4}\right)\frac{\partial u_3^d}{\partial x_2} - \frac{2x_3}{t_c}u_2^d + \frac{1}{t_c}\left(\frac{t_f^t + t_f^b}{2}\right)x_3\frac{\partial u_3^a}{\partial x_2}$$
$$+\frac{1}{t_c}\left(\frac{t_f^t - t_f^b}{2}\right)x_3\frac{\partial u_3^d}{\partial x_2} + \left(\frac{4x_3^2}{t_c^2} - 1\right)\Phi_2^c + \left(\frac{4x_3^2}{t_c^2} - 1\right)x_3\Omega_2^c \tag{8.8b}$$

$$v_3^c = u_3^a - \frac{2x_3}{t_c}u_3^d + \left(\frac{4x_3^2}{t_c^2} - 1\right)\Phi_3^c \tag{8.8c}$$

In the above displacement equations, the average and half-difference displacement function which were introduced in Chap. 3 are

$$u_i^a = \frac{1}{2}\left(u_i^t + u_i^b\right), \qquad u_i^d = \frac{1}{2}\left(u_i^t - u_i^b\right) \tag{8.9a, b}$$

ψ_α^a, ψ_α^d represent the rotation angles; while the core quantities Φ_1^c, Φ_2^c, Φ_3^c, Ω_1^c, and Ω_2^c represent the warping functions. In the following section, a stress-free initial geometric imperfection is introduced which remains constant during deformation. These geometric imperfections as before are introduced (see Hohe and Librescu (2003)) into the Green–Lagrange strain tensor as

$$\mathring{v}_3^t = \mathring{u}_3^a + \mathring{u}_3^d \tag{8.10}$$

$$\mathring{v}_3^b = \mathring{u}_3^a - \mathring{u}_3^d \tag{8.11}$$

$$\mathring{v}_3^c = \mathring{u}_3^a - \frac{2x_3}{t_c}\mathring{u}_3^d \tag{8.12}$$

8.3.2 Nonlinear Strain–Displacement Equations

The Green–Lagrange strain tensor with the geometric imperfections, from Eqs. (2.18)–(2.23), in conjunction with the von Kármán assumptions is (Librescu and Chang 1993, Amabili 2004)

$$\gamma_{11} = \frac{\partial v_1}{\partial x_1} - \frac{v_3}{R_1} + \frac{1}{2}\left(\frac{\partial v_3}{\partial x_1}\right)^2 + \frac{\partial v_3}{\partial x_1}\frac{\partial v_3^\circ}{\partial x_1} \tag{8.13a}$$

$$\gamma_{22} = \frac{\partial v_2}{\partial x_2} - \frac{v_3}{R_2} + \frac{1}{2}\left(\frac{\partial v_3}{\partial x_2}\right)^2 + \frac{\partial v_3}{\partial x_2}\frac{\partial v_3^\circ}{\partial x_2} \tag{8.13b}$$

$$\gamma_{12} = \frac{1}{2}\left(\frac{\partial v_2}{\partial x_1} + \frac{\partial v_1}{\partial x_2}\right) + \frac{1}{2}\frac{\partial v_3}{\partial x_1}\frac{\partial v_3}{\partial x_2} + \frac{1}{2}\frac{\partial v_3}{\partial x_1}\frac{\partial v_3^\circ}{\partial x_2} + \frac{1}{2}\frac{\partial v_3}{\partial x_2}\frac{\partial v_3^\circ}{\partial x_1} \tag{8.13c}$$

$$\gamma_{13} = \frac{1}{2}\left(\frac{\partial v_1}{\partial x_3} + \frac{\partial v_3}{\partial x_1}\right) + \frac{1}{2}\frac{\partial v_3}{\partial x_1}\frac{\partial v_3}{\partial x_3} + \frac{1}{2}\frac{\partial v_3}{\partial x_1}\frac{\partial v_3^\circ}{\partial x_3} + \frac{1}{2}\frac{\partial v_3}{\partial x_3}\frac{\partial v_3^\circ}{\partial x_1} \tag{8.13d}$$

$$\gamma_{23} = \frac{1}{2}\left(\frac{\partial v_2}{\partial x_3} + \frac{\partial v_3}{\partial x_2}\right) + \frac{1}{2}\frac{\partial v_3}{\partial x_2}\frac{\partial v_3}{\partial x_3} + \frac{1}{2}\frac{\partial v_3}{\partial x_2}\frac{\partial v_3^\circ}{\partial x_3} + \frac{1}{2}\frac{\partial v_3}{\partial x_3}\frac{\partial v_3^\circ}{\partial x_2} \tag{8.13e}$$

$$\gamma_{33} = \frac{\partial v_3}{\partial x_3} + \frac{1}{2}\left(\frac{\partial v_3}{\partial x_3}\right)^2 + \frac{\partial v_3}{\partial x_3}\frac{\partial v_3^\circ}{\partial x_3} \tag{8.13f}$$

Substituting the displacement equations, Eqs. (8.6a–c), (8.7a–c), and (8.8a–c) along with the introduction of the geometric imperfections, Eqs. (8.10)–(8.12) into the nonlinear strain–displacement relationships Eqs. (8.13a–f) gives for each layer

Top face sheets $\left(-\frac{t_c}{2} - t_f^t \le x_3 \le -\frac{t_c}{2}\right)$

$$\begin{aligned}
\gamma_{11}^t &= \overline{\gamma}_{11}^a + \overline{\gamma}_{11}^d + \left(x_3 + \frac{t_c + t_f^t}{2}\right)\kappa_{11}^a + \left(x_3 + \frac{t_c + t_f^t}{2}\right)\kappa_{11}^d \\
\gamma_{22}^t &= \overline{\gamma}_{22}^a + \overline{\gamma}_{22}^d + \left(x_3 + \frac{t_c + t_f^t}{2}\right)\kappa_{22}^a + \left(x_3 + \frac{t_c + t_f^t}{2}\right)\kappa_{22}^d \\
\gamma_{12}^t &= \overline{\gamma}_{12}^a + \overline{\gamma}_{12}^d + \left(x_3 + \frac{t_c + t_f^t}{2}\right)\kappa_{12}^a + \left(x_3 + \frac{t_c + t_f^t}{2}\right)\kappa_{12}^d
\end{aligned} \tag{8.14a – c}$$

Core $\left(-\frac{t_c}{2} \le x_3 \le \frac{t_c}{2}\right)$

$$\begin{aligned}
\gamma_{13}^{c} &= \overline{\gamma}^{c}{}_{13} + x_3\kappa_{13}^{c} + \left(x_3^2 - \frac{t_c^2}{4}\right)\eta_{13} + \left(x_3^2 - \frac{t_c^2}{4}\right)x_3\upsilon_{13} \\
\gamma_{23}^{c} &= \overline{\gamma}^{c}{}_{23} + x_3\kappa_{23}^{c} + \left(x_3^2 - \frac{t_c^2}{4}\right)\eta_{23} + \left(x_3^2 - \frac{t_c^2}{4}\right)x_3\upsilon_{23} \\
\gamma_{33}^{c} &= \overline{\gamma}^{c}{}_{33} + x_3\kappa_{33}^{c} + \left(x_3^2 - \frac{t_c^2}{4}\right)\eta_{33} + \left(x_3^2 - \frac{t_c^2}{4}\right)x_3\upsilon_{33}
\end{aligned} \tag{8.15a-c}$$

Bottom face sheets $\left(\frac{t_c}{2} \le x_3 \le \frac{t_c}{2} + t_f^b\right)$

$$\begin{aligned}
\gamma_{11}^{b} &= \overline{\gamma}_{11}^{a} - \overline{\gamma}_{11}^{d} + \left(x_3 - \frac{t_c + t_f^b}{2}\right)\kappa_{11}^{a} - \left(x_3 - \frac{t_c + t_f^b}{2}\right)\kappa_{11}^{d} \\
\gamma_{22}^{b} &= \overline{\gamma}_{22}^{a} - \overline{\gamma}_{22}^{d} + \left(x_3 - \frac{t_c + t_f^b}{2}\right)\kappa_{22}^{a} - \left(x_3 - \frac{t_c + t_f^b}{2}\right)\kappa_{22}^{d} \\
\gamma_{12}^{b} &= \overline{\gamma}_{12}^{a} - \overline{\gamma}_{12}^{d} + \left(x_3 - \frac{t_c + t_f^b}{2}\right)\kappa_{12}^{a} - \left(x_3 - \frac{t_c + t_f^b}{2}\right)\kappa_{12}^{d}
\end{aligned} \tag{8.16a-c}$$

where

$$\overline{\gamma}_{\alpha\beta}^{a} = \frac{1}{2}\left(\overline{\gamma}_{\alpha\beta}^{t} + \overline{\gamma}_{\alpha\beta}^{b}\right), \quad \overline{\gamma}_{\alpha\beta}^{d} = \frac{1}{2}\left(\overline{\gamma}_{\alpha\beta}^{t} - \overline{\gamma}_{\alpha\beta}^{b}\right) \tag{8.17a}$$

$$\kappa_{\alpha\beta}^{a} = \frac{1}{2}\left(\kappa_{\alpha\beta}^{t} + \kappa_{\alpha\beta}^{b}\right), \quad \kappa_{\alpha\beta}^{d} = \frac{1}{2}\left(\kappa_{\alpha\beta}^{t} - \kappa_{\alpha\beta}^{b}\right) \tag{8.17b}$$

In the above strain displacement equations, $\overline{\gamma}_{ij}^{(a,d)}$ are the in-plane average and the half difference of the tangential strains of the top and bottom facings, while $\kappa_{\alpha\beta}^{(a,d)}$ are the average and half difference bending strains of the top and bottom facings. For the core, $\overline{\gamma}_{i3}^{c}$ and κ_{i3}^{c} are the tangential and bending strains, respectively, while η_{i3}^{c} and υ_{i3}^{c} are higher-order strains. The expressions for these strains are given as

$$\overline{\gamma}_{11}^{a} = \frac{\partial u_1^a}{\partial x_1} - \frac{u_3^a}{R_1} + \frac{1}{2}\left(\frac{\partial u_3^a}{\partial x_1}\right)^2 + \frac{\partial u_3^a}{\partial x_1}\frac{\partial \mathring{u}_3^a}{\partial x_1} + \frac{1}{2}\left(\frac{\partial u_3^d}{\partial x_1}\right)^2 + \frac{\partial u_3^a}{\partial x_1}\frac{\partial \mathring{u}_3^d}{\partial x_1} \tag{8.18}$$

$$\overline{\gamma}_{22}^{a} = \frac{\partial u_2^a}{\partial x_2} - \frac{u_3^a}{R_2} + \frac{1}{2}\left(\frac{\partial u_3^a}{\partial x_2}\right)^2 + \frac{\partial u_3^a}{\partial x_2}\frac{\partial \mathring{u}_3^a}{\partial x_2} + \frac{1}{2}\left(\frac{\partial u_3^d}{\partial x_2}\right)^2 + \frac{\partial u_3^a}{\partial x_2}\frac{\partial \mathring{u}_3^d}{\partial x_2} \tag{8.19}$$

$$\overline{\gamma}_{12}^{a} = \frac{1}{2}\frac{\partial u_1^a}{\partial x_2} + \frac{1}{2}\frac{\partial u_2^a}{\partial x_1} + \frac{1}{2}\frac{\partial u_3^a}{\partial x_1}\frac{\partial u_3^a}{\partial x_2} + \frac{1}{2}\frac{\partial u_3^d}{\partial x_1}\frac{\partial u_3^d}{\partial x_2} + \frac{1}{2}\frac{\partial u_3^a}{\partial x_1}\frac{\partial \mathring{u}_3^a}{\partial x_2} + \frac{1}{2}\frac{\partial \mathring{u}_3^a}{\partial x_1}\frac{\partial u_3^a}{\partial x_2} + \frac{1}{2}\frac{\partial u_3^d}{\partial x_1}\frac{\partial \mathring{u}_3^d}{\partial x_2} + \frac{1}{2}\frac{\partial \mathring{u}_3^d}{\partial x_1}\frac{\partial u_3^d}{\partial x_2} \tag{8.20}$$

$$\overline{\gamma}_{11}^{d} = \frac{\partial u_1^d}{\partial x_1} - \frac{u_3^d}{R_1} + \frac{\partial u_3^a}{\partial x_1}\frac{\partial u_3^d}{\partial x_1} + \frac{\partial u_3^a}{\partial x_1}\frac{\partial \mathring{u}_3^d}{\partial x_1} + \frac{\partial \mathring{u}_3^a}{\partial x_1}\frac{\partial u_3^d}{\partial x_1} \tag{8.21}$$

$$\overline{\gamma}_{22}^{d} = \frac{\partial u_2^d}{\partial x_2} - \frac{u_3^d}{R_2} + \frac{\partial u_3^a}{\partial x_2}\frac{\partial u_3^d}{\partial x_2} + \frac{\partial u_3^a}{\partial x_2}\frac{\partial \mathring{u}_3^d}{\partial x_2} + \frac{\partial \mathring{u}_3^a}{\partial x_2}\frac{\partial u_3^d}{\partial x_2} \tag{8.22}$$

$$\overline{\gamma}_{12}^{d} = \frac{1}{2}\frac{\partial u_1^d}{\partial x_2} + \frac{1}{2}\frac{\partial u_2^d}{\partial x_1} + \frac{1}{2}\frac{\partial u_3^a}{\partial x_1}\frac{\partial u_3^d}{\partial x_2} + \frac{1}{2}\frac{\partial u_3^a}{\partial x_2}\frac{\partial u_3^d}{\partial x_1} + \frac{1}{2}\frac{\partial u_3^a}{\partial x_1}\frac{\partial \mathring{u}_3^d}{\partial x_2} + \frac{1}{2}\frac{\partial \mathring{u}_3^a}{\partial x_1}\frac{\partial u_3^d}{\partial x_2} + \frac{1}{2}\frac{\partial u_3^a}{\partial x_2}\frac{\partial \mathring{u}_3^d}{\partial x_1} + \frac{1}{2}\frac{\partial \mathring{u}_3^a}{\partial x_2}\frac{\partial u_3^d}{\partial x_1} \tag{8.23}$$

$$\kappa_{11}^{a} = -\frac{\partial^2 u_3^a}{\partial x_1^2} \tag{8.24}$$

$$\kappa_{22}^{a} = -\frac{\partial^2 u_3^a}{\partial x_2^2} \tag{8.25}$$

$$\kappa_{12}^{a} = -\frac{\partial^2 u_3^a}{\partial x_1 \partial x_2} \tag{8.26}$$

$$\kappa_{11}^{d} = -\frac{\partial^2 u_3^d}{\partial x_1^2} \tag{8.27}$$

$$\kappa_{22}^{d} = -\frac{\partial^2 u_3^d}{\partial x_2^2} \tag{8.28}$$

$$\kappa_{12}^{d} = -\frac{\partial^2 u_3^d}{\partial x_1 \partial x_2} \tag{8.29}$$

$$\overline{\gamma}_{33} = -\frac{2}{t_c}u_3^d + \frac{2}{t_c^2}\left(u_3^d\right)^2 + \frac{4}{t_c^2}u_3^d\left(u_3^\circ\right)^d + \frac{8}{t_c^2}\left(\Phi_3^c\right)^2 \tag{8.30}$$

$$\overline{\gamma}_{13}^{c} = -\frac{u_1^d}{t_c} + \left(\frac{1}{2} + \frac{1}{t_c}\left(\frac{t_f^t + t_f^b}{4}\right)\right)\frac{\partial u_3^a}{\partial x_1} + \frac{1}{t_c}\left(\frac{t_f^t - t_f^b}{4}\right)\frac{\partial u_3^d}{\partial x_1} + \frac{\Omega_1^c}{t_c} - \frac{u_3^d}{t_c}\frac{\partial u_3^a}{\partial x_1} - \frac{\mathring{u}_3^a}{t_c}\frac{\partial u_3^a}{\partial x_1} - \frac{u_3^d}{t_c}\frac{\partial \mathring{u}_3^a}{\partial x_1} - \frac{2\Phi_3^c}{t_c}\frac{\partial u_3^d}{\partial x_1} - \frac{2}{t_c}\Phi_3^c\frac{\partial \mathring{u}_3^d}{\partial x_1} \tag{8.31}$$

$$\overline{\gamma}_{23}^{c} = -\frac{u_2^d}{t_c} + \left(\frac{1}{2} + \frac{1}{t_c}\left(\frac{t_f^t + t_f^b}{4}\right)\right)\frac{\partial u_3^a}{\partial x_2} + \frac{1}{t_c}\left(\frac{t_f^t - t_f^b}{4}\right)\frac{\partial u_3^d}{\partial x_2} + \frac{\Omega_2^c}{t_c} - \frac{u_3^d}{t_c}\frac{\partial u_3^a}{\partial x_2} - \frac{\mathring{u}_3^a}{t_c}\frac{\partial u_3^a}{\partial x_2} - \frac{u_3^d}{t_c}\frac{\partial \mathring{u}_3^a}{\partial x_2} - \frac{2\Phi_3^c}{t_c}\frac{\partial u_3^d}{\partial x_2} - \frac{2}{t_c}\Phi_3^c\frac{\partial \mathring{u}_3^d}{\partial x_2} \tag{8.32}$$

$$\kappa_{33}^{c} = \frac{8}{t_c^2}\Phi_3^c - \frac{16}{t_c^3}u_3^d\Phi_3^c - \frac{16}{t_c^3}\Phi_3^c\left(\mathring{u}_3\right)^d \tag{8.33}$$

$$\kappa_{13}^{c} = \frac{4}{t_c^2}\Phi_1^c - \frac{1}{t_c}\frac{\partial u_3^d}{\partial x_1} + \frac{4}{t_c^2}\Phi_3^c\frac{\partial u_3^a}{\partial x_1} + \frac{2}{t_c^2}u_3^d\frac{\partial u_3^d}{\partial x_1} + \frac{2}{t_c^2}\left(\mathring{u}_3\right)^d\frac{\partial u_3^d}{\partial x_1} + \frac{2}{t_c^2}u_3^d\frac{\partial \mathring{u}_3^d}{\partial x_1} + \frac{4}{t_c^2}\Phi_3^c\frac{\partial \mathring{u}_3^a}{\partial x_1} \tag{8.34}$$

$$\kappa_{23}^{c} = \frac{4}{t_c^2}\Phi_2^c - \frac{1}{t_c}\frac{\partial u_3^d}{\partial x_2} + \frac{4}{t_c^2}\Phi_3^c\frac{\partial u_3^a}{\partial x_2} + \frac{2}{t_c^2}u_3^d\frac{\partial u_3^d}{\partial x_2} + \frac{2}{t_c^2}\left(\mathring{u}_3\right)^d\frac{\partial u_3^d}{\partial x_2} + \frac{2}{t_c^2}u_3^d\frac{\partial \mathring{u}_3^d}{\partial x_2} + \frac{4}{t_c^2}\Phi_3^c\frac{\partial \mathring{u}_3^a}{\partial x_2} \tag{8.35}$$

$$\eta_{33}^{c} = \frac{32}{t_c^4}\left(\Phi_3^c\right)^2 \tag{8.36}$$

$$\eta_{13}^{c} = \frac{6}{t_c^3}\Omega_1^c - \frac{8}{t_c^3}\Phi_3^c\frac{\partial u_3^d}{\partial x_1} - \frac{8}{t_c^3}\Phi_3^c\frac{\partial \mathring{u}_3^d}{\partial x_1} + \frac{2}{t_c^2}\frac{\partial \Phi_3^c}{\partial x_1} - \frac{4}{t_c^3}u_3^d\frac{\partial \Phi_3^c}{\partial x_1} - \frac{4}{t_c^3}\left(\mathring{u}_3\right)^d\frac{\partial \Phi_3^c}{\partial x_1} \tag{8.37}$$

$$\eta_{23}^{c} = \frac{6}{t_c^3}\Omega_2^c - \frac{8}{t_c^3}\Phi_3^c\frac{\partial u_3^d}{\partial x_2} - \frac{8}{t_c^3}\Phi_3^c\frac{\partial \mathring{u}_3^d}{\partial x_2} + \frac{2}{t_c^2}\frac{\partial \Phi_3^c}{\partial x_2} - \frac{4}{t_c^3}u_3^d\frac{\partial \Phi_3^c}{\partial x_2} - \frac{4}{t_c^3}\left(\mathring{u}_3\right)^d\frac{\partial \Phi_3^c}{\partial x_2} \tag{8.38}$$

$$\upsilon_{33}^{c} = 0 \tag{8.39}$$

$$\upsilon_{13}^{c} = \frac{16}{t_c^4}\Phi_3^c\frac{\partial \Phi_3^c}{\partial x_1} \tag{8.40}$$

$$\upsilon_{23}^{c} = \frac{16}{t_c^4}\Phi_3^c\frac{\partial \Phi_3^c}{\partial x_2} \tag{8.41}$$

8.3.3 Constitutive Equations

The top and bottom facings are considered to be constructed from unidirectional fiber-reinforced anisotropic laminated composites. The stress–strain relationships per lamina (see Reddy (2004) and Jones (1999)) of the facings is repeated here as

Top face sheets $\left(-\frac{t_c}{2} - t_f^t \leq x_3 \leq -\frac{t_c}{2}\right)$

$$\begin{pmatrix} \tau_{11}^t \\ \tau_{22}^t \\ \tau_{12}^t \end{pmatrix}_k = \begin{bmatrix} \widehat{Q}_{11}^t & \widehat{Q}_{11}^t & \widehat{Q}_{11}^t \\ & \widehat{Q}_{22}^t & \widehat{Q}_{26}^t \\ \text{Sym} & & \widehat{Q}_{66}^t \end{bmatrix}_k \begin{pmatrix} \gamma_{11}^t \\ \gamma_{22}^t \\ \gamma_{12}^t \end{pmatrix}_k - \begin{Bmatrix} \widehat{\lambda}_{11}^t \\ \widehat{\lambda}_{22}^t \\ \widehat{\lambda}_{12}^t \end{Bmatrix}_k \Delta T - \begin{Bmatrix} \widehat{\mu}_{11}^t \\ \widehat{\mu}_{22}^t \\ \widehat{\mu}_{12}^t \end{Bmatrix}_k \Delta M \quad (8.42)$$

where k represents the kth lamina in the facing and $\widehat{Q}_{ij}$ for $i, j = (1, 2, 6)$ are the transformed plane-stress reduced stiffness measures. These were given in Chap. 2.

Bottom face sheets $\left(\frac{t_c}{2} \leq x_3 \leq \frac{t_c}{2} + t_f^b\right)$

$$\begin{pmatrix} \tau_{11}^b \\ \tau_{22}^b \\ \tau_{12}^b \end{pmatrix}_k = \begin{bmatrix} \widehat{Q}_{11}^b & \widehat{Q}_{11}^b & \widehat{Q}_{11}^b \\ & \widehat{Q}_{22}^b & \widehat{Q}_{26}^b \\ \text{Sym} & & \widehat{Q}_{66}^b \end{bmatrix}_k \begin{pmatrix} \gamma_{11}^b \\ \gamma_{22}^b \\ \gamma_{12}^b \end{pmatrix}_k - \begin{Bmatrix} \widehat{\lambda}_{11}^b \\ \widehat{\lambda}_{22}^b \\ \widehat{\lambda}_{12}^b \end{Bmatrix}_k \Delta T - \begin{Bmatrix} \widehat{\mu}_{11}^b \\ \widehat{\mu}_{22}^b \\ \widehat{\mu}_{12}^b \end{Bmatrix}_k \Delta M \quad (8.43)$$

The stress–strain relationships for the orthotropic core with the geometrical and material axes coincident are expressed as

Core $\left(-\frac{t_c}{2} \leq x_3 \leq \frac{t_c}{2}\right)$

$$\begin{pmatrix} \tau_{33}^c \\ \tau_{23}^c \\ \tau_{13}^c \end{pmatrix} = \begin{bmatrix} Q_{33}^c & 0 & 0 \\ 0 & Q_{44}^c & 0 \\ 0 & 0 & Q_{55}^c \end{bmatrix} \begin{pmatrix} \gamma_{33}^c \\ \gamma_{23}^c \\ \gamma_{13}^c \end{pmatrix} \quad (8.44)$$

The core transverse and normal moduli are given as

$$Q_{33}^c = E^c, \quad Q_{44}^c = G_{23}^c, \quad Q_{55}^c = G_{13}^c \quad (8.45)$$

where E^c is the young's modulus for the core, and G_{13}^c, G_{23}^c are the shear moduli for the core.

8.4 Hamilton's Principle

An energy approach using Hamilton's Principle is used to derive the equations of motion. Letting U represent the strain energy, W represent the work done by external forces, and T represent the kinetic energy, Hamilton's Principle (see Soedel 2004) is expressed as

$$\delta J = \int_{t_0}^{t_1} (\delta U - \delta W - \delta T) dt = 0 \quad (8.46)$$

8.4.1 Strain Energy

Assuming a weak compressible core where the core carries only the transverse shear stresses, the variation in the strain energy is given by

$$\delta U = \int_A \left\{ \int_{\frac{t_c}{2}}^{\frac{t_c}{2}+t_f^b} (\tau_{11}^b \delta\gamma_{11}^b + \tau_{22}^b \delta\gamma_{22}^b + 2\tau_{12}^b \delta\gamma_{12}^b) dx_3 + \int_{-\frac{t_c}{2}}^{\frac{t_c}{2}} (2\tau_{13}^c \delta\gamma_{13}^c + 2\tau_{23}^c \delta\gamma_{23}^c + \tau_{33}^c \delta\gamma_{33}^c) dx_3 + \int_{-\frac{t_c}{2}-t_f^t}^{-\frac{t_c}{2}} (\tau_{11}^t \delta\gamma_{11}^t + \tau_{22}^t \delta\gamma_{22}^t + 2\tau_{12}^t \delta\gamma_{12}^t) dx_3 \right\} dA \tag{8.47}$$

where τ_{ij} are the tensorial components of the second Piola–Kirchhoff stress tensor, while A is attributed to the planar area of the sandwich shell. Substituting in the expressions for the strain relationships, Eqs. (8.14a–c)–(8.16a–c) results in

$$\begin{aligned}
\delta U = \int_A \Bigg\{ \int_{\frac{t_c}{2}}^{\frac{t_c}{2}+t_f^b} \Bigg[& \tau_{11}^b \left(\delta\overline{\gamma}_{11}^a - \delta\overline{\gamma}_{11}^d + \left(x_3 - \frac{t_c + t_f^b}{2} \right) \delta\kappa_{11}^a - \left(x_3 - \frac{t_c + t_f^b}{2} \right) \delta\kappa_{11}^d \right) \\
& + \tau_{22}^b \left(\delta\overline{\gamma}_{22}^a - \delta\overline{\gamma}_{22}^d + \left(x_3 - \frac{t_c + t_f^b}{2} \right) \delta\kappa_{22}^a - \left(x_3 - \frac{t_c + t_f^b}{2} \right) \delta\kappa_{22}^d \right) \\
& + 2\tau_{12}^b \left(\delta\overline{\gamma}_{12}^a - \delta\overline{\gamma}_{12}^d + \left(x_3 - \frac{t_c + t_f^b}{2} \right) \delta\kappa_{12}^a - \left(x_3 - \frac{t_c + t_f^b}{2} \right) \delta\kappa_{12}^d \right) \Bigg] dx_3 \\
& + \int_{-\frac{t_c}{2}}^{\frac{t_c}{2}} \Bigg[\tau_{33}^c \left(\delta\overline{\gamma}_{33}^c + x_3 \delta\kappa_{33}^c + \left(x_3^2 - \frac{t_c^2}{4} \right) \delta\eta_{33}^c \right) + 2\tau_{13}^c \Big(\delta\overline{\gamma}_{13}^c + x_3 \delta\kappa_{13}^c \\
& + \left(x_3^2 - \frac{t_c^2}{4} \right) \delta\eta_{13}^c + \left(x_3^2 - \frac{t_c^2}{4} \right) x_3 \delta\upsilon_{13}^c \Big) + 2\tau_{23}^c \Big(\delta\overline{\gamma}_{23}^c + x_3 \delta\kappa_{23}^c + \left(x_3^2 - \frac{t_c^2}{4} \right) \delta\eta_{23}^c \\
& + \left(x_3^2 - \frac{t_c^2}{4} \right) x_3 \delta\upsilon_{23}^c \Big) \Bigg] dx_3 + \int_{-\frac{t_c}{2}-t_f^t}^{-\frac{t_c}{2}} \Bigg[\tau_{11}^t \left(\delta\overline{\gamma}_{11}^a + \delta\overline{\gamma}_{11}^d + \left(x_3 + \frac{t_c + t_f^t}{2} \right) \delta\kappa_{11}^a + \right. \\
& \left. + \left(x_3 + \frac{t_c + t_f^b}{2} \right) \delta\kappa_{11}^d \right) + \tau_{22}^t \left(\delta\overline{\gamma}_{22}^a + \delta\overline{\gamma}_{22}^d + \left(x_3 + \frac{t_c + t_f^b}{2} \right) \delta\kappa_{22}^a \right. \\
& \left. + \left(x_3 + \frac{t_c + t_f^b}{2} \right) \delta\kappa_{22}^d \right) + 2\tau_{12}^t \left(\delta\overline{\gamma}_{12}^a + \delta\overline{\gamma}_{12}^d + \left(x_3 + \frac{t_c + t_f^b}{2} \right) \delta\kappa_{12}^a \right. \\
& \left. + \left(x_3 + \frac{t_c + t_f^b}{2} \right) \delta\kappa_{12}^d \right) \Bigg] dx_3 \Bigg\} dA
\end{aligned} \tag{8.48}$$

which can be written in terms of the local stress resultants and couples defined below in Eq. (8.50)–(8.52) as

$$\begin{aligned}\delta U = \int_A \{&N^t_{11}\delta\overline{\gamma}^a_{11} + N^t_{11}\delta\overline{\gamma}^d_{11} + M^t_{11}\delta\kappa^a_{11} + M^t_{11}\delta\kappa^d_{11} + N^t_{22}\delta\overline{\gamma}^a_{22} + N^t_{22}\delta\overline{\gamma}^d_{22}\\ &+M^t_{22}\delta\kappa^a_{22} + M^t_{22}\delta\kappa^d_{22} + 2N^t_{12}\delta\overline{\gamma}^a_{12} + 2N^t_{12}\delta\overline{\gamma}^d_{12} + 2M^t_{12}\delta\kappa^a_{12} + 2M^t_{12}\delta\kappa^d_{12}\\ &+N^c_{33}\delta\overline{\gamma}^c_{33} + M^c_{33}\delta\kappa^c_{33} + L^c_{33}\delta\eta^c_{33} + 2N^c_{13}\delta\overline{\gamma}^c_{13} + 2M^c_{13}\delta\kappa^c_{13} + 2L^c_{13}\delta\eta^c_{13}\\ &+2K^c_{13}\delta\upsilon^c_{13} + 2N^c_{23}\delta\overline{\gamma}^c_{23} + 2M^c_{23}\delta\kappa^c_{23} + 2L^c_{23}\delta\eta^c_{23} + 2K^c_{23}\delta\upsilon^c_{23} + N^b_{11}\delta\overline{\gamma}^a_{11}\\ &-N^b_{11}\delta\overline{\gamma}^d_{11} + M^b_{11}\delta\kappa^a_{11} - M^b_{11}\delta\kappa^d_{11} + N^b_{22}\delta\overline{\gamma}^a_{22} - N^b_{22}\delta\overline{\gamma}^d_{22} + M^b_{22}\delta\kappa^a_{22}\\ &-M^b_{22}\delta\kappa^d_{22} + 2N^b_{12}\delta\overline{\gamma}^a_{12} - 2N^b_{12}\delta\overline{\gamma}^d_{12} + 2M^b_{12}\delta\kappa^a_{12} - 2M^b_{12}\delta\kappa^d_{12}\}dA\end{aligned} \tag{8.49}$$

where the local stress resultants and stress couples are defined as

$$\left\{N^t_{\alpha\beta}, M^t_{\alpha\beta}\right\} = \int_{-\frac{t_c}{2}-t^t_f}^{-\frac{t_c}{2}} \tau^t_{\alpha\beta}\left\{1, \left(x_3 + \frac{t_c + t^t_f}{2}\right)\right\}dx_3 \tag{8.50}$$

$$\left\{N^b_{\alpha\beta}, M^b_{\alpha\beta}\right\} = \int_{\frac{t_c}{2}}^{\frac{t_c}{2}+t^b_f} \tau^b_{\alpha\beta}\left\{1, \left(x_3 - \frac{t_c + t^b_f}{2}\right)\right\}dx_3 \tag{8.51}$$

$$\left\{N^c_{i3}, M^c_{i3}, L^c_{i3}, K^c_{i3}\right\} = \int_{-\frac{t_c}{2}}^{\frac{t_c}{2}} \tau^c_{i3}\left(1, x_3, \left(x_3^2 - \frac{t_c^2}{4}\right), \left(x_3^2 - \frac{t_c^2}{4}\right)x_3\right)dx_3 \tag{8.52}$$

Combining like terms which contain the same variational displacements results in

$$\begin{aligned}\delta U = \int_A \{&(N^t_{11} + N^b_{11})\delta\overline{\gamma}^a_{11} + (N^t_{11} - N^b_{11})\delta\overline{\gamma}^d_{11} + (M^t_{11} + M^b_{11})\delta\kappa^a_{11} + (M^t_{11} - M^b_{11})\delta\kappa^d_{11} +\\ &(N^t_{22} + N^b_{22})\delta\overline{\gamma}^a_{22} + (N^t_{22} - N^b_{22})\delta\overline{\gamma}^d_{22} + (M^t_{22} + M^b_{22})\delta\kappa^a_{22} + (M^t_{22} - M^b_{22})\delta\kappa^d_{22} +\\ &2(N^t_{12} + N^b_{12})\delta\overline{\gamma}^a_{12} + 2(N^t_{12} - N^b_{12})\delta\overline{\gamma}^d_{12} + 2(M^t_{12} + M^b_{12})\delta\kappa^a_{12} + 2(M^t_{12} - M^b_{12})\delta\kappa^d_{12}\\ &+N^c_{33}\delta\overline{\gamma}^c_{33} + M^c_{33}\delta\kappa^c_{33} + L^c_{33}\delta\eta^c_{33} + 2N^c_{13}\delta\overline{\gamma}^c_{13} + 2M^c_{13}\delta\kappa^c_{13} + 2L^c_{13}\delta\eta^c_{13} + 2K^c_{13}\delta\upsilon^c_{13} +\\ &2N^c_{23}\delta\overline{\gamma}^c_{23} + 2M^c_{23}\delta\kappa^c_{23} + 2L^c_{23}\delta\eta^c_{23} + 2K^c_{23}\delta\upsilon^c_{23}\}dA\end{aligned} \tag{8.53}$$

Utilizing the definition of global stress resultants and global stress couples as defined below in Eqs. (8.55a and 8.55b), provides the variation in the strain energy as

$$\begin{aligned}\delta U=\int_A\big[&2N_{11}^a\delta\overline{\gamma}_{11}^a+2N_{11}^d\delta\overline{\gamma}_{11}^d+2M_{11}^a\delta\kappa_{11}^a+2M_{11}^d\delta\kappa_{11}^d+2N_{22}^a\delta\overline{\gamma}_{22}^a+2N_{22}^d\delta\overline{\gamma}_{22}^d+\\&2M_{22}^a\delta\kappa_{22}^a+2M_{22}^d\delta\kappa_{22}^d+4N_{12}^a\delta\overline{\gamma}_{12}^a+4N_{12}^d\delta\overline{\gamma}_{12}^d+4M_{12}^a\delta\kappa_{12}^a+4M_{12}^d\delta\kappa_{12}^d+\\&N_{33}^c\delta\overline{\gamma}_{33}^c+M_{33}^c\delta\kappa_{33}^c+L_{33}^c\delta\eta_{33}^c+2N_{13}^c\delta\overline{\gamma}_{13}^c+2M_{13}^c\delta\kappa_{13}^c+2L_{13}^c\delta\eta_{13}^c+2K_{13}^c\delta\upsilon_{13}^c+\\&2N_{23}^c\delta\overline{\gamma}_{23}^c+2M_{23}^c\delta\kappa_{23}^c+2L_{23}^c\delta\eta_{23}^c+2K_{23}^c\delta\upsilon_{23}^c\big]dA\end{aligned} \tag{8.54}$$

where the global stress resultants and global stress couples are defined as

$$\left(N_{\alpha\beta}^a,M_{\alpha\beta}^a\right)=\frac{1}{2}\left\{\left(N_{\alpha\beta}^t+N_{\alpha\beta}^b\right),\left(M_{\alpha\beta}^t+M_{\alpha\beta}^b\right)\right\} \tag{8.55a}$$

$$\left(N_{\alpha\beta}^d,M_{\alpha\beta}^d\right)=\frac{1}{2}\left\{\left(N_{\alpha\beta}^t-N_{\alpha\beta}^b\right),\left(M_{\alpha\beta}^t-M_{\alpha\beta}^b\right)\right\} \tag{8.55b}$$

Substituting in the strain–displacement equations, Eqs. (8.18)–(8.40) into Eq. (8.54) results in

$$\begin{aligned}\delta U=\int_A\Bigg[&2N_{11}^a\left\{\frac{\partial(\delta u_1^a)}{\partial x_1}-\frac{\delta u_3^a}{R_1}+\frac{\partial u_3^a}{\partial x_1}\frac{\partial(\delta u_3^a)}{\partial x_1}+\frac{\partial(\delta u_3^a)}{\partial x_1}\frac{\partial \mathring{u}_3^a}{\partial x_1}+\frac{\partial u_3^d}{\partial x_1}\frac{\partial(\delta u_3^d)}{\partial x_1}\right.\\&\left.+\frac{\partial(\delta u_3^a)}{\partial x_1}\frac{\partial \mathring{u}_3^d}{\partial x_1}\right\}+2N_{11}^d\left\{\frac{\partial(\delta u_1^d)}{\partial x_1}-\frac{\delta u_3^d}{R_1}+\frac{\partial u_3^d}{\partial x_1}\frac{\delta(u_3^a)}{\partial x_1}+\frac{\partial u_3^a}{\partial x_1}\frac{\partial(\delta u_3^d)}{\partial x_1}\right\}\\&+\frac{\partial(\delta u_3^a)}{\partial x_1}\frac{\partial \mathring{u}_3^d}{\partial x_1}+\frac{\partial(\delta u_3^d)}{\partial x_1}\frac{\partial \mathring{u}_3^a}{\partial x_1}-2M_{11}^a\frac{\partial^2(\delta u_3^a)}{\partial x_1^2}-2M_{11}^d\frac{\partial^2(\delta u_3^d)}{\partial x_1^2}\\&+2N_{22}^a\left\{\frac{\partial(\delta u_2^a)}{\partial x_2}-\frac{\delta u_3^a}{R_2}+\frac{\partial u_3^a}{\partial x_2}\frac{\partial(\delta u_3^a)}{\partial x_2}+\frac{\partial(\delta u_3^a)}{\partial x_2}\frac{\partial \mathring{u}_3^a}{\partial x_2}+\frac{\partial u_3^d}{\partial x_2}\frac{\partial(\delta u_3^d)}{\partial x_2}\right.\\&\left.+\frac{\partial(\delta u_3^a)}{\partial x_2}\frac{\partial \mathring{u}_3^d}{\partial x_2}\right\}+2N_{22}^d\left\{\frac{\partial(\delta u_2^d)}{\partial x_2}-\frac{\delta u_3^d}{R_2}+\frac{\partial u_3^d}{\partial x_2}\frac{\partial(\delta u_3^a)}{\partial x_2}+\frac{\partial u_3^a}{\partial x_2}\frac{\partial(\delta u_3^d)}{\partial x_2}+\right.\\&\left.+\frac{\partial(\delta u_3^a)}{\partial x_2}\frac{\partial \mathring{u}_3^d}{\partial x_2}+\frac{\partial(\delta u_3^d)}{\partial x_2}\frac{\partial \mathring{u}_3^a}{\partial x_2}\right\}-2M_{22}^a\frac{\partial^2(\delta u_3^a)}{\partial x_2^2}-2M_{22}^d\frac{\partial^2(\delta u_3^d)}{\partial x_2^2}\\&+4N_{12}^a\left\{\frac{\partial(\delta u_1^a)}{\partial x_2}++\frac{\partial(\delta u_2^a)}{\partial x_1}+\frac{\partial(\delta u_3^a)}{\partial x_1}\frac{\partial u_3^a}{\partial x_2}+\frac{\partial u_3^a}{\partial x_1}\frac{\partial(\delta u_3^a)}{\partial x_2}+\frac{\partial(\delta u_3^d)}{\partial x_1}\frac{\partial u_3^d}{\partial x_2}\right.\\&\left.+\frac{\partial u_3^d}{\partial x_1}\frac{\partial(\delta u_3^d)}{\partial x_2}+\frac{\partial(\delta u_3^a)}{\partial x_1}\frac{\partial \mathring{u}_3^a}{\partial x_2}+\frac{\partial \mathring{u}_3^a}{\partial x_1}\frac{\partial(\delta u_3^a)}{\partial x_2}+\frac{\partial(\delta u_3^d)}{\partial x_1}\frac{\partial \mathring{u}_3^a}{\partial x_2}+\frac{\partial \mathring{u}_3^d}{\partial x_1}\frac{\partial(\delta u_3^d)}{\partial x_2}\right\}\end{aligned}$$

$$+4N^d_{12}\left\{ \frac{\partial\left(\delta u_1^d\right)}{\partial x_2} + \frac{\partial\left(\delta u_2^d\right)}{\partial x_1} + \frac{\partial\left(\delta u_3^a\right)}{\partial x_1}\frac{\partial u_3^d}{\partial x_2} + \frac{\partial u_3^a}{\partial x_1}\frac{\partial\left(\delta u_3^d\right)}{\partial x_2} + \frac{\partial\left(\delta u_3^a\right)}{\partial x_2}\frac{\partial u_3^d}{\partial x_1}\right.$$

$$\left. + \frac{\partial u_3^a}{\partial x_2}\frac{\partial\left(\delta u_3^d\right)}{\partial x_1} + \frac{\partial\left(\delta u_3^a\right)}{\partial x_1}\frac{\partial \mathring{u}_3^d}{\partial x_2} + \frac{\partial\left(\delta u_3^d\right)}{\partial x_2}\frac{\partial\mathring{u}_3^a}{\partial x_1} + \frac{\partial\mathring{u}_3^d}{\partial x_1}\frac{\partial\left(\delta u_3^a\right)}{\partial x_2}\right\}$$

$$-4M^a_{12}\frac{\partial^2\left(\delta u_3^a\right)}{\partial x_1\partial x_2} - 4M^d_{12}\frac{\partial^2\left(\delta u_3^d\right)}{\partial x_1 \partial x_2} + N^c_{33}\left\{-\frac{2}{t_c}\delta u_3^d + \frac{4}{t_c^2}u_3^d\delta u_3^d + \right.$$

$$\left. +\frac{4}{t_c^2}\mathring{u}_3^d\delta u_3^d + \frac{16}{t_c^2}\Phi_3^c\delta\Phi_3^c\right\} + M^c_{33}\left\{\frac{8}{t_c^2}\delta\Phi_3^c - \frac{16}{t_c^3}\Phi_3^c\delta u_3^d - \frac{16}{t_c^3}u_3^d\delta\Phi_3^c\right.$$

$$\left. -\frac{16}{t_c^3}\mathring{u}_3^d\delta\Phi_3^c\right\} + L^c_{33}\left\{\frac{16}{t_c^4}\Phi_3^c\delta\Phi_3^c\right\} + 2N^c_{13}\left\{-\frac{1}{t_c}\delta u_1^d + \left(\frac{1}{2} + \frac{1}{t_c}\left(\frac{t_f^t+t_f^b}{4}\right)\right)\right.$$

$$\frac{\partial\left(\delta u_3^a\right)}{\partial x_1} + \frac{1}{t_c}\left(\frac{t_f^t - t_f^b}{4}\right)\frac{\partial\left(\delta u_3^d\right)}{\partial x_1} + \frac{1}{t_c}\delta\Omega_1^c - \frac{1}{t_c}\frac{\partial u_3^a}{\partial x_1}\delta u_3^d - \frac{1}{t_c}u_3^d\frac{\partial\left(\delta u_3^a\right)}{\partial x_1}$$

$$\left. -\frac{1}{t_c}\mathring{u}_3^a\frac{\partial\left(\delta u_3^a\right)}{\partial x_1} - \frac{1}{t_c}\frac{\partial\mathring{u}_3^a}{\partial x_1}\delta u_3^d - \frac{2}{t_c}\frac{\partial u_3^d}{\partial x_1}\delta\Phi_3^c - \frac{2}{t_c}\Phi_3^c\frac{\partial\left(\delta u_3^d\right)}{\partial x_1} - \frac{2}{t_c}\frac{\partial\mathring{u}_3^d}{\partial x_1}\delta\Phi_3^c\right\}$$

$$+2M^c_{13}\left\{\frac{4}{t_c^2}\delta\Phi_1^c - \frac{1}{t_c}\frac{\partial\left(\delta u_3^d\right)}{\partial x_1} + \frac{4}{t_c^2}\frac{\partial u_3^a}{\partial x_1}\delta\Phi_3^c + \frac{4}{t_c^2}\Phi_3^c\frac{\partial\left(\delta u_3^a\right)}{\partial x_1} + \frac{2}{t_c^2}\frac{\partial u_3^d}{\partial x_1}\delta u_3^d\right.$$

$$\left. +\frac{2}{t_c^2}u_3^d\frac{\partial\left(\delta u_3^d\right)}{\partial x_1} + \frac{2}{t_c^2}\mathring{u}_3^d\frac{\partial\left(\delta u_3^d\right)}{\partial x_1} + \frac{2}{t_c^2}\frac{\partial\mathring{u}_3^d}{\partial x_1}\delta u_3^d + \frac{4}{t_c^2}\frac{\partial\mathring{u}_3^a}{\partial x_1}\delta\Phi_3^c\right\} + 2L^c_{13}\left\{\frac{6}{t_c^3}\delta\Omega_1^c\right.$$

$$-\frac{8}{t_c^3}\frac{\partial u_3^d}{\partial x_1}\delta\Phi_3^c - \frac{8}{t_c^3}\Phi_3^c\frac{\partial\left(\delta u_3^d\right)}{\partial x_1} - \frac{8}{t_c^3}\frac{\partial\mathring{u}_3^d}{\partial x_1}\delta\Phi_3^c + \frac{2}{t_c^2}\frac{\partial\left(\delta\Phi_3^c\right)}{\partial x_1} - \frac{4}{t_c^3}\frac{\partial\Phi_3^c}{\partial x_1}\delta u_3^d$$

$$\left. -\frac{4}{t_c^3}u_3^d\frac{\partial\left(\delta\Phi_3^c\right)}{\partial x_1} - \frac{4}{t_c^3}\mathring{u}_3^d\frac{\partial\left(\delta\Phi_3^c\right)}{\partial x_1}\right\} + 2K^c_{13}\left\{\frac{16}{t_c^4}\frac{\partial\Phi_3^c}{\partial x_1}\delta\Phi_3^c + \frac{16}{t_c^4}\Phi_3^c\frac{\partial\left(\delta\Phi_3^c\right)}{\partial x_1}\right\}$$

$$+2N^c_{23}\left\{-\frac{1}{t_c}\delta u_2^d + \left(\frac{1}{2}+\frac{1}{t_c}\left(\frac{t_f^t+t_f^b}{4}\right)\right)\frac{\partial\left(\delta u_3^a\right)}{\partial x_2} + \frac{1}{t_c}\left(\frac{t_f^t-t_f^b}{4}\right)\frac{\partial\left(\delta u_3^d\right)}{\partial x_2}\right.$$

$$+\frac{1}{t_c}\delta\Omega_2^c - \frac{1}{t_c}\frac{\partial u_3^a}{\partial x_2}\delta u_3^d - \frac{1}{t_c}u_3^d\frac{\partial\left(\delta u_3^a\right)}{\partial x_2} - \frac{1}{t_c}\mathring{u}_3^a\frac{\partial\left(\delta u_3^a\right)}{\partial x_2} - \frac{1}{t_c}\frac{\partial\mathring{u}_3^a}{\partial x_2}\delta u_3^d$$

$$\left. -\frac{2}{t_c}\frac{\partial u_3^d}{\partial x_2}\delta\Phi_3^c - \frac{2}{t_c}\Phi_3^c\frac{\partial\left(\delta u_3^d\right)}{\partial x_2} - \frac{2}{t_c}\frac{\partial\mathring{u}_3^d}{\partial x_2}\delta\Phi_3^c\right\} + 2M^c_{23}\left\{\frac{4}{t_c^2}\delta\Phi_2^c - \frac{1}{t_c}\frac{\partial\left(\delta u_3^d\right)}{\partial x_2}\right.$$

$$+\frac{4}{t_c^2}\frac{\partial u_3^a}{\partial x_2}\delta\Phi_3^c+\frac{4}{t_c^2}\Phi_3^c\frac{\partial\left(\delta u_3^a\right)}{\partial x_2}+\frac{2}{t_c^2}\frac{\partial u_3^d}{\partial x_2}\delta u_3^d+\frac{2}{t_c^2}u_3^d\frac{\partial\left(\delta u_3^d\right)}{\partial x_2}+\frac{2}{t_c^2}\overset{\circ}{u}{}_3^d\frac{\partial\left(\delta u_3^d\right)}{\partial x_2}$$

$$+\frac{2}{t_c^2}\frac{\partial\overset{\circ}{u}{}_3^d}{\partial x_2}\delta u_3^d+\frac{4}{t_c^2}\frac{\partial\overset{\circ}{u}{}_3^a}{\partial x_2}\delta\Phi_3^c\Bigg\}+2L_{23}^c\left\{\frac{6}{t_c^3}\delta\Omega_2^c-\frac{8}{t_c^3}\frac{\partial u_3^d}{\partial x_2}\delta\Phi_3^c-\frac{8}{t_c^3}\Phi_3^c\frac{\partial\left(\delta u_3^d\right)}{\partial x_2}\right.$$

$$\left.-\frac{8}{t_c^3}\frac{\partial\overset{\circ}{u}{}_3^d}{\partial x_2}\delta\Phi_3^c+\frac{2}{t_c^2}\frac{\partial\left(\delta\Phi_3^c\right)}{\partial x_2}-\frac{4}{t_c^3}\frac{\partial\Phi_3^c}{\partial x_2}\delta u_3^d-\frac{4}{t_c^3}u_3^d\frac{\partial\left(\delta\Phi_3^c\right)}{\partial x_2}-\frac{4}{t_c^3}\overset{\circ}{u}{}_3^d\frac{\partial\left(\delta\Phi_3^c\right)}{\partial x_2}\right\}$$

$$+2K_{23}^c\left\{\frac{16}{t_c^4}\frac{\partial\Phi_3^c}{\partial x_2}\delta\Phi_3^c+\frac{16}{t_c^4}\Phi_3^c\frac{\partial\left(\delta\Phi_3^c\right)}{\partial x_2}\right\}\Bigg]dA \tag{8.56}$$

Integrating each appropriate term by parts, combining coefficients of identical variational displacements, and simplifying gives the variation in the strain energy as

$$\delta U=\int_0^{l_2}\int_0^{l_1}\Bigg\langle-2\left(\frac{\partial N_{11}^a}{\partial x_1}+\frac{\partial N_{12}^a}{\partial x_2}\right)\delta u_1^a-2\left(\frac{\partial N_{22}^a}{\partial x_2}+\frac{\partial N_{12}^a}{\partial x_1}\right)\delta u_2^a$$

$$-2\left(\frac{\partial N_{11}^a}{\partial x_1}+\frac{\partial N_{12}^a}{\partial x_2}+\frac{N_{13}^c}{t_c}\right)\delta u_1^d-2\left(\frac{\partial N_{22}^a}{\partial x_2}+\frac{\partial N_{12}^a}{\partial x_1}+\frac{N_{23}^c}{t_c}\right)\delta u_2^d$$

$$+\left(\frac{8}{t_c^2}M_{13}^c\right)\delta\Phi_1^c+\left(\frac{8}{t_c^2}M_{23}^c\right)\delta\Phi_2^c+2\left(\frac{N_{13}^c}{t_c}+\frac{6}{t_c^3}L_{13}^c\right)\delta\Omega_1^c$$

$$+2\left(\frac{N_{23}^c}{t_c}+\frac{6}{t_c^3}L_{23}^c\right)\delta\Omega_2^c+\Bigg\{-2N_{11}^a\left(\frac{\partial^2u_3^a}{\partial x_1^2}+\frac{\partial^2\overset{\circ}{u}{}^a}{\partial x_1^2}+\frac{1}{R_1}\right)$$

$$-4N_{12}^a\left(\frac{\partial^2u_3^a}{\partial x_1\partial x_2}+\frac{\partial^2\overset{\circ}{u}{}_3^a}{\partial x_1\partial x_2}\right)-2N_{22}^a\left(\frac{\partial^2u_3^a}{\partial x_2^2}+\frac{\partial^2\overset{\circ}{u}{}_3^a}{\partial x_2^2}+\frac{1}{R_2}\right)-2\frac{\partial^2M_{11}^a}{\partial x_1^2}$$

$$-4\frac{\partial^2M_{12}^a}{\partial x_1\partial x_2}-2\frac{\partial^2M_{22}^a}{\partial x_2^2}-2N_{11}^d\left(\frac{\partial^2u_3^a}{\partial x_1^2}+\frac{\partial^2\overset{\circ}{u}{}_3^a}{\partial x_1^2}\right)-4N_{12}^d\left(\frac{\partial^2u_3^d}{\partial x_1\partial x_2}+\frac{\partial^2\overset{\circ}{u}{}_3^d}{\partial x_1\partial x_2}\right)$$

$$-2N_{22}^d\left(\frac{\partial^2u_3^a}{\partial x_2^2}+\frac{\partial^2\overset{\circ}{u}{}_3^a}{\partial x_2^2}\right)-\frac{2}{t_c}\left(\frac{t_c}{2}+\frac{t_f^t+t_f^b}{4}-u_3^d-\overset{\circ}{u}{}_3^d\right)\left(\frac{\partial N_{13}^c}{\partial x_1}++\frac{\partial N_{23}^c}{\partial x_2}\right)$$

$$+\frac{4}{t_c}\left(\frac{\partial u_3^d}{\partial x_2}+\frac{\partial\overset{\circ}{u}{}_3^d}{\partial x_2}\right)N_{23}^c+\frac{4}{t_c}\left(\frac{\partial u_3^d}{\partial x_1}+\frac{\partial\overset{\circ}{u}{}_3^d}{\partial x_1}\right)N_{13}^c-2\left(\frac{\partial u_3^a}{\partial x_1}+\frac{\partial\overset{\circ}{u}{}_3^a}{\partial x_1}\right)$$

$$\left(\frac{\partial N_{11}^a}{\partial x_1}+\frac{\partial N_{12}^a}{\partial x_2}\right)-2\left(\frac{\partial u_3^a}{\partial x_2}+\frac{\partial\overset{\circ}{u}{}_3^a}{\partial x_2}\right)\left(\frac{\partial N_{22}^a}{\partial x_2}+\frac{\partial N_{12}^a}{\partial x_1}\right)-2\left(\frac{\partial u_3^d}{\partial x_1}+\frac{\partial\overset{\circ}{u}{}_3^d}{\partial x_1}\right)$$

$$\left(\frac{\partial N_{11}^d}{\partial x_1}+\frac{\partial N_{12}^d}{\partial x_2}\right)-2\left(\frac{\partial u_3^d}{\partial x_2}+\;+\frac{\partial \overset{\circ}{u}{}_3^d}{\partial x_2}\right)\left(\frac{\partial N_{22}^d}{\partial x_2}+\frac{\partial N_{12}^d}{\partial x_1}\right)$$

$$-\frac{4}{t_c^2}\Phi_3^c\left(\frac{\partial M_{13}^c}{\partial x_1}+\frac{\partial M_{23}^c}{\partial x_2}\right)-\frac{4}{t_c^2}\frac{\partial \Phi_3^c}{\partial x_1}M_{13}^c-\frac{4}{t_c^2}\frac{\partial \Phi_3^c}{\partial x_1}M_{13}^c\Bigg\}\delta u_3^a+$$

$$\Bigg\{-2N_{11}^d\left(\frac{\partial^2 u_3^a}{\partial x_1^2}+\frac{\partial^2 \overset{\circ}{u}{}_3^a}{\partial x_1^2}+\frac{1}{R_1}\right)-4N_{12}^d\left(\frac{\partial^2 u_3^a}{\partial x_1\partial x_2}+\frac{\partial^2 \overset{\circ}{u}{}_3^a}{\partial x_1\partial x_2}\right)$$

$$-2N_{22}^d\left(\frac{\partial^2 u_3^a}{\partial x_2^2}+\frac{\partial^2 \overset{\circ}{u}{}_3^a}{\partial x_2^2}+\frac{1}{R_2}\right)-2\frac{\partial^2 M_{11}^d}{\partial x_1^2}-4\frac{\partial^2 M_{12}^d}{\partial x_1\partial x_2}-2\frac{\partial^2 M_{22}^d}{\partial x_2^2}$$

$$-2N_{11}^a\left(\frac{\partial^2 u_3^d}{\partial x_1^2}+\frac{\partial^2 \overset{\circ}{u}{}_3^d}{\partial x_1^2}\right)-4N_{12}^a\left(\frac{\partial^2 u_3^d}{\partial x_1\partial x_2}+\frac{\partial^2 \overset{\circ}{u}{}_3^d}{\partial x_1\partial x_2}\right)-$$

$$-2N_{22}^a\left(\frac{\partial^2 u_3^d}{\partial x_2^2}+\frac{\partial^2 \overset{\circ}{u}{}_3^d}{\partial x_2^2}\right)-\frac{4}{t_c}\left(\frac{t_c}{2}-u_3^d-\overset{\circ}{u}{}_3^d\right)N_{33}^c+\frac{4}{t_c}\left(\Phi_3^c+\frac{t_f^t-t_f^b}{16}\right)$$

$$\left(\frac{\partial N_{13}^c}{\partial x_1}+\frac{\partial N_{23}^c}{\partial x_2}\right)++\frac{4}{t_c}\frac{\partial \Phi_3^c}{\partial x_1}N_{13}^c+\frac{4}{t_c}\frac{\partial \Phi_3^c}{\partial x_2}N_{23}^c+\frac{16}{t_c^3}\Phi_3^c\left(\frac{\partial L_{13}^c}{\partial x_1}+\frac{\partial L_{23}^c}{\partial x_2}\right)$$

$$+\frac{8}{t_c^3}\frac{\partial \Phi_3^c}{\partial x_1}L_{13}^c+\frac{8}{t_c^3}\frac{\partial \Phi_3^c}{\partial x_2}L_{23}^c+\frac{16}{t_c^3}\Phi_3^c M_{33}^c-2\left(\frac{\partial u_3^d}{\partial x_1}+\frac{\partial \overset{\circ}{u}{}_3^d}{\partial x_1}\right)\left(\frac{\partial N_{11}^a}{\partial x_1}+\frac{\partial N_{12}^a}{\partial x_2}\right)$$

$$-2\left(\frac{\partial u_3^d}{\partial x_2}+\frac{\partial \overset{\circ}{u}{}_3^d}{\partial x_2}\right)\left(\frac{\partial N_{22}^a}{\partial x_2}+\frac{\partial N_{12}^a}{\partial x_1}\right)-2\left(\frac{\partial u_3^a}{\partial x_1}+\frac{\partial \overset{\circ}{u}{}_3^a}{\partial x_1}\right)\left(\frac{\partial N_{11}^d}{\partial x_1}+\frac{\partial N_{12}^d}{\partial x_2}+\frac{N_{13}^c}{t_c}\right)$$

$$-2\left(\frac{\partial u_3^a}{\partial x_2}+\frac{\partial \overset{\circ}{u}{}_3^a}{\partial x_2}\right)\left(\frac{\partial N_{22}^d}{\partial x_2}+\frac{\partial N_{12}^d}{\partial x_1}+\frac{N_{23}^c}{t_c}\right)+\frac{4}{t_c^2}\left(\frac{t_c}{2}-u_3^d-\overset{\circ}{u}{}_3^d\right)\left(\frac{\partial M_{13}^c}{\partial x_1}+\right.$$

$$\left.+\frac{\partial M_{23}^c}{\partial x_2}\right)\Bigg\}\delta u_3^d+\Bigg\{-\frac{4}{t_c}\left(\frac{\partial u_3^d}{\partial x_1}+\frac{\partial \overset{\circ}{u}{}_3^d}{\partial x_1}\right)N_{13}^c-\frac{4}{t_c}\left(\frac{\partial u_3^d}{\partial x_2}+\frac{\partial \overset{\circ}{u}{}_3^d}{\partial x_2}\right)N_{23}^c$$

$$-\frac{8}{t_c^3}\left(\frac{\partial L_{13}^c}{\partial x_1}+\frac{\partial L_{23}^c}{\partial x_2}\right)\left(\frac{t_c}{2}-u_3^d-\overset{\circ}{u}{}_3^d\right)-\frac{8}{t_c^3}L_{13}^c\left(\frac{\partial u_3^d}{\partial x_1}+\frac{\partial \overset{\circ}{u}{}_3^d}{\partial x_1}\right)$$

$$-\frac{8}{t_c^3}L_{23}^c\left(\frac{\partial u_3^d}{\partial x_2}+\frac{\partial \overset{\circ}{u}{}_3^d}{\partial x_2}\right)+\frac{16}{t_c^2}\Phi_3^c N_{33}^c+\frac{16}{t_c^3}\left(\frac{t_c}{2}-u_3^d-\overset{\circ}{u}{}_3^d\right)M_{33}^c+\frac{64}{t_c^4}\Phi_3^c L_{33}^c$$

$$
-\frac{32}{t_c^4}\Phi_3^c\left(\frac{\partial K_{13}^c}{\partial x_1}+\frac{\partial K_{23}^c}{\partial x_2}\right)+\frac{8}{t_c^2}M_{13}^c\left(\frac{\partial u_3^a}{\partial x_1}+\frac{\partial \overset{\circ}{u}{}_3^a}{\partial x_1}\right)
$$

$$
+\frac{8}{t_c^2}M_{23}^c\left(\frac{\partial u_3^a}{\partial x_2}+\frac{\partial \overset{\circ}{u}{}_3^a}{\partial x_2}\right)\Bigg\}\delta\Phi_3^c\Bigg\rangle dx_1dx_2+
$$

$$
\int_0^{l_2}\Bigg\langle 2N_{11}^a\delta u_1^a+2N_{12}^a\delta u_2^a+2N_{11}^d\delta u_2^d+2N_{12}^d\delta u_2^d+\Bigg\{2N_{11}^a\left(\frac{\partial u_3^a}{\partial x_1}+\frac{\partial \overset{\circ}{u}{}_3^a}{\partial x_1}\right)+
$$

$$
+2N_{12}^a\left(\frac{\partial u_3^a}{\partial x_2}+\frac{\partial \overset{\circ}{u}{}_3^a}{\partial x_2}\right)+2N_{11}^d\left(\frac{\partial u_3^d}{\partial x_1}+\frac{\partial \overset{\circ}{u}{}_3^d}{\partial x_1}\right)+2N_{12}^d\left(\frac{\partial u_3^d}{\partial x_2}+\frac{\partial \overset{\circ}{u}{}_3^d}{\partial x_2}\right)+2\frac{\partial M_{11}^a}{\partial x_1}
$$

$$
+4\frac{\partial M_{12}^a}{\partial x_2}+\frac{2}{t_c}\left[\frac{t_c}{2}+\frac{t_f^t+t_f^b}{4}-u_3^d-\overset{\circ}{u}{}_3^d\right]N_{13}^c++\frac{8}{t_c^2}M_{13}^c\Phi_3^c\Bigg\}\delta u_3^a
$$

$$
+\Bigg\{2N_{11}^a\left(\frac{\partial u_3^d}{\partial x_1}+\frac{\partial \overset{\circ}{u}{}_3^d}{\partial x_1}\right)+2N_{11}^d\left(\frac{\partial u_3^a}{\partial x_1}+\frac{\partial \overset{\circ}{u}{}_3^a}{\partial x_1}\right)+2N_{12}^a\left(\frac{\partial u_3^d}{\partial x_2}+\frac{\partial \overset{\circ}{u}{}_3^d}{\partial x_2}\right)
$$

$$
+2N_{12}^d\left(\frac{\partial u_3^a}{\partial x_2}+\frac{\partial \overset{\circ}{u}{}_3^a}{\partial x_2}\right)+2\frac{\partial M_{11}^d}{\partial x_1}+4\frac{\partial M_{12}^d}{\partial x_2}-\frac{8}{t_c^3}L_{13}^c\Phi_3^c-\frac{2}{t_c}\left(\Phi_3^c-\frac{t_f^t-t_f^b}{8}\right)N_{13}^c
$$

$$
-\frac{2}{t_c^2}\left(\frac{t_c}{2}-u_3^d-\overset{\circ}{u}{}_3^d\right)M_{13}^c\Bigg\}\delta u_3^d+\Bigg\{\frac{8}{t_c^3}\left(\frac{t_c}{2}-u_3^d-\overset{\circ}{u}{}_3^d\right)L_{13}^c+\frac{32}{t_c^4}\Phi_3^cK_{13}^c\Bigg\}\delta\Phi_3^c
$$

$$
+2M_{11}^a\delta\left(\frac{\partial u_3^a}{\partial x_1}\right)+2M_{11}^d\delta\left(\frac{\partial u_3^d}{\partial x_1}\right)\Bigg\rangle\Bigg|_0^{l_1}dx_2+
$$

$$
\int_0^{l_1}\Bigg\langle 2N_{22}^a\delta u_2^a+2N_{21}^a\delta u_1^a+2N_{22}^d\delta u_1^d+2N_{21}^d\delta u_1^d+\Bigg\{2N_{22}^a\left(\frac{\partial u_3^a}{\partial x_2}+\frac{\partial \overset{\circ}{u}{}_3^a}{\partial x_2}\right)
$$

$$
+2N_{21}^a\left(\frac{\partial u_3^a}{\partial x_1}+\frac{\partial \overset{\circ}{u}{}_3^a}{\partial x_1}\right)++2N_{22}^d\left(\frac{\partial u_3^d}{\partial x_2}+\frac{\partial \overset{\circ}{u}{}_3^d}{\partial x_2}\right)+2N_{21}^d\left(\frac{\partial u_3^d}{\partial x_1}+\frac{\partial \overset{\circ}{u}{}_3^d}{\partial x_1}\right)
$$

$$
+2\frac{\partial M_{22}^a}{\partial x_2}+4\frac{\partial M_{21}^a}{\partial x_1}+\frac{2}{t_c}\left[\frac{t_c}{2}+\frac{t_f^t+t_f^b}{4}-u_3^d-\overset{\circ}{u}{}_3^d\right]N_{23}^c+\frac{8}{t_c^2}M_{23}^c\Phi_3^c\Bigg\}\delta u_3^a
$$

$$
+\Bigg\{2N_{22}^a\left(\frac{\partial u_3^d}{\partial x_2}+\frac{\partial \overset{\circ}{u}{}_3^d}{\partial x_2}\right)+2N_{22}^d\left(\frac{\partial u_3^a}{\partial x_2}+\frac{\partial \overset{\circ}{u}{}_3^a}{\partial x_2}\right)+2N_{21}^a\left(\frac{\partial u_3^d}{\partial x_1}+\frac{\partial \overset{\circ}{u}{}_3^d}{\partial x_1}\right)
$$

$$+2N^d_{21}\left(\frac{\partial u^a_3}{\partial x_1}+\frac{\partial \overset{\circ}{u}{}^a_3}{\partial x_1}\right)+2\frac{\partial M^d_{22}}{\partial x_2}+4\frac{\partial M^d_{21}}{\partial x_1}-\frac{8}{t_c^3}L^c_{23}\Phi^c_3-\frac{2}{t_c}\left(\Phi^c_3-\frac{t^t_f-t^b_f}{8}\right)N^c_{23}$$

$$-\frac{2}{t_c^2}\left(\frac{t_c}{2}-u^d_3-\overset{\circ}{u}{}^d_3\right)M^c_{23}\Bigg\}\delta u^d_3+\left\{\frac{8}{t_c^3}\left(\frac{t_c}{2}-u^d_3-\overset{\circ}{u}{}^d_3\right)L^c_{23}+\frac{32}{t_c^4}\Phi^c_3K^c_{23}\right\}\delta\Phi^c_3$$

$$+2M^a_{22}\delta\left(\frac{\partial u^a_3}{\partial x_2}\right)+2M^d_{22}\delta\left(\frac{\partial u^d_3}{\partial x_2}\right)\Bigg\rangle\Bigg|_0^{l_2}dx_1+ \tag{8.57}$$

8.4.2 Work Done by External Loads

The total work consists of the work due to body forces, edge loads, surface tractions, and damping. The work due to each of these is presented mathematically as

- *Work due to body forces*

$$\delta W_b=\int_\sigma\left\{\int_{\bar{h}}^{\bar{h}+h'}\rho' H'_i\delta V'_i dx_3+\int_{-\bar{h}}^{\bar{h}}\bar{\rho}\bar{H}_i\delta\bar{V}_i dx_3+\int_{-\bar{h}-h''}^{-\bar{h}}\rho'' H''_i\delta V''_i dx_3\right\}d\sigma \tag{8.58}$$

where ρ is the mass density and H_i is the body force vector. The body forces will not be included in the following developments. Typical body forces are gravity, electrical and magnetic forces which will be neglected.

- *Work due to edge loads*

$$W_{\text{edge loads}}=W^t_{\text{edge loads}}+W^c_{\text{edge loads}}+W^b_{\text{edge loads}} \tag{8.59}$$

Considering the contribution from each layer, the total work due to edge loads along each boundary is the total summation of the work of each layer. The work due to edge loads is by definition

$$\delta W_{el}=\int_{x_1}\left(\int_{\frac{-t_c}{2}-t'_f}^{\frac{-t_c}{2}}\left(\hat{\tau}^t_{22}\delta v^t_2+\hat{\tau}^t_{21}\delta v^t_1\right)dx_3+\int_{\frac{-t_c}{2}}^{\frac{t_c}{2}}\hat{\tau}^c_{23}\delta v^c_3 dx_3+\int_{\frac{t_c}{2}}^{\frac{t_c}{2}+t^b_f}\left(\hat{\tau}^b_{22}\delta v^b_2+\hat{\tau}^b_{21}\delta v^b_1\right)dx_3\right)dx_1$$
$$+\int_{x_2}\left(\int_{\frac{-t_c}{2}-t'_f}^{\frac{-t_c}{2}}\left(\hat{\tau}^t_{11}\delta v^t_1+\hat{\tau}^t_{12}\delta v^t_2\right)dx_3+\int_{\frac{-t_c}{2}}^{\frac{t_c}{2}}\hat{\tau}^c_{13}\delta v^c_3 dx_3+\int_{\frac{t_c}{2}}^{\frac{t_c}{2}+t^b_f}\left(\hat{\tau}^b_{11}\delta v^b_1+\hat{\tau}^b_{12}\delta v^b_2\right)dx_3\right)dx_2 \tag{8.60}$$

Substituting in the pertinent displacement equations, Eq. (8.6a–c)–(8.8a–c) into Eq. (8.60) using the expressions for the local stress resultants and stress couples defined earlier in Eqs. (8.50)–(8.52) gives

$$\begin{aligned}
\delta W_{el} = & \int_{x_1} \left(\widehat{N}^t_{22}\delta u^a_2 + \widehat{N}^t_{22}\delta u^d_2 - \widehat{M}^t_{22}\delta\left(\frac{\partial u^a_3}{\partial x_2}\right) - \widehat{M}^t_{22}\delta\left(\frac{\partial u^d_3}{\partial x_2}\right) + \widehat{N}^t_{21}\delta u^a_1 + \widehat{N}^t_{21}\delta u^d_1 + \right. \\
& -\widehat{M}^t_{21}\delta\left(\frac{\partial u^a_3}{\partial x_1}\right) - \widehat{M}^t_{21}\delta\left(\frac{\partial u^d_3}{\partial x_1}\right) + \widehat{N}^c_{23}\delta u^a_3 - \frac{2}{t_c}\widehat{M}^c_{23}\delta u^d_3 + \frac{4}{t_c^2}\widehat{L}^c_{23}\delta\Phi^c_3 + \widehat{N}^b_{22}\delta u^a_2 \\
& -\widehat{N}^b_{22}\delta u^d_2 - \widehat{M}^b_{22}\delta\left(\frac{\partial u^a_3}{\partial x_2}\right) - \widehat{M}^b_{22}\delta\left(\frac{\partial u^d_3}{\partial x_2}\right) + \widehat{N}^b_{21}\delta u^a_1 - \widehat{N}^b_{21}\delta u^d_1 - \underline{\widehat{M}^b_{21}\delta\left(\frac{\partial u^a_3}{\partial x_1}\right) +} \\
& \left. \underline{-\widehat{M}^b_{21}\delta\left(\frac{\partial u^d_3}{\partial x_1}\right)} \right) dx_1 + \\
& \int_{x_2} \left(\widehat{N}^t_{11}\delta u^a_1 + \widehat{N}^t_{11}\delta u^d_1 - \widehat{M}^t_{11}\delta\left(\frac{\partial u^a_3}{\partial x_1}\right) - \widehat{M}^t_{11}\delta\left(\frac{\partial u^d_3}{\partial x_1}\right) + \widehat{N}^t_{12}\delta u^a_2 + \widehat{N}^t_{12}\delta u^d_2 + \right. \\
& -\widehat{M}^t_{12}\delta\left(\frac{\partial u^a_3}{\partial x_2}\right) - \widehat{M}^t_{12}\delta\left(\frac{\partial u^d_3}{\partial x_2}\right) + \widehat{N}^c_{13}\delta u^a_3 - \frac{2}{t_c}\widehat{M}^c_{13}\delta u^d_3 + \frac{4}{t_c^2}\widehat{L}^c_{13}\delta\Phi^c_3 + \widehat{N}^b_{11}\delta u^a_1 \\
& -\widehat{N}^b_{11}\delta u^d_1 - \widehat{M}^b_{11}\delta\left(\frac{\partial u^a_3}{\partial x_1}\right) - \widehat{M}^b_{11}\delta\left(\frac{\partial u^d_3}{\partial x_1}\right) + \widehat{N}^b_{12}\delta u^a_2 - \widehat{N}^b_{12}\delta u^d_2 - \underline{\widehat{M}^b_{12}\delta\left(\frac{\partial u^a_3}{\partial x_2}\right) +} \\
& \left. \underline{-\widehat{M}^b_{12}\delta\left(\frac{\partial u^d_3}{\partial x_2}\right)} \right) dx_2
\end{aligned} \tag{8.61}$$

Integrating the underlined terms, simplifying, and gathering like terms results in

$$\begin{aligned}
W_{el} = & \int_{x_1} \left\{ 2\widehat{N}^a_{22}\delta u^a_2 + 2\widehat{N}^d_{22}\delta u^d_2 + 2\widehat{N}^a_{21}\delta u^a_1 + 2\widehat{N}^d_{21}\delta u^d_1 + \left(2\frac{\partial\widehat{M}^a_{21}}{\partial x_1} + \widehat{N}^c_{23}\right)\delta u^a_3 + \right. \\
& \left. \left(2\frac{\partial\widehat{M}^d_{21}}{\partial x_1} - \frac{\widehat{M}^c_{23}}{t_c}\right)\delta u^d_3 + 2\widehat{M}^a_{22}\delta\left(\frac{\partial u^a_3}{\partial x_2}\right) + 2\widehat{M}^d_{22}\delta\left(\frac{\partial u^d_3}{\partial x_2}\right) + \frac{4}{t_c^2}\widehat{L}^c_{23}\delta\Phi^c_3 \right\} dx_1 + \\
& \int_{x_2} \left\{ 2\widehat{N}^a_{11}\delta u^a_1 + 2\widehat{N}^d_{11}\delta u^d_1 + 2\widehat{N}^a_{12}\delta u^a_2 + 2\widehat{N}^d_{12}\delta u^d_2 + \left(2\frac{\partial\widehat{M}^a_{12}}{\partial x_2} + \widehat{N}^c_{13}\right)\delta u^a_3 + \right. \\
& \left. \left(2\frac{\partial\widehat{M}^d_{12}}{\partial x_2} - \frac{\widehat{M}^c_{13}}{t_c}\right)\delta u^d_3 + 2\widehat{M}^a_{11}\delta\left(\frac{\partial u^a_3}{\partial x_1}\right) + 2\widehat{M}^d_{11}\delta\left(\frac{\partial u^d_3}{\partial x_1}\right) + \frac{4}{t_c^2}\widehat{L}^c_{13}\delta\Phi^c_3 \right\} dx_2
\end{aligned} \tag{8.62}$$

where the defined global stress resultants and stress couples from Eqs. (8.55a and 8.55b) have been utilized.

- *Work due to surface tractions*

The expression for the work done by lateral vertical loading such as an external pressure is by definition given as

$$W_{st} = \int_A \left(\widehat{q}_3^t \delta v_3^t + \widehat{q}_3^b \delta v_3^b \right) dA \tag{8.63}$$

Substituting Eqs. (8.6c) and (8.7c) into Eq. (8.63) gives

$$W_{st} = \int_A \left(\widehat{q}_3^t \left(\delta u_3^a + \delta u_3^d \right) + \widehat{q}_3^b \left(\delta u_3^a - \delta u_3^d \right) \right) dA \tag{8.64}$$

Simplifying results in

$$W_{st} = \int_A \left\{ \left(\widehat{q}_3^t + \widehat{q}_3^b \right) \delta u_3^a + \left(\widehat{q}_3^t - \widehat{q}_3^b \right) \delta u_3^d \right\} dA = 2 \int_A \left(\widehat{q}_3^a \delta u_3^a + \widehat{q}_3^d \delta u_3^d \right) dA \tag{8.65}$$

where

$$\widehat{q}_3^a = \frac{\widehat{q}_3^t + \widehat{q}_3^b}{2} \tag{8.66}$$

$$\widehat{q}_3^d = \frac{\widehat{q}_3^t - \widehat{q}_3^b}{2} \tag{8.67}$$

- *Work due to damping*

By definition, the work due to damping for the transverse direction is given by

$$W_d = \int_A \left(C^t \dot{v}_3^t \delta v_3^t + C^c \dot{v}_3^c \delta v_3^c + C^b \dot{v}_3^b \delta v_3^b \right) dA \tag{8.68}$$

where C^t, C^c and C^b are the structural damping coefficients per unit area of the facings and the core. Substituting in Eqs. (8.6)–(8.8) gives

$$W_d = \int_A \left(C^t \left(\dot{u}_3^a + \dot{u}_3^d \right) \left(\delta u_3^a + \delta u_3^d \right) + C^c \dot{u}_3^a \delta u_3^a + C^b \left(\dot{u}_3^a - \dot{u}_3^d \right) \left(\delta u_3^a - \delta u_3^d \right) \right) dA \tag{8.69}$$

Assuming that the damping is constant throughout the thickness of the core, simplifying, and gathering like variational terms gives

$$W_d = \int_A \Big[\left(C^t + C^b + C^c \right) \dot{u}_3^a \delta u_3^a + \left(C^t - C^b \right) \dot{u}_3^a \delta u_3^d + \left(C^t - C^b \right) \dot{u}_3^d \delta u_3^a + \\ \left(C^t + C^b \right) \dot{u}_3^d \delta u_3^d \Big] dA \tag{8.70}$$

which can be written as

$$W_d = 2\int_A \left[\left(\left(C^a + \frac{C^c}{2} \right) \dot{u}_3^a + C^d \dot{u}_3^d \right) \delta u_3^a + \left(C^d \dot{u}_3^a + C^a \dot{u}_3^d \right) \delta u_3^d \right] dA \tag{8.71}$$

where

$$C^a = \frac{C^t + C^b}{2}, \qquad C^d = \frac{C^t - C^b}{2} \tag{8.72}$$

This expression is substituted into energy function Eq. (8.46).

8.4.3 Kinetic Energy

By definition, the kinetic energy is given by

$$T = \frac{1}{2} \int_{x_3} \int_{x_2} \int_{x_1} \rho \left[\left(\frac{\partial V_1}{\partial t} \right)^2 + \left(\frac{\partial V_2}{\partial t} \right)^2 \left(\frac{\partial V_3}{\partial t} \right)^2 \right] dx_1 dx_2 dx_3 \tag{8.73}$$

Taking a variation in the expression for the kinetic energy and integrating with respect to time gives

$$\begin{aligned} \int_{t_0}^{t_1} \delta T dt = \int_{t_0}^{t_1} \int_\sigma \Bigg[& \int_{\frac{t_c}{2}}^{\frac{t_c}{2}+t^b_f} \rho^b \left(\ddot{V}_1^b \delta V_1^b + \ddot{V}_2^b \delta V_2^b + \ddot{V}_3^b \delta V_3^b \right) dx_3 \\ & + \int_{-\frac{t_c}{2}}^{\frac{t_c}{2}} \rho^c \left(\ddot{V}_1^c \delta V_1^c + \ddot{V}_2^c \delta V_2^c + \ddot{V}_3^c \delta V_3^c \right) dx_3 \\ & + \int_{-\frac{t_c}{2}-t^t_f}^{-\frac{t_c}{2}} \rho^t \left(\ddot{V}_1^t \delta V_1^t + \ddot{V}_2^t \delta V_2^t + \ddot{V}_3^t \delta V_3^t \right) dx_3 \Bigg] d\sigma dt \end{aligned} \tag{8.74}$$

If the in-plane inertias are neglected and only the transverse inertias are considered, the kinetic energy can be expressed as

$$\int_{t_0}^{t_1} \delta T dt = \int_{t_0}^{t_1}\int_A \left[\int_{\frac{t_c}{2}}^{\frac{t_c}{2}+t_f^b} \rho^b \ddot{V}_3^b \delta V_3^b dx_3 + \int_{-\frac{t_c}{2}}^{\frac{t_c}{2}} \rho^c \ddot{V}_3^c \delta V_3^c dx_3 + \int_{-\frac{t_c}{2}-t_f^t}^{-\frac{t_c}{2}} \rho^t \ddot{V}_3^t \delta V_3^t dx_3\right] dAdt \tag{8.75}$$

Using the expressions for the displacements for each respective layer, from Eqs. (8.6)–(8.8), gives

$$\int_{t_0}^{t_1} \delta T dt = \int_{t_0}^{t_1}\int_A \left\{\int_{-\frac{t_c}{2}-t_f^t}^{-\frac{t_c}{2}} \rho^t \left(\ddot{u}_3^a + \ddot{u}_3^d\right)\left(\delta u_3^a + \delta u_3^d\right) dx_3 + \int_{-\frac{t_c}{2}}^{\frac{t_c}{2}} \rho^c \left(\ddot{u}_3^a - \frac{2}{t_c} x_3 \ddot{u}_3^d + \left(\frac{4x_3^2}{t_c^2} - 1\right)\ddot{\Phi}_3^c\right)\right.$$
$$\left.\cdot\left(\delta u_3^a - \frac{2}{t_c} x_3 \delta u_3^d + \left(\frac{4x_3^2}{t_c^2} - 1\right)\delta\Phi_3^c\right) dx_3 + \int_{\frac{t_c}{2}}^{\frac{t_c}{2}+t_f^b} \rho^b \left(\ddot{u}_3^a - \ddot{u}_3^d\right)\left(\delta u_3^a - \delta u_3^d\right) dx_3\right) dAdt \tag{8.76}$$

Integrating and simplifying gives

$$\int_{t_0}^{t_1} \delta T dt = \int_{t_0}^{t_1}\int_A \left[\left\{2\left(m^a + \frac{m_c}{2}\right)\ddot{u}_3^a + 2m^d \ddot{u}_3^d - \frac{2m_c}{3}\ddot{\Phi}_3^c\right\}\delta u_3^a + \left\{2\left(m^a + \frac{m_c}{6}\right)\ddot{u}_3^d + \right.\right.$$
$$\left.\left. + 2m^d \ddot{u}_3^a\right\}\delta u_3^d - \left\{\frac{2m_c}{3}\ddot{u}_3^a + \frac{6m_c}{15}\ddot{\Phi}_3^c\right\}\delta\Phi_3^c\right] dAdt \tag{8.77}$$

where

$$m^a = \frac{m_f^t + m_f^b}{2}, \qquad m^d = \frac{m_f^t - m_f^b}{2} \tag{8.78a, b}$$

and

$$m_f^t = \int_{-\frac{t_c}{2}-t_f^t}^{-\frac{t_c}{2}} \rho_{(k)}^t dx_3, \qquad m_f^b = \int_{\frac{t_c}{2}}^{\frac{t_c}{2}+t_f^b} \rho_{(k)}^b dx_3, \qquad m_c = \rho^c t_c \tag{8.79a – c}$$

8.5 Governing System

8.5.1 Equations of Motion

Substituting δU, δW, and the δT back into Hamilton's Equation, Eq. (8.46) gives

$$\int_{t_0}^{t_1}\left\langle\int_0^{l_2}\int_0^{l_1}\left\langle -2\left(\frac{\partial N_{11}^a}{\partial x_1}+\frac{\partial N_{12}^a}{\partial x_2}\right)\delta u_1^a-2\left(\frac{\partial N_{22}^a}{\partial x_2}+\frac{\partial N_{12}^a}{\partial x_1}\right)\delta u_2^a-2\left(\frac{\partial N_{11}^a}{\partial x_1}+\frac{\partial N_{12}^a}{\partial x_2}\right.\right.\right.$$

$$\left.+\frac{N_{13}^c}{t_c}\right)\delta u_1^d-2\left(\frac{\partial N_{22}^a}{\partial x_2}+\frac{\partial N_{12}^a}{\partial x_1}+\frac{N_{23}^c}{t_c}\right)\delta u_2^d+\left(\frac{8}{t_c^2}M_{13}^c\right)\delta\Phi_1^c+\left(\frac{8}{t_c^2}M_{23}^c\right)\delta\Phi_2^c$$

$$+2\left(\frac{N_{13}^c}{t_c}+\frac{6}{t_c^3}L_{13}^c\right)\delta\Omega_1^c+2\left(\frac{N_{23}^c}{t_c}+\frac{6}{t_c^3}L_{23}^c\right)\delta\Omega_2^c+\left\{-2N_{11}^a\left(\frac{\partial^2u_3^a}{\partial x_1^2}+\frac{\partial^2\overset{\circ}{u}{}_3^a}{\partial x_1^2}+\frac{1}{R_1}\right)\right.$$

$$-4N_{12}^a\left(\frac{\partial^2u_3^a}{\partial x_1\partial x_2}+\frac{\partial^2\overset{\circ}{u}{}_3^a}{\partial x_1\partial x_2}\right)-2N_{22}^a\left(\frac{\partial^2u_3^a}{\partial x_2^2}+\frac{\partial^2\overset{\circ}{u}{}_3^a}{\partial x_2^2}+\frac{1}{R_2}\right)-2\frac{\partial^2M_{11}^a}{\partial x_1^2}$$

$$-4\frac{\partial^2M_{12}^a}{\partial x_1\partial x_2}-2\frac{\partial^2M_{22}^a}{\partial x_2^2}-2N_{11}^d\left(\frac{\partial^2u_3^a}{\partial x_1^2}+\frac{\partial^2\overset{\circ}{u}{}_3^a}{\partial x_1^2}\right)-4N_{12}^d\left(\frac{\partial^2u_3^d}{\partial x_1\partial x_2}+\frac{\partial^2\overset{\circ}{u}{}_3^d}{\partial x_1\partial x_2}\right)$$

$$-2N_{22}^d\left(\frac{\partial^2u_3^a}{\partial x_2^2}+\frac{\partial^2\overset{\circ}{u}{}_3^a}{\partial x_2^2}\right)-\frac{2}{t_c}\left(\frac{t_c}{2}+\frac{t_f^t+t_f^b}{4}-u_3^d-\overset{\circ}{u}{}_3^d\right)\left(\frac{\partial N_{13}^c}{\partial x_1}+\frac{\partial N_{23}^c}{\partial x_2}\right)$$

$$+\frac{4}{t_c}\left(\frac{\partial u_3^d}{\partial x_2}+\frac{\partial\overset{\circ}{u}{}_3^d}{\partial x_2}\right)N_{23}^c+\frac{4}{t_c}\left(\frac{\partial u_3^d}{\partial x_1}+\frac{\partial\overset{\circ}{u}{}_3^d}{\partial x_1}\right)N_{13}^c-2\left(\frac{\partial u_3^a}{\partial x_1}+\frac{\partial\overset{\circ}{u}{}_3^a}{\partial x_1}\right)$$

$$\left(\frac{\partial N_{11}^a}{\partial x_1}+\frac{\partial N_{12}^a}{\partial x_2}\right)-2\left(\frac{\partial u_3^a}{\partial x_2}+\frac{\circ\,\partial u_3^a}{\partial x_2}\right)\left(\frac{\partial N_{22}^a}{\partial x_2}+\frac{\partial N_{12}^a}{\partial x_1}\right)-2\left(\frac{\partial u_3^d}{\partial x_1}+\frac{\circ\,\partial u_3^d}{\partial x_1}\right)$$

$$\left(\frac{\partial N_{11}^d}{\partial x_1}+\frac{\partial N_{12}^d}{\partial x_2}\right)-2\left(\frac{\partial u_3^d}{\partial x_2}++\frac{\circ\,\partial u_3^d}{\partial x_2}\right)\left(\frac{\partial N_{22}^d}{\partial x_2}+\frac{\partial N_{12}^d}{\partial x_1}\right)$$

$$-\frac{4}{t_c^2}\Phi_3^c\left(\frac{\partial M_{13}^c}{\partial x_1}+\frac{\partial M_{23}^c}{\partial x_2}\right)-\frac{4}{t_c^2}\frac{\partial\Phi_3^c}{\partial x_1}M_{13}^c-\frac{4}{t_c^2}\frac{\partial\Phi_3^c}{\partial x_1}M_{13}^c-2\widehat{q}_3^a$$

$$\left.-2(C^a+C^c/2)\dot{u}_3^a-2C^d\dot{u}_3^d-2\left(m^a+\frac{m_c}{2}\right)\ddot{u}_3^a-2m^d\ddot{u}_3^d+\frac{2m_c}{3}\ddot{\Phi}_3^c\right\}\delta u_3^a+$$

$$\left\{-2N_{11}^d\left(\frac{\partial^2u_3^a}{\partial x_1^2}+\frac{\partial^2\overset{\circ}{u}{}_3^a}{\partial x_1^2}+\frac{1}{R_1}\right)-4N_{12}^d\left(\frac{\partial^2u_3^a}{\partial x_1\partial x_2}+\frac{\partial^2\overset{\circ}{u}{}_3^a}{\partial x_1\partial x_2}\right)\right.$$

$$-2N_{22}^d\left(\frac{\partial^2u_3^a}{\partial x_2^2}+\frac{\partial^2\overset{\circ}{u}{}_3^a}{\partial x_2^2}+\frac{1}{R_2}\right)-2\frac{\partial^2M_{11}^d}{\partial x_1^2}-4\frac{\partial^2M_{12}^d}{\partial x_1\partial x_2}-2\frac{\partial^2M_{22}^d}{\partial x_2^2}$$

$$-2N_{11}^a\left(\frac{\partial^2u_3^d}{\partial x_1^2}+\frac{\partial^2\overset{\circ}{u}{}_3^d}{\partial x_1^2}\right)-4N_{12}^a\left(\frac{\partial^2u_3^d}{\partial x_1\partial x_2}+\frac{\partial^2\overset{\circ}{u}{}_3^d}{\partial x_1\partial x_2}\right)-2N_{22}^a\left(\frac{\partial^2u_3^d}{\partial x_2^2}+\frac{\partial^2\overset{\circ}{u}{}_3^d}{\partial x_2^2}\right)$$

$$-\frac{4}{t_c}\left(\frac{t_c}{2}-u_3^d-\overset{\circ}{u}{}_3^d\right)N_{33}^c+\frac{4}{t_c}\left(\Phi_3^c+\frac{t_f^t-t_f^b}{16}\right)\left(\frac{\partial N_{13}^c}{\partial x_1}+\frac{\partial N_{23}^c}{\partial x_2}\right)+$$

$$+\frac{4}{t_c}\frac{\partial \Phi_3^c}{\partial x_1}N_{13}^c+\frac{4}{t_c}\frac{\partial \Phi_3^c}{\partial x_2}N_{23}^c+\frac{16}{t_c^3}\Phi_3^c\left(\frac{\partial L_{13}^c}{\partial x_1}+\frac{\partial L_{23}^c}{\partial x_2}\right)+\frac{8}{t_c^3}\frac{\partial \Phi_3^c}{\partial x_1}L_{13}^c+\frac{8}{t_c^3}\frac{\partial \Phi_3^c}{\partial x_2}L_{23}^c$$

$$+\frac{16}{t_c^3}\Phi_3^c M_{33}^c-2\left(\frac{\partial u_3^d}{\partial x_1}+\frac{\partial \mathring{u}_3^d}{\partial x_1}\right)\left(\frac{\partial N_{11}^a}{\partial x_1}+\frac{\partial N_{12}^a}{\partial x_2}\right)-2\left(\frac{\partial u_3^d}{\partial x_2}+\frac{\partial \mathring{u}_3^d}{\partial x_2}\right)$$

$$\left(\frac{\partial N_{22}^a}{\partial x_2}+\frac{\partial N_{12}^a}{\partial x_1}\right)-2\left(\frac{\partial u_3^a}{\partial x_1}+\frac{\partial \mathring{u}_3^a}{\partial x_1}\right)\left(\frac{\partial N_{11}^d}{\partial x_1}+\frac{\partial N_{12}^d}{\partial x_2}+\frac{N_{13}^c}{t_c}\right)-2\left(\frac{\partial u_3^a}{\partial x_2}+\frac{\partial \mathring{u}_3^a}{\partial x_2}\right)$$

$$\left(\frac{\partial N_{22}^d}{\partial x_2}+\frac{\partial N_{12}^d}{\partial x_1}+\frac{N_{23}^c}{t_c}\right)+\frac{4}{t_c^2}\left(\frac{t_c}{2}-u_3^d-\mathring{u}_3^d\right)\left(\frac{\partial M_{13}^c}{\partial x_1}++\frac{\partial M_{23}^c}{\partial x_2}\right)-2\widehat{q}_3^d-2C^d\dot{u}_3^a$$

$$-2C^a\dot{u}_3^d-2\left(m^a+\frac{m_c}{6}\right)\ddot{u}_3^d-2m^d\ddot{u}_3^a\Bigg\}\delta u_3^d+\Bigg\{-\frac{4}{t_c}\left(\frac{\partial u_3^d}{\partial x_1}+\frac{\partial \mathring{u}_3^d}{\partial x_1}\right)N_{13}^c$$

$$-\frac{4}{t_c}\left(\frac{\partial u_3^d}{\partial x_2}+\frac{\partial \mathring{u}_3^d}{\partial x_2}\right)N_{23}^c-\frac{8}{t_c^3}\left(\frac{\partial L_{13}^c}{\partial x_1}+\frac{\partial L_{23}^c}{\partial x_2}\right)\left(\frac{t_c}{2}-u_3^d-\mathring{u}_3^d\right)$$

$$-\frac{8}{t_c^3}L_{13}^c\left(\frac{\partial u_3^d}{\partial x_1}+\frac{\partial \mathring{u}_3^d}{\partial x_1}\right)-\frac{8}{t_c^3}L_{23}^c\left(\frac{\partial u_3^d}{\partial x_2}+\frac{\partial \mathring{u}_3^d}{\partial x_2}\right)+\frac{16}{t_c^2}\Phi_3^c N_{33}^c$$

$$+\frac{16}{t_c^3}\left(\frac{t_c}{2}-u_3^d-\mathring{u}_3^d\right)M_{33}^c+\frac{64}{t_c^4}\Phi_3^c L_{33}^c-\frac{32}{t_c^4}\Phi_3^c\left(\frac{\partial K_{13}^c}{\partial x_1}+\frac{\partial K_{23}^c}{\partial x_2}\right)+\frac{8}{t_c^2}M_{13}^c$$

$$\left(\frac{\partial u_3^a}{\partial x_1}+\frac{\partial \mathring{u}_3^a}{\partial x_1}\right)+\frac{8}{t_c^2}M_{23}^c\left(\frac{\partial u_3^a}{\partial x_2}+\frac{\partial \mathring{u}_3^a}{\partial x_2}\right)+\frac{2m_c}{3}\ddot{u}_3^a+\frac{6m_c}{15}\ddot{\Phi}_3^c\Bigg\}\delta\Phi_3^c\Bigg\rangle dx_1dx_2dt$$

$$+\int_{t_0}^{t_1}\left[\int_0^{l_2}\Bigg\langle 2\left(N_{11}^a-\widehat{N}_{11}^a\right)\delta u_1^a+2\left(N_{12}^a-\widehat{N}_{12}^a\right)\delta u_2^a+2\left(N_{11}^d-\widehat{N}_{11}^d\right)\delta u_2^d\right.$$

$$+2\left(N_{12}^d-\widehat{N}_{12}^d\right)\delta u_2^d+\Bigg\{2N_{11}^a\left(\frac{\partial u_3^a}{\partial x_1}+\frac{\partial \mathring{u}_3^a}{\partial x_1}\right)+2N_{12}^a\left(\frac{\partial u_3^a}{\partial x_2}+\frac{\partial \mathring{u}_3^a}{\partial x_2}\right)$$

$$+2N_{11}^d\left(\frac{\partial u_3^d}{\partial x_1}+\frac{\partial \mathring{u}_3^d}{\partial x_1}\right)+2N_{12}^d\left(\frac{\partial u_3^d}{\partial x_2}+\frac{\partial \mathring{u}_3^d}{\partial x_2}\right)+2\frac{\partial M_{11}^a}{\partial x_1}+4\frac{\partial M_{12}^a}{\partial x_2}$$

$$+\frac{2}{t_c}\left[\frac{t_c}{2}+\frac{t_f^t+t_f^b}{4}-u_3^d-\mathring{u}_3^d\right]N_{13}^c+\frac{8}{t_c^2}M_{13}^c\Phi_3^c-2\frac{\partial \widehat{M}_{12}^a}{\partial x_2}-\widehat{N}_{13}^c\Bigg\}\delta u_3^a+$$

$$+\Bigg\{2N_{11}^a\left(\frac{\partial u_3^d}{\partial x_1}+\frac{\partial \mathring{u}_3^d}{\partial x_1}\right)+2N_{11}^d\left(\frac{\partial u_3^a}{\partial x_1}+\frac{\partial \mathring{u}_3^a}{\partial x_1}\right)+2N_{12}^a\left(\frac{\partial u_3^d}{\partial x_2}+\frac{\partial \mathring{u}_3^d}{\partial x_2}\right)$$

$$+2N_{12}^d\left(\frac{\partial u_3^a}{\partial x_2}+\frac{\partial \overset{\circ}{u}{}_3^a}{\partial x_2}\right)+2\frac{\partial M_{11}^d}{\partial x_1}+4\frac{\partial M_{12}^d}{\partial x_2}-\frac{8}{t_c^3}L_{13}^c\Phi_3^c-\frac{2}{t_c}\left(\Phi_3^c-\frac{t_f^t-t_f^b}{8}\right)N_{13}^c$$

$$-\frac{2}{t_c^2}\left(\frac{t_c}{2}-u_3^d-\overset{\circ}{u}{}_3^d\right)M_{13}^c-2\frac{\partial \widehat{M}_{12}^d}{\partial x_2}+\frac{\widehat{M}_{13}^c}{t_c}\Bigg\}\delta u_3^d$$

$$+\left\{\frac{8}{t_c^3}\left(\frac{t_c}{2}-u_3^d-\overset{\circ}{u}{}_3^d\right)L_{13}^c+\frac{32}{t_c^4}\Phi_3^cK_{13}^c-\frac{4}{t_c^2}\widehat{L}_{13}^c\right\}\delta\Phi_3^c+2\left(M_{11}^a-\widehat{M}_{11}^a\right)\delta\left(\frac{\partial u_3^a}{\partial x_1}\right)$$

$$+2\left(M_{11}^d-\widehat{M}_{11}^d\right)\delta\left(\frac{\partial u_3^d}{\partial x_1}\right)\Bigg]dx_2dt+\int_{t_0}^{t_1}\left[\left[\int_0^{l_1}\left\langle 2\left(N_{22}^a-\widehat{N}_{22}^a\right)\delta u_2^a\right.\right.\right.$$

$$+2\left(N_{21}^a-\widehat{N}_{21}^a\right)\delta u_1^a+2\left(N_{22}^d-\widehat{N}_{22}^d\right)\delta u_1^d+2\left(N_{21}^d-\widehat{N}_{21}^d\right)\delta u_1^d+$$

$$\Bigg\{2N_{22}^a\left(\frac{\partial u_3^a}{\partial x_2}+\frac{\partial \overset{\circ}{u}{}_3^a}{\partial x_2}\right)+2N_{21}^a\left(\frac{\partial u_3^a}{\partial x_1}+\frac{\partial \overset{\circ}{u}{}_3^a}{\partial x_1}\right)+2N_{22}^d\left(\frac{\partial u_3^d}{\partial x_2}+\frac{\partial \overset{\circ}{u}{}_3^d}{\partial x_2}\right)$$

$$+2N_{21}^d\left(\frac{\partial u_3^d}{\partial x_1}+\frac{\partial \overset{\circ}{u}{}_3^d}{\partial x_1}\right)+2\frac{\partial M_{22}^a}{\partial x_2}+4\frac{\partial M_{21}^a}{\partial x_1}+\frac{2}{t_c}\left[\frac{t_c}{2}+\frac{t_f^t+t_f^b}{4}-u_3^d-\overset{\circ}{u}{}_3^d\right]N_{23}^c$$

$$+\frac{8}{t_c^2}M_{23}^c\Phi_3^c-2\frac{\partial \widehat{M}_{21}^a}{\partial x_1}-\widehat{N}_{23}^c\Bigg\}\delta u_3^a+\Bigg\{2N_{22}^a\left(\frac{\partial u_3^d}{\partial x_2}+\frac{\partial \overset{\circ}{u}{}_3^d}{\partial x_2}\right)$$

$$+2N_{22}^d\left(\frac{\partial u_3^a}{\partial x_2}+\frac{\partial \overset{\circ}{u}{}_3^a}{\partial x_2}\right)+2N_{21}^a\left(\frac{\partial u_3^d}{\partial x_1}+\frac{\partial \overset{\circ}{u}{}_3^d}{\partial x_1}\right)+2N_{21}^d\left(\frac{\partial u_3^a}{\partial x_1}+\frac{\partial \overset{\circ}{u}{}_3^a}{\partial x_1}\right)+2\frac{\partial M_{22}^d}{\partial x_2}$$

$$+4\frac{\partial M_{21}^d}{\partial x_1}-\frac{8}{t_c^3}L_{23}^c\Phi_3^c-\frac{2}{t_c}\left(\Phi_3^c-\frac{t_f^t-t_f^b}{8}\right)N_{23}^c-\frac{2}{t_c^2}\left(\frac{t_c}{2}-u_3^d-\overset{\circ}{u}{}_3^d\right)M_{23}^c$$

$$-2\frac{\partial \widehat{M}_{21}^d}{\partial x_1}+\frac{\widehat{M}_{23}^c}{t_c}\Bigg\}\delta u_3^d+\left\{\frac{8}{t_c^3}\left(\frac{t_c}{2}-u_3^d-\overset{\circ}{u}{}_3^d\right)L_{23}^c+\frac{32}{t_c^4}\Phi_3^cK_{23}^c-\frac{4}{t_c^2}\widehat{L}_{23}^c\right\}\delta\Phi_3^c$$

$$+2\left(M_{22}^a-\widehat{M}_{22}^a\right)\delta\left(\frac{\partial u_3^a}{\partial x_2}\right)+2\left(M_{22}^d-\widehat{M}_{22}^d\right)\delta\left(\frac{\partial u_3^d}{\partial x_2}\right)\Bigg]dx_1dt=0 \quad (8.80)$$

This above equation can only be satisfied if the integrals are equal to zero each individually. This implies that, since the variational displacements are arbitrary, the coefficients of the variational displacements are zero. This results in the appearance of the equations of motion and the corresponding boundary conditions. The equations of motion appear as

$$\delta u_1^a: \qquad \frac{\partial N_{11}^a}{\partial x_1}+\frac{\partial N_{12}^a}{\partial x_2}=0 \quad (8.81)$$

$$\delta u_2^a: \qquad \frac{\partial N_{12}^a}{\partial x_1} + \frac{\partial N_{22}^a}{\partial x_2} = 0 \tag{8.82}$$

$$\delta u_1^d: \qquad \frac{\partial N_{11}^d}{\partial x_1} + \frac{\partial N_{12}^d}{\partial x_2} + \frac{N_{13}^c}{t_c} = 0 \tag{8.83}$$

$$\delta u_2^d: \qquad \frac{\partial N_{12}^d}{\partial x_1} + \frac{\partial N_{22}^d}{\partial x_2} + \frac{N_{23}^c}{t_c} = 0 \tag{8.84}$$

$$\delta \Phi_1^c: \qquad M_{13}^c = 0 \tag{8.85}$$

$$\delta \Phi_2^c: \qquad M_{23}^c = 0 \tag{8.86}$$

$$\delta \Omega_1^c: \qquad N_{13}^c + \frac{6}{t_c^2} L_{13}^c = 0 \tag{8.87}$$

$$\delta \Omega_2^c: \qquad N_{23}^c + \frac{6}{t_c^2} L_{23}^c = 0 \tag{8.88}$$

δu_3^a :

$$\begin{aligned}
&N_{11}^a \left(\frac{\partial^2 u_3^a}{\partial x_1^2} + \frac{\partial^2 \overset{\circ}{u}{}_3^a}{\partial x_1^2} + \frac{1}{R_1} \right) + 2N_{12}^a \left(\frac{\partial^2 u_3^a}{\partial x_1 \partial x_2} + \frac{\partial^2 \overset{\circ}{u}{}_3^a}{\partial x_1 \partial x_2} \right) + N_{22}^a \left(\frac{\partial^2 u_3^a}{\partial x_2^2} + \frac{\partial^2 \overset{\circ}{u}{}_3^a}{\partial x_2^2} + \frac{1}{R_2} \right) \\
&+ \frac{\partial^2 M_{11}^a}{\partial x_1^2} + 2\frac{\partial^2 M_{12}^a}{\partial x_1 \partial x_2} + \frac{\partial^2 M_{22}^a}{\partial x_2^2} + N_{11}^d \left(\frac{\partial^2 u_3^a}{\partial x_1^2} + \frac{\partial^2 \overset{\circ}{u}{}_3^a}{\partial x_1^2} \right) + 2N_{12}^d \left(\frac{\partial^2 u_3^d}{\partial x_1 \partial x_2} + \frac{\partial^2 \overset{\circ}{u}{}_3^d}{\partial x_1 \partial x_2} \right) \\
&+ N_{22}^d \left(\frac{\partial^2 u_3^a}{\partial x_2^2} + \frac{\partial^2 \overset{\circ}{u}{}_3^a}{\partial x_2^2} \right) + \frac{1}{t_c} \left(\frac{t_c}{2} + \frac{t_f^t + t_f^b}{4} - u_3^d - \overset{\circ}{u}{}_3^d \right) \left(\frac{\partial N_{13}^c}{\partial x_1} + \frac{\partial N_{23}^c}{\partial x_2} \right) \\
&- \frac{2}{t_c} \left(\frac{\partial u_3^d}{\partial x_2} + \frac{\circ \partial u_3^d}{\partial x_2} \right) N_{23}^c - \frac{2}{t_c} \left(\frac{\partial u_3^d}{\partial x_1} + \frac{\circ \partial u_3^d}{\partial x_1} \right) N_{13}^c \\
&+ \underline{\left(\frac{\partial u_3^a}{\partial x_1} + \frac{\circ \partial u_3^a}{\partial x_1} \right) \left(\frac{\partial N_{11}^a}{\partial x_1} + \frac{\partial N_{12}^a}{\partial x_2} \right)} + \underline{\left(\frac{\partial u_3^a}{\partial x_2} + \frac{\circ \partial u_3^a}{\partial x_2} \right) \left(\frac{\partial N_{22}^a}{\partial x_2} + \frac{\partial N_{12}^a}{\partial x_1} \right)} \\
&+ \underline{\left(\frac{\partial u_3^d}{\partial x_1} + \frac{\circ \partial u_3^d}{\partial x_1} \right) \left(\frac{\partial N_{11}^d}{\partial x_1} + \frac{\partial N_{12}^d}{\partial x_2} \right) \left(\frac{\partial u_3^d}{\partial x_2} + \frac{\circ \partial u_3^d}{\partial x_2} \right) \left(\frac{\partial N_{22}^d}{\partial x_2} + \frac{\partial N_{12}^d}{\partial x_1} \right)} \\
&+ \underline{\frac{2}{t_c^2} \Phi_3^c \left(\frac{\partial M_{13}^c}{\partial x_1} + \frac{\partial M_{23}^c}{\partial x_2} \right)} + \underline{\frac{2}{t_c^2} \frac{\partial \Phi_3^c}{\partial x_1} M_{13}^c} + \underline{\frac{2}{t_c^2} \frac{\partial \Phi_3^c}{\partial x_1} M_{13}^c} + \widehat{q}_3^a \\
&+ \left(C^a + \frac{C^c}{2} \right) \dot{u}_3^a + C^d \dot{u}_3^d + \left(m^a + \frac{m_c}{2} \right) \ddot{u}_3^a + m^d \ddot{u}_3^d - \frac{m_c}{3} \ddot{\Phi}_3^c = 0
\end{aligned} \tag{8.89}$$

The underlined terms in the ninth equation of motion vanish with the use of Eqs. (8.81), (8.82), (8.85), and (8.86). Utilizing these equations simplifies the ninth equation of motion to the following form.

$$
\begin{aligned}
&N_{11}^{a}\left(\frac{\partial^{2}u_{3}^{a}}{\partial x_{1}^{2}}+\frac{\partial^{2}\overset{\circ}{u}_{3}^{a}}{\partial x_{1}^{2}}+\frac{1}{R_{1}}\right)+2N_{12}^{a}\left(\frac{\partial^{2}u_{3}^{a}}{\partial x_{1}\partial x_{2}}+\frac{\partial^{2}\overset{\circ}{u}_{3}^{a}}{\partial x_{1}\partial x_{2}}\right)+N_{22}^{a}\left(\frac{\partial^{2}u_{3}^{a}}{\partial x_{2}^{2}}+\frac{\partial^{2}\overset{\circ}{u}_{3}^{a}}{\partial x_{2}^{2}}+\frac{1}{R_{2}}\right)\\
&+\frac{\partial^{2}M_{11}^{a}}{\partial x_{1}^{2}}+2\frac{\partial^{2}M_{12}^{a}}{\partial x_{1}\partial x_{2}}+\frac{\partial^{2}M_{22}^{a}}{\partial x_{2}^{2}}+N_{11}^{d}\left(\frac{\partial^{2}u_{3}^{a}}{\partial x_{1}^{2}}+\frac{\partial^{2}\overset{\circ}{u}_{3}^{a}}{\partial x_{1}^{2}}\right)+2N_{12}^{d}\left(\frac{\partial^{2}u_{3}^{d}}{\partial x_{1}\partial x_{2}}+\frac{\partial^{2}\overset{\circ}{u}_{3}^{d}}{\partial x_{1}\partial x_{2}}\right)\\
&+N_{22}^{d}\left(\frac{\partial^{2}u_{3}^{a}}{\partial x_{2}^{2}}+\frac{\partial^{2}\overset{\circ}{u}_{3}^{a}}{\partial x_{2}^{2}}\right)+\frac{1}{t_{c}}\left(\frac{t_{c}}{2}+\frac{t_{f}^{t}+t_{f}^{b}}{4}-u_{3}^{d}-\overset{\circ}{u}_{3}^{d}\right)\left(\frac{\partial N_{13}^{c}}{\partial x_{1}}+\frac{\partial N_{23}^{c}}{\partial x_{2}}\right)\\
&-\frac{2}{t_{c}}\left(\frac{\partial u_{3}^{d}}{\partial x_{2}}+\frac{\partial\overset{\circ}{u}_{3}^{d}}{\partial x_{2}}\right)N_{23}^{c}-\frac{2}{t_{c}}\left(\frac{\partial u_{3}^{d}}{\partial x_{1}}+\frac{\partial\overset{\circ}{u}_{3}^{d}}{\partial x_{1}}\right)N_{13}^{c}\widehat{q}_{3}^{a}+\left(C^{a}+\frac{C^{c}}{2}\right)\dot{u}_{3}^{a}\\
&+C^{d}\dot{u}_{3}^{d}+\left(m^{a}+\frac{m_{c}}{2}\right)\ddot{u}_{3}^{a}+m^{d}\ddot{u}_{3}^{d}-\frac{m_{c}}{3}\ddot{\Phi}_{3}^{c}=0
\end{aligned}
\tag{8.90}
$$

The tenth equation of motion appears as

δu_{3}^{d} :

$$
\begin{aligned}
&N_{11}^{d}\left(\frac{\partial^{2}u_{3}^{a}}{\partial x_{1}^{2}}+\frac{\partial^{2}\overset{\circ}{u}_{3}^{a}}{\partial x_{1}^{2}}+\frac{1}{R_{1}}\right)+2N_{12}^{d}\left(\frac{\partial^{2}u_{3}^{a}}{\partial x_{1}\partial x_{2}}+\frac{\partial^{2}\overset{\circ}{u}_{3}^{a}}{\partial x_{1}\partial x_{2}}\right)+N_{22}^{d}\left(\frac{\partial^{2}u_{3}^{a}}{\partial x_{2}^{2}}+\frac{\partial^{2}\overset{\circ}{u}_{3}^{a}}{\partial x_{2}^{2}}+\frac{1}{R_{2}}\right)\\
&+\frac{\partial^{2}M_{11}^{d}}{\partial x_{1}^{2}}+2\frac{\partial^{2}M_{12}^{d}}{\partial x_{1}\partial x_{2}}+\frac{\partial^{2}M_{22}^{d}}{\partial x_{2}^{2}}+N_{11}^{a}\left(\frac{\partial^{2}u_{3}^{d}}{\partial x_{1}^{2}}+\frac{\partial^{2}\overset{\circ}{u}_{3}^{d}}{\partial x_{1}^{2}}\right)+2N_{12}^{a}\left(\frac{\partial^{2}u_{3}^{d}}{\partial x_{1}\partial x_{2}}+\frac{\partial^{2}\overset{\circ}{u}_{3}^{d}}{\partial x_{1}\partial x_{2}}\right)\\
&+N_{22}^{a}\left(\frac{\partial^{2}u_{3}^{d}}{\partial x_{2}^{2}}+\frac{\partial^{2}\overset{\circ}{u}_{3}^{d}}{\partial x_{2}^{2}}\right)+\frac{2}{t_{c}}\left(\frac{t_{c}}{2}-u_{3}^{d}-\overset{\circ}{u}_{3}^{d}\right)N_{33}^{c}-\frac{2}{t_{c}}\left(\Phi_{3}^{c}+\frac{t_{f}^{t}-t_{f}^{b}}{16}\right)\left(\frac{\partial N_{13}^{c}}{\partial x_{1}}+\frac{\partial N_{23}^{c}}{\partial x_{2}}\right)\\
&-\frac{2}{t_{c}}\frac{\partial\Phi_{3}^{c}}{\partial x_{1}}N_{13}^{c}-\frac{2}{t_{c}}\frac{\partial\Phi_{3}^{c}}{\partial x_{2}}N_{23}^{c}-\frac{8}{t_{c}^{3}}\Phi_{3}^{c}\left(\frac{\partial L_{13}^{c}}{\partial x_{1}}+\frac{\partial L_{23}^{c}}{\partial x_{2}}\right)-\frac{4}{t_{c}^{3}}\frac{\partial\Phi_{3}^{c}}{\partial x_{1}}L_{13}^{c}-\frac{4}{t_{c}^{3}}\frac{\partial\Phi_{3}^{c}}{\partial x_{2}}L_{23}^{c}\\
&-\frac{8}{t_{c}^{3}}\Phi_{3}^{c}M_{33}^{c}+\underline{\left(\frac{\partial u_{3}^{d}}{\partial x_{1}}+\frac{\partial\overset{\circ}{u}_{3}^{d}}{\partial x_{1}}\right)\left(\frac{\partial N_{11}^{a}}{\partial x_{1}}+\frac{\partial N_{12}^{a}}{\partial x_{2}}\right)}+\underline{\left(\frac{\partial u_{3}^{d}}{\partial x_{2}}+\frac{\partial\overset{\circ}{u}_{3}^{d}}{\partial x_{2}}\right)\left(\frac{\partial N_{22}^{a}}{\partial x_{2}}+\frac{\partial N_{12}^{a}}{\partial x_{1}}\right)}\\
&+\underline{\left(\frac{\partial u_{3}^{a}}{\partial x_{1}}+\frac{\partial\overset{\circ}{u}_{3}^{a}}{\partial x_{1}}\right)\left(\frac{\partial N_{11}^{d}}{\partial x_{1}}+\frac{\partial N_{12}^{d}}{\partial x_{2}}+\frac{N_{13}^{c}}{t_{c}}\right)}+\underline{\left(\frac{\partial u_{3}^{a}}{\partial x_{2}}+\frac{\partial\overset{\circ}{u}_{3}^{a}}{\partial x_{2}}\right)\left(\frac{\partial N_{22}^{d}}{\partial x_{2}}+\frac{\partial N_{12}^{d}}{\partial x_{1}}+\frac{N_{23}^{c}}{t_{c}}\right)}\\
&\underline{-\frac{2}{t_{c}^{2}}\left(\frac{t_{c}}{2}-u_{3}^{d}-\overset{\circ}{u}_{3}^{d}\right)\left(\frac{\partial M_{13}^{c}}{\partial x_{1}}+\frac{\partial M_{23}^{c}}{\partial x_{2}}\right)}+\widehat{q}_{3}^{d}+C^{d}\dot{u}_{3}^{a}+C^{a}\dot{u}_{3}^{d}+\left(m^{a}+\frac{m_{c}}{6}\right)\ddot{u}_{3}^{d}+m^{d}\ddot{u}_{3}^{a}\\
&=0
\end{aligned}
\tag{8.91}
$$

The underlined terms in the tenth equation of motion vanish by using Eqs. (8.81)–(8.86). The tenth equation of motion simplifies to

$$N_{11}^{d}\left(\frac{\partial^2 u_3^a}{\partial x_1^2}+\frac{\partial^2 \overset{\circ}{u}{}_3^a}{\partial x_1^2}+\frac{1}{R_1}\right)+2N_{12}^{d}\left(\frac{\partial^2 u_3^a}{\partial x_1\partial x_2}+\frac{\partial^2 \overset{\circ}{u}{}_3^a}{\partial x_1\partial x_2}\right)+N_{22}^{d}\left(\frac{\partial^2 u_3^a}{\partial x_2^2}+\frac{\partial^2 \overset{\circ}{u}{}_3^a}{\partial x_2^2}+\frac{1}{R_2}\right)$$
$$+\frac{\partial^2 M_{11}^d}{\partial x_1^2}+2\frac{\partial^2 M_{12}^d}{\partial x_1\partial x_2}+\frac{\partial^2 M_{22}^d}{\partial x_2^2}+N_{11}^{a}\left(\frac{\partial^2 u_3^d}{\partial x_1^2}+\frac{\partial^2 \overset{\circ}{u}{}_3^d}{\partial x_1^2}\right)+2N_{12}^{a}\left(\frac{\partial^2 u_3^d}{\partial x_1\partial x_2}+\frac{\partial^2 \overset{\circ}{u}{}_3^d}{\partial x_1\partial x_2}\right)$$
$$+N_{22}^{a}\left(\frac{\partial^2 u_3^d}{\partial x_2^2}+\frac{\partial^2 \overset{\circ}{u}{}_3^d}{\partial x_2^2}\right)+\frac{2}{t_c}\left(\frac{t_c}{2}-u_3^d-\overset{\circ}{u}{}_3^d\right)N_{33}^c-\frac{2}{t_c}\left(\Phi_3^c+\frac{t_f^t-t_f^b}{16}\right)\left(\frac{\partial N_{13}^c}{\partial x_1}+\frac{\partial N_{23}^c}{\partial x_2}\right)$$
$$-\frac{2}{t_c}\frac{\partial \Phi_3^c}{\partial x_1}N_{13}^c-\frac{2}{t_c}\frac{\partial \Phi_3^c}{\partial x_2}N_{23}^c-\underline{\underline{\frac{8}{t_c^3}\Phi_3^c\left(\frac{\partial L_{13}^c}{\partial x_1}+\frac{\partial L_{23}^c}{\partial x_2}\right)}}-\underline{\underline{\frac{4}{t_c^3}\frac{\partial \Phi_3^c}{\partial x_1}L_{13}^c}}-\underline{\underline{\frac{4}{t_c^3}\frac{\partial \Phi_3^c}{\partial x_2}L_{23}^c}}$$
$$-\frac{8}{t_c^3}\Phi_3^c M_{33}^c+\widehat{q}_3^d+C^d\dot{u}_3^a+C^a\dot{u}_3^d+\left(m^a+\frac{m_c}{6}\right)\ddot{u}_3^d+m^d\ddot{u}_3^a=0 \tag{8.92}$$

The above tenth equation of motion can be further simplified through the use of Eqs. (8.87) and (8.88) by expressing L_{13}^c and L_{23}^c in terms of N_{13}^c and N_{23}^c as shown below. The three double underlined terms expressed in terms of N_{13}^c and N_{23}^c become

1. $$\frac{8}{t_c^3}\Phi_3^c\left(\frac{\partial L_{13}^c}{\partial x_1}+\frac{\partial L_{23}^c}{\partial x_2}\right)=-\frac{4}{3t_c}\Phi_3^c\left(\frac{\partial N_{13}^c}{\partial x_1}+\frac{\partial N_{23}^c}{\partial x_2}\right) \tag{8.93}$$

2. $$\frac{4}{t_c^3}\frac{\partial \Phi_3^c}{\partial x_1}L_{13}^c=-\frac{2}{3t_c}\frac{\partial \Phi_3^c}{\partial x_1}N_{13}^c \tag{8.94}$$

3. $$\frac{4}{t_c^3}\frac{\partial \Phi_3^c}{\partial x_2}L_{23}^c=-\frac{2}{3t_c}\frac{\partial \Phi_3^c}{\partial x_2}N_{23}^c \tag{8.95}$$

Utilizing these expressions in terms of N_{13}^c and N_{23}^c in Eq. (8.92) simplifies the tenth equation of motion further while eliminating the variables L_{13}^c and L_{23}^c. The tenth equation of motion is now expressed as

$$N_{11}^d\left(\frac{\partial^2 u_3^a}{\partial x_1^2}+\frac{\partial^2 \overset{\circ}{u}{}_3^a}{\partial x_1^2}+\frac{1}{R_1}\right)+2N_{12}^d\left(\frac{\partial^2 u_3^a}{\partial x_1 \partial x_2}+\frac{\partial^2 \overset{\circ}{u}{}_3^a}{\partial x_1 \partial x_2}\right)+N_{22}^d\left(\frac{\partial^2 u_3^a}{\partial x_2^2}+\frac{\partial^2 \overset{\circ}{u}{}_3^a}{\partial x_2^2}+\frac{1}{R_2}\right)$$
$$+\frac{\partial^2 M_{11}^d}{\partial x_1^2}+2\frac{\partial^2 M_{12}^d}{\partial x_1 \partial x_2}+\frac{\partial^2 M_{22}^d}{\partial x_2^2}+N_{11}^a\left(\frac{\partial^2 u_3^d}{\partial x_1^2}+\frac{\partial^2 \overset{\circ}{u}{}_3^d}{\partial x_1^2}\right)+2N_{12}^a\left(\frac{\partial^2 u_3^d}{\partial x_1 \partial x_2}+\frac{\partial^2 \overset{\circ}{u}{}_3^d}{\partial x_1 \partial x_2}\right)$$
$$+N_{22}^a\left(\frac{\partial^2 u_3^d}{\partial x_2^2}+\frac{\partial^2 \overset{\circ}{u}{}_3^d}{\partial x_2^2}\right)+\frac{2}{t_c}\left(\frac{t_c}{2}-u_3^d-\overset{\circ}{u}{}_3^d\right)N_{33}^c-\frac{2}{3t_c}\left(\Phi_3^c+\frac{t_f^t-t_f^b}{16}\right)\left(\frac{\partial N_{13}^c}{\partial x_1}+\frac{\partial N_{23}^c}{\partial x_2}\right)$$
$$-\frac{4}{3t_c}\frac{\partial \Phi_3^c}{\partial x_1}N_{13}^c-\frac{4}{3t_c}\frac{\partial \Phi_3^c}{\partial x_2}N_{23}^c-\frac{8}{t_c^3}\Phi_3^c M_{33}^c+\widehat{q}_3^d+C^d\dot{u}_3^a+C^a\dot{u}_3^d+\left(m^a+\frac{m_c}{6}\right)\ddot{u}_3^d$$
$$+m^d\ddot{u}_3^a=0 \tag{8.96}$$

The final eleventh equation of motion appears as

$\delta\Phi_3^c$:

$$-\frac{4}{t_c}\left(\frac{\partial u_3^d}{\partial x_1}+\frac{\partial \overset{\circ}{u}{}_3^d}{\partial x_1}\right)N_{13}^c-\frac{4}{t_c}\left(\frac{\partial u_3^d}{\partial x_2}+\frac{\partial \overset{\circ}{u}{}_3^d}{\partial x_2}\right)N_{23}^c-\underline{\frac{8}{t_c^3}\left(\frac{\partial L_{13}^c}{\partial x_1}+\frac{\partial L_{23}^c}{\partial x_2}\right)\left(\frac{t_c}{2}-u_3^d-\overset{\circ}{u}{}_3^d\right)}$$
$$\underline{-\frac{8}{t_c^3}L_{13}^c\left(\frac{\partial u_3^d}{\partial x_1}+\frac{\partial \overset{\circ}{u}{}_3^d}{\partial x_1}\right)}-\underline{\frac{8}{t_c^3}L_{23}^c\left(\frac{\partial u_3^d}{\partial x_2}+\frac{\partial \overset{\circ}{u}{}_3^d}{\partial x_2}\right)}+\frac{16}{t_c^2}\Phi_3^c N_{33}^c+\frac{16}{t_c^3}\left(\frac{t_c}{2}-u_3^d-\overset{\circ}{u}{}_3^d\right)M_{33}^c$$
$$+\frac{64}{t_c^4}\Phi_3^c L_{33}^c-\frac{32}{t_c^4}\Phi_3^c\left(\frac{\partial K_{13}^c}{\partial x_1}+\frac{\partial K_{23}^c}{\partial x_2}\right)+\underline{\frac{8}{t_c^2}M_{13}^c\left(\frac{\partial u_3^a}{\partial x_1}+\frac{\partial \overset{\circ}{u}{}_3^a}{\partial x_1}\right)}$$
$$\underline{+\frac{8}{t_c^2}M_{23}^c\left(\frac{\partial u_3^a}{\partial x_2}+\frac{\partial \overset{\circ}{u}{}_3^a}{\partial x_2}\right)}+\frac{2m_c}{3}\ddot{u}_3^a+\frac{6m_c}{15}\ddot{\Phi}_3^c=0 \tag{8.97}$$

Just as was done previously, expressing L_{13}^c and L_{23}^c in terms of N_{13}^c and N_{23}^c with the use of Eqs. (8.87) and (8.88) for the first and second underlined terms in the above equation along with utilizing Eqs. (8.85) and (8.86) for the third and fourth underlined terms in the above equation results in the eleventh equation of motion being simplified to

$$\frac{4}{3t_c}\left(\frac{\partial u_3^d}{\partial x_1}+\frac{\partial \overset{\circ}{u}{}_3^d}{\partial x_1}\right)N_{13}^c+\frac{4}{3t_c}\left(\frac{\partial u_3^d}{\partial x_2}+\frac{\partial \overset{\circ}{u}{}_3^d}{\partial x_2}\right)N_{23}^c+\frac{8}{t_c^2}\Phi_3^c N_{33}^c-\frac{8}{t_c^3}\left(\frac{t_c}{2}-u_3^d-\overset{\circ}{u}{}_3^d\right)M_{33}^c$$
$$-\frac{32}{t_c^4}\Phi_3^c L_{33}^c-\frac{2}{3t_c}\left(\frac{t_c}{2}-u_3^d-\overset{\circ}{u}{}_3^d\right)\left(\frac{\partial N_{13}^c}{\partial x_1}+\frac{\partial N_{23}^c}{\partial x_2}\right)+\frac{16}{t_c^4}\Phi_3^c\left(\frac{\partial K_{13}^c}{\partial x_1}+\frac{\partial K_{23}^c}{\partial x_2}\right)$$
$$-\frac{m_c}{3}\ddot{u}_3^a-\frac{3m_c}{15}\ddot{\Phi}_3^c=0 \tag{8.98}$$

In summary, the final form of the eleven equations of motion are

$$\frac{\partial N_{11}^{a}}{\partial x_1}+\frac{\partial N_{12}^{a}}{\partial x_2}=0 \tag{8.99}$$

$$\frac{\partial N_{12}^{a}}{\partial x_1}+\frac{\partial N_{22}^{a}}{\partial x_2}=0 \tag{8.100}$$

$$\frac{\partial N_{11}^{d}}{\partial x_1}+\frac{\partial N_{12}^{d}}{\partial x_2}+\frac{N_{13}^{c}}{t_c}=0 \tag{8.101}$$

$$\frac{\partial N_{12}^{d}}{\partial x_1}+\frac{\partial N_{22}^{d}}{\partial x_2}+\frac{N_{23}^{c}}{t_c}=0 \tag{8.102}$$

$$M_{13}^{c}=0 \tag{8.103}$$

$$M_{23}^{c}=0 \tag{8.104}$$

$$N_{13}^{c}+\frac{6}{t_c^2}L_{13}^{c}=0 \tag{8.105}$$

$$N_{23}^{c}+\frac{6}{t_c^2}L_{23}^{c}=0 \tag{8.106}$$

$$\begin{aligned}
&N_{11}^{a}\left(\frac{\partial^2 u_3^a}{\partial x_1^2}+\frac{\partial^2 \overset{\circ}{u}{}_3^{a}}{\partial x_1^2}+\frac{1}{R_1}\right)+2N_{12}^{a}\left(\frac{\partial^2 u_3^a}{\partial x_1\partial x_2}+\frac{\partial^2 \overset{\circ}{u}{}_3^{a}}{\partial x_1\partial x_2}\right)+N_{22}^{a}\left(\frac{\partial^2 u_3^a}{\partial x_2^2}+\frac{\partial^2 \overset{\circ}{u}{}_3^{a}}{\partial x_2^2}+\frac{1}{R_2}\right)\\
&+\frac{\partial^2 M_{11}^{a}}{\partial x_1^2}+2\frac{\partial^2 M_{12}^{a}}{\partial x_1\partial x_2}+\frac{\partial^2 M_{22}^{a}}{\partial x_2^2}+N_{11}^{d}\left(\frac{\partial^2 u_3^a}{\partial x_1^2}+\frac{\partial^2 \overset{\circ}{u}{}_3^{a}}{\partial x_1^2}\right)+2N_{12}^{d}\left(\frac{\partial^2 u_3^d}{\partial x_1\partial x_2}+\frac{\partial^2 \overset{\circ}{u}{}_3^{d}}{\partial x_1\partial x_2}\right)\\
&+N_{22}^{d}\left(\frac{\partial^2 u_3^a}{\partial x_2^2}+\frac{\partial^2 \overset{\circ}{u}{}_3^{a}}{\partial x_2^2}\right)+\frac{1}{t_c}\left(\frac{t_c}{2}+\frac{t_f^t+t_f^b}{4}-u_3^d-\overset{\circ}{u}{}_3^{d}\right)\left(\frac{\partial N_{13}^{c}}{\partial x_1}+\frac{\partial N_{23}^{c}}{\partial x_2}\right)\\
&-\frac{2}{t_c}\left(\frac{\partial u_3^d}{\partial x_2}+\frac{\circ\,\partial u_3^d}{\partial x_2}\right)N_{23}^{c}-\frac{2}{t_c}\left(\frac{\partial u_3^d}{\partial x_1}+\frac{\circ\,\partial u_3^d}{\partial x_1}\right)N_{13}^{c}\widehat{q}_3^{a}+\left(C^a+\frac{C^c}{2}\right)\dot{u}_3^a+C^d\dot{u}_3^d\\
&+\left(m^a+\frac{m_c}{2}\right)\ddot{u}_3^a+m^d\ddot{u}_3^d-\frac{m_c}{3}\ddot{\Phi}_3^{c}=0
\end{aligned} \tag{8.107}$$

$$
N_{11}^{d}\left(\frac{\partial^{2}u_{3}^{a}}{\partial x_{1}^{2}}+\frac{\partial^{2}\overset{\circ}{u}{}_{3}^{a}}{\partial x_{1}^{2}}+\frac{1}{R_{1}}\right)+2N_{12}^{d}\left(\frac{\partial^{2}u_{3}^{a}}{\partial x_{1}\partial x_{2}}+\frac{\partial^{2}\overset{\circ}{u}{}_{3}^{a}}{\partial x_{1}\partial x_{2}}\right)+N_{22}^{d}\left(\frac{\partial^{2}u_{3}^{a}}{\partial x_{2}^{2}}+\frac{\partial^{2}\overset{\circ}{u}{}_{3}^{a}}{\partial x_{2}^{2}}+\frac{1}{R_{2}}\right)
$$
$$
+\frac{\partial^{2}M_{11}^{d}}{\partial x_{1}^{2}}+2\frac{\partial^{2}M_{12}^{d}}{\partial x_{1}\partial x_{2}}+\frac{\partial^{2}M_{22}^{d}}{\partial x_{2}^{2}}+N_{11}^{a}\left(\frac{\partial^{2}u_{3}^{d}}{\partial x_{1}^{2}}+\frac{\partial^{2}\overset{\circ}{u}{}_{3}^{d}}{\partial x_{1}^{2}}\right)+2N_{12}^{a}\left(\frac{\partial^{2}u_{3}^{d}}{\partial x_{1}\partial x_{2}}+\frac{\partial^{2}\overset{\circ}{u}{}_{3}^{d}}{\partial x_{1}\partial x_{2}}\right)
$$
$$
+N_{22}^{a}\left(\frac{\partial^{2}u_{3}^{d}}{\partial x_{2}^{2}}+\frac{\partial^{2}\overset{\circ}{u}{}_{3}^{d}}{\partial x_{2}^{2}}\right)+\frac{2}{t_{c}}\left(\frac{t_{c}}{2}-u_{3}^{d}-\overset{\circ}{u}{}_{3}^{d}\right)N_{33}^{c}-\frac{2}{3t_{c}}\left(\Phi_{3}^{c}+\frac{t_{f}^{t}-t_{f}^{b}}{16}\right)\left(\frac{\partial N_{13}^{c}}{\partial x_{1}}+\frac{\partial N_{23}^{c}}{\partial x_{2}}\right)
$$
$$
-\frac{4}{3t_{c}}\frac{\partial\Phi_{3}^{c}}{\partial x_{1}}N_{13}^{c}-\frac{4}{3t_{c}}\frac{\partial\Phi_{3}^{c}}{\partial x_{2}}N_{23}^{c}-\frac{8}{t_{c}^{3}}\Phi_{3}^{c}M_{33}^{c}+\widehat{q}_{3}^{d}+C^{d}\dot{u}_{3}^{a}+C^{a}\dot{u}_{3}^{d}+\left(m^{a}+\frac{m_{c}}{6}\right)\ddot{u}_{3}^{d}
$$
$$
+m^{d}\ddot{u}_{3}^{a}=0 \tag{8.108}
$$

$$
\frac{4}{3t_{c}}\left(\frac{\partial u_{3}^{d}}{\partial x_{1}}+\frac{\partial\overset{\circ}{u}{}_{3}^{d}}{\partial x_{1}}\right)N_{13}^{c}+\frac{4}{3t_{c}}\left(\frac{\partial u_{3}^{d}}{\partial x_{2}}+\frac{\partial\overset{\circ}{u}{}_{3}^{d}}{\partial x_{2}}\right)N_{23}^{c}+\frac{8}{t_{c}^{2}}\Phi_{3}^{c}N_{33}^{c}-\frac{8}{t_{c}^{3}}\left(\frac{t_{c}}{2}-u_{3}^{d}-\overset{\circ}{u}{}_{3}^{d}\right)M_{33}^{c}
$$
$$
-\frac{32}{t_{c}^{4}}\Phi_{3}^{c}L_{33}^{c}-\frac{2}{3t_{c}}\left(\frac{t_{c}}{2}-u_{3}^{d}-\overset{\circ}{u}{}_{3}^{d}\right)\left(\frac{\partial N_{13}^{c}}{\partial x_{1}}+\frac{\partial N_{23}^{c}}{\partial x_{2}}\right)+\frac{16}{t_{c}^{4}}\Phi_{3}^{c}\left(\frac{\partial K_{13}^{c}}{\partial x_{1}}+\frac{\partial K_{23}^{c}}{\partial x_{2}}\right)
$$
$$
-\frac{m_{c}}{3}\ddot{u}_{3}^{a}-\frac{3m_{c}}{15}\ddot{\Phi}_{3}^{c}=0 \tag{8.109}
$$

Setting the boundary integrals to zero provides the associated boundary conditions along the edges $x_n = \text{const}(n = 1, 2)$

$$
u_{n}^{a}=\widehat{u}_{n}^{a}\quad\text{or}\quad N_{nn}^{a}=\widehat{N}_{nn}^{a} \tag{8.110}
$$

$$
u_{t}^{a}=\widehat{u}_{t}^{a}\quad\text{or}\quad N_{nt}^{a}=\widehat{N}_{nt}^{a} \tag{8.111}
$$

$$
u_{n}^{d}=\widehat{u}_{n}^{d}\quad\text{or}\quad N_{nn}^{d}=\widehat{N}_{nn}^{d} \tag{8.112}
$$

$$
u_{t}^{d}=\widehat{u}_{t}^{d}\quad\text{or}\quad N_{nt}^{d}=\widehat{N}_{nt}^{d} \tag{8.113}
$$

$$u_3^a = \widehat{u}_3^a \quad \text{or} \quad \left(\frac{\partial u_3^a}{\partial x_n} + \frac{\partial \overset{\circ}{u}{}_3^a}{\partial x_n}\right) N_{nn}^a + \left(\frac{\partial u_3^a}{\partial x_t} + \frac{\partial \overset{\circ}{u}{}_3^a}{\partial x_t}\right) N_{nt}^a + \left(\frac{\partial u_3^d}{\partial x_n} + \frac{\partial \overset{\circ}{u}{}_3^d}{\partial x_n}\right) N_{nn}^d +$$
$$\left(\frac{\partial u_3^d}{\partial x_t} + \frac{\partial \overset{\circ}{u}{}_3^d}{\partial x_t}\right) N_{nt}^d + \frac{\partial M_{nn}^a}{\partial x_n} + 2\frac{\partial M_{nt}^a}{\partial x_t} + \frac{1}{t_c}\left(\frac{t_c}{2} + \frac{t_f^t + t_f^b}{4} - u_3^d - \overset{\circ}{u}{}_3^d\right) N_{13}^c$$
$$= \frac{\partial \widehat{M}_{nt}^a}{\partial x_t} + \frac{1}{2}\widehat{N}_{n3}^c \tag{8.114}$$

$$u_3^d = \widehat{u}_3^d \quad \text{or} \quad \left(\frac{\partial u_3^d}{\partial x_n} + \frac{\partial \overset{\circ}{u}{}_3^d}{\partial x_n}\right) N_{nn}^a + \left(\frac{\partial u_3^d}{\partial x_t} + \frac{\partial \overset{\circ}{u}{}_3^d}{\partial x_t}\right) N_{nt}^a + \left(\frac{\partial u_3^a}{\partial x_n} + \frac{\partial \overset{\circ}{u}{}_3^a}{\partial x_n}\right) N_{nn}^d +$$
$$\left(\frac{\partial u_3^a}{\partial x_t} + \frac{\partial \overset{\circ}{u}{}_3^a}{\partial x_t}\right) N_{nt}^d + \frac{\partial M_{nn}^d}{\partial x_n} + 2\frac{\partial M_{nt}^d}{\partial x_t} - \frac{2}{t_c}\left(\frac{\Phi_3^c}{3} - \frac{t_f^t - t_f^b}{8}\right) N_{n3}^c =$$
$$\frac{\partial \widehat{M}_{nt}^d}{\partial x_t} - \frac{\widehat{M}_{n3}^c}{t_c} \tag{8.115}$$

$$\Phi_3^c = \widehat{\Phi}_3^c \quad \text{or} \quad \frac{2}{t_c}\left(\frac{t_c}{2} - u_3^d - \circ \quad u_3^d\right) N_{n3}^c - \frac{48}{t_c^4}\Phi_3^c K_{n3}^c = \widehat{N}_{n3}^c \tag{8.116}$$

$$\frac{\partial u_3^a}{\partial x_n} = \frac{\partial \widehat{u}_3^a}{\partial x_n} \quad \text{or} \quad M_{nn}^a = \widehat{M}_{nn}^a \tag{8.117}$$

$$\frac{\partial u_3^d}{\partial x_n} = \frac{\partial \widehat{u}_3^d}{\partial x_n} \quad \text{or} \quad M_{nn}^d = \widehat{M}_{nn}^d \tag{8.118}$$

where n and t are the normal and tangential directions to the boundary When $n = 1$, $t = 2$ and when $n = 2$, $t = 1$. It should be mentioned that the sixth and seventh boundary conditions have been simplified from their original forms just as was done for the ninth, tenth, and the eleventh equations of motion. The terms that contained L_{13}^c and L_{23}^c were expressed in terms of N_{13}^c and N_{23}^c with the use of the pertinent equations of motion. In addition, terms containing M_{13}^c and M_{23}^c were set to zero. For the case of a symmetric sandwich panel, these boundary conditions and equations of motion agree with Hohe and Librescu (2006).

8.5.2 Boundary Conditions

For the case of **simply supported** boundary conditions the following conditions must be satisfied.

Along the edges $x_n = 0,\ L_n$

$$N^a_{nn} = N^d_{nn} = N^a_{nt} = N^d_{nt} = M^a_{nn} = M^d_{nn} = u^a_3 = u^d_3 = 0 \tag{8.119}$$

For the case of **clamped** boundary conditions, the following conditions must be satisfied.

$$u^a_n = u^a_t = u^d_n = u^d_t = u^a_3 = u^d_3 = \frac{\partial u^a_3}{\partial x_n} = \frac{\partial u^d_3}{\partial x_n} = 0 \tag{8.120}$$

8.6 The Stress Resultants, Stress Couples, and Stiffnesses

The local stress resultants and stress couples can be expressed in terms of the strains by substituting Eqs. (8.42)–(8.44) into Eqs. (8.50)–(8.52). For the top face, the matrix form (see Hause 2012) becomes

$$\begin{pmatrix} N^t_{11} \\ N^t_{22} \\ N^t_{12} \\ M^t_{11} \\ M^t_{22} \\ M^t_{12} \end{pmatrix} = \begin{pmatrix} A^t_{11} & A^t_{12} & A^t_{16} & B^t_{11} & B^t_{12} & B^t_{16} \\ & A^t_{22} & A^t_{26} & B^t_{12} & B^t_{22} & B^t_{26} \\ & & A^t_{66} & B^t_{16} & B^t_{26} & B^t_{66} \\ & & & D^t_{11} & D^t_{12} & D^t_{16} \\ & & & & D^t_{22} & D^t_{26} \\ \text{Sym} & & & & & D^t_{66} \end{pmatrix} \begin{pmatrix} \overline{\gamma}^t_{11} \\ \overline{\gamma}^t_{22} \\ 2\overline{\gamma}^t_{12} \\ \kappa^t_{11} \\ \kappa^t_{22} \\ 2\kappa^t_{12} \end{pmatrix} - \begin{pmatrix} N^t_{11} \\ N^t_{22} \\ N^t_{12} \\ M^t_{11} \\ M^t_{22} \\ M^t_{12} \end{pmatrix}^T - \begin{pmatrix} N^t_{11} \\ N^t_{22} \\ N^t_{12} \\ M^t_{11} \\ M^t_{22} \\ M^t_{12} \end{pmatrix}^m \tag{8.121}$$

where the local stiffnesses $A^t_{ij}, B^t_{ij}, D^t_{ij}$ are defined as

$$\left(A^t_{ij}, B^t_{ij}, D^t_{ij}\right) = \int_{-\frac{t_c}{2}-t^t_f}^{-\frac{t_c}{2}} \widehat{Q}^t_{ij} \left\{ 1, \left(x_3 + \frac{t_c + t^t_f}{2}\right), \left(x_3 + \frac{t_c + t^t_f}{2}\right)^2 \right\} dx_3 \tag{8.122}$$

For the bottom face, the local stress resultants and couples in matrix form become

$$\begin{pmatrix} N_{11}^b \\ N_{22}^b \\ N_{12}^b \\ M_{11}^b \\ M_{22}^b \\ M_{12}^b \end{pmatrix} = \begin{pmatrix} A_{11}^b & A_{12}^b & A_{16}^b & B_{11}^b & B_{12}^b & B_{16}^b \\ & A_{22}^b & A_{26}^b & B_{12}^b & B_{22}^b & B_{26}^b \\ & & A_{66}^b & B_{16}^b & B_{26}^b & B_{66}^b \\ & & & D_{11}^b & D_{12}^b & D_{16}^b \\ & & & & D_{22}^b & D_{26}^b \\ \text{Sym} & & & & & D_{66}^b \end{pmatrix} \begin{pmatrix} \bar{\gamma}_{11}^b \\ \bar{\gamma}_{22}^b \\ 2\bar{\gamma}_{12}^b \\ \kappa_{11}^b \\ \kappa_{22}^b \\ 2\kappa_{12}^b \end{pmatrix} - \begin{pmatrix} N_{11}^b \\ N_{22}^b \\ N_{12}^b \\ M_{11}^b \\ M_{22}^b \\ M_{12}^b \end{pmatrix}^T - \begin{pmatrix} N_{11}^b \\ N_{22}^b \\ N_{12}^b \\ M_{11}^b \\ M_{22}^b \\ M_{12}^b \end{pmatrix}^m \tag{8.123}$$

where the local stiffnesses $A_{ij}^b, B_{ij}^b, D_{ij}^b$ are defined as

$$\left(A_{ij}^b, B_{ij}^b, D_{ij}^b\right) = \int_{\frac{t_c}{2}}^{\frac{t_c}{2}+t_f^b} \widehat{Q}_{ij}^b \left\{1, \left(x_3 - \frac{t_c + t_f^b}{2}\right), \left(x_3 - \frac{t_c + t_f^b}{2}\right)^2\right\} dx_3 \tag{8.124}$$

If the structural tailoring of the faces is such that the stacking sequence is symmetric with respect to its local mid-surface, then $\left[B_{ij}^{(t,b)}\right] = 0$. The stress resultants and stress couples for the core are expressed as

$$N_{33}^c = A_{33}^c \bar{\gamma}_{33} + \left(D_{33}^c - \frac{t_c^2}{4} A_{33}^c\right) \eta_{33}^c \tag{8.125}$$

$$N_{23}^c = 2A_{44}^c \bar{\gamma}_{23} + \left(D_{44}^c - \frac{t_c^2}{4} A_{44}^c\right) 2\eta_{23}^c \tag{8.126}$$

$$N_{13}^c = 2A_{55}^c \bar{\gamma}_{13} + \left(D_{55}^c - \frac{t_c^2}{4} A_{55}^c\right) 2\eta_{13}^c \tag{8.127}$$

$$M_{33}^c = D_{33}^c \kappa_{33}^c \tag{8.128}$$

$$M_{23}^c = 2D_{44}^c \kappa_{23}^c + 2\left(F_{44}^c + \frac{t_c^2}{4} D_{44}^c - \frac{t_c^4}{16} A_{44}^c\right) \upsilon_{23}^c \tag{8.129}$$

$$M_{13}^c = 2D_{55}^c \kappa_{13}^c + 2\left(F_{55}^c + \frac{t_c^2}{4} D_{55}^c - \frac{t_c^4}{16} A_{55}^c\right) \upsilon_{13}^c \tag{8.130}$$

$$L_{33}^c = \left(D_{33}^c - \frac{t_c^2}{4} A_{33}^c\right) \bar{\gamma}_{33}^c + F_{33}^c \eta_{33}^c \tag{8.131}$$

$$L_{23}^c = 2\left(D_{44}^c - \frac{t_c^2}{4} A_{44}^c\right) \bar{\gamma}_{23}^c + 2F_{44}^c \eta_{23}^c \tag{8.132}$$

$$L_{13}^{c}=2\left(D_{55}^{c}-\frac{t_{c}^{2}}{4}A_{55}^{c}\right)\overline{\gamma}_{13}^{c}+2F_{55}^{c}\eta_{13}^{c} \tag{8.133}$$

$$K_{23}^{c}=2\left(F_{44}^{c}+\frac{t_{c}^{2}}{4}D_{44}^{c}-\frac{t_{c}^{4}}{16}A_{44}^{c}\right)\kappa_{23}^{c}+2H_{44}^{c}\upsilon_{23}^{c} \tag{8.1314}$$

$$K_{13}^{c}=2\left(F_{55}^{c}+\frac{t_{c}^{2}}{4}D_{55}^{c}-\frac{t_{c}^{4}}{16}A_{55}^{c}\right)\kappa_{13}^{c}+2H_{55}^{c}\upsilon_{13}^{c} \tag{8.135}$$

where the core stiffnesses are given by

$$\left(A_{ii}^{c},D_{ii}^{c},F_{ii}^{c},H_{ii}^{c}\right)=\int_{-\frac{t_c}{2}}^{\frac{t_c}{2}}Q_{ii}^{c}\left(1,x_3^2,\left(x_3^2-\frac{t_c^2}{4}\right)^2,\left(x_3^2-\frac{t_c^2}{4}\right)^2x_3^2\right)dx_3,\quad i=(3,4,5) \tag{8.136}$$

Note: No summation with respect to the repeated indices is assumed. The core transverse and normal moduli are given as

8.7 Comments

In Chap. 7, an extremely advanced theory of doubly curved sandwich panels considering a transversely compressible core was introduced where a first-order power series was considered for the core transverse displacement function. In contrast with the present case, a second-order power series was considered for the core transverse displacement. This enhanced the first theory on a higher level providing an additional level of fidelity. Again, the following application addressing the buckling and post-buckling response only provides the solution methodology to demonstrate how to apply and solve the equations. There are several results by Hohe and Librescu (2003, 2006) that the reader is referred to.

8.8 Application – Buckling/Post-Buckling

8.8.1 *Comments*

In relation to the following application, the following assumptions are made to enable a more favorable solution. The face sheets are symmetric with respect to their local and global mid-surfaces. The face sheets are assumed to contain orthotropic laminae where the axes of orthotropy coincide with the geometrical axes. In this way, the stiffness terms A_{16} and A_{26} vanish. Additionally, it is assumed that the core is orthotropic and the axes of orthotropy coincide with the geometrical axes as well. Finally, the Kirchhoff assumptions are assumed to apply in the facings.

8.8.2 Governing Equations

From Eqs. (8.99)–(8.109) the governing equations are

- *Equations of motion*

$$\frac{\partial N_{11}^a}{\partial x_1} + \frac{\partial N_{12}^a}{\partial x_2} = 0 \tag{8.137}$$

$$\frac{\partial N_{12}^a}{\partial x_1} + \frac{\partial N_{22}^a}{\partial x_2} = 0 \tag{8.138}$$

$$\frac{\partial N_{11}^d}{\partial x_1} + \frac{\partial N_{12}^d}{\partial x_2} + \frac{N_{13}^c}{t_c} = 0 \tag{8.139}$$

$$\frac{\partial N_{12}^d}{\partial x_1} + \frac{\partial N_{22}^d}{\partial x_2} + \frac{N_{23}^c}{t_c} = 0 \tag{8.140}$$

$$M_{13}^c = 0 \tag{8.141}$$

$$M_{23}^c = 0 \tag{8.142}$$

$$N_{13}^c + \frac{6}{t_c^2} L_{13}^c = 0 \tag{8.143}$$

$$N_{23}^c + \frac{6}{t_c^2} L_{23}^c = 0 \tag{8.144}$$

$$
\begin{aligned}
&N_{11}^{a}\left(\frac{\partial^{2}u_{3}^{a}}{\partial x_{1}^{2}}+\frac{\partial^{2}\overset{\circ a}{u_{3}}}{\partial x_{1}^{2}}+\frac{1}{R_{1}}\right)+2N_{12}^{a}\left(\frac{\partial^{2}u_{3}^{a}}{\partial x_{1}\partial x_{2}}+\frac{\partial^{2}\overset{\circ a}{u_{3}}}{\partial x_{1}\partial x_{2}}\right)+N_{22}^{a}\left(\frac{\partial^{2}u_{3}^{a}}{\partial x_{2}^{2}}+\frac{\partial^{2}\overset{\circ a}{u_{3}}}{\partial x_{2}^{2}}+\frac{1}{R_{2}}\right)\\
&+\frac{\partial^{2}M_{11}^{a}}{\partial x_{1}^{2}}+2\frac{\partial^{2}M_{12}^{a}}{\partial x_{1}\partial x_{2}}+\frac{\partial^{2}M_{22}^{a}}{\partial x_{2}^{2}}+N_{11}^{d}\left(\frac{\partial^{2}u_{3}^{a}}{\partial x_{1}^{2}}+\frac{\partial^{2}\overset{\circ a}{u_{3}}}{\partial x_{1}^{2}}\right)+2N_{12}^{d}\left(\frac{\partial^{2}u_{3}^{d}}{\partial x_{1}\partial x_{2}}+\frac{\partial^{2}\overset{\circ d}{u_{3}}}{\partial x_{1}\partial x_{2}}\right)\\
&+N_{22}^{d}\left(\frac{\partial^{2}u_{3}^{a}}{\partial x_{2}^{2}}+\frac{\partial^{2}\overset{\circ a}{u_{3}}}{\partial x_{2}^{2}}\right)+\frac{1}{t_{c}}\left(\frac{t_{c}}{2}+\frac{t_{f}^{t}+t_{f}^{b}}{4}-u_{3}^{d}-\overset{\circ d}{u_{3}}\right)\left(\frac{\partial N_{13}^{c}}{\partial x_{1}}+\frac{\partial N_{23}^{c}}{\partial x_{2}}\right)\\
&-\frac{2}{t_{c}}\left(\frac{\partial u_{3}^{d}}{\partial x_{2}}+\frac{\circ\partial u_{3}^{d}}{\partial x_{2}}\right)N_{23}^{c}-\frac{2}{t_{c}}\left(\frac{\partial u_{3}^{d}}{\partial x_{1}}+\frac{\circ\partial u_{3}^{d}}{\partial x_{1}}\right)N_{13}^{c}\widehat{q}_{3}^{a}+\left(C^{a}+\frac{C^{c}}{2}\right)\dot{u}_{3}^{a}+C^{d}\dot{u}_{3}^{d}\\
&+\left(m^{a}+\frac{m_{c}}{2}\right)\ddot{u}_{3}^{a}+m^{d}\ddot{u}_{3}^{d}-\frac{m_{c}}{3}\ddot{\Phi}_{3}^{c}=0
\end{aligned}
\tag{8.145}
$$

$$
\begin{aligned}
&N_{11}^{d}\left(\frac{\partial^{2}u_{3}^{a}}{\partial x_{1}^{2}}+\frac{\partial^{2}\overset{\circ a}{u_{3}}}{\partial x_{1}^{2}}+\frac{1}{R_{1}}\right)+2N_{12}^{d}\left(\frac{\partial^{2}u_{3}^{a}}{\partial x_{1}\partial x_{2}}+\frac{\partial^{2}\overset{\circ a}{u_{3}}}{\partial x_{1}\partial x_{2}}\right)+N_{22}^{d}\left(\frac{\partial^{2}u_{3}^{a}}{\partial x_{2}^{2}}+\frac{\partial^{2}\overset{\circ a}{u_{3}}}{\partial x_{2}^{2}}+\frac{1}{R_{2}}\right)\\
&+\frac{\partial^{2}M_{11}^{d}}{\partial x_{1}^{2}}+2\frac{\partial^{2}M_{12}^{d}}{\partial x_{1}\partial x_{2}}+\frac{\partial^{2}M_{22}^{d}}{\partial x_{2}^{2}}+N_{11}^{a}\left(\frac{\partial^{2}u_{3}^{d}}{\partial x_{1}^{2}}+\frac{\partial^{2}\overset{\circ d}{u_{3}}}{\partial x_{1}^{2}}\right)+2N_{12}^{a}\left(\frac{\partial^{2}u_{3}^{d}}{\partial x_{1}\partial x_{2}}+\frac{\partial^{2}\overset{\circ d}{u_{3}}}{\partial x_{1}\partial x_{2}}\right)\\
&+N_{22}^{a}\left(\frac{\partial^{2}u_{3}^{d}}{\partial x_{2}^{2}}+\frac{\partial^{2}\overset{\circ d}{u_{3}}}{\partial x_{2}^{2}}\right)+\frac{2}{t_{c}}\left(\frac{t_{c}}{2}-u_{3}^{d}-\overset{\circ d}{u_{3}}\right)N_{33}^{c}-\frac{2}{3t_{c}}\left(\Phi_{3}^{c}+\frac{t_{f}^{t}-t_{f}^{b}}{16}\right)\left(\frac{\partial N_{13}^{c}}{\partial x_{1}}+\frac{\partial N_{23}^{c}}{\partial x_{2}}\right)\\
&-\frac{4}{3t_{c}}\frac{\partial\Phi_{3}^{c}}{\partial x_{1}}N_{13}^{c}-\frac{4}{3t_{c}}\frac{\partial\Phi_{3}^{c}}{\partial x_{2}}N_{23}^{c}-\frac{8}{t_{c}^{3}}\Phi_{3}^{c}M_{33}^{c}+\widehat{q}_{3}^{d}+C^{d}\dot{u}_{3}^{a}+C^{a}\dot{u}_{3}^{d}+\left(m^{a}+\frac{m_{c}}{6}\right)\ddot{u}_{3}^{d}\\
&+m^{d}\ddot{u}_{3}^{a}=0
\end{aligned}
\tag{8.146}
$$

$$
\begin{aligned}
&\frac{4}{3t_{c}}\left(\frac{\partial u_{3}^{d}}{\partial x_{1}}+\frac{\partial\overset{\circ d}{u_{3}}}{\partial x_{1}}\right)N_{13}^{c}+\frac{4}{3t_{c}}\left(\frac{\partial u_{3}^{d}}{\partial x_{2}}+\frac{\partial\overset{\circ d}{u_{3}}}{\partial x_{2}}\right)N_{23}^{c}+\frac{8}{t_{c}^{2}}\Phi_{3}^{c}N_{33}^{c}-\frac{8}{t_{c}^{3}}\left(\frac{t_{c}}{2}-u_{3}^{d}-\overset{\circ d}{u_{3}}\right)M_{33}^{c}\\
&-\frac{32}{t_{c}^{4}}\Phi_{3}^{c}L_{33}^{c}-\frac{2}{3t_{c}}\left(\frac{t_{c}}{2}-u_{3}^{d}-\overset{\circ d}{u_{3}}\right)\left(\frac{\partial N_{13}^{c}}{\partial x_{1}}+\frac{\partial N_{23}^{c}}{\partial x_{2}}\right)+\frac{16}{t_{c}^{4}}\Phi_{3}^{c}\left(\frac{\partial K_{13}^{c}}{\partial x_{1}}+\frac{\partial K_{23}^{c}}{\partial x_{2}}\right)\\
&-\frac{m_{c}}{3}\ddot{u}_{3}^{a}-\frac{3m_{c}}{15}\ddot{\Phi}_{3}^{c}=0
\end{aligned}
\tag{8.147}
$$

- *Boundary conditions*

In addition, as part of the governing system of equations, simply supported boundary conditions are considered and are given as

Along the edges $x_1 = 0,\ L_1$

$$N_{11}^a = N_{11}^d = N_{12}^a = N_{12}^d = M_{11}^a = M_{11}^d = u_3^a = u_3^d = \Phi_3^c = 0 \qquad (8.148\text{a} - \text{h})$$

Along the edges $x_2 = 0,\ L_2$

$$N_{22}^a = N_{22}^d = N_{21}^a = N_{21}^d = M_{22}^a = M_{22}^d = u_3^a = u_3^d = \Phi_3^c = 0 \qquad (8.148\text{i} - \text{p})$$

With this governing system of equations in hand, the following solution methodology is discussed pertaining to buckling and post-buckling of flat and curved sandwich panels.

8.8.3 Solution Methodology

At this point there are 11 equations of motion and 9 boundary conditions along each edge which need to be fulfilled. To facilitate the solution methodology, a stress potential function is introduced which satisfies the first two equations of motion. This stress potential function is given as

$$N_{11}^a = \frac{\partial^2 \phi}{\partial x_2^2}, \qquad N_{22}^a = \frac{\partial^2 \phi}{\partial x_1^2}, \qquad N_{12}^a = -\frac{\partial^2 \phi}{\partial x_1 \partial x_2} \qquad (8.149\text{a} - \text{c})$$

With the introduction of the stress potential function, another variable is introduced into the system at the cost of eliminating two equations of motion. Because of this new variable, another equation in terms of this new variable is needed. This will come from a compatibility equation. This compatibility equation can be arrived at by eliminating the in-plane displacements between the tangential strains. The result is

$$\frac{\partial^2 \overline{\gamma}_{11}^a}{\partial x_2^2} - 2\frac{\partial^2 \overline{\gamma}_{12}^a}{\partial x_1 \partial x_2} + \frac{\partial^2 \overline{\gamma}_{22}^a}{\partial x_1^2} = -\frac{1}{R_2}\frac{\partial^2 u_3^a}{\partial x_1^2} - \frac{1}{R_1}\frac{\partial^2 u_3^a}{\partial x_2^2} + \left(\frac{\partial u_3^a}{\partial x_1 \partial x_2}\right)^2 + \left(\frac{\partial u_3^d}{\partial x_1 \partial x_2}\right)^2 -$$
$$-\frac{\partial^2 u_3^a}{\partial x_1^2}\frac{\partial^2 u_3^a}{\partial x_2^2} - \frac{\partial^2 u_3^d}{\partial x_1^2}\frac{\partial^2 u_3^d}{\partial x_2^2} + 2\frac{\partial u_3^a}{\partial x_1 \partial x_2}\frac{\partial \left(\overset{\circ}{u}_3\right)^a}{\partial x_1 \partial x_2} - \frac{\partial^2 \left(\overset{\circ}{u}_3\right)^a}{\partial x_1^2}\frac{\partial^2 u_3^a}{\partial x_2^2} - \frac{\partial^2 u_3^a}{\partial x_1^2}\frac{\partial^2 \left(\overset{\circ}{u}_3\right)^a}{\partial x_2^2}$$
$$+2\frac{\partial u_3^d}{\partial x_1 \partial x_2}\frac{\partial \left(\overset{\circ}{u}_3\right)^d}{\partial x_1 \partial x_2} - \frac{\partial^2 \left(\overset{\circ}{u}_3\right)^d}{\partial x_1^2}\frac{\partial^2 u_3^d}{\partial x_2^2} - \frac{\partial^2 u_3^d}{\partial x_1^2}\frac{\partial^2 \left(\overset{\circ}{u}_3\right)^d}{\partial x_2^2}$$
$$(8.150)$$

In essence the governing equations have been reduced by one. Two have been eliminated while one has been added. With the first two equations of motion out of the way attention is turned to the fifth through the eight equations of motion, Eqs. (7.5-58)–(7.5-61). Expressing $N^c_{\alpha 3}$, $M^c_{\alpha 3}$, $L^c_{\alpha 3}$ $(\alpha = 1, 2)$ in terms of the strain–displacement relations, $\overline{\gamma}^c_{i3}$, κ^c_{i3}, η^c_{13}, and υ^c_{13} through the use of Eqs. (7.99)–(7.6-15) and Eqs. (7.19)–(7.28) and substituting the result into the fifth through the eighth equations of motion gives four algebraic equations in terms of Φ^c_1, Φ^c_2, Ω^c_1, and Ω^c_2. Solving these four simultaneous algebraic equations gives

$$\Phi^c_\alpha = \frac{t_c}{4}\frac{\partial u^d_3}{\partial x_\alpha} - \frac{1}{2}u^d_3\frac{\partial u^d_3}{\partial x_\alpha} - \frac{1}{2}u^d_3\left(\frac{\partial \overset{\circ}{u}_3}{\partial x_\alpha}\right)^d - \frac{1}{2}\left(\overset{\circ}{u}_3\right)^d\frac{\partial u^d_3}{\partial x_\alpha} - \Phi^c_3\frac{\partial u^d_3}{\partial x_\alpha} - \Phi^c_3\left(\frac{\partial \overset{\circ}{u}_3}{\partial x_\alpha}\right)^d + \frac{2}{5}\Phi^c_3\frac{\partial \Phi^c_3}{\partial x_\alpha} \tag{8.151a}$$

$$\Omega^c_\alpha = -\frac{t_c}{6}\frac{\partial \Phi^c_3}{\partial x_\alpha} + \frac{2}{3}\Phi_3\frac{\partial u^d_3}{\partial x_\alpha} + \frac{2}{3}\Phi_3\left(\frac{\partial \overset{\circ}{u}_3}{\partial x_\alpha}\right)^d + \frac{1}{3}\frac{\partial \Phi^c_3}{\partial x_\alpha}u^d_3 + \frac{1}{3}\frac{\partial \Phi^c_3}{\partial x_\alpha}\left(\overset{\circ}{u}_3\right)^d \tag{8.151b}$$

Next the transverse displacements and the geometric imperfections can be represented by

$$u^a_3 = w^a_{mn}\sin\left(\lambda^a_m x_1\right)\sin\left(\mu^a_n x_2\right), \qquad u^d_3 = w^d_{pq}\sin\left(\lambda^d_p x_1\right)\sin\left(\mu^d_q x_2\right) \tag{8.152a, b}$$

$$\overset{\circ}{u}{}^a_3 = \overset{\circ}{w}{}^a_{mm}\sin\left(\lambda^a_m x_1\right)\sin\left(\mu^a_n x_2\right), \qquad \overset{\circ}{u}{}^d_3 = \overset{\circ}{w}{}^d_{pq}\sin\left(\lambda^d_p x_1\right)\sin\left(\mu^d_q x_2\right) \tag{8.153a, b}$$

where

$$\lambda^a_m = \frac{m\pi}{L_1}, \quad \mu^a_n = \frac{n\pi}{L_2}, \qquad \lambda^d_p = \frac{p\pi}{L_1}, \quad \mu^d_q = \frac{q\pi}{L_2} \tag{8.154a, b}$$

where m, n, p, q are the number of half sine waves while w^a_{mn} and w^d_{pq} are the modal amplitudes. Substituting these expressions into the compatibility equation and following a similar procedure as was carried out with the compatibility equation in Chap. 4 results in an inhomogeneous partial differential equation in terms of the stress potential function. The solution to this partial differential equation can be assumed in the following form

$$\begin{aligned}\varphi(x_1, x_2, t) &= \frac{1}{2}\overline{N}^a_{11}x^2_2 + \frac{1}{2}\overline{N}^a_{22}x^2_1 + \left(\left(w^a_{mn}\right)^2 + 2w^a_{mn}\overset{\circ}{w}{}^a_{mn}\right)C_1\cos 2\lambda^a_m x_1 \\ &+ \left(\left(w^a_{mn}\right)^2 + 2w^a_{mn}\overset{\circ}{w}{}^a_{mn}\right)C_2\cos 2\mu^a_n x_2 + w^a_{mn}C_3\sin\left(\lambda^a_m x_1\right)\sin\left(\mu^a_n x_2\right) + \\ &\left(\left(w^d_{pq}\right)^2 + 2w^d_{pq}\overset{\circ}{w}{}^d_{pq}\right)C_4\cos 2\lambda^d_p x_1 + \left(\left(w^d_{pq}\right)^2 + 2w^d_{pq}\overset{\circ}{w}{}^d_{pq}\right)C_5\cos 2\mu^d_q x_2\end{aligned} \tag{8.155}$$

Substituting this assumed solution, Eq. (8.155) into the governing partial differential equation in terms of the stress potential and comparing coefficients provides the constants C1–C5.

Continuing with the third and fourth equations of equilibrium by expressing them in terms of displacements results in two coupled inhomogeneous differential equations in terms of u_1^d and u_2^d. The expressions for these displacement quantities can be assumed in the following form

$$
\begin{aligned}
u_1^d &= w_{mn}^a A_1 \cos\left(\lambda_m^a x_1\right) \sin\left(\mu_n^a x_2\right) + w_{pq}^d A_2 \cos\left(\lambda_p^d x_1\right) \sin\left(\mu_q^d x_2\right) \\
&+ \left(w_{mn}^a w_{pq}^d + \left(w_{mn}^\circ\right)^a w_{pq}^d + w_{mn}^a \left(w_{pq}^\circ\right)^d \right) A_3 \cos\left(\lambda_m^a x_1\right) \sin\left(\mu_n^a x_2\right) \sin\left(\lambda_p^d x_1\right) \sin\left(\mu_q^d x_2\right) \\
&+ \left(w_{mn}^a w_{pq}^d + \left(w_{mn}^\circ\right)^a w_{pq}^d + w_{mn}^a \left(w_{pq}^\circ\right)^d \right) A_4 \sin\left(\lambda_m^a x_1\right) \cos\left(\mu_n^a x_2\right) \cos\left(\lambda_p^d x_1\right) \cos\left(\mu_q^d x_2\right) \\
&+ \left(w_{mn}^a w_{pq}^d + \left(w_{mn}^\circ\right)^a w_{pq}^d + w_{mn}^a \left(w_{pq}^\circ\right)^d \right) A_5 \sin\left(\lambda_m^a x_1\right) \sin\left(\mu_n^a x_2\right) \cos\left(\lambda_p^d x_1\right) \sin\left(\mu_q^d x_2\right) \\
&+ \left(w_{mn}^a w_{pq}^d + \left(w_{mn}^\circ\right)^a w_{pq}^d + w_{mn}^a \left(w_{pq}^\circ\right)^d \right) A_6 \cos\left(\lambda_m^a x_1\right) \cos\left(\mu_n^a x_2\right) \sin\left(\lambda_p^d x_1\right) \cos\left(\mu_q^d x_2\right)
\end{aligned}
\tag{8.156}
$$

$$
\begin{aligned}
u_2^d &= w_{mn}^a B_1 \sin\left(\lambda_m^a x_1\right) \cos\left(\mu_n^a x_2\right) + w_{pq}^d B_2 \sin\left(\lambda_p^d x_1\right) \cos\left(\mu_q^d x_2\right) \\
&+ \left(w_{mn}^a w_{pq}^d + \left(w_{mn}^\circ\right)^a w_{pq}^d + w_{mn}^a \left(w_{pq}^\circ\right)^d \right) B_3 \sin\left(\lambda_m^a x_1\right) \cos\left(\mu_n^a x_2\right) \sin\left(\lambda_p^d x_1\right) \sin\left(\mu_q^d x_2\right) \\
&+ \left(w_{mn}^a w_{pq}^d + \left(w_{mn}^\circ\right)^a w_{pq}^d + w_{mn}^a \left(w_{pq}^\circ\right)^d \right) B_4 \cos\left(\lambda_m^a x_1\right) \sin\left(\mu_n^a x_2\right) \cos\left(\lambda_p^d x_1\right) \cos\left(\mu_q^d x_2\right) \\
&+ \left(w_{mn}^a w_{pq}^d + \left(w_{mn}^\circ\right)^a w_{pq}^d + w_{mn}^a \left(w_{pq}^\circ\right)^d \right) B_5 \sin\left(\lambda_m^a x_1\right) \sin\left(\mu_n^a x_2\right) \sin\left(\lambda_p^d x_1\right) \cos\left(\mu_q^d x_2\right) \\
&+ \left(w_{mn}^a w_{pq}^d + \left(w_{mn}^\circ\right)^a w_{pq}^d + w_{mn}^a \left(w_{pq}^\circ\right)^d \right) B_6 \cos\left(\lambda_m^a x_1\right) \cos\left(\mu_n^a x_2\right) \cos\left(\lambda_p^d x_1\right) \sin\left(\mu_q^d x_2\right)
\end{aligned}
\tag{8.157}
$$

Substituting these expressions back into the third and fourth equilibrium equations and comparing coefficients provides the constants $A_1 - A_6$ and $B_1 - B_6$. Up to this point, eight equations of equilibrium are satisfied and all of the displacement quantities are known except u_1^a and u_2^a. Considering Eqs. (8.149a–c) with Eq. (8.155) and substituting into the constitutive equations, Eqs. (8.121) and (8.123) along with the strain–displacement relationships and the transverse displacement relationships, Eqs. (8.152a, b) and (8.153a, b) result in two coupled inhomogeneous partial differential equations in terms of u_1^a and u_2^a. Looking to these two coupled differential equations, u_1^a and u_2^a can be assumed in the following form

$$
\begin{aligned}
u_1^a = & \left(\left(w_{mn}^a\right)^2 + 2w_{mn}^a \circ w_{mn}^a\right) D_1 x_1 + \left(\left(w_{mn}^a\right)^2 + 2w_{mn}^a \circ w_{mn}^a\right) D_2 \sin\left(2\lambda_m^a x_1\right) + \\
& + \left(\left(w_{mn}^a\right)^2 + 2w_{mn}^a \circ w_{mn}^a\right) D_3 \sin\left(2\lambda_m^a x_1\right) \cos\left(2\mu_n^a x_2\right) \\
& + \left(\left(w_{pq}^d\right)^2 + 2w_{pq}^d \circ w_{pq}^d\right) D_4 x_1 + \left(\left(w_{pq}^d\right)^2 + 2w_{pq}^d \circ w_{pq}^d\right) D_5 \sin\left(2\lambda_p^d x_1\right) \\
& + \left(\left(w_{pq}^d\right)^2 + 2w_{pq}^d \circ w_{pq}^d\right) D_6 \sin\left(2\lambda_p^d x_1\right) \cos\left(2\mu_q^d x_2\right) \\
& + w_{mn}^a D_7 \cos\left(\lambda_m^a x_1\right) \sin\left(\mu_n^a x_2\right) + D_8 \overline{N}_{11} x_1 + D_9 \overline{N}_{22} x_1
\end{aligned}
\tag{8.158}
$$

$$
\begin{aligned}
u_2^a = & \left(\left(w_{mn}^a\right)^2 + 2w_{mn}^a \circ w_{mn}^a\right) E_1 x_2 + \left(\left(w_{mn}^a\right)^2 + 2w_{mn}^a \circ w_{mn}^a\right) E_2 \sin\left(2\mu_n^a x_2\right) + \\
& + \left(\left(w_{mn}^a\right)^2 + 2w_{mn}^a \circ w_{mn}^a\right) E_3 \cos\left(2\lambda_m^a x_1\right) \sin\left(2\mu_n^a x_2\right) \\
& + \left(\left(w_{pq}^d\right)^2 + 2w_{pq}^d \circ w_{pq}^d\right) E_4 x_2 + \left(\left(w_{pq}^d\right)^2 + 2w_{pq}^d \circ w_{pq}^d\right) E_5 \sin\left(2\mu_q^d x_2\right) \\
& + \left(\left(w_{pq}^d\right)^2 + 2w_{pq}^d \circ w_{pq}^d\right) E_6 \cos\left(2\lambda_p^d x_1\right) \sin\left(2\mu_q^d x_2\right) \\
& + w_{mn}^a E_7 \sin\left(\lambda_m^a x_1\right) \cos\left(\mu_n^a x_2\right) + E_8 \overline{N}_{11} x_2 + E_9 \overline{N}_{22} x_2
\end{aligned}
\tag{8.159}
$$

Where the constants can be determined by substituting these assumed forms for the corresponding displacements into the two partial differential equations and comparing coefficients. At this point, eight equations of equilibrium, the boundary conditions pertaining to the transverse displacements and the stress couples M_{nn}^a and M_{nn}^d are satisfied, while all of the displacement quantities except the modal amplitudes are known. The eleventh equation of motion becomes immaterial due to the transverse warping of the core vanishing. The ninth and tenth equations of motion along with the unfulfilled boundary conditions can be expressed in terms of the modal amplitudes and retained in the energy functional where application of the extended Galerkin method is carried out. What results are two independent cubic algebraic equations in terms of the geometric imperfections, the modal amplitudes w_{mn}^a and w_{pq}^d, the prescribed edge loads $\overline{N}_{11}^a$ and $\overline{N}_{22}^a$, and the transverse pressure $\widehat{q}_3^a$ and $\widehat{q}_3^d$ which can be solved via Newton's Method.

References

Amabili, M. (2004). *Nonlinear vibrations and stability of shells and plates*. Cambridge University Press.

Barut, A., Madenci, E., Heinrich, J., & Tessler, A. (2001). Analysis of thick sandwich construction by a {3, 2}-order theory. *International Journal of Solids and Structures, 38*, 6063–6077.

Hause, T. (2012). Elastic structural response of anisotropic sandwich plates with a first-order compressible core impacted by a Friedlander-type shock loading. *Composite Structures, 94*, 1634–1645.

Hohe, J., & Librescu, L. (2003). A nonlinear theory for doubly curved anisotropic sandwich shells with transversely compressible core. *International Journal of Solids and Structures, 40*, 1059–1088.

Hohe, J., & Librescu, L. (2006). Dynamic buckling of flat and curved sandwich panels with transversely compressible core. *Composite Structures, 74*, 10–24.

Jones, R. M. (1999). *Mechanics of composite materials* (2nd ed.). New York/London: Taylor and Francis.

Librescu, L., & Chang, M. J. (1993). Effects of geometric imperfections on vibration of compressed shear deformable composite curved panels. *Acta Mechanica, 96*, 203–224.

Reddy, J. N. (2004). *Mechanics of laminated composite plates and shells-theory and analysis* (2nd ed.). Boca Raton: CRC Press.

Soedel, W. (2004). *Vibrations of shells and plates* (3rd ed.). New York: Marcel Dekker, Inc.

Chapter 9
Theory of Functionally Graded Sandwich Plates and Shells

Abstract The foundation of functionally graded doubly curved sandwich structures is presented. The theory considers a comprehensive nonlinear theory with two types of variations of the constituent materials. The theory considers the first-order shear deformation construct at the cost of only a slight decrease in accuracy. An energy approach is taken in the formulation of the governing equations. Within the context of this theoretical model, the work due to edge loads, lateral loads, and surface tractions is included.

9.1 Introduction

The foundation of functionally graded doubly curved sandwich structures is presented. The theory considers a comprehensive nonlinear theory with two types of variations of the constituent materials. These are as follows:

1. The faces sheets are functionally graded while the core is homogeneous.
2. The face-sheets are homogeneous while the core is functionally graded.

Within the theory, the first-order deformation theory is chosen over the third-order deformation theory due to only a slight increase in accuracy at an increased computational effort. The governing equations are derived using Hamilton's Equation with the work due to edge loads, lateral loads, or surface tractions included.

9.2 Preliminaries and Basic Assumptions

The shell mid-surface is referred to a curvilinear orthogonal system of coordinates (x_1, x_2, x_3), where x_3 is the thickness coordinate measured in the transverse direction which is measured positive in the upward direction from the mid-surface of the shell. The nonlinear elastic theory of functionally graded shells is adopted for the First-Order Shear Deformation Theory (FOSDT). It should be noted that with higher-order shear deformation theories it has been found that only a slight increase in

T. J. Hause, *Sandwich Structures: Theory and Responses*,
https://doi.org/10.1007/978-3-030-71895-4_9

accuracy has resulted at the cost of computational effort. Although not exact, the theory gives an approximate estimate of the thermomechanical behavior of these functionally graded structures. All three layers are considered to be heterogeneous in material composition. Three types of functionally graded shells will be considered.

Based on the rules of mixtures, the material properties such as Young's Modulus, Density, and Poisson's Ratio are assumed to vary across the wall thickness as

$$P(x_3) = P_c V_c(x_3) + P_m V_m(x_3) \tag{9.1}$$

where $P(x_3)$ is the effective material property of each layer. P_c and P_m denote the temperature-dependent material properties of the ceramic and metal phases of the plate, respectively, and may be expressed as a function of temperature (see Touloukian 1967, Huang and Shen 2004) as

$$P = P_0\left(P_{-1}T^{-1} + P_1 T + P_2 T^2 + P_3 T^3\right). \tag{9.2}$$

where, $P_0,\ P_{-1}, P_1, P_2$, and P_3 are the coefficients which depend on temperature and are unique to the constituent materials. $V_c(x_3)$ and $V_m(x_3)$ are the volume fractions of the ceramic and metal phases, respectively. The volume fractions must obey the following relationship.

$$V_c(x_3) + V_m(x_3) = 1 \tag{9.3}$$

By using Eq. (9.3), Eq. (9.1) can be expressed as

$$P(x_3, T) = [P_c(T) - P_m(T)]V_c(x_3) + P_m(T) \tag{9.4}$$

It can be deduced by observation that for $V_c(x_3) = 0,\ P(x_3, T) = P_m(T)$ and for $V_c(x_3) = 1$, $P(x_3, T) = P_c(T)$. This implies that $V_c(x_3) \in [0, 1]$.

In general, $P(x_3, T)$ can be expressed as

$$P^{(n)}(x_3, T) = [P_1(T) - P_2(T)]V^{(n)}(x_3) + P_2(T). \tag{9.5}$$

where $P^{(n)}(x_3, T)$ is the effective material property for the FGM of Layer n. For Case (1), P_1 and P_2 are the properties of each of the top and bottom faces of layer 1 and layer 3, respectively. For case (2), P_1 and P_2 are the properties of layer 1 and layer 3, respectively.

Case (1) – FGM Face-Sheets and Homogeneous Core

In this case, the faces are functionally graded, and the core is homogeneous. The interfaces between the core and the face sheets are either metal or ceramic to fulfill continuity at the core material. This power law is expressed as

$$\begin{cases} V^{(1)} = \left(\frac{x_3-h_1}{h_2-h_1}\right)^N, & x_3 \in [h_1, h_2] \\ V^{(2)} = 1, & x_3 \in [h_2, h_3] \\ V^{(3)} = \left(\frac{x_3-h_3}{h_3-h_4}\right)^N, & x_3 \in [h_3, h_4] \end{cases} \tag{9.6}$$

Here $V^{(n)}(n = 1, 2, 3)$ denotes the volume fraction of layer n and N denotes the volume fraction index.

Case (2) – Homogeneous Face-Sheets and FGM Core
This case exhibits homogeneous face sheets and a functionally graded core. The face sheets are opposite in the material type. So, if one face is ceramic, the other face is metal, and vice versa. The volume fraction function can be expressed as

$$\begin{cases} V^{(1)} = 0, & x_3 \in [h_1, h_2] \\ V^{(2)} = \left(\frac{x_3-h_2}{h_3-h_2}\right)^N, & x_3 \in [h_2, h_3] \\ V^{(3)} = 1, & x_3 \in [h_3, h_4] \end{cases} \tag{9.7}$$

As in Case (2), the situation could be reversed with ceramic on the bottom face and metal on the top face (see Abdelaziz et al. 2011 and Ashraf and Alghamdi 2008)

9.3 Basic Equations

9.3.1 Displacement Field

Consistent with plate and shell theory (Reddy 2004), the 3D displacement field through the wall thickness can be expressed as

$$u(x_1, x_2, x_3, t) = u_0(x_1, x_2, t) + x_3\psi_1(x_1, x_2, t) \tag{9.8a}$$

$$v(x_1, x_2, x_3, t) = v_0(x_1, x_2, t) + x_3\psi_2(x_1, x_2, t) \tag{9.8b}$$

$$w(x_1, x_2, x_3, t) = w_0(x_1, x_2, t) \tag{9.8c}$$

where u_0, v_0, w_0 are the 2D mid-surface displacements of the core, while ψ_1, ψ_2 are the rotation angles of the mid-surface. **Note:** In this chapter, as compared to the theory in the past two chapters, the displacement field is unified throughout the entire structure, where before, each layer had its own defined displacement field. For functionally graded materials one equivalent displacement field will be utilized to describe the entire distribution of the displacement quantities through the thickness.

9.3.2 Nonlinear Strain–Displacement Equations

The nonlinear strain displacement relationships (Shen 2009, Reddy 2004) for doubly curved shells assuming the von Kármán assumptions across the plate thickness at distance from the mid-surface are expressed as

$$\varepsilon_{11} = \frac{\partial u}{\partial x_1} - \frac{w}{R_1} + \frac{1}{2}\left(\frac{\partial w}{\partial x_1}\right)^2 \tag{9.9a}$$

$$\varepsilon_{22} = \frac{\partial v}{\partial x_2} - \frac{w}{R_2} + \frac{1}{2}\left(\frac{\partial w}{\partial x_2}\right)^2 \tag{9.9b}$$

$$\varepsilon_{12} = \frac{1}{2}\left(\frac{\partial v}{\partial x_1} + \frac{\partial u}{\partial x_2}\right) + \frac{1}{2}\frac{\partial w}{\partial x_1}\frac{\partial w}{\partial x_2} \tag{9.9c}$$

$$\varepsilon_{13} = \frac{1}{2}\left(\frac{\partial u}{\partial x_3} + \frac{\partial w}{\partial x_1}\right) + \frac{1}{2}\frac{\partial w}{\partial x_1}\frac{\partial w}{\partial x_3} \tag{9.9d}$$

$$\varepsilon_{23} = \frac{1}{2}\left(\frac{\partial v}{\partial x_3} + \frac{\partial w}{\partial x_2}\right) + \frac{1}{2}\frac{\partial w}{\partial x_2}\frac{\partial w}{\partial x_3} \tag{9.9e}$$

$$\varepsilon_{33} = \frac{\partial w}{\partial x_3} + \frac{1}{2}\left(\frac{\partial w}{\partial x_3}\right)^2 \tag{9.9f}$$

Substituting Eqs. (9.8a–c) into Eqs. (9.9a–f) gives the 3D strain displacement equations in terms of the 2D displacement quantities of the mid-surface of the shell. These are expressed as

$$\begin{aligned} \varepsilon_{11} &= \frac{\partial u_0}{\partial x_1} + \frac{1}{2}\left(\frac{\partial w_0}{\partial x_1}\right)^2 - \frac{w_0}{R_1} + x_3\frac{\partial \psi_1}{\partial x_1}, \\ \varepsilon_{22} &= \frac{\partial v_0}{\partial x_2} + \frac{1}{2}\left(\frac{\partial w_0}{\partial x_2}\right)^2 - \frac{w_0}{R_2} + x_3\frac{\partial \psi_2}{\partial x_2} \end{aligned} \tag{9.10a, b}$$

$$\gamma_{12} = \frac{\partial v_0}{\partial x_1} + \frac{\partial u_0}{\partial x_2} + \frac{\partial w_0}{\partial x_1}\frac{\partial w_0}{\partial x_1} + x_3\left(\frac{\partial \psi_1}{\partial x_2} + \frac{\partial \psi_2}{\partial x_1}\right) \tag{9.10c}$$

$$\gamma_{23} = \psi_2 + \frac{\partial w_0}{\partial x_2}, \quad \gamma_{13} = \psi_1 + \frac{\partial w_0}{\partial x_1}, \quad \varepsilon_{33} = 0 \tag{9.10d, e}$$

These expressions can be put into matrix form as

$$\begin{Bmatrix} \varepsilon_{11} \\ \varepsilon_{22} \\ \gamma_{23} \\ \gamma_{13} \\ \gamma_{12} \end{Bmatrix} = \begin{Bmatrix} \varepsilon_{11}^{(0)} \\ \varepsilon_{22}^{(0)} \\ \gamma_{23}^{(0)} \\ \gamma_{13}^{(0)} \\ \gamma_{12}^{(0)} \end{Bmatrix} + x_3 \begin{Bmatrix} \varepsilon_{11}^{(1)} \\ \varepsilon_{22}^{(1)} \\ \gamma_{23}^{(1)} \\ \gamma_{13}^{(1)} \\ \gamma_{12}^{(1)} \end{Bmatrix} \tag{9.11}$$

where

$$\begin{aligned}
&\varepsilon_{11}^{(0)} = \frac{\partial u_0}{\partial x_1} + \frac{1}{2}\left(\frac{\partial w_0}{\partial x_1}\right)^2 - \frac{w_0}{R_1}, \quad \varepsilon_{22}^{(0)} = \frac{\partial v_0}{\partial x_2} + \frac{1}{2}\left(\frac{\partial w_0}{\partial x_2}\right)^2 - \frac{w_0}{R_2} \\
&\gamma_{23}^{(0)} = \psi_2 + \frac{\partial w_0}{\partial x_2}, \quad \gamma_{13}^{(0)} = \psi_1 + \frac{\partial w_0}{\partial x_1}, \quad \gamma_{12}^{0} = \frac{\partial v_0}{\partial x_1} + \frac{\partial u_0}{\partial x_2} + \frac{\partial w_0}{\partial x_1}\frac{\partial w_0}{\partial x_2} \\
&\varepsilon_{11}^{(1)} = \frac{\partial \psi_1}{\partial x_1}, \quad \varepsilon_{22}^{(1)} = \frac{\partial \psi_2}{\partial x_2}, \quad \gamma_{12}^{(1)} = \frac{\partial \psi_1}{\partial x_2} + \frac{\partial \psi_2}{\partial x_1} \\
&\gamma_{23}^{(1)} = 0, \quad \gamma_{13}^{(1)} = 0
\end{aligned} \tag{9.12a – j}$$

9.3.3 Constitutive Equations

For an elastic and isotropic body, the stress–strain relationships are governed by

$$\begin{Bmatrix} \sigma_{11} \\ \sigma_{22} \\ \tau_{12} \end{Bmatrix}^{(n)} = \begin{bmatrix} Q_{11} & Q_{12} & 0 \\ Q_{12} & Q_{22} & 0 \\ 0 & 0 & Q_{66} \end{bmatrix}^{(n)} \begin{Bmatrix} \varepsilon_{11} - \alpha(x_3, T)\Delta T \\ \varepsilon_{22} - \alpha(x_3, T)\Delta T \\ \gamma_{12} \end{Bmatrix}^{(n)} \tag{9.13}$$

$$\begin{Bmatrix} \tau_{23} \\ \tau_{13} \end{Bmatrix}^{(n)} = K \begin{bmatrix} Q_{44} & 0 \\ 0 & Q_{55} \end{bmatrix}^{(n)} \begin{Bmatrix} \gamma_{23} \\ \gamma_{13} \end{Bmatrix}^{(n)} \tag{9.14}$$

The material stiffnesses, $Q_{ij}(x_3)$, $(i = 1, 2, 6)$ are given by

$$\begin{aligned}
&Q_{11}^{(n)} = Q_{22}^{(n)} = \frac{E^{(n)}(x_3, T)}{1 - \left(\nu^{(n)}(T)\right)^2}, \quad Q_{12} = \frac{\nu^{(n)}(T)E^{(n)}(x_3, T)}{1 - \left(\nu^{(n)}(T)\right)^2} \\
&Q_{44}^{(n)} = Q_{55}^{(n)} = Q_{66}^{(n)} = \frac{E^{(n)}(x_3, T)}{2[1 + \nu^{(n)}(T)]}
\end{aligned} \tag{9.15}$$

Consistent with the standard plate and shell theory the stress resultants and stress couples are defined as

$$\left(N_{\alpha\beta}, M_{\alpha\beta}\right) = \sum_{n=1}^{3} \int_{h_n}^{h_{n+1}} \sigma_{\alpha\beta}^{(n)}(1, x_3)dx_3, \qquad (\alpha = 1, 2) \tag{9.16a}$$

$$N_{\alpha 3} = \sum_{n=1}^{3} \int_{h_n}^{h_{n+1}} \tau_{\alpha 3}^{(n)} dx_3 \tag{9.16b}$$

where K is the shear correction factor. With the use of Eqs. (9.13), (9.14) and (9.16a, b), the stress resultants and couples can be written in matrix form as

$$\begin{pmatrix} N_{11} \\ N_{22} \\ N_{12} \\ M_{11} \\ M_{22} \\ M_{12} \end{pmatrix} = \begin{pmatrix} A_{11} & A_{12} & A_{16} & B_{11} & B_{12} & B_{16} \\ A_{12} & A_{22} & A_{26} & B_{12} & B_{22} & B_{26} \\ A_{16} & A_{26} & A_{66} & B_{16} & B_{26} & B_{66} \\ B_{11} & B_{12} & B_{16} & D_{11} & D_{12} & D_{16} \\ B_{12} & B_{22} & B_{26} & D_{12} & D_{22} & D_{26} \\ B_{16} & B_{26} & B_{66} & D_{16} & D_{26} & D_{66} \end{pmatrix} \begin{pmatrix} \varepsilon_{11}^{(0)} \\ \varepsilon_{22}^{(0)} \\ \gamma_{12}^{(0)} \\ \varepsilon_{11}^{(1)} \\ \varepsilon_{22}^{(1)} \\ \gamma_{12}^{(1)} \end{pmatrix} - \begin{pmatrix} N_{11}^{T} \\ N_{22}^{T} \\ N_{12}^{T} \\ M_{11}^{T} \\ M_{22}^{T} \\ M_{12}^{T} \end{pmatrix} \tag{9.17a}$$

$$\begin{pmatrix} N_{23} \\ N_{13} \end{pmatrix} = K \begin{bmatrix} A_{44} & \\ & A_{55} \end{bmatrix} \begin{pmatrix} \gamma_{23}^{(0)} \\ \gamma_{13}^{(0)} \end{pmatrix} \tag{9.17b}$$

or in compact form Eq. (9.17a) can be expressed as

$$\begin{bmatrix} \{N\} \\ \{M\} \end{bmatrix} = \begin{bmatrix} [A] & [B] \\ [B] & [D] \end{bmatrix} \begin{bmatrix} \{\varepsilon^{(0)}\} \\ \{\varepsilon^{(1)}\} \end{bmatrix} - \begin{bmatrix} \{N^{T}\} \\ \{M^{T}\} \end{bmatrix} \tag{9.18}$$

where the global stiffnesses A_{ij}, B_{ij}, and D_{ij}, $(i, j = 1, 2, 6)$ are given by

$$\left(A_{ij}, B_{ij}, D_{ij}\right) = \sum_{n=1}^{3} \int_{h_n}^{h_{n+1}} Q_{ij}^{n}\left(1, x_3, x_3^2\right)dx_3, \qquad (i, j = 1, 2, 6) \tag{9.19a}$$

$$A_{IJ} = \sum_{n=1}^{3} \int_{h_n}^{h_{n+1}} Q_{IJ}^{(n)} dx_3 \qquad (I, J = 4, 5) \tag{9.19b}$$

9.4 Hamilton's Principle

As in the other chapters, an energy approach will be taken to derive the equations of motion. For convenience Hamilton's Principle (Soedel 2004) is expressed as

$$\int_{t_0}^{t_1} (\delta U - \delta W - \delta T)dt = 0 \tag{9.20}$$

9.4.1 *Strain Energy*

Assuming an inextensible core and faces, the variation in the strain energy is given by

$$\delta U = \int_0^{l_2}\int_0^{l_1}\left\langle \int_{-\frac{h}{2}}^{\frac{h}{2}} \left\{\sigma_{11}^{(n)}\delta\varepsilon_{11} + \sigma_{22}^{(n)}\delta\varepsilon_{22} + \tau_{12}^{(n)}\delta\gamma_{12} + \tau_{13}^{(n)}\delta\gamma_{13} + \tau_{23}^{(n)}\delta\gamma_{23}\right\}dx_3\right\rangle dx_1 dx_2 \tag{9.21}$$

Taking a variation in the strains from Eqs. (9.11) and substituting into Eq. (9.21) gives

$$\begin{aligned}\delta U = \int_0^{l_2}\int_0^{l_1}\left\langle \int_{-\frac{h}{2}}^{\frac{h}{2}} \left\{\sigma_{11}^{(n)}\left(\delta\varepsilon_{11}^{(0)} + x_3\delta\varepsilon_{11}^{(1)}\right) + \sigma_{22}^{(n)}\left(\delta\varepsilon_{22}^{(0)} + x_3\delta\varepsilon_{22}^{(1)}\right)\right.\right.\\ \left.\left. +\tau_{12}^{(n)}\left(\delta\gamma_{12}^{(0)} + x_3\delta\gamma_{12}^{(1)}\right) + \tau_{13}^{(n)}\delta\gamma_{13}^{(0)} + \tau_{23}^{(n)}\delta\gamma_{23}^{(0)}\right\}dx_3\right\rangle dx_1 dx_2\end{aligned} \tag{9.22}$$

which can be expressed as

$$\begin{aligned}\delta U = \int_0^{l_2}\int_0^{l_1}\left\{N_{11}\delta\varepsilon_{11}^{(0)} + M_{11}\delta\varepsilon_{11}^{(1)} + N_{22}\delta\varepsilon_{22}^{(0)} + M_{22}\delta\varepsilon_{22}^{(1)} + N_{12}\delta\gamma_{12}^{(0)} + M_{12}\delta\gamma_{12}^{(1)} + \right.\\ \left. N_{13}\delta\gamma_{13}^{(0)} + N_{23}\delta\gamma_{23}^{(0)}\right\}dx_1 dx_2\end{aligned} \tag{9.23}$$

Using the mid-surface strain–displacement relationships, Eq. (9.12a–j) in Eq. (9.23) gives the variation in the strain energy in terms of the variation in the displacement quantities which results in

$$\begin{aligned}\delta U = \int_0^{l_2}\int_0^{l_1} \Bigg\langle & N_{11}\left\{\frac{\partial(\delta u_0)}{\partial x_1} + \frac{\partial w_0}{\partial x_1}\frac{\partial(\delta w_0)}{\partial x_1} - \frac{\delta w_0}{R_1}\right\} + M_{11}\frac{\partial(\delta\psi_1)}{\partial x_1} \\ & + N_{22}\left\{\frac{\partial(\delta v_0)}{\partial x_2} + \frac{\partial w_0}{\partial x_2}\frac{\partial(\delta w_0)}{\partial x_2} - \frac{\delta w_0}{R_2}\right\} + M_{22}\frac{\partial(\delta\psi_2)}{\partial x_2} \\ & + N_{12}\left\{\frac{\partial(\delta v_0)}{\partial x_1} + \frac{\partial(\delta u_0)}{\partial x_2} + \frac{\partial(\delta w_0)}{\partial x_1}\frac{\partial w_0}{\partial x_2} + \frac{\partial w_0}{\partial x_1}\frac{\partial(\delta w_0)}{\partial x_2}\right\} \\ & + M_{12}\left\{\frac{\partial(\delta\psi_1)}{\partial x_2} + \frac{\partial(\delta\psi_2)}{\partial x_1}\right\} + N_{13}\left\{\delta\psi_1 + \frac{\partial(\delta w_0)}{\partial x_1}\right\} \\ & + N_{23}\left\{\delta\psi_2 + \frac{\partial(\delta w_0)}{\partial x_2}\right\}\Bigg\rangle dx_1 dx_2\end{aligned} \tag{9.24}$$

Integrating Eq. (9.24) by parts gives

$$\begin{aligned}\delta U = \int_0^{l_2}\int_0^{l_1} \Bigg\langle & -\left(\frac{\partial N_{11}}{\partial x_1} + \frac{\partial N_{12}}{\partial x_2}\right)\delta u_0 - \left(\frac{\partial N_{22}}{\partial x_2} + \frac{\partial N_{12}}{\partial x_1}\right)\delta v_0 \\ & - \left(\frac{\partial M_{11}}{\partial x_1} + \frac{\partial M_{12}}{\partial x_2} - N_{13}\right)\delta\psi_1 - \left(\frac{\partial M_{22}}{\partial x_2} + \frac{\partial M_{12}}{\partial x_1} - N_{23}\right)\delta\psi_2 \\ & + \Bigg\{-\left(\frac{\partial N_{22}}{\partial x_2} + \frac{\partial N_{12}}{\partial x_1}\right)\frac{\partial w_0}{\partial x_2} - \left(\frac{\partial N_{11}}{\partial x_1} + \frac{\partial N_{12}}{\partial x_2}\right)\frac{\partial w_0}{\partial x_1} - N_{11}\frac{\partial^2 w_0}{\partial x_1^2} \\ & - 2N_{12}\frac{\partial^2 w_0}{\partial x_1 \partial x_2} - N_{22}\frac{\partial^2 w_0}{\partial x_2^2} - \frac{N_{11}}{R_1} - \frac{N_{22}}{R_2} - \left(\frac{\partial N_{13}}{\partial x_1} + \frac{\partial N_{23}}{\partial x_2}\right)\Bigg\}\delta w_0\Bigg\rangle dx_1 dx_2 \\ & + \int_0^{l_2}\left\{N_{11}\delta u_0 + N_{12}\delta v_0 + M_{11}\delta\psi_1 + M_{12}\delta\psi_2 + \left(N_{11}\frac{\partial w_0}{\partial x_1} + N_{12}\frac{\partial w_0}{\partial x_2} + N_{13}\right)\delta w_0\right\}\Bigg|_0^{l_1} dx_2 \\ & + \int_0^{l_1}\left\{N_{12}\delta u_0 + N_{22}\delta v_0 + M_{12}\delta\psi_1 + M_{22}\delta\psi_2 + \left(N_{12}\frac{\partial w_0}{\partial x_1} + N_{22}\frac{\partial w_0}{\partial x_2} + N_{23}\right)\delta w_0\right\}\Bigg|_0^{l_2} dx\end{aligned} \tag{9.25}$$

9.4.2 *Work Done by External Loads*

As before the external work acting on the structure is due to edge loading, lateral loading, body forces, and damping. The work due to body forces is provided below but is neglected in the theoretical developments.

- *Work due to body forces*

$$\delta W_b = \int_\sigma \left\{ \int_{\bar{h}}^{\bar{h}+h^/} \rho^/ H_i^/ \delta V_i^/ dx_3 + \int_{-\bar{h}}^{\bar{h}} \bar{\rho} \bar{H}_i \delta \bar{V}_i dx_3 + \int_{-\bar{h}-h^{//}}^{-\bar{h}} \rho^{//} H_i^{//} \delta V_i^{//} dx_3 \right\} d\sigma \tag{9.26}$$

Where ρ is the mass density and H_i is the body force vector. The body forces are neglected such that gravitational, electrical, and magnetic forces are irrelevant.

- *Work due to surface tractions*

$$\delta W_{st} = \int_0^{l_2} \int_0^{l_1} q_3(x_1, x_2) \delta w_0 dx_1 dx_2 \tag{9.27}$$

- *Work due to edge loads*

$$\begin{aligned} \delta W_{el} = & \int_0^{l_1} \left\{ \int_{-\frac{h}{2}}^{\frac{h}{2}} \left(\widetilde{\sigma}_{22}^{(n)} \delta v + \widetilde{\sigma}_{21}^{(n)} \delta u + \widetilde{\sigma}_{23}^{(n)} \delta w \right) dx_3 \right\} dx_1 \\ & + \int_0^{l_2} \left\{ \int_{-\frac{h}{2}}^{\frac{h}{2}} \left(\widetilde{\sigma}_{11}^{(n)} \delta u + \widetilde{\sigma}_{12}^{(n)} \delta v + \widetilde{\sigma}_{13}^{(n)} \delta w \right) dx_3 \right\} dx_2 \end{aligned} \tag{9.28}$$

Substituting the displacement relationships into Eq. (9.28) gives

$$\begin{aligned} \delta W_{el} = & \int_0^{l_1} \left\{ \int_{-\frac{h}{2}}^{\frac{h}{2}} \left(\widetilde{\sigma}_{22}^{(n)} (\delta v_0 + x_3 \delta \psi_2) + \widetilde{\sigma}_{21}^{(n)} (\delta u_0 + x_3 \delta \psi_1) + \widetilde{\sigma}_{23}^{(n)} \delta w_0 \right) dx_3 \right\} dx_1 + \\ & + \int_0^{l_2} \left\{ \int_{-\frac{h}{2}}^{\frac{h}{2}} \left(\widetilde{\sigma}_{11}^{(n)} (\delta u_0 + x_3 \delta \psi_1) + \widetilde{\sigma}_{12}^{(n)} (\delta v_0 + x_3 \delta \psi_2) + \widetilde{\sigma}_{13}^{(n)} \delta w_0 \right) dx_3 \right\} dx_2 \end{aligned} \tag{9.29}$$

Using the definition of stress resultants and couples gives

$$\begin{aligned} \delta W_{el} = & \int_0^{l_1} \left(\widetilde{N}_{21} \delta u_0 + \widetilde{N}_{22} \delta v_0 + \widetilde{M}_{21} \delta \psi_1 + \widetilde{M}_{22} \delta \psi_2 + \widetilde{N}_{23} \delta w_0 \right) dx_1 \\ & \int_0^{l_2} \left(\widetilde{N}_{11} \delta u_0 + \widetilde{N}_{12} \delta v_0 + \widetilde{M}_{11} \delta \psi_1 + \widetilde{M}_{12} \delta \psi_2 + \widetilde{N}_{13} \delta w_0 \right) dx_2 \end{aligned} \tag{9.30}$$

- *Work due to damping*

The work due to damping for the transverse direction is given by

$$W_d = -\int_A C^{(ave)} \dot{w}_0 \delta w_0 dA \tag{9.31}$$

where $C^{(a)}$ is the average structural damping coefficient per unit area among the three layers (the facings and the core).

9.4.3 Kinetic Energy

As from the previous chapters, by definition the kinetic energy of an elastic body is given by

$$T = \frac{1}{2}\int_{x_3}\int_{x_2}\int_{x_1} \rho\left[\left(\frac{\partial V_1}{\partial t}\right)^2 + \left(\frac{\partial V_2}{\partial t}\right)^2 \left(\frac{\partial V_3}{\partial t}\right)^2\right] dx_1 dx_2 dx_3 \tag{9.32}$$

Taking the variation in the kinetic energy and integrating by parts with respect to time gives

$$\int_{t_0}^{t_1} \delta T dt = -\int_{t_0}^{t_1}\int_A\left\{\int_{-\frac{h}{2}}^{\frac{h}{2}} \rho^{(n)}(x_3)\left(\ddot{V}_1\delta V_1 + \ddot{V}_2\delta V_2 + \ddot{V}_3\delta V_3\right)dx_3\right\} dA dt \tag{9.33}$$

Substituting the expressions for the displacements from Eqs. (9.8a–c) into Eq. (9.33) results in

$$\begin{aligned}\int_{t_0}^{t_1} \delta T dt = -\int_{t_0}^{t_1}\int_A\Bigg\{\int_{-\frac{h}{2}}^{\frac{h}{2}} \rho^{(n)}(x_3)[(\ddot{u}_0 + x_3\ddot{\psi}_1)(\delta u_0 + x_3\delta\psi_1)(\ddot{v}_0 + x_3\ddot{\psi}_2)\\ +(\delta v_0 + x_3\delta\psi_2) + \ddot{w}_0\delta w_0]dx_3\Bigg\} dA dt\end{aligned} \tag{9.34}$$

Simplifying and combining like terms gives

$$\begin{aligned}\int_{t_0}^{t_1} \delta T dt = -\int_{t_0}^{t_1}\int_A\Big\{(m_0\ddot{u}_0 + m_1\ddot{\psi}_1)\delta u_0 + (m_0\ddot{v}_0 + m_1\ddot{\psi}_2)\delta v_0 + (m_1\ddot{u}_0 + m_2\ddot{\psi}_1)\delta\psi_1 +\\ (m_1\ddot{v}_0 + m_2\ddot{\psi}_2)\delta\psi_2 + m_0\ddot{w}_0\delta w_0\Big\} dA dt\end{aligned} \tag{9.35}$$

where

$$
\begin{aligned}
m_0 &= \sum_{n=1}^{3}\int_{h_n}^{h_{n+1}} \rho^{(n)}(x_3)dx_3, \qquad m_1 = \sum_{n=1}^{3}\int_{h_n}^{h_{n+1}} \rho^{(n)}(x_3)x_3dx_3,\\
m_2 &= \sum_{n=1}^{3}\int_{h_n}^{h_{n+1}} \rho^{(n)}(x_3)x_3^2dx_3
\end{aligned} \tag{9.36}
$$

9.5 Equations of Motion

9.5.1 Equations of Motion – Mixed Formulation

Substituting δU, δW, δT back into Hamilton's Equation gives

$$
\begin{aligned}
&\int_{t_0}^{t_1}\Big\langle\int_0^{l_2}\int_0^{l_1}\Big\{-\left(\frac{\partial N_{11}}{\partial x_1}+\frac{\partial N_{12}}{\partial x_2}-m_0\ddot{u}_0-m_1\ddot{\psi}_1\right)\delta u_0-\left(\frac{\partial N_{22}}{\partial x_2}+\frac{\partial N_{12}}{\partial x_1}-m_0\ddot{v}_0-m_1\ddot{\psi}_2\right)\delta v_0-\\
&\quad\left(\frac{\partial M_{11}}{\partial x_1}+\frac{\partial M_{12}}{\partial x_2}-N_{13}-m_1\ddot{u}_0-m_2\ddot{\psi}_1\right)\delta\psi_1-\left(\frac{\partial M_{22}}{\partial x_2}+\frac{\partial M_{12}}{\partial x_1}-N_{23}-m_1\ddot{v}_0\right.\\
&\quad\left.-m_2\ddot{\psi}_2\right)\delta\psi_2+\Big\{-\left(\frac{\partial N_{22}}{\partial x_2}+\frac{\partial N_{12}}{\partial x_1}\right)\frac{\partial w_0}{\partial x_2}-\left(\frac{\partial N_{11}}{\partial x_1}+\frac{\partial N_{12}}{\partial x_2}\right)\frac{\partial w_0}{\partial x_1}-N_{11}\frac{\partial^2 w_0}{\partial x_1^2}-\\
&\quad 2N_{12}\frac{\partial^2 w_0}{\partial x_1\partial x_2}-N_{22}\frac{\partial^2 w_0}{\partial x_2^2}-\frac{N_{11}}{R_1}-\frac{N_{22}}{R_2}-\left(\frac{\partial N_{13}}{\partial x_1}+\frac{\partial N_{23}}{\partial x_2}\right)+q_3-C^{(ave)}\dot{w}_0\\
&\quad -m_0\ddot{w}_0\Big\}\delta w_0\Big\}dx_1dx_2\Big\rangle dt+\\
&\int_{t_0}^{t_1}\int_0^{l_2}\Big\{(N_{11}-\widetilde{N}_{11})\delta u_0+(N_{12}-\widetilde{N}_{12})\delta v_0+(M_{11}-\widetilde{M}_{11})\delta\psi_1+(M_{12}-\widetilde{M}_{12})\delta\psi_2+\\
&\quad\left(N_{11}\frac{\partial w_0}{\partial x_1}+N_{12}\frac{\partial w_0}{\partial x_2}+N_{13}-\widetilde{N}_{13}\right)\delta w_0\Big\}\Big|_0^{l_1}dx_2dt+\\
&\int_{t_0}^{t_1}\int_0^{l_1}\Big\{(N_{12}-\widetilde{N}_{12})\delta u_0+(N_{22}-\widetilde{N}_{22})\delta v_0+(M_{12}-\widetilde{M}_{12})\delta\psi_1+(M_{22}-\widetilde{M}_{22})\delta\psi_2+\\
&\quad\left(N_{12}\frac{\partial w_0}{\partial x_1}+N_{22}\frac{\partial w_0}{\partial x_2}+N_{23}-\widetilde{N}_{23}\right)\delta w_0\Big\}\Big|_0^{l_2}dx_1dt=0
\end{aligned} \tag{9.37}
$$

This equation can only be satisfied by setting the coefficients of the variational displacements equal to zero since the variational displacements are arbitrary. Setting the coefficients of δu_0, δv_0, $\delta\psi_1$, $\delta\psi_2$, δw_0 to zero provides the equations of motion along with the boundary conditions. These are provided as

$$
\delta u_0: \qquad \frac{\partial N_{11}}{\partial x_1}+\frac{\partial N_{12}}{\partial x_2}-m_0\ddot{u}_0-m_1\ddot{\psi}_1=0 \tag{9.38a}
$$

$$\delta v_0 : \quad \frac{\partial N_{22}}{\partial x_2} + \frac{\partial N_{12}}{\partial x_1} - m_0 \ddot{v}_0 - m_1 \ddot{\psi}_2 = 0 \tag{9.38b}$$

$$\delta \psi_1 : \quad \frac{\partial M_{11}}{\partial x_1} + \frac{\partial M_{12}}{\partial x_2} - N_{13} - m_1 \ddot{u}_0 - m_2 \ddot{\psi}_1 = 0 \tag{9.38c}$$

$$\delta \psi_2 : \quad \frac{\partial M_{22}}{\partial x_2} + \frac{\partial M_{12}}{\partial x_1} - N_{23} - m_1 \ddot{v}_0 - m_2 \ddot{\psi}_2 = 0 \tag{9.38d}$$

$$\begin{aligned} \delta w_0 : \quad & N_{11} \frac{\partial^2 w_0}{\partial x_1^2} + 2N_{12} \frac{\partial^2 w_0}{\partial x_1 \partial x_2} + N_{22} \frac{\partial^2 w_0}{\partial x_2^2} + \frac{N_{11}}{R_1} + \frac{N_{22}}{R_2} + \left(\frac{\partial N_{13}}{\partial x_1} + \frac{\partial N_{23}}{\partial x_2} \right) + C^{(a)} \dot{w}_0 \\ & + m_0 \ddot{w}_0 - q = 0 \end{aligned} \tag{9.38e}$$

The boundary conditions become
Along the edges $x_1 = 0,\ L_1$

$$\begin{aligned} & N_{11} = \widetilde{N}_{11} \quad \text{or} \quad u_0 = \widetilde{u}_0 \\ & N_{12} = \widetilde{N}_{12} \quad \text{or} \quad v_0 = \widetilde{v}_0 \\ & M_{11} = \widetilde{M}_{11} \quad \text{or} \quad \psi_1 = \widetilde{\psi}_1 \\ & M_{12} = \widetilde{M}_{12} \quad \text{or} \quad \psi_2 = \widetilde{\psi}_2 \\ & N_{11} \frac{\partial w_0}{\partial x_1} + N_{12} \frac{\partial w_0}{\partial x_2} + N_{13} = \widetilde{N}_{13} \quad \text{or} \quad w_0 = \widetilde{w}_0 \end{aligned} \tag{9.39a – e}$$

Along the edges $x_2 = 0,\ L_2$

$$\begin{aligned} & N_{12} = \widetilde{N}_{12} \quad \text{or} \quad u_0 = \widetilde{u}_0 \\ & N_{22} = \widetilde{N}_{22} \quad \text{or} \quad v_0 = \widetilde{v}_0 \\ & M_{12} = \widetilde{M}_{12} \quad \text{or} \quad \psi_1 = \widetilde{\psi}_1 \\ & M_{22} = \widetilde{M}_{22} \quad \text{or} \quad \psi_2 = \widetilde{\psi}_2 \\ & N_{12} \frac{\partial w_0}{\partial x_1} + N_{22} \frac{\partial w_0}{\partial x_2} + N_{23} = \widetilde{N}_{23} \quad \text{or} \quad w_0 = \widetilde{w}_0 \end{aligned} \tag{9.40a – e}$$

9.5.2 The Equations of Motion – The Displacement Formulation

The above equations of motion, which are expressed in terms of the mixed formulations, can be expressed in terms of displacements by using Eqs. (9.17a, b) with Eqs. (9.12a–j). With this in hand, the above equations of motion can be expressed in terms of displacements as

$$\delta u_0:\ A_{11}\left(\frac{\partial^2 u_0}{\partial x_1^2}+\frac{\partial w_0}{\partial x_1}\frac{\partial^2 w_0}{\partial x_1^2}-\frac{1}{R_1}\frac{\partial w_0}{\partial x_1}\right)+A_{12}\left(\frac{\partial^2 v_0}{\partial x_1\partial x_2}+\frac{\partial w_0}{\partial x_2}\frac{\partial^2 w_0}{\partial x_1\partial x_2}-\frac{1}{R_2}\frac{\partial w_0}{\partial x_1}\right)$$
$$+A_{16}\left(\frac{\partial^2 v_0}{\partial x_1^2}+2\frac{\partial^2 u_0}{\partial x_1\partial x_2}+\frac{\partial^2 w_0}{\partial x_1^2}\frac{\partial w_0}{\partial x_2}+2\frac{\partial w_0}{\partial x_1}\frac{\partial^2 w_0}{\partial x_1\partial x_2}-\frac{1}{R_1}\frac{\partial w_0}{\partial x_2}\right)$$
$$+A_{26}\left(\frac{\partial^2 v_0}{\partial x_2^2}+\frac{\partial w_0}{\partial x_2}\frac{\partial^2 w_0}{\partial x_2^2}-\frac{1}{R_2}\frac{\partial w_0}{\partial x_2}\right)$$
$$+A_{66}\left(\frac{\partial^2 v_0}{\partial x_1\partial x_2}+\frac{\partial^2 u_0}{\partial x_2^2}+\frac{\partial^2 w_0}{\partial x_1\partial x_2}\frac{\partial w_0}{\partial x_2}+\frac{\partial w_0}{\partial x_1}\frac{\partial^2 w_0}{\partial x_2^2}\right)$$
$$+B_{11}\frac{\partial^2\psi_1}{\partial x_1^2}+B_{12}\frac{\partial^2\psi_2}{\partial x_1\partial x_2}+B_{16}\left(2\frac{\partial^2\psi_1}{\partial x_1\partial x_2}+\frac{\partial^2\psi_2}{\partial x_1^2}\right)$$
$$+B_{26}\frac{\partial^2\psi_2}{\partial x_2^2}+B_{66}\left(\frac{\partial^2\psi_1}{\partial x_2^2}+\frac{\partial^2\psi_2}{\partial x_1\partial x_2}\right)-\frac{\partial N_{11}^T}{\partial x_1}-\frac{\partial N_{12}^T}{\partial x_2}=m_0\frac{\partial^2 u_0}{\partial t^2}+m_1\frac{\partial^2\psi_1}{\partial t^2}$$
(9.41a)

$$\delta v_0:\ A_{22}\left(\frac{\partial^2 v_0}{\partial x_2^2}+\frac{\partial w_0}{\partial x_2}\frac{\partial^2 w_0}{\partial x_2^2}-\frac{1}{R_2}\frac{\partial w_0}{\partial x_2}\right)+A_{12}\left(\frac{\partial^2 u_0}{\partial x_1\partial x_2}+\frac{\partial w_0}{\partial x_1}\frac{\partial^2 w_0}{\partial x_1\partial x_2}-\frac{1}{R_1}\frac{\partial w_0}{\partial x_2}\right)$$
$$+A_{16}\left(\frac{\partial^2 u_0}{\partial x_2^2}+\frac{\partial w_0}{\partial x_1}\frac{\partial^2 w_0}{\partial x_1^2}-\frac{1}{R_1}\frac{\partial w_0}{\partial x_1}\right)+A_{26}\left(\frac{\partial^2 u_0}{\partial x_1^2}+2\frac{\partial^2 v_0}{\partial x_1\partial x_2}+\frac{\partial^2 w_0}{\partial x_2^2}\frac{\partial w_0}{\partial x_1}\right.$$
$$\left.+2\frac{\partial w_0}{\partial x_2}\frac{\partial^2 w_0}{\partial x_1\partial x_2}-\frac{1}{R_2}\frac{\partial w_0}{\partial x_1}\right)+A_{66}\left(\frac{\partial^2 u_0}{\partial x_1\partial x_2}+\frac{\partial^2 v_0}{\partial x_1^2}+\frac{\partial^2 w_0}{\partial x_1\partial x_2}\frac{\partial w_0}{\partial x_1}+\frac{\partial w_0}{\partial x_2}\frac{\partial^2 w_0}{\partial x_1^2}\right)$$
$$+B_{22}\frac{\partial^2\psi_2}{\partial x_2^2}+B_{12}\frac{\partial^2\psi_1}{\partial x_1\partial x_2}+B_{16}\frac{\partial^2\psi_1}{\partial x_1^2}+B_{26}\left(2\frac{\partial^2\psi_2}{\partial x_1\partial x_2}+\frac{\partial^2\psi_1}{\partial x_2^2}\right)$$
$$+B_{66}\left(\frac{\partial^2\psi_2}{\partial x_1^2}+\frac{\partial^2\psi_1}{\partial x_1\partial x_2}\right)-\frac{\partial N_{22}^T}{\partial x_2}-\frac{\partial N_{12}^T}{\partial x_1}=m_0\frac{\partial^2 v_0}{\partial t^2}+m_1\frac{\partial^2\psi_2}{\partial t^2}$$
(9.41b)

$$\delta\psi_1:\ B_{11}\left(\frac{\partial^2 u_0}{\partial x_1^2}+\frac{\partial w_0}{\partial x_1}\frac{\partial^2 w_0}{\partial x_1^2}-\frac{1}{R_1}\frac{\partial w_0}{\partial x_1}\right)+B_{12}\left(\frac{\partial^2 v_0}{\partial x_1\partial x_2}+\frac{\partial w_0}{\partial x_2}\frac{\partial^2 w_0}{\partial x_1\partial x_2}-\frac{1}{R_2}\frac{\partial w_0}{\partial x_1}\right)$$
$$+B_{16}\left(\frac{\partial^2 v_0}{\partial x_1^2}+2\frac{\partial^2 u_0}{\partial x_1\partial x_2}+\frac{\partial^2 w_0}{\partial x_1^2}\frac{\partial w_0}{\partial x_2}+2\frac{\partial w_0}{\partial x_1}\frac{\partial^2 w_0}{\partial x_1\partial x_2}-\frac{1}{R_1}\frac{\partial w_0}{\partial x_2}\right)+B_{26}\left(\frac{\partial^2 v_0}{\partial x_2^2}+\right.$$
$$\left.\frac{\partial w_0}{\partial x_2}\frac{\partial^2 w_0}{\partial x_2^2}-\frac{1}{R_2}\frac{\partial w_0}{\partial x_2}\right)+B_{66}\left(\frac{\partial^2 v_0}{\partial x_1\partial x_2}+\frac{\partial^2 u_0}{\partial x_2^2}+\frac{\partial^2 w_0}{\partial x_1\partial x_2}\frac{\partial w_0}{\partial x_2}+\frac{\partial w_0}{\partial x_1}\frac{\partial^2 w_0}{\partial x_2^2}\right)+$$
$$D_{11}\frac{\partial^2\psi_1}{\partial x_1^2}+D_{12}\frac{\partial^2\psi_2}{\partial x_1\partial x_2}+D_{16}\left(2\frac{\partial^2\psi_1}{\partial x_1\partial x_2}+\frac{\partial^2\psi_2}{\partial x_1^2}\right)+D_{26}\frac{\partial^2\psi_2}{\partial x_2^2}$$
$$+D_{66}\left(\frac{\partial^2\psi_1}{\partial x_2^2}+\frac{\partial^2\psi_2}{\partial x_1\partial x_2}\right)-A_{55}\left(\psi_1+\frac{\partial w_0}{\partial x_1}\right)-\frac{\partial M_{11}^T}{\partial x_1}-\frac{\partial M_{12}^T}{\partial x_2}$$
$$=m_1\frac{\partial^2 u_0}{\partial t^2}+m_2\frac{\partial^2\psi_1}{\partial t^2}$$
(9.41c)

$$\delta\psi_2 : \; B_{22}\left(\frac{\partial^2 v_0}{\partial x_2^2} + \frac{\partial w_0}{\partial x_2}\frac{\partial^2 w_0}{\partial x_2^2} - \frac{1}{R_2}\frac{\partial w_0}{\partial x_2}\right) + B_{12}\left(\frac{\partial^2 u_0}{\partial x_1 \partial x_2} + \frac{\partial w_0}{\partial x_1}\frac{\partial^2 w_0}{\partial x_1 \partial x_2} - \frac{1}{R_1}\frac{\partial w_0}{\partial x_2}\right)$$
$$+B_{16}\left(\frac{\partial^2 u_0}{\partial x_2^2} + \frac{\partial w_0}{\partial x_1}\frac{\partial^2 w_0}{\partial x_1^2} - \frac{1}{R_1}\frac{\partial w_0}{\partial x_1}\right) + B_{26}\left(\frac{\partial^2 u_0}{\partial x_1^2} + 2\frac{\partial^2 v_0}{\partial x_1 \partial x_2} + \frac{\partial^2 w_0}{\partial x_2^2}\frac{\partial w_0}{\partial x_1} + \right.$$
$$\left. 2\frac{\partial w_0}{\partial x_2}\frac{\partial^2 w_0}{\partial x_1 \partial x_2} - \frac{1}{R_2}\frac{\partial w_0}{\partial x_1}\right) + B_{66}\left(\frac{\partial^2 u_0}{\partial x_1 \partial x_2} + \frac{\partial^2 v_0}{\partial x_1^2} + \frac{\partial^2 w_0}{\partial x_1 \partial x_2}\frac{\partial w_0}{\partial x_1} + \frac{\partial w_0}{\partial x_2}\frac{\partial^2 w_0}{\partial x_1^2}\right)$$
$$+D_{22}\frac{\partial^2 \psi_2}{\partial x_2^2} + D_{12}\frac{\partial^2 \psi_1}{\partial x_1 \partial x_2} + D_{16}\frac{\partial^2 \psi_1}{\partial x_1^2} + D_{26}\left(2\frac{\partial^2 \psi_2}{\partial x_1 \partial x_2} + \frac{\partial^2 \psi_1}{\partial x_2^2}\right) + D_{66}\left(\frac{\partial^2 \psi_2}{\partial x_1^2} + \right.$$
$$\left.\frac{\partial^2 \psi_1}{\partial x_1 \partial x_2}\right) - A_{44}\left(\psi_2 + \frac{\partial w_0}{\partial x_2}\right) - \frac{\partial M_{22}^T}{\partial x_2} - \frac{\partial M_{12}^T}{\partial x_1} = m_1\frac{\partial^2 v_0}{\partial t^2} + m_2\frac{\partial^2 \psi_2}{\partial t^2} \tag{9.41d}$$

δw_0 :

$$A_{11}\left\{\frac{\partial u_0}{\partial x_1}\frac{\partial^2 w_0}{\partial x_1^2} + \frac{1}{2}\left(\frac{\partial w_0}{\partial x_1}\right)^2\frac{\partial^2 w_0}{\partial x_1^2} + \frac{1}{R_1}\left(\frac{\partial u_0}{\partial x_1} + \frac{1}{2}\left(\frac{\partial w_0}{\partial x_1}\right)^2 - w_0\frac{\partial^2 w_0}{\partial x_1^2} - \frac{w_0}{R_1}\right)\right\} +$$
$$A_{12}\left\{\frac{\partial u_0}{\partial x_1}\frac{\partial^2 w_0}{\partial x_2^2} + \frac{\partial v_0}{\partial x_2}\frac{\partial^2 w_0}{\partial x_1^2} + \frac{1}{2}\left(\frac{\partial w_0}{\partial x_1}\right)^2\frac{\partial^2 w_0}{\partial x_2^2} + \frac{1}{2}\left(\frac{\partial w_0}{\partial x_2}\right)^2\frac{\partial^2 w_0}{\partial x_1^2} + \frac{1}{R_1}\left(\frac{\partial v_0}{\partial x_2} + \right.\right.$$
$$\left.\left.\frac{1}{2}\left(\frac{\partial w_0}{\partial x_2}\right)^2 - w_0\frac{\partial^2 w_0}{\partial x_2^2} - \frac{w_0}{R_2}\right) + \frac{1}{R_2}\left(\frac{\partial u_0}{\partial x_1} + \frac{1}{2}\left(\frac{\partial w_0}{\partial x_1}\right)^2 - w_0\frac{\partial^2 w_0}{\partial x_1^2} - \frac{w_0}{R_1}\right)\right\} +$$
$$A_{22}\left\{\frac{\partial v_0}{\partial x_2}\frac{\partial^2 w_0}{\partial x_2^2} + \frac{1}{2}\left(\frac{\partial w_0}{\partial x_2}\right)^2\frac{\partial^2 w_0}{\partial x_2^2} + \frac{1}{R_2}\left(\frac{\partial v_0}{\partial x_2} + \frac{1}{2}\left(\frac{\partial w_0}{\partial x_2}\right)^2 - w_0\frac{\partial^2 w_0}{\partial x_2^2} - \frac{w_0}{R_2}\right)\right\} +$$
$$2A_{66}\left\{\frac{\partial v_0}{\partial x_1}\frac{\partial^2 w_0}{\partial x_1 \partial x_2} + \frac{\partial u_0}{\partial x_2}\frac{\partial^2 w_0}{\partial x_1 \partial x_2} + \frac{\partial w_0}{\partial x_1}\frac{\partial w_0}{\partial x_2}\frac{\partial^2 w_0}{\partial x_1 \partial x_2}\right\} + A_{16}\left\{\frac{\partial v_0}{\partial x_1}\frac{\partial^2 w_0}{\partial x_1^2} + \frac{\partial u_0}{\partial x_2}\frac{\partial^2 w_0}{\partial x_1^2}\right.$$
$$+2\frac{\partial u_0}{\partial x_1}\frac{\partial^2 w_0}{\partial x_1 \partial x_2} + \left(\frac{\partial w_0}{\partial x_1}\right)^2\frac{\partial^2 w_0}{\partial x_1 \partial x_2} + \frac{\partial w_0}{\partial x_1}\frac{\partial w_0}{\partial x_2}\frac{\partial^2 w_0}{\partial x_1^2} + \frac{1}{R_1}\left(\frac{\partial v_0}{\partial x_1} + \frac{\partial u_0}{\partial x_2} + \frac{\partial w_0}{\partial x_1}\frac{\partial w_0}{\partial x_2} - \right.$$
$$\left.\left. 2w_0\frac{\partial^2 w_0}{\partial x_1 \partial x_2}\right)\right\} + A_{26}\left\{\frac{\partial v_0}{\partial x_1}\frac{\partial^2 w_0}{\partial x_2^2} + \frac{\partial u_0}{\partial x_2}\frac{\partial^2 w_0}{\partial x_2^2} + 2\frac{\partial v_0}{\partial x_2}\frac{\partial^2 w_0}{\partial x_1 \partial x_2} + \left(\frac{\partial w_0}{\partial x_2}\right)^2\frac{\partial^2 w_0}{\partial x_1 \partial x_2} + \right.$$
$$\left.\frac{\partial w_0}{\partial x_1}\frac{\partial w_0}{\partial x_2}\frac{\partial^2 w_0}{\partial x_2^2} + \frac{1}{R_2}\left(\frac{\partial v_0}{\partial x_1} + \frac{\partial u_0}{\partial x_2} + \frac{\partial w_0}{\partial x_1}\frac{\partial w_0}{\partial x_2} - 2w_0\frac{\partial^2 w_0}{\partial x_1 \partial x_2}\right)\right\} + B_{11}\left(\frac{\partial \psi_1}{\partial x_1}\frac{\partial^2 w_0}{\partial x_1^2} + \right.$$
$$\left.\frac{1}{R_1}\frac{\partial \psi_1}{\partial x_1}\right) + B_{12}\left\{\frac{\partial \psi_1}{\partial x_1}\frac{\partial^2 w_0}{\partial x_2^2} + \frac{\partial \psi_2}{\partial x_2}\frac{\partial^2 w_0}{\partial x_1^2} + \frac{1}{R_1}\frac{\partial \psi_2}{\partial x_2} + \frac{1}{R_2}\frac{\partial \psi_1}{\partial x_1}\right\} + B_{22}\left(\frac{\partial \psi_2}{\partial x_2}\frac{\partial^2 w_0}{\partial x_2^2} + \right.$$

$$\left.\frac{1}{R_2}\frac{\partial\psi_2}{\partial x_2}\right) + 2B_{66}\left(\frac{\partial\psi_1}{\partial x_2}\frac{\partial^2 w_0}{\partial x_1\partial x_2} + \frac{\partial\psi_2}{\partial x_1}\frac{\partial^2 w_0}{\partial x_1\partial x_2}\right) + B_{16}\left\{\frac{\partial\psi_1}{\partial x_2}\frac{\partial^2 w_0}{\partial x_1^2} + \frac{\partial\psi_2}{\partial x_1}\frac{\partial^2 w_0}{\partial x_1^2} + \right.$$

$$\left.2\frac{\partial\psi_1}{\partial x_1}\frac{\partial^2 w_0}{\partial x_1\partial x_2} + \frac{1}{R_1}\left(\frac{\partial\psi_1}{\partial x_2} + \frac{\partial\psi_2}{\partial x_1}\right)\right\} + B_{26}\left\{\frac{\partial\psi_1}{\partial x_2}\frac{\partial^2 w_0}{\partial x_2^2} + \frac{\partial\psi_2}{\partial x_1}\frac{\partial^2 w_0}{\partial x_2^2} + 2\frac{\partial\psi_2}{\partial x_2}\frac{\partial^2 w_0}{\partial x_1\partial x_2}\right.$$

$$\left.+\frac{1}{R_2}\left(\frac{\partial\psi_1}{\partial x_2} + \frac{\partial\psi_2}{\partial x_1}\right)\right\} + KA_{55}\left(\frac{\partial\psi_1}{\partial x_1} + \frac{\partial^2 w_0}{\partial x_1^2}\right) + KA_{44}\left(\frac{\partial\psi_2}{\partial x_2} + \frac{\partial^2 w_0}{\partial x_2^2}\right)$$

$$-N_{11}^T\left(\frac{\partial^2 w_0}{\partial x_1^2} + \frac{1}{R_1}\right) - 2N_{12}^T\frac{\partial^2 w_0}{\partial x_1\partial x_2} - N_{22}^T\left(\frac{\partial^2 w_0}{\partial x_2^2} + \frac{1}{R_2}\right) = 0 \tag{9.41e}$$

9.5.3 The Boundary Conditions

As described in the previous chapters, n and t are the normal and tangential directions to the boundary when $n = 1,\ t = 2$ and when $n = 2,\ t = 1$. For the case of **simply supported** boundary conditions the following conditions must be satisfied.

Along the edges $x_n = 0,\ L_n$

$$N_{nn} = N_{nt} = M_{nn} = M_{nt} = w_0 = 0 \tag{9.42a, b}$$

For the case of **clamped** boundary conditions, the following conditions must be satisfied.

$$N_{nn} = N_{nt} = \psi_n = \psi_t = w_0 = 0 \tag{9.43a, b}$$

9.6 Final Comments

A basic theoretical foundation concerning functionally graded doubly curved sandwich panels has been presented. Two cases were introduced. One case considered that the facings were homogeneous while the core was functionally graded, and vice versa for the second case. This chapter solely presented the theoretical equations necessary to introduce the topic. There is a plethora of results in the literature upon which the reader is referred to. The reader is referred to such authors as Shen (2009),

Abdelaziz et al. (2011), and Ashraf and Alghamdi (2008) for additional information and results on the topic.

References

Abdelaziz, H. H., Atmane, H. A., Mechab, I., Boumia, L., Tounsi, A., & El Abbas, A. B. (2011). Static analysis of functionally graded sandwich plates using an efficient and simple refined theory. *Chinese Journal of Aeronautics, 24*, 434–448.

Ashraf, Z. M., & Alghamdi, N. A. (2008). Thermoelastic bending analysis of functionally graded sandwich plates. *Journal of Material Science, 43*, 2574–2589.

Huang, X. L., & Shen, H. S. (2004). Nonlinear vibration and dynamic response of functionally graded plates in thermal environments. *International Journal of Solids and Structures, 41*, 2403–2427.

Reddy, J. N. (2004). *Mechanics of laminated composite plates and shells-theory and analysis* (2nd ed.). Boca Raton: CRC Press.

Shen, H. S. (2009). *Functionally graded materials-nonlinear analysis of plates and shells*. Boca Raton: CRC Press.

Soedel, W. (2004). *Vibrations of shells and plates* (3rd ed.). New York: Marcel Dekker, Inc.

Touloukian, Y. S. (1967). *Thermophysical properties of high temperature solid materials*. New York: Macmillan.

Index

T. J. Hause, *Sandwich Structures: Theory and Responses*,
https://doi.org/10.1007/978-3-030-71895-4